Rice Genomics, Genetics and Breeding

Rice Genomics, Genetics and Breeding

Editor: Abbie Chavez

R CALLISTO REFERENCE

www.callistoreference.com

Callisto Reference,
118-35 Queens Blvd., Suite 400,
Forest Hills, NY 11375, USA

Visit us on the World Wide Web at:
www.callistoreference.com

ISBN: 978-1-64116-527-3 (Hardback)

Cataloging-in-Publication Data

Rice genomics, genetics and breeding / edited by Abbie Chavez.
   p. cm.
Includes bibliographical references and index.
ISBN 978-1-64116-527-3
1. Rice. 2. Rice--Genetics. 3. Rice--Breeding. I. Chavez, Abbie.
SB191.R5 R53 2022
633.18--dc23

Table of Contents

Preface

Rice is the seed of primarily two species of grass, namely, Oryza sativa and Oryza glaberrima. Oryza sativa is popularly known as Asian rice while the common name for Oryza glaberrima is African rice. The study of rice genomics deals with the entirety of genes present in it. Rice genetics focuses on the transfer of DNA from one generation to the other. Rice breeding is involved in a number of objectives such as increasing the quality, yield and resistance to biotic and abiotic stress. This book unfolds the innovative aspects of genomics, genetics and breeding of rice which will be crucial for the progress of this field in the future. It consists of contributions made by international experts. The book is appropriate for students seeking detailed information in this area as well as for experts.

The researches compiled throughout the book are authentic and of high quality, combining several disciplines and from very diverse regions from around the world. Drawing on the contributions of many researchers from diverse countries, the book's objective is to provide the readers with the latest achievements in the area of research. This book will surely be a source of knowledge to all interested and researching the field.

In the end, I would like to express my deep sense of gratitude to all the authors for meeting the set deadlines in completing and submitting their research chapters. I would also like to thank the publisher for the support offered to us throughout the course of the book. Finally, I extend my sincere thanks to my family for being a constant source of inspiration and encouragement.

Editor

Rice Improvement Through Genome-Based Functional Analysis and Molecular Breeding in India

Pinky Agarwal[1†], Swarup K. Parida[1†], Saurabh Raghuvanshi[2], Sanjay Kapoor[2], Paramjit Khurana[2], Jitendra P. Khurana[2] and Akhilesh K. Tyagi[1,2*]

Abstract

Rice is one of the main pillars of food security in India. Its improvement for higher yield in sustainable agriculture system is also vital to provide energy and nutritional needs of growing world population, expected to reach more than 9 billion by 2050. The high quality genome sequence of rice has provided a rich resource to mine information about diversity of genes and alleles which can contribute to improvement of useful agronomic traits. Defining the function of each gene and regulatory element of rice remains a challenge for the rice community in the coming years. Subsequent to participation in IRGSP, India has continued to contribute in the areas of diversity analysis, transcriptomics, functional genomics, marker development, QTL mapping and molecular breeding, through national and multi-national research programs. These efforts have helped generate resources for rice improvement, some of which have already been deployed to mitigate loss due to environmental stress and pathogens. With renewed efforts, Indian researchers are making new strides, along with the international scientific community, in both basic research and realization of its translational impact.

Keywords: Gene function, Genome, India, Marker-assisted selection, Rice, *Oryza sativa*, Transcriptomics

Introduction

Rice feeds a large part of the world's population, including Indians, and hence is an extremely important cereal crop. Sustained efforts have been made to improve its yield to meet the ever-increasing global demand. This has been possible largely through studies carried out on its agronomy, physiology, genetics and molecular biology. Since the genome sequence of an organism opens up new vistas to improve its performance, multiple efforts were simultaneously made to sequence *indica* and *japonica* subspecies of rice (*Oryza sativa*) genomes culminating in the availability of a map-based sequence of the genome from rice (International Rice Genome Sequencing Project 2005; reviewed in Vij et al. 2006). To this international initiative, India contributed by sequencing and assembly of the long arm of chromosome 11 and participation in mapping/annotation of the rice genome. This was achieved by the collaborative efforts of the researchers at the Department of Plant Molecular Biology, University of Delhi, South Campus, and National Research Center on Plant Biotechnology, Indian Council of Agricultural Research (Chen et al. 2002; International Rice Genome Sequencing Project 2005; The Rice Chromosomes 11 and 12 Sequencing Consortia 2005; Rice Annotation Project 2007, 2008). Recently, remarkable progress has been made in understanding genetic and functional diversity in rice by sequencing *Oryza glaberrima* (Wang et al. 2014) and 3000 other globally distributed accessions of rice (Li et al. 2014).

An annotated genome enlists the putative genes and their sequence catalogs. Subsequently, the encoded proteins by such genes can be classified into families depending on the presence of a conserved domain or motif. Once the locus IDs of genes are delineated, gene organization and structure can be analyzed, to determine

* Correspondence: akhilesh@genomeindia.org
†Equal contributors
[1]National Institute of Plant Genome Research (NIPGR), Aruna Asaf Ali Marg, New Delhi 110067, India
[2]Interdisciplinary Centre for Plant Genomics and Department of Plant Molecular Biology, University of Delhi, South Campus, New Delhi 110021, India

unique features of genes and their products. Similarly, evolutionary and phylogenetic analyses help classify the family members into distinct classes and define their origin. Hence, the availability of the annotated rice genome has paved the way for the identification of members of multifarious gene families, India being a major contributor to this analysis. Many of these have been linked with the Rice Genome Annotation Project (RGAP) and are available at the community annotation project, while others are yet to be added (http://rice.plantbiology.msu.edu/annotation_community_families.shtml).

The next imperative step is to gain knowledge about the expression patterns of genes (Agarwal et al. 2014). For this, techniques like microarrays, RNA sequencing, microRNA sequencing and downstream analyses have been used to study transcriptomes, and their regulation by miRNA in multifarious developmental and stress conditions related to rice. Differentially regulated genes have been selected as targets for functional validation with the eventual aim to raise improved rice plants. This has been accomplished by generating transgenic plants with altered expression of the genes, followed by their detailed analysis. In this way, Indian researchers have been able to identify genes involved in development and stress response and elucidate downstream genes and pathways. For the larger benefit of the scientific community, the data generated has been deposited in international public repositories. To this effect, databases as well as bioinformatic tools are being created to facilitate easy access and seamless analysis of large data sets.

Both annotated genes and intergenic regions can be used to define functions of genes/quantitative trait loci (QTLs) by forward genetic approaches. Indian labs are making use of rich diversity of the rice genome for such purposes. The available genome/transcriptome sequences of diverse rice genotypes have been used to generate enormous resources in the form of genomic (genic) SSR (simple sequence repeat) and SNP (single nucleotide polymorphism) markers at a genome-wide scale especially in *indica* and aromatic rice (Parida et al. 2009; Jain et al. 2014). The large-scale validation and high-throughput genotyping of these genome/gene-based markers in diverse natural Indian germplasm (core and mini-core) collections, advanced generation mapping and mutant populations of rice are underway. These efforts assisted in mining of novel natural/functional allelic variants, understanding of molecular diversity and domestication pattern, construction of high-density genetic linkage maps and apply genetic/association mapping to identify potential genomic loci associated with complex quantitative traits of agronomic importance in rice (Sharma et al. 2005; Ngangkham et al. 2010; Marathi et al. 2012; Kumar et al. 2015). Marker-assisted selection (MAS)

to introgress and pyramid superior functional genes (alleles)/QTLs regulating complex yield/quality component and stress tolerance traits into diverse Indian rice genotypes for their genetic enhancement has been attempted (Singh et al. 2011a). In this article, an attempt has been made to highlight some key investigations from India in the post-genomic era that help gain knowledge about molecular biology and genetics along with its applications to improve the rice crop. Several of these findings were also presented during the 11th International Symposium on Rice Functional Genomics held in India in 2013.

Review
Loci for 3394 Genes from 50 Families Annotated

Transcription factors (TFs) are integral components of any regulatory and/or metabolic pathway. Comprehensive genome wide analyses of their gene structure and expression have revealed interesting aspects about their regulation during development and stress and provided leads for their functional characterization. Amongst the first detailed analysis of large TF gene families, the C_2H_2 zinc-finger family, encoded by 189 *ZOS* genes was found to have 97 new members, including 10 previously unannotated members. Also, new types of zinc fingers were discovered, apart from the cannonical ones. These genes were found to differentially express during reproductive development and abiotic stresses (Agarwal et al. 2007). The rice genome was found to encode for 75 MADS box TFs. Nine members belonging to the Mβ group of the MADS-family, which was considered absent in monocots, were also identified (Arora et al. 2007). Eighteen new bZIP motif-encoding genes were identified, making the total to 89 genes for this class of TFs. Sixty-two genes out of 72 intron-containing genes had the intron within the bZIP region (Nijhawan et al. 2008). The analysis of homeobox gene family showed that HD-ZIP III sub family was intron-rich, resulting in a high number of alternatively spliced genes. Also, 50 % of the genes had evolved due to duplication events (Jain et al. 2008c). Most of the HSFs were found to be induced by heat (Mittal et al. 2009), apart from other developmental and stress conditions. Splicing of a member resulted in the exclusion of its NES (Chauhan et al. 2011). Many DOF genes were found to be expressed during various stages of seed development (Gaur et al. 2011) and it is known that DOFs play an important role in the process (Agarwal et al. 2011). The promoters of some of the *MYB* genes were found to have novel *cis* elements and lower arms of chromosomes had relatively more *MYB* genes (Katiyar et al. 2012). Thus, a comprehensive analysis of members of each TF family brings to light certain unique features, which remain unnoticed during whole genome annotation.

Besides TFs, a number of gene families coding for various signal transduction components (STCs) have also been characterized for clues to their structure-function relationships. Amongst hormone signaling pathways, 31 *OsAux/IAA* genes were identified and most of them were found to be up-regulated upon auxin treatment, six of them at much higher levels (Jain et al. 2006a). Another auxin responsive family, *GH3*, encoding amido synthetases involved in catalyzing the synthesis of IAA-amino acid conjugates, was found to have 12 members in rice. Group III genes of this family are apparently absent not only in rice but other monocots too (Jain et al. 2006b). The early auxin responsive intronless gene family *SAUR* has 58 members, 17 of which are clustered on chromosome 9 (Jain et al. 2006d). Kinases and phosphatases are essential parts of any signaling pathway. Rice genome has been found to code for 31 CDPKs, two of which were newly identified (Ray et al. 2007). The identification protocol led to 192 and 70 previously unidentified genes as coding for RLCKs (Vij et al. 2008) and MAPKKKs, respectively (Rao et al. 2010). Most of the 11 SERK/SERLs had been previously annotated incorrectly and were found to express in tissues other than somatic embryos as well (Singla et al. 2009). Also, 173 members for three classes of LecRLKs were identified (Vaid et al. 2012). For ten CBLs, 33 interacting CIPKs were found. Many of these interactions were found to be conserved from *Arabidopsis* (Kanwar et al. 2014). Five families of phosphatases were found to have been encoded by 132 genes (Singh et al. 2010a). Also, 51 genes were found to code for 73 TCS proteins, including 14 histidine kinases, 5 phosphotransfer proteins and 32 response regulator encoding genes. They were found to be clustered at various regions on rice chromosomes (Pareek et al. 2006). Amongst other STCs, 687 F-box protein encoding genes were identified in rice, many of which had other domains as well (Jain et al. 2007). Another class of STCs comprises of known stress inducible genes. Rice genome was found to have 18 abiotic stress inducible *SAP* genes (Vij and Tyagi 2006), heat stress and other abiotic stress inducible 23 sHsps (Sarkar et al. 2009) and only five *Hsp100* genes (Batra et al. 2007). J-proteins are co-chaperones to Hsp70 and rice has 104 such genes, coding for proteins with high molecular weights. Five such genes could even rescue the corresponding yeast mutant under heat stress (Sarkar et al. 2013). Amongst transporters, rice genome was found to code for 33 calcium transporters, which included one channel, 14/15 ATPases and 16 exchangers (Goel et al. 2011; Kamrul Huda et al. 2013), 133 ABC transporters (Saha et al. 2015) and 14 sulphate transporters. The sulphate transporters are expressed during sulphate starvation, heavy metal and abiotic stress as well as reproductive development (Kumar et al. 2011). Apart from

these, phospholipase A and C components have also been identified from rice genome (Singh et al. 2013a; Singh et al. 2012a).

The identification of genes encoding transcriptional regulators in rice has its share of contribution from India. In effect, eight, 19 and five genes coding for Dicer-like proteins, Argonautes and RNA-dependent RNA polymerases have been identified (Kapoor et al. 2008). Apart from this, 115 RNA helicases and 31 DNA helicases have also been identified (Umate et al. 2010). Rice genome has been found to code for 51 Mediator complex protein coding genes, which include all subunits identified so far in various organisms (Mathur et al. 2011). Eleven regulators belonging to various families have the KIX domain, which is responsible for protein-protein interactions (Thakur et al. 2013). Gene family analysis often brings out interesting aspects of the genome. A pair of chymotrypsin protease inhibitor encoding genes, out of a total of 17, shares a bidirectional promoter (Singh et al. 2009). Also, genes coding for an assortment of other proteins have been identified. This includes 48 genes coding for the enzyme glutaredoxin (Garg et al. 2010), 11 for peroxiredoxins (Umate 2010), 14 for lipoxygenase (Umate 2011; Marla and Singh 2012), 11 for Class I metallothioneins induced under heavy metal stress (Gautam et al. 2012), 28 for cyclophilins (Trivedi et al. 2012), 14 for glyoxalases I and II (Mustafiz et al. 2011), 16 carotenoid biosynthesis genes (Chaudhary et al. 2010) and 158 *ARMADILLO* genes, which have an ARM repeat (Sharma et al. 2014a). Fifty-nine CBS domain containing proteins have also been found to contain other domains, and hence may perform diverse functions (Kushwaha et al. 2009). Thus, in totality, locus IDs have been assigned to the members of several important gene families along with insights into their expression profiles during development and abiotic stress conditions. Also, the interacting partners and/or promoter sequences have been analyzed for a few members. The general points of interest include (i) possibility of identifying new genes and clades for a family on the basis of molecular model-based comparisons, (ii) conservation in position of introns, (iii) compensating evolutionary selection of mutations in protein structure (iv) identification of state-specific or inducible genes of a family and (v) identification of gene functions that have evolved specifically in monocots.

Databases and Tools are Being Used for Storing and Mining Information

As more information is getting accumulated in plant biology, databases are being created to make the information publically available, for easy access and analysis. These databases have a user friendly interface for easy data retrieval (Agarwal et al. 2014). Indian rice scientists have developed various databases on varied aspects. For QTLs related to abiotic stress, QlicRice provides not only the genomic locations, but also the responsible locus IDs (Smita et al. 2011). STIFDB2 documents all stress responsive TFs in both *indica* and *japonica* rice. Apart from an extensive *cis* regulatory element analysis, it can be used for network prediction as well (Naika et al. 2013). Similarly, RiceSRTFDB has records of stress-responsive TFs along with their expression patterns, mutants and *cis* element analysis (Priya and Jain 2013).

A new concept for literature based 'Manually Curated Database of Rice Proteins' has been developed based on novel data curation methods that enable digitization and semantic integration of experimental data. The current release of database has data of 2401 rice proteins manually curated from 538 research articles. Over 800 phenotypic/biochemical traits have been curated along with their associated genes (Gour et al. 2014). The availability of the genome sequence allows researchers to perform multiple analysis, such as elucidation of intronless genes (Jain et al. 2008a) and prediction of miRNA targets (Archak and Nagaraju 2007). Various tools have been developed and validated using rice genome sequence, such as RetroPred for prediction of non-LTR retrotransposons (Naik et al. 2008), a machine learning tool to predict infection by rice blast fungus depending on the weather conditions (Kaundal et al. 2006), MirtronPred to predict plant mirtrons (Joshi et al. 2012) and pTAREF for prediction of miRNA targets (Jha and Shankar 2011). The URLs for all the webservers and softwares have been mentioned in Table 1.

Transcriptome Analysis Presents a Snapshot of Participating Genes

Prior to the advent of microarray and RNA seq, other strategies including subtractive hybridization were used to determine the transcript levels of multiple genes (Agarwal et al. 2014). After the sequences of rice BAC and PAC clones were available, genes were identified from drought-stressed ESTs, some of which were novel (Babu et al. 2002; Gorantla et al. 2007; Reddy et al. 2002). In order to identify genes responsible for salt tolerance, subtractive cDNA libraries have been compared between tolerant (Pokkali) and susceptible (IR64) genotypes (Kumari et al. 2009). The samples that have undergone transcriptome analysis till date fall into various categories. Multiple developmental stages, tissues subjected to a/biotic or nutrient stress, developmental stages under stress, and resistant genotypes have been used to elucidate various genes/pathways operating in the desired situation. We have performed microarray analysis on 19 stages of rice development, including both vegetative and reproductive phases, and the data have been deposited at the Gene Expression Omnibus of NCBI. Along with this, microarray data from three stages of abiotic stresses, namely, cold, dehydration and salt, have also been deposited (Agarwal et al. 2007; Arora et al. 2007; Ray et al. 2011; Sharma et al. 2012). Also, the transcriptome dynamics during entire anther development from pre-pollen mother cell stage to mature anthers with tri-nucleate pollen have been analyzed by using microarrays to reveal genes specifically expressing during meiosis and playing potential roles in sporophyte to gametophyte switching during male gametophyte development (Deveshwar et al. 2011). Thidiazuron

Table 1 Details of databases and web tools developed by Indian rice scientists

Sl. no.	Name of database/ tool	URL	Information generated	Reference
1	Rice blast infection prediction	http://www.imtech.res.in/raghava/rbpred/	Predicts chances of rice blast infection by using weather conditions as an input.	Kaundal et al. 2006
2	RetroPred	http://www.juit.ac.in/attachments/RetroPred/home.html	A downloadable tool which detects repeat regions in genomic sequences and classifies them as LINEs and SINEs.	Naik et al. 2008
3	pTAREF	http://scbb.ihbt.res.in/new/p-taref/form1.html	A downloadable tool to predict miRNA targets in transcriptome data.	Jha and Shankar 2011
4	QlicRice	http://nabg.iasri.res.in:8080/qlic-rice/	A database of 974 abiotic stress-related QTLs connected with 460 locus IDs, with their physical and genetic data, which can be easily queried.	Smita et al. 2011
5	MirtronPred	http://203.92.44.117/mirtronPred/prediction.php	Uses intronic sequences as an input to predict mirtrons.	Joshi et al. 2012
6	STIFDB2	http://caps.ncbs.res.in/stifdb2/	Encompasses over 5000 stress responsive genes from Arabidopsis and both *indica* and *japonica* rice, with a list of putative TF binding sites in their promoters.	Naika et al. 2013
7	RiceSRTFDB	http://www.nipgr.res.in/RiceSRTFDB.html	Salt and drought stress-responsive TFs from 99 Affymetrix chips can be queried.	Priya and Jain 2013
8	MCDRP	http://www.genomeindia.org/biocuration/	Manually curated data from all published rice genes, including ontology, functions, pathways and interactions.	Gour et al. 2014

treated rice callus has also been analyzed by microarray to identify genes related to differentiation (Chakrabarty et al. 2009). The gene expression analysis of rice roots subjected to heavy metal stress helped in the identification of genes that can be modulated for detoxification (Dubey et al. 2010; Dubey et al. 2014). In different studies, microarray analyses have been performed on root, leaf and panicle under drought stress (Smita et al. 2013) and on two contrasting genotypes under drought stress (Lenka et al. 2011), rice samples subjected to heat and/or oxidative stresses (Mittal et al. 2012a, 2012b) and heat stress followed by recovery (Sarkar et al. 2014). With respect to biotic stress, microarray has been done for *Xanthomonas* infected rice plants (Grewal et al. 2012). Amongst nutrient stress, microarray analysis of calcium and phosphate starved plants shows involvement of various biosynthetic pathway genes such as carbohydrate, phosphate, lipid and nitrogen metabolism (Shankar et al. 2013; Shankar et al. 2014). Thus, whole genome transcriptome analyses have been done for diverse abiotic and biotic stress situations and various developmental stages in rice and they have helped in delineating essential genes and the associated pathways. These high quality transcriptome analyses have added to the efforts of the international community of rice researchers in generating an indispensible resource for future studies.

Applications of Genome Sequencing for Functional and Regulatory Aspects

There are many scientific groups in India working on understanding regulation of various aspects of rice development and stress responses; they have been able to pin point the specific function of important genes by raising and analyzing transgenic plants with modified expression patterns. *RFL*, a rice ortholog of *LEAFY* from Arabidopsis, controls flowering time and architecture (Deshpande et al. 2015; Rao et al. 2008; Table 2). Further, transcriptome analysis of rice plants with altered expression of a desired gene by overexpression or silencing or mutation, along with phenotypic analysis, identifies the functional relevance of the gene. The downstream genes and networks can also be elucidated by such an analysis (Agarwal et al. 2014). A mutant *ewst1* in Nagina22 results in enhanced water tolerance and genes related to production of osmoprotectants and secondary metabolites are up regulated in it (Lima et al. 2015). Ectopic expression of *OsiSAP1* or *OsSAP11* or *OsRLCK53* results in up regulation of genes imparting drought tolerance (Giri et al. 2011; Dansana et al. 2014; Mukhopadhyay et al. 2004). Plants overexpressing *OsMKK6* have also been analyzed by microarray, indicating the role of this gene in stress regulation (Kumar and Sinha 2014). Microarray analysis of plants with overexpression and RNAi constructs of *OsMADS29* showed that the gene

controls cytokinin mediated starch biosynthesis during seed development (Nayar et al. 2013). Nuclear translocation of this TF has been shown to be regulated by way of homodimerization and interaction with at least 19 other seed-expressing MADS-box proteins (Nayar et al. 2014). The gene *Pi54* in Taipei309 imparts tolerance to *Magnaporthea* infection by activating defense responsive pathways (Gupta et al. 2012). Thus, downstream operable genes and pathways have been elucidated. This has given valuable information on various genes and their related pathways. Table 2 enlists rice genes for which transgenic plants have been made in a homologous system.

Many genes of rice have been characterized by their altered expression in a heterologous system as Arabidopsis and tobacco. Although these model plants are adopted for the ease with which the phenotype can be easily scored, in most cases, the results can be extrapolated to rice. For example, not only rice plants overexpressing OsDREB2A are abiotic stress tolerant (Mallikarjuna et al. 2011), the gene OsDREB1B confers similar phenotype in tobacco as well (Gutha and Reddy 2008). Even in the case of promoters, rice sucrose synthase1, *RSs1* promoter confers phloem-specific expression in Arabidopsis (Saha et al. 2007). Such promoter sequences can be used for targeted gene expression. Ectopic expression of many abiotic stress responsive rice genes in Arabidopsis and/or tobacco makes plants tolerant to abiotic stress responses. Prominent amongst these are stress associated proteins, SAP1/11, receptor-like cytoplasmic kinase, OsRLCK253 (Giri et al. 2011), topoisomerases OsTOP6A1, OsTOP6A3 and OsTOP6B (Jain et al. 2006c, 2008b), OsTCP19 (Mukhopadhyay and Tyagi 2015), a bZIP TF in the Saltol QTL, OsHBP1b (Lakra et al. 2015), cyclin, OsCyp2-P (Kumari et al. 2014), protein phosphatase 2C, OsPP108 (Singh et al. 2015a), glutathione S-transferases, OsGSTU4 and OsGSTL2 (Sharma et al. 2014b; Kumar et al. 2013), glutaredoxin, OsGRX8 (Sharma et al. 2013), cystathionine β-synthase, OsCBSX4 (Singh et al. 2012c), Ca(2+)ATPase, OSACA6 (Kamrul Huda et al. 2014; Huda et al. 2013), metallothionein, OsMT1e-P (Kumar et al. 2012), myoinositol phosphate synthase from wild rice, PcINO1 (Patra et al. 2010; Das-Chatterjee et al. 2006; Majee et al. 2004), LEA protein, Rab16A (RoyChoudhury et al. 2007) and chymotrypsin protease inhibitor, OCPI2 (Tiwari et al. 2015b). On the biotic stress front, the promoter of *CYP76M7* is *Magnaporthea* inducible and can be used for expression of defense related genes (Vijayan et al. 2015). *OsSAP1* enhances tolerance to pathogen infection (Tyagi et al. 2014). Proteins with multiple roles have also been characterized in heterologous systems. Multidrug and toxic compound extrusion proteins, OsMATE1 and 2 have roles in both abiotic and biotic stresses and development (Tiwari et al. 2014). Germin-like protein1, OsGLP1,

Table 2 Some rice genes and promoters analyzed to elucidate their activities in transgenic rice

Sl. no.	Gene	Locus ID[a]	Construct	Function	Reference
1	*ANS* (Anthocyanidin synthase)	LOC_Os01g27490	OE	Increase in flavonoids and anthocyanins and hence oxidation potential	Reddy et al. (2007)
2	*ETHE1 (ETHYLMALONIC ENCEPHALOPATHY PROTEIN 1)*	LOC_Os01g47690	promoter	Stress and calcium responsive	Kaur et al. (2014)
3	*GLYOXALASEII*	LOC_Os09g34100	OE	Confers salt stress tolerance	Singla-Pareek et al. (2008)
4	*IPK1* (inositol 1,3,4,5,6-pentakisphosphate 2-kinase)	LOC_Os04g56580	RNAi	Silencing of *IPK1* decreases the antinutrient phytic acid	Ali et al. (2013a)
5	*MIPS* (myo-inositol-3-phosphate synthase)	LOC_Os03g09250	RNAi	Decrease in phytate and increase in inorganic phosphate and iron levels	Ali et al. (2013b)
6	*NBS-Str1* (nucleotide binding site-LRR) *and BLEC-Str8* (β-lectin domain protein)	LOC_Os02g09790 and LOC_Os04g01950	Promoter	Cause increased expression under stress	Ray et al. (2012)
7	*orfB*	LOC_Os12g34018	RNAi	Unedited mitochondrial *orfB* gene causes male sterility by decreasing ATP synthase activity	Chakraborty et al. (2015)
8	*OsbHLH, OsFbox, OsiPK*	LOC_Os01g18870, LOC_Os05g46050, LOC_Os12g12860	Promoter	Responsible for anther-specific expression	Khurana et al. (2013)
9	*OsCDR1* (aspartate protease)	LOC_Os01g08330	OE	Provides resistance against fungal and bacterial pathogens	Prasad et al. (2009)
10	*OsClpB-cyt* (ClpB-cytoplasmic /Hsp100)	LOC_Os05g44340	Promoter	Classical heat shock element supresses expression under unstressed conditions	Singh et al. (2012b)
11	*OsCPK31* (CDPK)	AK110341[b]	OE, RNAi	Helps in rice grain filling	Manimaran et al. (2015)
12	*OsDREB2A* (Dehydration-responsive element binding transcription factor)	LOC_Os01g07120	OE	Osmotic, dehydration and salt stress tolerance	Mallikarjuna et al. (2011)
13	*OsGLP1* (germin-like protein)	LOC_Os08g35760	RNAi	Controls plant height and disease resistance	Banerjee and Maiti (2010)
14	*OsGLYII-2* (*GLYOXALASEII*)	LOC_Os03g21460	OE	Provides salinity and decarbonyl stress tolerance	Ghosh et al. (2014)
15	*OsGRX8* (glutaredoxin)	LOC_Os02g30850	RNAi	Abiotic stress response	Sharma et al. (2013)
16	*OSIAGP* (Arabinogalactan)	LOC_Os06g21410	Promoter	Provides pollen-preferential expression	Anand and Tyagi (2010)
17	*OsiPA*	LOC_Os10g40090	Promoter	Confers pollen-specific expression	Swapna et al. (2011)
18	*OsiSAP1*	LOC_Os09g31200	OE	Positively regulates water deficit stress by affecting gene expression leading to loss of membrane damage and lipid peroxidation	Dansana et al. (2014)
19	*OsiSAP8*	LOC_Os06g41010	OE	Abiotic stress tolerance	Kanneganti and Gupta (2008)
20	*OsiWAK1* (wall associated kinase)	LOC_Os11g46860	RNAi	Plant growth and development	Kanneganti and Gupta (2011)
21	*OsMADS1*	LOC_Os03g11614	OE	Involved in panicle, lemma and palea development	Prasad et al. (2001)
22	*OsMADS29*	LOC_Os02g07430	RNAi, OE, Protein-protein interaction	Regulates starch biosynthesis during seeds development by regulating active cytokinin levels	Nayar et al. (2013); Nayar et al. (2014)
22	*OsMKK6*	LOC_Os01g32660	OE	Participates in salt stress signaling and regulates phytoalexin biosynthesis due to UV exposure	Kumar and Sinha (2013); Wankhede et al. (2013)
23	*Osoxo4* (oxalate oxidase)	LOC_Os03g48780	OE	Resistance to sheath blight	Molla et al. (2013)
24	*OsSAG12-1* (senescence associated gene)	LOC_Os01g67980	RNAi	Negatively regulates cell death	Singh et al. (2013d)

Table 2 Some rice genes and promoters analyzed to elucidate their activities in transgenic rice *(Continued)*

25	*OsSUV3* (*SUPPRESSOR OF VAR3*, DNA and RNA helicase, ATPase)	LOC_Os03g53500	OE, RNAi	Provides tolerance against salt stress by upregulation of multiple hormones, resistance against metal stress	Sahoo et al. (2014a); Sahoo et al. (2014b); Tuteja et al. (2013)
26	*Pi54*	LOC_Os11g42010	OE	Provides resistance to *Magnaporthea* irrespective of insertion position	Arora et al. (2015)
27	*Rab16A*, 4X ABRE, 2X ABRC	LOC_Os11g26790	Promoter	Responsible for ABA responsive expression under stress and development	Ganguly et al. (2011)
28	*RFL* (*RICE FLORICULA/ LEAFY*)	LOC_Os04g51000	promoter and intron	Introns have regulatory regions	Prasad et al. (2003)
29	*RFL* (*RICE FLORICULA/ LEAFY*)	LOC_Os04g51000	RNAi, OE	Controls flowering, promotes axillary meristem initiation and hence tillering through auxin and strigolactone signaling	Deshpande et al. (2015); Rao et al. (2008)
30	*Ubi1* (Ubiquitin) from rice	LOC_Os06g46770	Promoter	Confers high levels of constitutive expression	Bhattacharyya et al. (2012)

[a]based on RGAP version 7 (http://rice.plantbiology.msu.edu/)
[b]represents cDNA clone from KOME (http://cdna01.dna.affrc.go.jp/cgi-bin/sogo.cgi?pj=598&class=598&page=cDNA) as no RGAP locus ID has been found

provides tolerance to both abiotic and biotic stress (Banerjee and Maiti 2010; Banerjee et al. 2010).

Likewise, functions of few miRNAs have also been elucidated. Since miRNAs cause degradation of their target mRNAs, the two express inversely (Raghuram et al. 2014). Osa-MIR414, osa-MIR164e and osa-MIR408 have been found to regulate the expression of *OsABP, OsDBH* and *OsDSHCT*. These miRNAs are down regulated during salinity stress (Macovei and Tuteja 2012). Thus, during salt stress, miRNAs targeting DEAD box helicases are down regulated (Macovei and Tuteja 2012; Umate and Tuteja 2010). Micro RNAs are also being identified by sequencing, such as those involved in tungro virus infection and salt stress (Sanan-Mishra et al. 2009). Drought tolerant genotype, Nagina 22, shows a variety of miRNAs differentially expressed during 'anthesis' stage drought, which may be responsible for its tolerance (Kansal et al. 2015). Further, miR408 which targets several plantacyanin genes is regulated differentially in Nagina 22 as compared to drought sensitive rice variety during drought stress (Mutum et al. 2013). Another study on salt-responsive miRNA markers has highlighted the differences in miRNAs amongst a sensitive and tolerant genotype of rice (Mondal and Ganie 2014). Moreover, over half of the miRNA targets have been found to be conserved amongst *indica* and *japonica* genotypes (Archak and Nagaraju 2007). miRNAs are also responsible for low N-tolerant genotypes (Nischal et al. 2012). The antagonistic effects of arsenic and selenium are controlled by the miRNA population (Pandey et al. 2015). Also, varieties showing variance in tolerance to arsenite stress show differences in miRNA accumulation (Sharma et al. 2015). Eleven TFs, controlled by miRNAs, regulate abiotic stress responsive genes (Nigam et al. 2015). Non-conserved miRNAs have also been predicted in rice (Kumar et al. 2014). Rice has been found to have polycistronic *MIR166*s (Barik et al. 2014). Submergence

responsive miRNA targets (Paul and Chakraborty 2013) and TF binding motifs in miRNAs (Devi et al. 2013) have been predicted. miRNA species are differentially and preferentially expressed amongst various developmental stages (Mittal et al. 2013) and callus differentiation (Chakrabarty et al. 2010). They are even responsible for the cross talk amongst biotic and abiotic stresses (Sanan-Mishra et al. 2009). The studies discussed above provide evidence how the annotation of the rice genome has been useful in determining the gene networks and also unraveling the function of genes involved in various aspects of development in rice and when it is exposed in unfavorable conditions.

Genome-Wide Development and Use of Informative Genetic Markers

The availability of gold standard reference genome sequence of *japonica* rice cv. Nipponbare (International Rice Genome Sequencing Project 2005) has propelled the genome resequencing and transcriptome sequencing of diverse rice genotypes in recent years by use of NGS (next-generation sequencing) approaches in India. This in turn led to the development of enormous resources in the form of genomic (genic) simple sequence repeat (SSR) and single nucleotide polymorphism (SNP) markers at a genome-wide scale in rice. Despite availability of 2240 RM (rice microsatellite) and 18,828 hyper-variable class I SSR markers from rice draft and/ or whole genome sequences (McCouch et al. 2002; International Rice Genome Sequencing Project 2005), many initiatives have been undertaken by Indian researchers to develop multiple kinds of novel concept-based informative genomic and genic SSR markers at a genome-wide scale for expediting high-throughput genetic analysis in rice. Preliminary efforts have been made to utilize 33,722 rice unigene sequences (48.8 Mb) for developing a novel class of non-redundant 13,230 genic microsatellite markers designated as UniGene derived MicroSatellite (UGMS) markers, which are bin-mapped across 12 rice chromosomes (Parida et al. 2006). Considering the added advantages of these markers in assaying allelic variation in transcribed genic sequence components of the genome, these have been placed in public domain as a web-based freely accessible relational database, *UgMicroSatdb* (*U*nigene *M*icro*S*atellite database) for unrestricted use (Aishwarya and Sharma 2007). In another study, 17,966 novel "GNMS (genic non-coding microsatellite)" markers targeting different upstream regulatory (5′UTRs and promoters) and non-coding (3′UTRs and intronic) sequence components of protein coding genes annotated from Nipponbare rice genome were developed, which are bin-mapped across 12 rice chromosomes (Parida et al. 2009). This further led to the identification of 112 orthologous and paralogous "CNMS (Conserved Non-coding MicroSatellite)" markers from the putative rice promoter sequences for comparative genome mapping, and understanding of evolutionary and gene regulatory complexities among rice and other members of grass family (Parida et al. 2009). The GNMS and CNMS markers with their many desirable genetic attributes, including high polymorphic potential and functional significance have efficiency to serve as candidate gene-based microsatellite markers in diverse genomics-assisted breeding applications in rice. Subsequently, utilizing the complete rice genome sequence information, 436 well-validated HvSSR (highly variable SSR) markers with repeat-length of 51–70 bp have been developed at a genome-wide scale (Singh et al. 2010b). The efficacy of these markers for exhibiting consistent PCR amplification and detecting higher allelic polymorphism among accessions even by a cost-effective agarose gel, have demonstrated their suitability in large-scale genotyping applications in rice. In order to develop trait-specific genetic markers, salinity-responsive microRNAs (miRNAs) have been targeted to develop 130 miRNA-SSR markers for their efficient utilization in salinity stress tolerance marker-assisted genetic enhancement of rice (Mondal and Ganie 2014). The genetic variation and evolutionary dynamics between *indica* cv. 93–11 and *japonica* cv. Nipponbare have been understood by identifying and analysing the repeat-units mutability of ~50,000 *in silico* polymorphic SSR markers occurring in diverse coding and non-coding sequence components of rice genome (Grover et al. 2007).

In addition to SSR markers, the whole genome resequences/pseudomolecules and transcriptomic sequences of multiple rice genotypes enabled rapid discovery and development of genomic and genic SNP and InDel (insertion-deletion) markers *in silico* at a genome-wide

scale. Recently, an international initiative, i.e., "The 3000 rice genome sequence project" has been completed to generate 20 million SNPs by genome resequencing of 3000 rice genotypes (Alexandrov et al. 2015). Similar efforts also have been made by Indian scientists to generate genome/gene-derived SNP and InDel markers by low-coverage resequencing specifically of diverse *indica* genotypes, including aromatic rice. For instance, non-redundant 2495052 SNP and 324034 InDel markers have been discovered by comparing the NGS-based whole-genome resequencing data of six elite *indica* inbred lines (three of each cytoplasmic male sterile and restorer lines) to accelerate genomics-assisted breeding for hybrid performance in rice (Subbaiyan et al. 2012). Subsequently, the whole genome resequencing of three drought/salinity tolerant (Nagina 22 and Pokkali) and sensitive (IR64) rice accessions identified non-redundant 1784583 SNPs and 154275 InDels between reference Nipponbare and three resequenced rice accessions. Based on this outcome, genome-wide 401683 SNPs between IR64 and Pokkali and 662509 SNPs between IR64 and Nagina 22 that are well-distributed across coding and non-coding regions of these sequenced genomes were discovered with the eventual aim to deploy them in marker-assisted breeding for abiotic stress tolerance in rice (Jain et al. 2014). More recently, the comparison of whole genome resequencing data of a widely cultivated low glycemic index-containing *indica* rice variety, Swarna, with reference genome Nipponbare, identified 1,149,698 SNPs (65,984 non-synonymous SNPs) and 104,163 InDels for deciphering the genetic basis of complex glycemic index quantitative trait in rice (Rathinasabapathi et al. 2015).

Essentially, the genomic and genic SSR and SNP markers developed especially by Indian initiatives are amenable to large-scale validation and high-throughput genotyping at the whole genome level. These informative markers are thus suitable for multi-dimensional genomics-assisted breeding applications.

Characterization of Core/Mini-Core Germplasm and Mutant Resource for Rice Genetic Enhancement

Rice is grown in the wide range of agro-ecological conditions and adapted to diverse sociocultural traditions prevalent in India. Thus, Indian rice germplasm lines are rich in trait diversity. The rice germplasm resources, including cultivated varieties, breeding lines, landraces, wild accessions representing diverse agro-climatic regions of India as well as the world, have been conserved efficiently in different National germplasm repository centers like National Bureau of Plant Genetic Resources (NBPGR, New Delhi). The phenotypic and molecular characterization of the huge Indian rice germplasm collections is, therefore, vital in order to use the natural allelic diversity information in genomics-assisted rice crop improvement program. Considering the economic importance of aromatic rice in Basmati trade and commerce, significant efforts have been made for varietal identification and understanding the genetic diversity and domestication patterns majorly among different traditional (landraces) and evolved (elite) long- and short-grained Basmati and between Basmati and non-Basmati *indica* rice accessions (Nagaraju et al. 2002; Aggarwal et al. 2002; Sajib et al. 2012; Meti et al. 2013; Vhora et al. 2013). For example, the distinctness, uniformity and stability (DUS) in an aromatic rice variety, Pusa Basmati1, has been established by its precise phenotyping for diverse aroma and grain quality traits and large-scale genotyping of genome-wide SSR markers (Singh et al. 2004). The efficiency of genome-wide SSR markers and five mitochondrial gene-specific CAPS (cleaved amplified polymorphic sequences) markers for testing the genetic purity of Pusa6A parental lines during seed production of Pusa Rice Hybrid10 has been demonstrated. This would be of immense use in unambiguous identification and protection of rice hybrids, including marker-assisted production of pure hybrid rice seeds in India (Ngangkham et al. 2010; Anand et al. 2012). More recently, the efficacy of gene-based 50 K SNP chip in molecular diversity and evolutionary studies among cultivated and wild rice accessions was demonstrated (Singh et al. 2015b). The large-scale diversity analysis of Indian rice germplasm will also be useful in identifying important hot-spot (rich in trait and/or allelic diversity) rice producing regions of the country, like Eastern Indo-gangetic planes of Uttar Pradesh and Bihar for wild rice accessions as well as Eastern and North-eastern states of India for distinctive landraces (Das et al. 2013; Singh et al. 2013b, 2013c). Considering the difficulties involved in genotypic and phenotypic characterization of the huge set of available germplasm resources of rice, efforts have been made currently in India to constitute the core and mini-core collections in rice by identifying the largest amount of genetic diversity with a minimum number of accessions. In order to constitute core/minicore germplasm collections, diverse popularly grown and rarely cultivated traditional varieties and landraces belonging to *indica*, which constitute about 80 % of total cultivated Indian rice, have been assessed for phenotypic and genotypic diversity analysis (Prashanth et al. 2002). To constitute a trait-specific mini-core in rice, diversity analysis of germplasm lines have also been performed targeting various yield-component, biotic stress tolerance and grain quality traits (grain color and micronutrient content) (Banumathy et al. 2010; Singh et al. 2011a; Prasad et al. 2013; Patel et al. 2014). With the multi-institutional efforts of Indian scientists, recently a set of 98 accessions belonging to a core/

mini-core collection representing 94 % allelic diversity of the total 6912 rice germplasm lines has been constituted utilizing both marker-based genotyping and phenotyping strategies and different precise statistical measures (Tiwari et al. 2015a). Similarly, a core set of 701 accessions representing 99.9 % allelic diversity of total 6984 rice germplasm lines belonging to North Eastern region (considered as hot-spot region) of India has been developed (Choudhury et al. 2014). These readily available core/mini-core germplasm resources of rice have been phenotyped for diverse agronomic traits, including yield component and abiotic/biotic stress tolerance traits at different geographical locations (multi-environment) and hot-spot regions of India for multiple years in field.

Based on phenotypic and genotypic characterization of germplasm lines, accessions contrasting for major yield component and stress tolerance traits have been selected and utilized as parents for generation of advanced bi-parental and back-cross mapping populations such as RILs (recombinant inbred lines) and NILs (near isogenic lines) in rice. One of the upland *indica* (*aus*) rice accession, Nagina 22, has been induced with EMS (ethyl methane sulfonate) mutagen to generate and characterize 22,292 mutant lines through an Indian National initiative involving several research institutes (Mohapatra et al. 2014). These mutant lines were phenotyped with a collaborative effort for a range of traits, which led to identify mutants for flowering, maturity, grain number and size, plant growth and architecture, yield, resistance to blast and bacterial leaf blight diseases, tolerance to drought, heat, salinity and herbicide and phosphorus use efficiency (Ashokkumar et al. 2013; Kulkarni et al. 2013; Poli et al. 2013; Mohapatra et al. 2014; Panigrahy et al. 2014; Lima et al. 2015). These mutant repositories could serve as a valuable resource for mining of novel functional alleles regulating qualitative and quantitative traits for genetic improvement in rice.

Genome-Wide Scanning of Trait-Associated Functionally Relevant Molecular Tags

Investigations by Indian researchers in the context of high-throughput SSR and SNP marker-based genotyping in advanced generation bi-parental mapping population enabled to construct high-density genetic linkage and functional transcript maps and hastened identification and mapping of genes/QTLs associated with agronomic traits in rice. For instance, about 300 QTLs governing growth and grain yield contributing traits (panicle length, days to heading/flowering, plant height, grain yield, grain weight, grain size and grain length), various quality component traits (amylose content, cooked kernel elongation ratio, aroma, grain physico-chemical and cooking quality traits, and iron and zinc concentration nutritional quality traits) and stress tolerance traits (sheath blight, drought and salinity resistance) have been identified and mapped on high-density SSR and SNP marker-based genetic linkage maps derived from diverse *indica* and aromatic rice based mapping population of rice (Marri et al. 2005; Ammar et al. 2009; Amarawathi et al. 2008; Channamallikarjuna et al. 2010; Vikram et al. 2011; Salunkhe et al. 2011; Anand et al. 2012, 2013; Anuradha et al. 2012; Guleria et al. 2012; Marathi et al. 2012; Meenakshisundaram et al. 2011; Shanmugavadivel et al. 2013). The constructed high-density genetic linkage maps have been integrated with sequence-based physical map and improved the resolution and accuracy of trait-specific genes/QTLs identification. Utilizing such map-based cloning strategy, genes harbouring major QTLs associated with diverse agronomic traits, including *Pi-K^h* (*Pi54*) and PPR (pentatricopeptide repeat) genes regulating blast resistance and fertility restoration have been identified in rice (Sharma et al. 2005; Reddy et al. 2008; Ngangkham et al. 2010). Recently, the integration of QTL mapping and microarray-based genome-wide transcriptome profiling of parents and bulks of homozygous RILs has been found to be a powerful approach to narrow down the number of candidate genes underlying the QTLs of interest and for isolating the possible genes regulating the traits. Using such strategy, a candidate gene belonging to glycosyl hydrolase family regulating the number of grains per panicle in a major QTL region (*qGN4-1*) has been mapped on chromosome 4, and two genes encoding integral transmembrane protein DUF6 and cation chloride cotransporter co-localized in the significant QTL intervals on chromosomes 1, 8, and 12 for salt ion concentrations and two known genes, *badh1* and *badh2* (betaine aldehyde dehydrogenase) at QTL interval governing aroma have been identified in rice (Deshmukh et al. 2010; Pandit et al. 2010; Sharma et al. 2011; Pachauri et al. 2014).

Efforts have been made by Indian scientists to mine novel allelic variants in the known cloned genes such as *Pi54* and *Pita* regulating blast resistance, *Xa21*, *Xa26* and *xa5* for bacterial blight resistance, *OsDREB1F* for drought tolerance, *GW2* for grain size/weight and nine candidate stress responsive genes governing abiotic and biotic stress tolerance traits in rice (Singh et al. 2010c; Das et al. 2012; Parida et al. 2012; Dixit et al. 2013; Kumari et al. 2013; Devanna et al. 2014; Ramkumar et al. 2014; Bimolata et al. 2015; Singh et al. 2015c, 2015d; Thakur et al. 2015). In addition, diverse mutant populations and natural germplasm collections (core and mini-core) available for rice have been assayed through TILLING (targeting induced local lesions in genomes) and EcoTILLING to mine novel functional allelic variants in the known/candidate genes associated with various agronomic traits (Ashokkumar et al. 2013; Kulkarni et al. 2013, 2014; Poli et al. 2013; Mohapatra

et al. 2014; Panigrahy et al. 2014; Lima et al. 2015). More recently, GWAS (genome-wide association study) using the genotyping information of custom designed Illumina Infinium array based on 6000 SNPs (present in many stress responsive genes distributed across 12 chromosomes with average distance of <100 kb between SNP loci) assayed in 220 accessions (association panel), has been performed to identify 20 SNPs in known/candidate genes significantly associated with Na^+/K^+ ratio and 44 with other salinity stress tolerance contributing traits in rice (Kumar et al. 2015).

Marker-Assisted Breeding for Rice Crop Improvement

Many successful endeavors have been made by Indian rice molecular breeders to introgress and pyramid the superior functional genes and major QTLs/alleles regulating complex yield/quality component and stress tolerance traits into diverse rice genotypes especially using marker-assisted selection (MAS) for their genetic enhancement. The genetic improvement of Basmati rice for yield, quality and resistance to bacterial leaf blight (*Xa21*, *xa4*, *xa13*, *xa5*, *Xa33t*, *xa34t* and *Xa38*) and blast (*Pi1*, *Pi2*, *Pi5*, *Pi9*, *Pi54*, *Pib*, *Piz*, *Piz5*, *Pi-ta* and *Pi54/ Pi-K^h*), brown plant hopper [*Bph-3/17/18/20/21* and *Bph18(t)*], sheath blight (*qSHB*) and gall midge (*Gm4* and *Gm8*) diseases has been performed by pyramiding the multiple genes/QTLs through marker-assisted backcrossing (MABC)/marker-assisted foreground and background selection (Joseph et al. 2004; Sharma et al. 2005; Cheema et al. 2008; Sundaram et al. 2008, 2009; Gopalakrishnan et al. 2008; Himabindu et al. 2010; Basavaraj et al. 2010; Madhavi et al. 2011; Sama et al. 2010, 2012; Hari et al. 2011,2013; Singh et al. 2011b; Natarajkumar et al. 2012; Sujatha et al. 2010, 2013; Pandey et al. 2013; Pradhan et al. 2015). A multi-institutional National network project with an objective to introgress known cloned QTLs regulating drought (*DTY1.1*, *DTY2.1*, *DTY2.2*, *DTY3.1*, *DTY3.2*, *DTY9.1* and *DTY12.1*), flood (sub-mergence) (*Sub1*) and salinity stress (*Saltol*) tolerance (http:// india.irri.org/mega-projects-in-india, Singh et al. 2015b) into high-yielding mega rice varieties (ADT46, Bahadur, MTU1075, Pooja, Rajendra, Mahsuri, Ranjit, ADT39, Pusa44, ADT45, Gayatri and Savitri) of India through MAS for their genetic enhancement for target traits is under progress. This eventually may lead to development of certain diverse genetically-tailored high-yielding and climate resilient early maturing Indian rice varieties for sustaining food security.

Conclusions

The rice genome sequence has served as a catalyst to accelerate efforts on the functional analysis of genes/ QTLs by reverse and forward genetics in India. This is coupled with use of genomics-assisted breeding of rice to improve traits such as yield, abiotic stress tolerance and biotic stress resistance. Diverse varieties of rice are grown in different regions of India and concerted efforts are required to introgress genes/QTLs for desirable traits into them. Some of these efforts have already led to release of improved varieties for submergence tolerance and biotic stress resistance. Indian rice researchers have highly benefited from new knowledge generated worldwide and from their own collaborative efforts in crop improvement programs. It is hoped that appropriate regulatory process will help move transgenic rice also to the field level evaluation and deregulation in due course. In the meantime, several projects related to molecular dissection of desirable agronomic traits of rice and introgression of appropriate genes/QTLs have been initiated. India is also poised to contribute to international projects in the area of functional genomics of rice and looks forward to launch of activities like riceENCODE. It is hoped that the outcome of rice genomics would go a long way to influence research in other related crops as well.

Competing Interests
The authors declare that they have no competing interest.

Authors' Contributions
PA and SKP did literature survey and compiled the information. All authors contributed to the writing and finalization of the manuscript. AKT was responsible for the overall concept. All authors read and approved the final manuscript.

Acknowledgements
Investigations in authors' labs are supported by the Department of Biotechnology and the Department of Science and Technology, Government of India.

References
Agarwal P, Arora R, Ray S, Singh AK, Singh VP, Takatsuji H, Kapoor S, Tyagi AK (2007) Genome-wide identification of C2H2 zinc-finger gene family in rice and their phylogeny and expression analysis. Plant Mol Biol 65:467–485

Agarwal P, Kapoor S, Tyagi AK (2011) Transcription factors regulating the progression of monocot and dicot seed development. Bioessays 33:189–202

Agarwal P, Parida SK, Mahto A, Das S, Mathew IE, Malik N, Tyagi AK (2014) Expanding frontiers in plant transcriptomics in aid of functional genomics and molecular breeding. Biotechnol J 9:1480–1492

Aggarwal K, Shenoy V, Ramadevi J, Rajkumar R, Singh L (2002) Molecular characterization of some Indian Basmati and other elite rice genotypes using fluorescent-AFLP. Theor Appl Genet 105:680–690

Aishwarya V, Sharma PC (2007) UgMicroSatdb: database for mining microsatellites from unigenes. Nucleic Acids Res 36:D53–D56

Alexandrov N, Tai S, Wang W, Mansueto L, Palis K, Fuentes RR, Ulat VJ, Chebotarov D, Zhang G, Li Z, Mauleon R, Hamilton RS, McNally KL (2015) SNP-seek database of SNPs derived from 3000 rice genomes. Nucleic Acids Res 43:D1023–D1027

Ali N, Paul S, Gayen D, Sarkar SN, Datta K, Datta SK (2013a) Development of low phytate rice by RNAi mediated seed-specific silencing of inositol 1,3,4,5,6-pentakisphosphate 2-kinase gene (IPK1). PLoS One 8:e68161

Ali N, Paul S, Gayen D, Sarkar SN, Datta SK, Datta K (2013b) RNAi mediated down regulation of myo-inositol-3-phosphate synthase to generate low phytate rice. Rice 6:12

Amarawathi Y, Singh R, Singh AK, Singh VP, Mahopatra T, Sharma TR, Singh NK (2008) Mapping of quantitative trait loci for Basmati quality traits in rice (Oryza sativa L.). Mol Breed 21:49–65

Ammar MHM, Pandit A, Singh RK, Sameena S, Chauhan MS, Singh AK, Sharma PC, Gaikwad K, Sharma TR, Mohapatra T, Singh NK (2009) Mapping of QTLs controlling Na+, K+ and Cl− ion concentrations in salt tolerant indica rice variety CSR27. J Plant Biochem Biotechnol 18:139–150

Anand S, Tyagi AK (2010) Characterization of a pollen-preferential gene OSIAGP from rice (Oryza sativa L. subspecies indica) coding for an arabinogalactan protein homologue, and analysis of its promoter activity during pollen development and pollen tube growth. Transgenic Res 19:385–397

Anand D, Prabhu KV, Singh AK (2012) Analysis of molecular diversity and fingerprinting of commercially grown Indian rice hybrids. J Plant Biochem Biotechnol 21:173–179

Anand D, Baunthiyal M, Singh A, Gopalakrishnan S, Singh NK, Prabhu KV (2013) Validation of gene based marker-QTL association for grain dimension traits in rice. J Plant Biochem Biotechnol 22:467–473

Anuradha K, Agarwal S, Rao YV, Rao KV, Viraktamath BC, Sarla N (2012) Mapping QTLs and candidate genes for iron and zinc concentrations in unpolished rice of Madhukar x Swarna RILs. Gene 508:233–240

Archak S, Nagaraju J (2007) Computational prediction of rice (Oryza sativa) miRNA targets. Genomics Proteomics Bioinformatics 5:196–206

Arora R, Agarwal P, Ray S, Singh AK, Singh VP, Tyagi AK, Kapoor S (2007) MADS-box gene family in rice: genome-wide identification, organization and expression profiling during reproductive development and stress. BMC Genomics 8:242

Arora K, Rai AK, Gupta SK, Singh PK, Narula A, Sharma TR (2015) Phenotypic expression of blast resistance gene Pi54 is not affected by its chromosomal position. Plant Cell Rep 34:63–70

Ashokkumar K, Raveendran M, Senthil N, Vijayalaxmi D, Sowmya M, Sharma RP, Robin S (2013) Isolation and characterization of altered root growth behavior and salinity tolerant mutants in rice. Afr J Biotechnol 12:5852–5859

Babu PR, Sekhar AC, Ithal N, Markandeya G, Reddy AR (2002) Annotation and BAC/PAC localization of nonredundant ESTs from drought-stressed seedlings of an indica rice. J Genet 81:25–44

Banerjee J, Maiti MK (2010) Functional role of rice germin-like protein1 in regulation of plant height and disease resistance. Biochem Biophys Res Commun 394:178–183

Banerjee J, Das N, Dey P, Maiti MK (2010) Transgenically expressed rice germin-like protein1 in tobacco causes hyper-accumulation of H2O2 and reinforcement of the cell wall components. Biochem Biophys Res Commun 402:637–643

Banumathy S, Manimaran R, Sheeba A, Manivannan N, Ramya B, Kumar D, Ramasubramanian GV (2010) Genetic diversity analysis of rice germplasm lines for yield attributing traits. J Plant Breed 1:500–504

Barik S, SarkarDas S, Singh A, Gautam V, Kumar P, Majee M, Sarkar AK (2014) Phylogenetic analysis reveals conservation and diversification of micro RNA166 genes among diverse plant species. Genomics 103:114–121

Basavaraj SH, Singh VK, Singh A, Singh A, Singh A, Anand D, Yadav S, Ellur RK, Singh D, Gopalakrishnan S, Nagarajan M, Mohapatra T, Prabhu KV, Singh AK (2010) Marker-assisted improvement of bacterial blight resistance in parental lines of Pusa RH10, a superfine grain aromatic rice hybrid. Mol Breed 26:293–305

Batra G, Chauhan VS, Singh A, Sarkar NK, Grover A (2007) Complexity of rice Hsp100 gene family: lessons from rice genome sequence data. J Biosci 32:611–619

Bhattacharyya J, Chowdhury AH, Ray S, Jha JK, Das S, Gayen S, Chakraborty A, Mitra J, Maiti MK, Basu A, Sen SK (2012) Native polyubiquitin promoter of rice provides increased constitutive expression in stable transgenic rice plants. Plant Cell Rep 31:271–279

Bimolata W, Kumar A, Reddy MSK, Sundaram RM, Laha GS, Qureshi IA, Ghazi IA (2015) Nucleotide diversity analysis of three major bacterial blight resistance genes in rice. PLoS One 10:e0120186

Chakrabarty D, Trivedi PK, Misra P, Tiwari M, Shri M, Shukla D, Kumar S, Rai A, Pandey A, Nigam D, Tripathi RD, Tuli R (2009) Comparative transcriptome analysis of arsenate and arsenite stresses in rice seedlings. Chemosphere 74:688–702

Chakrabarty D, Trivedi PK, Shri M, Misra P, Asif MH, Dubey S, Kumar S, Rai A, Tiwari M, Shukla D, Pandey A, Nigam D, Tripathi RD, Tuli R (2010) Differential transcriptional expression following thidiazuron-induced callus differentiation developmental shifts in rice. Plant Biol 12:46–59

Chakraborty A, Mitra J, Bhattacharyya J, Pradhan S, Sikdar N, Das S, Chakraborty S, Kumar S, Lakhanpaul S, Sen SK (2015) Transgenic expression of an unedited mitochondrial *orfB* gene product from wild abortive (WA) cytoplasm of rice (*Oryza sativa* L.) generates male sterility in fertile rice lines. Planta 241:1463–1479

Channamallikarjuna V, Sonah H, Prasad M, Rao GJN, Chand S, Upreti HC, Singh NK, Sharma TR (2010) Identification of major quantitative trait loci *qSBR11-1* for sheath blight resistance in rice. Mol Breed 25:155–166

Chaudhary N, Nijhawan A, Khurana JP, Khurana P (2010) Carotenoid biosynthesis genes in rice: structural analysis, genome-wide expression profiling and phylogenetic analysis. Mol Genet Genomics 283:13–33

Chauhan H, Khurana N, Agarwal P, Khurana P (2011) Heat shock factors in rice (*Oryza sativa* L.): genome-wide expression analysis during reproductive development and abiotic stress. Mol Genet Genomics 286:171–187

Cheema K, Grewal N, Vikal Y, Sharma R, Lore JS, Das A, Bhatia D, Mahajan R, Gupta V, Bharaj TS, Singh K (2008) A novel bacterial blight resistance gene from *Oryza nivara* mapped to 38 kb region on chromosome 4 L and transferred to *Oryza sativa* L. Genet Res 90:397–407

Chen M, Presting G, Barbazuk WB, Goicoechea JL, Blackmon B, Fang G, Kim H, Frisch D, Yu Y, Sun S, Higingbottom S, Phimphilai J, Phimphilai D, Thurmond S, Gaudette B, Li P, Liu J, Hatfield J, Main D, Farrar K, Henderson C, Barnett L, Costa R, Williams B, Walser S, Atkins M, Hall C, Budiman MA, Tomkins JP, Luo M, Bancroft I, Salse J, Regad F, Mohapatra T, Singh NK, Tyagi AK, Soderlund C, Dean RA, Wing RA (2002) An integrated physical and genetic map of the rice genome. Plant Cell 14:537–545

Choudhury DR, Singh N, Singh AK, Kumar S, Srinivasan K, Tyagi RK, Ahmad A, Singh NK, Singh R (2014) Analysis of genetic diversity and population structure of rice germplasm from North-eastern region of India and development of a core germplasm set. PLoS One 9:e113094

Dansana PK, Kothari KS, Vij S, Tyagi AK (2014) *OsiSAP1* overexpression improves water-deficit stress tolerance in transgenic rice by affecting expression of endogenous stress-related genes. Plant Cell Rep 33:1425–1440

Das A, Soubam D, Singh PK, Thakur S, Singh NK, Sharma TR (2012) A novel blast resistance gene, *Pi54rh* cloned from wild species of rice, *Oryza rhizomatis* confers broad spectrum resistance to *Magnaporthe oryzae*. Funct Integr Genomics 12:215–228

Das B, Sengupta S, Parida SK, Roy B, Ghosh M, Prasad M, Ghose TK (2013) Genetic diversity and population structure of rice landraces from Eastern and North Eastern states of India. BMC Genet 14:71

Das-Chatterjee A, Goswami L, Maitra S, Dastidar KG, Ray S, Majumder AL (2006) Introgression of a novel salt-tolerant L-myo-inositol 1-phosphate synthase from *Porteresia coarctata* (Roxb.) Tateoka (*PcINO1*) confers salt tolerance to evolutionary diverse organisms. FEBS Lett 580:3980–3988

Deshmukh R, Singh A, Jain N, Anand S, Gacche R, Singh A, Gaikwad K, Sharma T, Mohapatra T, Singh N (2010) Identification of candidate genes for grain number in rice (*Oryza sativa* L.). Funct Integr Genomics 10:339–347

Deshpande GM, Ramakrishna K, Chongloi GL, Vijayraghavan U (2015) Functions for rice *RFL* in vegetative axillary meristem specification and outgrowth. J Exp Bot 66:2773–2784

Devanna NB, Vijayan J, Sharma TR (2014) The blast resistance gene *Pi54 of* cloned from *Oryza officinalis* interacts with Avr-Pi54 through its novel non-LRR domains. PLoS One 9:e104840

Deveshwar P, Bovill WD, Sharma R, Able JA, Kapoor S (2011) Analysis of anther transcriptomes to identify genes contributing to meiosis and male gametophyte development in rice. BMC Plant Biol 11:78

Devi SJ, Madhav MS, Kumar GR, Goel AK, Umakanth B, Jahnavi B, Viraktamath BC (2013) Identification of abiotic stress miRNA transcription factor binding motifs (TFBMs) in rice. Gene 531:15–22

Dixit N, Dokku P, Amitha Mithra SV, Parida SK, Singh AK, Singh NK, Mohapatra T (2013) Haplotype structure in grain weight gene *GW2* and its association with grain characteristics in rice. Euphytica 192:55–61

Dubey S, Misra P, Dwivedi S, Chatterjee S, Bag SK, Mantri S, Asif MH, Rai A, Kumar S, Shri M, Tripathi P, Tripathi RD, Trivedi PK, Chakrabarty D, Tuli R (2010) Transcriptomic and metabolomic shifts in rice roots in response to Cr (VI) stress. BMC Genomics 11:648

Dubey S, Shri M, Misra P, Lakhwani D, Bag SK, Asif MH, Trivedi PK, Tripathi RD, Chakrabarty D (2014) Heavy metals induce oxidative stress and genome-wide modulation in transcriptome of rice root. Funct Integr Genomics 14:401–417

Ganguly M, Roychoudhury A, Sarkar SN, Sengupta DN, Datta SK, Datta K (2011) Inducibility of three salinity/abscisic acid-regulated promoters in transgenic rice with *gusA* reporter gene. Plant Cell Rep 30:1617–1625

Garg R, Jhanwar S, Tyagi AK, Jain M (2010) Genome-wide survey and expression analysis suggest diverse roles of glutaredoxin gene family members during development and response to various stimuli in rice. DNA Res 17:353–367

Gaur VS, Singh US, Kumar A (2011) Transcriptional profiling and *in silico* analysis of Dof transcription factor gene family for understanding their regulation during seed development of rice *Oryza sativa* L. Mol Biol Rep 38:2827–2848

Gautam N, Verma PK, Verma S, Tripathi RD, Trivedi PK, Adhikari B, Chakrabarty D (2012) Genome-wide identification of rice class I metallothionein gene: tissue expression patterns and induction in response to heavy metal stress. Funct Integr Genomics 12:635–647

Ghosh A, Pareek A, Sopory SK, Singla-Pareek SL (2014) A glutathione responsive rice glyoxalase II, *OsGLYII-2*, functions in salinity adaptation by maintaining better photosynthesis efficiency and anti-oxidant pool. Plant J 80:93–105

Giri J, Vij S, Dansana PK, Tyagi AK (2011) Rice A20/AN1 zinc-finger containing stress-associated proteins (SAP1/11) and a receptor-like cytoplasmic kinase (OsRLCK253) interact via A20 zinc-finger and confer abiotic stress tolerance in transgenic Arabidopsis plants. New Phytol 191:721–732

Goel A, Taj G, Pandey D, Gupta S, Kumar A (2011) Genome-wide comparative *in silico* analysis of calcium transporters of rice and sorghum. Genomics Proteomics Bioinformatics 9:138–150

Gopalakrishnan S, Sharma RK, Rajkumar KA, Joseph M, Singh VP, Singh AK, Bhat KV, Singh NK, Mohapatra T (2008) Integrating marker-assisted background analysis with foreground selection for identification of superior bacterial blight resistant recombinants in Basmati rice. Plant Breed 127:131–139

Gorantla M, Babu PR, Lachagari VB, Reddy AM, Wusirika R, Bennetzen JL, Reddy AR (2007) Identification of stress-responsive genes in an *indica* rice (*Oryza sativa* L.) using ESTs generated from drought-stressed seedlings. J Exp Bot 58:253–265

Gour P, Garg P, Jain R, Joseph SV, Tyagi AK, Raghuvanshi S (2014) Manually curated database of rice proteins. Nucleic Acids Res 42:D1214–D1221

Grewal RK, Gupta S, Das S (2012) *Xanthomonas oryzae* pv oryzae triggers immediate transcriptomic modulations in rice. BMC Genomics 13:49

Grover A, Aishwarya V, Sharma PC (2007) Biased distribution of microsatellite motifs in the rice genome. Mol Genet Genomics 277:469–480

Guleria S, Sharma V, Marathi B, Anand S, Singh S, Singh NK, Mohapatra T, Gopala S, Prabhu KV, Singh AK (2012) Molecular mapping of grain physico-chemical and cooking quality traits using recombinant inbred lines in rice (*Oryza sativa* L.). J Plant Biochem Biotechnol 21:1–10

Gupta SK, Rai AK, Kanwar SS, Chand D, Singh NK, Sharma TR (2012) The single functional blast resistance gene *Pi54* activates a complex defence mechanism in rice. J Exp Bot 63:757–772

Gutha LR, Reddy AR (2008) Rice DREB1B promoter shows distinct stress-specific responses, and the overexpression of cDNA in tobacco confers improved abiotic and biotic stress tolerance. Plant Mol Biol 68:533–555

Hari Y, Srinivasarao K, Viraktamath BC, Hariprasad AS, Laha GS, Ahmed MI, Natarajkumar P, Ramesha MS, Neeraja CN, Balachandran SM, Rani NS, Suresh PB, Sujatha K, Pandey M, Reddy AG, Madhav MS, Sundaram RM (2011) Marker-assisted improvement of a stable restorer line, KMR-3R and its derived hybrid KRH2 for bacterial blight resistance and grain quality. Plant Breed 130:608–616

Hari Y, Srinivasarao K, Viraktamath BC, Hariprasad AS, Laha GS, Ahmed MI, Natarajkumar P, Sujatha K, Prasad MS, Pandey M, Ramesha MS, Neeraja CN, Balachandran SM, Shobharani N, Kemparaju B, Madhanmohan K, Sama VSAK, Hajira SK, Baachiranjeevi CH, Pranathi K, Reddy AG, Madhav MS, Sundaram RM (2013) Marker-assisted introgression of bacterial blight and blast resistance into IR 58025B, an elite maintainer line of rice. Plant Breed 132:586–594

Himabindu K, Suneetha K, Sama V, Bentur JS (2010) A new rice gall midge resistance gene in the breeding line CR57-MR1523, mapping with flanking markers and development of NILs. Euphytica 174:179–187

Huda KM, Banu MS, Garg B, Tula S, Tuteja R, Tuteja N (2013) OsACA6, a P-type IIB Ca^{2+} ATPase promotes salinity and drought stress tolerance in tobacco by ROS scavenging and enhancing the expression of stress-responsive genes. Plant J 76:997–1015

International Rice Genome Sequencing Project (2005) The map-based sequence of the rice genome. Nature 436:793–800

Jain M, Kaur N, Garg R, Thakur JK, Tyagi AK, Khurana JP (2006a) Structure and expression analysis of early auxin-responsive Aux/IAA gene family in rice (*Oryza sativa*). Funct Integr Genomics 6:47–59

Jain M, Kaur N, Tyagi AK, Khurana JP (2006b) The auxin-responsive GH3 gene family in rice (*Oryza sativa*). Funct Integr Genomics 6:36–46

Jain M, Tyagi AK, Khurana JP (2006c) Overexpression of putative topoisomerase 6 genes from rice confers stress tolerance in transgenic Arabidopsis plants. FEBS J 273:5245–5260

Jain M, Tyagi AK, Khurana JP (2006d) Genome-wide analysis, evolutionary expansion, and expression of early auxin-responsive SAUR gene family in rice (Oryza sativa). Genomics 88:360–371

Jain M, Nijhawan A, Arora R, Agarwal P, Ray S, Sharma P, Kapoor S, Tyagi AK, Khurana JP (2007) F-box proteins in rice. Genome-wide analysis, classification, temporal and spatial gene expression during panicle and seed development, and regulation by light and abiotic stress. Plant Physiol 143:1467–1483

Jain M, Khurana P, Tyagi AK, Khurana JP (2008a) Genome-wide analysis of intronless genes in rice and Arabidopsis. Funct Integr Genomics 8:69–78

Jain M, Tyagi AK, Khurana JP (2008b) Constitutive expression of a meiotic recombination protein gene homolog, OsTOP6A1, from rice confers abiotic stress tolerance in transgenic Arabidopsis plants. Plant Cell Rep 27:767–778

Jain M, Tyagi AK, Khurana JP (2008c) Genome-wide identification, classification, evolutionary expansion and expression analyses of homeobox genes in rice. FEBS J 275:2845–2861

Jain M, Moharana KC, Shankar R, Kumari R, Garg R (2014) Genomewide discovery of DNA polymorphisms in rice cultivars with contrasting drought and salinity stress response and their functional relevance. Plant Biotechnol J 12:253–264

Jha A, Shankar R (2011) Employing machine learning for reliable miRNA target identification in plants. BMC Genomics 12:636

Joseph M, Gopalakrishnan S, Sharma RK, Singh AK, Singh VP, Singh NK, Mohapatra T (2004) Combining bacterial blight resistance and Basmati quality characteristics by phenotypic and molecular marker assisted selection in rice. Mol Breed 13:377–387

Joshi PK, Gupta D, Nandal UK, Khan Y, Mukherjee SK, Sanan-Mishra N (2012) Identification of mirtrons in rice using MirtronPred: a tool for predicting plant mirtrons. Genomics 99:370–375

Kamrul Huda KM, Yadav S, Akhter Banu MS, Trivedi DK, Tuteja N (2013) Genome-wide analysis of plant-type II Ca²⁺ ATPases gene family from rice and Arabidopsis: potential role in abiotic stresses. Plant Physiol Biochem 65:32–47

Kamrul Huda KM, Akhter Banu MS, Yadav S, Sahoo RK, Tuteja R, Tuteja N (2014) Salinity and drought tolerant OsACA6 enhances cold tolerance in transgenic tobacco by interacting with stress-inducible proteins. Plant Physiol Biochem 82:229–238

Kanneganti V, Gupta AK (2008) Overexpression of OsiSAP8, a member of stress associated protein (SAP) gene family of rice confers tolerance to salt, drought and cold stress in transgenic tobacco and rice. Plant Mol Biol 66:445–462

Kanneganti V, Gupta AK (2011) RNAi mediated silencing of a wall associated kinase, OsiWAK1 in Oryza sativa results in impaired root development and sterility due to anther indehiscence: Wall Associated Kinases from Oryza sativa. Physiol Mol Biol Plant 17:65–77

Kansal S, Devi RM, Balyan SC, Arora MK, Singh AK, Mathur S, Raghuvanshi S (2015) Unique miRNome during anthesis in drought-tolerant indica rice var. Nagina 22. Planta 241:1543–1559

Kanwar P, Sanyal SK, Tokas I, Yadav AK, Pandey A, Kapoor S, Pandey GK (2014) Comprehensive structural, interaction and expression analysis of CBL and CIPK complement during abiotic stresses and development in rice. Cell Calcium 56:81–95

Kapoor M, Arora R, Lama T, Nijhawan A, Khurana JP, Tyagi AK, Kapoor S (2008) Genome-wide identification, organization and phylogenetic analysis of Dicer-like, Argonaute and RNA-dependent RNA Polymerase gene families and their expression analysis during reproductive development and stress in rice. BMC Genomics 9:451

Katiyar A, Smita S, Lenka SK, Rajwanshi R, Chinnusamy V, Bansal KC (2012) Genome-wide classification and expression analysis of MYB transcription factor families in rice and Arabidopsis. BMC Genomics 13:544

Kaundal R, Kapoor AS, Raghava GP (2006) Machine learning techniques in disease forecasting: a case study on rice blast prediction. BMC Bioinformatics 7:485

Kaur C, Mustafiz A, Sarkar AK, Ariyadasa TU, Singla-Pareek SL, Sopory SK (2014) Expression of abiotic stress inducible ETHE1-like protein from rice is higher in roots and is regulated by calcium. Physiol Plant 152:1–16

Khurana R, Kapoor S, Tyagi AK (2013) Spatial and temporal activity of upstream regulatory regions of rice anther-specific genes in transgenic rice and Arabidopsis. Transgenic Res 22:31–46

Kulkarni KP, Vishwakarma C, Sahoo SP, Lima JM, Nath M, Dokku P, Gacche RN, Mohapatra T, Robin S, Sarla N, Seshashayee M, Singh AK, Singh K, Singh NK, Sharma RP (2013) Phenotypic characterization and genetic analysis of dwarf and early flowering mutants of rice variety Nagina22. Oryza 50:18–25

Kulkarni KP, Vishwakarma C, Sahoo SP, Lima JM, Nath M, Dokku P, Gacche RN, Mohapatra T, Robin S, Sarla N, Seshashayee M, Singh AK, Singh K, Singh NK, Sharma RP (2014) A substitution mutation in OsCCD7 cosegregates with dwarf and increased tillering phenotype in rice. J Genet 93:389–401

Kumar K, Sinha AK (2013) Overexpression of constitutively active mitogen activated protein kinase kinase 6 enhances tolerance to salt stress in rice. Rice 6:25

Kumar K, Sinha AK (2014) Genome-wide transcriptome modulation in rice transgenic lines expressing engineered mitogen activated protein kinase kinase 6. Plant Signal Behav 9:e28502

Kumar S, Asif MH, Chakrabarty D, Tripathi RD, Trivedi PK (2011) Differential expression and alternative splicing of rice sulphate transporter family members regulate sulphur status during plant growth, development and stress conditions. Funct Integr Genomics 11:259–273

Kumar G, Kushwaha HR, Panjabi-Sabharwal V, Kumari S, Joshi R, Karan R, Mittal S, Pareek SL, Pareek A (2012) Clustered metallothionein genes are co-regulated in rice and ectopic expression of OsMT1e-P confers multiple abiotic stress tolerance in tobacco via ROS scavenging. BMC Plant Biol 12:107

Kumar S, Asif MH, Chakrabarty D, Tripathi RD, Dubey RS, Trivedi PK (2013) Expression of a rice Lambda class of glutathione S-transferase, OsGSTL2, in Arabidopsis provides tolerance to heavy metal and other abiotic stresses. J Hazard Mater 248–249:228–237

Kumar SP, Pandya HA, Jasrai YT (2014) A computational model for non-conserved mature miRNAs from the rice genome. SAR QSAR Environ Res 25:205–220

Kumar V, Singh A, Amitha Mithra SV, Krishnamurthy SL, Parida SK, Jain S, Tiwari KK, Kumar P, Rao AR, Sharma SK, Khurana JP, Singh NK, Mohapatra T (2015) Genome-wide association mapping of salinity tolerance in rice (Oryza sativa). DNA Res 22:133–145

Kumari S, Sabharwal VP, Kushwaha HR, Sopory SK, Singla-Pareek SL, Pareek A (2009) Transcriptome map for seedling stage specific salinity stress response indicates a specific set of genes as candidate for saline tolerance in Oryza sativa L. Funct Integr Genomics 9:109–123

Kumari A, Das A, Devanna B, Thakur S, Singh P, Singh N, Sharma T (2013) Mining of rice blast resistance gene Pi54 shows effect of single nucleotide polymorphisms on phenotypic expression of the alleles. Eur J Plant Pathology 137:55–65

Kumari S, Joshi R, Singh K, Roy S, Tripathi AK, Singh P, Singla-Pareek SL, Pareek A (2015) Expression of a cyclophilin OsCyp2-P isolated from a salt-tolerant landrace of rice in tobacco alleviates stress via ion homeostasis and limiting ROS accumulation. Funct Integr Genomics. 15:395–412

Kushwaha HR, Singh AK, Sopory SK, Singla-Pareek SL, Pareek A (2009) Genome wide expression analysis of CBS domain containing proteins in Arabidopsis thaliana (L.) Heynh and Oryza sativa L. reveals their developmental and stress regulation. BMC Genomics 10:200

Lakra N, Nutan KK, Das P, Anwar K, Singla-Pareek SL, Pareek A (2015) A nuclear-localized histone-gene binding protein from rice (OsHBP1b) functions in salinity and drought stress tolerance by maintaining chlorophyll content and improving the antioxidant machinery. J Plant Physiol 176:36–46

Lenka SK, Katiyar A, Chinnusamy V, Bansal KC (2011) Comparative analysis of drought-responsive transcriptome in indica rice genotypes with contrasting drought tolerance. Plant Biotechnol J 9:315–327

Li JY, Wang J, Zeigler RS (2014) The 3,000 rice genomes project: new opportunities and challenges for future rice research. Gigascience 3:8

Lima JM, Nath M, Dokku P, Raman KV, Kulkarni KP, Vishwakarma C, Sahoo SP, Mohapatra UB, Amitha Mithra SV, Chinnusamy V, Robin S, Sarla N, Seshashayee M, Singh K, Singh AK, Singh NK, Sharma RP, Mohapatra T (2015) Physiological, anatomical and transcriptional alterations in a rice mutant leading to enhanced water stress tolerance. AoB Plants. 7:plv023

Macovei A, Tuteja N (2012) microRNAs targeting DEAD-box helicases are involved in salinity stress response in rice (Oryza sativa L.). BMC Plant Biol 12:183

Madhavi KR, Prasad MS, Laha GS, Mohan KM, Madhav MS, Viraktamath BC (2011) Combining blast and bacterial blight resistance in rice cultivar, improved Samba Mahsuri. Indian J Plant Protect 39:124–129

Majee M, Maitra S, Dastidar KG, Pattnaik S, Chatterjee A, Hait NC, Das KP, Majumder AL (2004) A novel salt-tolerant L-myo-inositol-1-phosphate synthase from Porteresia coarctata (Roxb.) Tateoka, a halophytic wild rice: molecular cloning, bacterial overexpression, characterization, and functional introgression into tobacco-conferring salt tolerance phenotype. J Biol Chem 279:28539–28552

Mallikarjuna G, Mallikarjuna K, Reddy MK, Kaul T (2011) Expression of OsDREB2A transcription factor confers enhanced dehydration and salt stress tolerance in rice (Oryza sativa L.). Biotechnol Lett 33:1689–1697

Manimaran P, Mangrauthia SK, Sundaram RM, Balachandran SM (2015) Constitutive expression and silencing of a novel seed specific calcium dependent protein kinase gene in rice reveals its role in grain filling. J Plant Physiol 174:41–48

Marathi B, Guleria S, Mohapatra T, Parsad R, Mariappan N, Kurungara VK, Atwal SS, Prabhu KV, Singh NK, Singh AK (2012) QTL analysis of novel genomic regions associated with yield and yield related traits in new plant type based recombinant inbred lines of rice (Oryza sativa L.). BMC Plant Biol 12:137–137

Marla SS, Singh VK (2012) LOX genes in blast fungus (Magnaporthe grisea) resistance in rice. Funct Integr Genomics 12:265–275

Marri PR, Sarla N, Reddy LV, Siddiq EA (2005) Identification and mapping of yield and yield related QTLs from an Indian accession of Oryza rufipogon. BMC Genet 6:33

Mathur S, Vyas S, Kapoor S, Tyagi AK (2011) The Mediator complex in plants: structure, phylogeny, and expression profiling of representative genes in a dicot (Arabidopsis) and a monocot (rice) during reproduction and abiotic stress. Plant Physiol 157:1609–1627

McCouch SR, Teytelman L, Xu Y, Lobos KB, Clare K, Walton M, Fu B, Maghirang R, Li Z, Xing Y, Zhang Q, Kono I, Yano M, Fjellstrom R, DeClerck G, Schneider D, Cartinhour S, Ware D, Stein L (2002) Development of 2240 new SSR markers for rice (Oryza sativa L.). DNA Res 9:199–207

Meenakshisundaram P, Patel SB, Sudha M, Geethanjali S, Vinod KK, Selvaraju K, Govindaraj P, Arumugachamy S, Shanmugasundaram P, Maheswaran M (2011) Microsatellite marker based linkage map construction and mapping of granule bound starch synthase (GBSS) in rice using recombinant inbred lines of the cross Basmati 370/ASD16. Crop Improv 38:155–162

Meti N, Samal KC, Bastia DN, Rout GR (2013) Genetic diversity analysis in aromatic rice genotypes using microsatellite based simple sequence repeats (SSR) marker. Afr J Biotechnol 12:4238–4250

Mittal D, Chakrabarti S, Sarkar A, Singh A, Grover A (2009) Heat shock factor gene family in rice: genomic organization and transcript expression profiling in response to high temperature, low temperature and oxidative stresses. Plant Physiol Biochem 47:785–795

Mittal D, Madhyastha DA, Grover A (2012a) Gene expression analysis in response to low and high temperature and oxidative stresses in rice: combination of stresses evokes different transcriptional changes as against stresses applied individually. Plant Sci 197:102–113

Mittal D, Madhyastha DA, Grover A (2012b) Genome-wide transcriptional profiles during temperature and oxidative stress reveal coordinated expression patterns and overlapping regulons in rice. PLoS One 7:e40899

Mittal D, Mukherjee SK, Vasudevan M, Mishra NS (2013) Identification of tissue-preferential expression patterns of rice miRNAs. J Cell Biochem 114:2071–2081

Mohapatra T, Robin S, Sarla N, Sheshashayee M, Singh AK, Singh K, Singh NK, Amitha Mithra SV, Sharma RP (2014) EMS induced mutants of upland rice variety Nagina22: generation and characterization. Proc Indian Natl Sci Acad 80:163–172

Molla KA, Karmakar S, Chanda PK, Ghosh S, Sarkar SN, Datta SK, Datta K (2013) Rice oxalate oxidase gene driven by green tissue-specific promoter increases tolerance to sheath blight pathogen (Rhizoctonia solani) in transgenic rice. Mol Plant Pathol 14:910–922

Mondal TK, Ganie SA (2014) Identification and characterization of salt responsive miRNA-SSR markers in rice (Oryza sativa). Gene 535:204–209

Mukhopadhyay P, Tyagi AK (2015) OsTCP19 influences developmental and abiotic stress signaling by modulating ABI4-mediated pathways. Sci Rep 29:9998

Mukhopadhyay P, Vij S, Tyagi AK (2004) Overexpression of a zinc-finger protein gene from rice confers tolerance to cold, dehydration, and salt stress in transgenic tobacco. Proc Natl Acad Sci U S A 101:6309–6314

Mustafiz A, Singh AK, Pareek A, Sopory SK, Singla-Pareek SL (2011) Genome-wide analysis of rice and Arabidopsis identifies two glyoxalase genes that are highly expressed in abiotic stresses. Funct Integr Genomics 11:293–305

Mutum RD, Balyan SC, Kansal S, Agarwal P, Kumar S, Kumar M, Raghuvanshi S (2013) Evolution of variety-specific regulatory schema for expression of osa-miR408 in indica rice varieties under drought stress. FEBS J 280:1717–1730

Nagaraju J, Kathirvel M, Kumar RR, Siddiq EA, Hasnain SE (2002) Genetic analysis of traditional and evolved Basmati and non-Basmati rice varieties by using fluorescence-based ISSR-PCR and SSR markers. Proc Natl Acad Sci U S A 99:5836–5841

Naik PK, Mittal VK, Gupta S (2008) RetroPred: A tool for prediction, classification and extraction of non-LTR retrotransposons (LINEs & SINEs) from the genome by integrating PALS, PILER, MEME and ANN. Bioinformation 2:263–270

Naika M, Shameer K, Mathew OK, Gowda R, Sowdhamini R (2013) STIFDB2: an updated version of plant stress-responsive transcription factor database with additional stress signals, stress-responsive transcription factor binding sites and stress-responsive genes in Arabidopsis and rice. Plant Cell Physiol 54:e8

Natarajkumar P, Sujatha K, Laha GS, Rao KS, Mishra B, Viraktamath BC, Hari Y, Reddy CS, Balachandran SM, Ram T, Madhav MS, Rani NS, Neeraja CN, Reddy GA, Shaik H, Sundaram RM (2012) Identification and fine-mapping of Xa33, a novel gene for resistance to Xanthomonas oryzae pv. oryzae. Phytopathology 102:222–228

Nayar S, Sharma R, Tyagi AK, Kapoor S (2013) Functional delineation of rice MADS29 reveals its role in embryo and endosperm development by affecting hormone homeostasis. J Exp Bot 64:4239–4253

Nayar S, Kapoor M, Kapoor S (2014) Post-translational regulation of rice MADS29 function: homodimerization or binary interactions with other seed-expressed MADS proteins modulate its translocation into the nucleus. J Exp Bot 65:5339–53350

Ngangkham U, Parida SK, De S, Kumar KAR, Singh AK, Singh NK, Mohapatra T (2010) Genic markers for wild abortive (WA) cytoplasm based male sterility and its fertility restoration in rice. Mol Breed 26:275–292

Nigam D, Kumar S, Mishra DC, Rai A, Smita S, Saha A (2015) Synergistic regulatory networks mediated by microRNAs and transcription factors under drought, heat and salt stresses in Oryza sativa spp. Gene 555:127–139

Nijhawan A, Jain M, Tyagi AK, Khurana JP (2008) Genomic survey and gene expression analysis of the basic leucine zipper transcription factor family in rice. Plant Physiol 146:333–350

Nischal L, Mohsin M, Khan I, Kardam H, Wadhwa A, Abrol YP, Iqbal M, Ahmad A (2012) Identification and comparative analysis of microRNAs associated with low-N tolerance in rice genotypes. PLoS One 7:e50261

Pachauri V, Mishra V, Mishra P, Singh AK, Singh S, Singh R, Singh NK (2014) Identification of candidate genes for rice grain aroma by combining QTL mapping and transcriptome profiling approaches. Cereal Res Comm 42:376–388

Pandey MK, Rani NS, Sundaram RM, Laha GS, Madhav MS, Rao KS, Sudharshan I, Hari Y, Varaprasad GS, Rao LVS, Suneetha K, Sivaranjani AKP, Viraktamath BC (2013) Improvement of two traditional Basmati rice varieties for bacterial blight resistance and plant stature through morphological and marker-assisted selection. Mol Breed 31:239–246

Pandey C, Raghuram B, Sinha AK, Gupta M (2015) miRNA plays a role in the antagonistic effect of selenium on arsenic stress in rice seedlings. Metallomics 7:857–866

Pandit A, Rai V, Bal S, Sinha S, Kumar V, Chauhan M, Gautam RK, Singh R, Sharma PC, Singh AK, Gaikwad K, Sharma TR, Mohapatra T, Singh NK (2010) Combining QTL mapping and transcriptome profiling of bulked RILs for identification of functional polymorphism for salt tolerance genes in rice (Oryza sativa L.). Mol Genet Genomics 284:121–136

Panigrahy M, Rao DN, Yugandhar P, Raju NS, Krishnamurthya P, Voletia SR, Reddy GA, Mohapatra T, Robin S, Singh AK, Singh K, Sheshshayeef M, Sharma RP, Sarla N (2014) Hydroponic experiment for identification of tolerance traits developed by rice Nagina22 mutants to low-phosphorus in field condition. Arch Agron Soil Sci 60:565–576

Pareek A, Singh A, Kumar M, Kushwaha HR, Lynn AM, Singla-Pareek SL (2006) Whole-genome analysis of Oryza sativa reveals similar architecture of two-component signaling machinery with Arabidopsis. Plant Physiol 142:380–397

Parida SK, Kumar KA, Dalal V, Singh NK, Mohaptra T (2006) Unigene derived microsatellite markers for the cereal genomes. Theor Appl Genet 112:808–817

Parida SK, Dalal V, Singh AK, Singh NK, Mohaptra T (2009) Genic non-coding microsatellites in the rice genome: characterization, marker design and use in assessing genetic and evolutionary relationships among domesticated groups. BMC Genomics 10:140

Parida SK, Mukerji M, Singh AK, Singh NK, Mohapatra T (2012) SNPs in stress-responsive rice genes: validation, genotyping, functional relevance and population structure. BMC Genomics 13:426

Patel S, Ravikiran R, Chakraborty S, Macwana S, Sasidharan N, Trivedi R, Aher B (2014) Genetic diversity analysis of colored and white rice genotypes using microsatellite (SSR) and Insertion-Deletion (INDEL) markers. Emir J Food Agric 26:497–507

Patra B, Ray S, Richter A, Majumder AL (2010) Enhanced salt tolerance of transgenic tobacco plants by co-expression of PcINO1 and McIMT1 is accompanied by increased level of myo-inositol and methylated inositol. Protoplasma 245:143–152

Paul P, Chakraborty S (2013) Computational prediction of submergence responsive microRNA and their binding position within the genome of *Oryza sativa*. Bioinformation 9:858–863

Poli Y, Basava RK, Panigrahy M, Vinukonda VP, Dokula NR, Voleti SR, Desiraju S, Neelamraju S (2013) Characterization of a Nagina22 rice mutant for heat tolerance and mapping of yield traits. Rice 6:36

Pradhan SK, Nayak DK, Mohanty S, Behera L, Barik SR, Pandit E, Lenka S, Anandan A (2015) Pyramiding of three bacterial blight resistance genes for broad-spectrum resistance in deepwater rice variety, Jalmagna. Rice 8:19

Prasad K, Sriram P, Kumar CS, Kushalappa K, Vijayraghavan U (2001) Ectopic expression of rice OsMADS1 reveals a role in specifying the lemma and palea, grass floral organs analogous to sepals. Dev Genes Evol 211:281–290

Prasad K, Kushalappa K, Vijayraghavan U (2003) Mechanism underlying regulated expression of RFL, a conserved transcription factor, in the developing rice inflorescence. Mech Dev 120:491–502

Prasad BD, Creissen G, Lamb C, Chattoo BB (2009) Overexpression of rice (*Oryza sativa* L.) OsCDR1 leads to constitutive activation of defense responses in rice and Arabidopsis. Mol Plant Microbe Interact 22:1635–1644

Prasad SG, Sujatha M, Rao SLV, Chaitanya U (2013) Screening of *indica* rice (*Oryza sativa* L.) genotypes against low temperature stress using diversity analysis. Helix 2:280–283

Prashanth SR, Parani M, Mohanty BP, Talame V, Tuberosa R, Parida A (2002) Genetic diversity in cultivars and landraces of *Oryza sativa* subsp. *indica* as revealed by AFLP markers. Genome 45:451–459

Priya P, Jain M (2013) RiceSRTFDB: a database of rice transcription factors containing comprehensive expression, cis-regulatory element and mutant information to facilitate gene function analysis. Database (Oxford). 2013: bat027. doi:10.1093/database/bat027

Raghuram B, Sheikh AH, Sinha AK (2014) Regulation of MAP kinase signaling cascade by microRNAs in *Oryza sativa*. Plant Signal Behav 9:e972130

Ramkumar G, Madhav MS, Devi SJSR, Manimaran P, Mohan KM, Prasad MS, Balachandran SM, Neeraja CN, Sundaram RM, Viraktamath BC (2014) Nucleotide diversity of *Pita*, a major blast resistance gene and identification of its minimal promoter. Gene 546:250–256

Rao NN, Prasad K, Kumar PR, Vijayraghavan U (2008) Distinct regulatory role for RFL, the rice LFY homolog, in determining flowering time and plant architecture. Proc Natl Acad Sci U S A 105:3646–3651

Rao KP, Richa T, Kumar K, Raghuram B, Sinha AK (2010) *In silico* analysis reveals 75 members of mitogen-activated protein kinase kinase kinase gene family in rice. DNA Res 17:139–153

Rathinasabapathi P, Purushothaman N, Ramprasad VL, Parani M (2015) Whole genome sequencing and analysis of Swarna, a widely cultivated *indica* rice variety with low glycemic index. Sci Rep 5:11303

Ray S, Agarwal P, Arora R, Kapoor S, Tyagi AK (2007) Expression analysis of calcium-dependent protein kinase gene family during reproductive development and abiotic stress conditions in rice (*Oryza sativa* L. ssp. *indica*). Mol Genet Genomics 278:493–505

Ray S, Dansana PK, Giri J, Deveshwar P, Arora R, Agarwal P, Khurana JP, Kapoor S, Tyagi AK (2011) Modulation of transcription factor and metabolic pathway genes in response to water-deficit stress in rice. Funct Integr Genomics 11:157–178

Ray S, Kapoor S, Tyagi AK (2012) Analysis of transcriptional and upstream regulatory sequence activity of two environmental stress-inducible genes, *NBS-Str1* and *BLEC-Str8*, of rice. Transgenic Res 21:351–366

Reddy AR, Ramakrishna W, Sekhar AC, Ithal N, Babu PR, Bonaldo MF, Soares MB, Bennetzen JL (2002) Novel genes are enriched in normalized cDNA libraries from drought-stressed seedlings of rice (*Oryza sativa* L. subsp. *indica* cv. Nagina 22). Genome 45:204–211

Reddy AM, Reddy VS, Scheffler BE, Wienand U, Reddy AR (2007) Novel transgenic rice overexpressing anthocyanidin synthase accumulates a mixture of flavonoids leading to an increased antioxidant potential. Metab Eng 9:95–111

Reddy BPN, Deshmukh RK, Gupta B, Deshmukh NK, Bhaganagare G, Shivraj SM, Singh T, Kotashane AS (2008) Identification of candidate genes for bacterial leaf blight resistance in rice by integration of genetic QTL map with the physical map. Asian J Bio Sci 3:24–29

Rice Annotation Project (2007) Curated genome annotation of *Oryza sativa* ssp. *japonica* and comparative genome analysis with *Arabidopsis thaliana*. Genome Res 17:175–183

Rice Annotation Project (2008) The Rice Annotation Project Database (RAP-DB): 2008 update. Nucleic Acids Res 36:D1028–D1033

Rice Chromosomes 11 and 12 Sequencing Consortia (2005) The sequence of rice chromosomes 11 and 12, rich in disease resistance genes and recent gene duplications. BMC Biol 3:20

RoyChoudhury A, Roy C, Sengupta DN (2007) Transgenic tobacco plants overexpressing the heterologous *lea* gene *Rab16A* from rice during high salt and water deficit display enhanced tolerance to salinity stress. Plant Cell Rep 26:1839–1859

Saha P, Chakraborti D, Sarkar A, Dutta I, Basu D, Das S (2007) Characterization of vascular-specific *RSs1* and *rolC* promoters for their utilization in engineering plants to develop resistance against hemipteran insect pests. Planta 226:429–442

Saha J, Sengupta A, Gupta K, Gupta B (2015) Molecular phylogenetic study and expression analysis of ATP-binding cassette transporter gene family in *Oryza sativa* in response to salt stress. Comput Biol Chem 54:18–32

Sahoo RK, Ansari MW, Pradhan M, Dangar TK, Mohanty S, Tuteja N (2014a) Phenotypic and molecular characterization of native *Azospirillum* strains from rice fields to improve crop productivity. Protoplasma 251:943–953

Sahoo RK, Ansari MW, Tuteja R, Tuteja N (2014b) OsSUV3 transgenic rice maintains higher endogenous levels of plant hormones that mitigates adverse effects of salinity and sustains crop productivity. Rice 7:17

Sajib AM, Hossain MM, Mosnaz ATMJ, Hossain H, Islam MM, Ali MS, Prodhan SH (2012) SSR marker-based molecular characterization and genetic diversity analysis of aromatic landraces of rice (*Oryza sativa* L.). J BioSci Biotechnol 1:107–116

Salunkhe AS, Poornima R, Prince KS, Kanagaraj P, Sheeba JA, Amudha K, Suji KK, Senthil A, Babu RC (2011) Fine mapping QTL for drought resistance traits in rice (*Oryza sativa* L.) using bulk segregant analysis. Mol Biotechnol 49:90–95

Sama VSAK, Himabindu K, Sundaram RM, Viraktamath BC, Bentur JS (2010) Mapping of a rice gall midge resistance gene, *gm3*, in RP 2068-18-3-5 and *in silico* identification of candidate gene(s). Rice Genet Newsl 25:76

Sama VSAK, Himabindu K, Naik SB, Sundaram RM, Viraktamath BC, Bentur JS (2012) Mapping and MAS breeding of an allelic gene to the *Gm8* for resistance to Asian rice gall midge. Euphytica 187:393–400

Sanan-Mishra N, Kumar V, Sopory SK, Mukherjee SK (2009) Cloning and validation of novel miRNA from Basmati rice indicates cross talk between abiotic and biotic stresses. Mol Genet Genomics 282:463–474

Sarkar NK, Kim YK, Grover A (2009) Rice sHsp genes: genomic organization and expression profiling under stress and development. BMC Genomics 10:393

Sarkar NK, Kundnani P, Grover A (2013) Functional analysis of Hsp70 superfamily proteins of rice (*Oryza sativa*). Cell Stress Chaperones 18:427–437

Sarkar NK, Kim YK, Grover A (2014) Coexpression network analysis associated with call of rice seedlings for encountering heat stress. Plant Mol Biol 84:125–143

Shankar A, Singh A, Kanwar P, Srivastava AK, Pandey A, Suprasanna P, Kapoor S, Pandey GK (2013) Gene expression analysis of rice seedling under potassium deprivation reveals major changes in metabolism and signaling components. PLoS One 8:e70321

Shankar A, Srivastava AK, Yadav AK, Sharma M, Pandey A, Raut VV, Das MK, Suprasanna P, Pandey GK (2014) Whole genome transcriptome analysis of rice seedling reveals alterations in Ca²⁺ ion signaling and homeostasis in response to Ca²⁺ deficiency. Cell Calcium 55:155–165

Shanmugavadivel PS, Mithra SVA, Dokku P, Kumar KAR, Rao GJN, Singh VP, Singh AK, Singh NK, Mohapatra T (2013) Mapping quantitative trait loci (QTL) for grain size in rice using a RIL population from Basmati x *indica* cross showing high segregation distortion. Euphytica 194:401–416

Sharma TR, Madhav MS, Singh BK, Shanker P, Jana TK, Dalal V, Pandit A, Singh A, Gaikwad K, Upreti HC, Singh NK (2005) High-resolution mapping, cloning and molecular characterization of the gene of rice, which confers resistance to rice blast. Mol Genet Genomics 274:569–578

Sharma A, Deshmukh RK, Jain N, Singh NK (2011) Combining QTL mapping and transcriptome profiling for an insight into genes for grain number in rice (*Oryza sativa* L.). Indian J Genet 71:115–119

Sharma R, Agarwal P, Ray S, Deveshwar P, Sharma P, Sharma N, Nijhawan A, Jain M, Singh AK, Singh VP, Khurana JP, Tyagi AK, Kapoor S (2012) Expression dynamics of metabolic and regulatory components across stages of panicle and seed development in *indica* rice. Funct Integr Genomics 12:229–248

Sharma R, Priya P, Jain M (2013) Modified expression of an auxin-responsive rice CC-type glutaredoxin gene affects multiple abiotic stress responses. Planta 238:871–884

Sharma M, Singh A, Shankar A, Pandey A, Baranwal V, Kapoor S, Tyagi AK, Pandey GK (2014a) Comprehensive expression analysis of rice Armadillo gene family during abiotic stress and development. DNA Res 21:267–283

Sharma R, Sahoo A, Devendran R, Jain M (2014b) Over-expression of a rice tau class glutathione s-transferase gene improves tolerance to salinity and oxidative stresses in Arabidopsis. PLoS One 9:e92900

Sharma D, Tiwari M, Lakhwani D, Tripathi RD, Trivedi PK (2015) Differential expression of microRNAs by arsenate and arsenite stress in natural accessions of rice. Metallomics 7:174–187

Singh RK, Sharma RK, Singh AK, Singh VP, Singh NK, Tiwari SP, Mohapatra T (2004) Suitability of mapped sequence tagged microsatellite site markers for establishing distinctness, uniformity and stability in aromatic rice. Euphytica 135:135–143

Singh A, Sahi C, Grover A (2009) Chymotrypsin protease inhibitor gene family in rice: Genomic organization and evidence for the presence of a bidirectional promoter shared between two chymotrypsin protease inhibitor genes. Gene 428:9–19

Singh A, Giri J, Kapoor S, Tyagi AK Pandey GK (2010a) Protein phosphatase complement in rice: genome-wide identification and transcriptional analysis under abiotic stress conditions and reproductive development. BMC Genomics 11:435

Singh A, Singh PK, Singh R, Pandit A, Mahato AK, Gupta DK, Tyagi K, Singh AK, Singh NK, Sharma TR (2010b) SNP haplotypes of the BADH1 gene and their association with aroma in rice (Oryza sativa L.). Mol Breed 26:325–338

Singh H, Deshmukh RK, Singh A, Singh AK, Gaikwad K, Sharma TR, Mohapatra T, Singh NK (2010c) Highly variable SSR markers suitable for rice genotyping using agarose gels. Mol Breed 25:359–364

Singh AK, Gopalakrishnan S, Singh VP, Prabhu KV, Mohapatra T, Singh NK, Sharma TR, Nagarajan M, Vinod KK, Singh D, Singh UD, Subhash C, Atwal SS, Rakesh S, Singh VK, Ranjith KE, Singh A, Deepti A, Apurva K, Sheel Y, Nitika G, Singh A, Shikari AB, Singh A, Marath B (2011a) Marker assisted selection: a paradigm shift in Basmati breeding. Indian J Genet 71:1–9

Singh SK, Singh CM, Lal GM (2011b) Assessment of genetic variability for yield and its component characters in rice (Oryza sativa L.). Res Plant Biol 1:73–76

Singh A, Baranwal V, Shankar A, Kanwar P, Ranjan R, Yadav S, Pandey A, Kapoor S, Pandey GK (2012a) Rice phospholipase A superfamily: organization, phylogenetic and expression analysis during abiotic stresses and development. PLoS One 7:e30947

Singh A, Mittal D, Lavania D, Agarwal M, Mishra RC, Grover A (2012b) OsHsfA2c and OsHsfB4b are involved in the transcriptional regulation of cytoplasmic OsClpB (Hsp100) gene in rice (Oryza sativa L.). Cell Stress Chaperones 17:243–254

Singh AK, Kumar R, Pareek A, Sopory SK, Singla-Pareek SL (2012c) Overexpression of rice CBS domain containing protein improves salinity, oxidative, and heavy metal tolerance in transgenic tobacco. Mol Biotechnol 52:205–216

Singh A, Kanwar P, Pandey A, Tyagi AK, Sopory SK, Kapoor S, Pandey GK (2013a) Comprehensive genomic analysis and expression profiling of phospholipase C gene family during abiotic stresses and development in rice. PLoS One 8:e62494

Singh A, Singh B, Panda K, Rai VP, Singh AK, Singh SP, Chouhan SK, Vandna R, Kumar P, Singh NK (2013b) Wild rices of Eastern Indo-gangetic plains of India constitute two sub-populations harbouring rich genetic diversity. J Plant Mol Biol Omics 62:121–127

Singh N, Choudhury DR, Singh AK, Kumar S, Srinivasan K, Tyagi RK, Singh NK, Singh R (2013c) Comparison of SSR and SNP markers in estimation of genetic diversity and population structure of Indian rice varieties. PLoS One 8:e84136

Singh S, Giri MK, Singh PK, Siddiqui A, Nandi AK (2013d) Down-regulation of OsSAG12-1 results in enhanced senescence and pathogen-induced cell death in transgenic rice plants. J Biosci 38:583–592

Singh A, Jha SK, Bagri J, Pandey GK (2015a) ABA inducible rice protein phosphatase 2C confers ABA insensitivity and abiotic stress tolerance in Arabidopsis. PLoS One 10:e0125168

Singh BP, Jayaswal PK, Singh B, Singh PK, Kumar V, Mishra S, Singh N, Panda K, Singh NK (2015b) Natural allelic diversity in OsDREB1F gene in the Indian wild rice germplasm led to ascertain its association with drought tolerance. Plant Cell Rep 34:993–1004

Singh N, Jayaswal PK, Panda K, Mandal P, Kumar V, Singh B, Mishra S, Singh Y, Singh R, Rai V, Gupta A, Sharma TR, Singh NK (2015c) Single-copy gene based 50 K SNP chip for genetic studies and molecular breeding in rice. Sci Rep 5:11600

Singh S, Chand S, Singh NK, Sharma TR (2015d) Genome-wide distribution, organisation and functional characterization of disease resistance and defence response genes across rice species. PLoS One 10:e0125964

Singla B, Khurana JP, Khurana P (2009) Structural characterization and expression analysis of the SERK/SERL gene family in rice (Oryza sativa). Int J Plant Genomics 2009:539402

Singla-Pareek SL, Yadav SK, Pareek A, Reddy MK, Sopory SK (2008) Enhancing salt tolerance in a crop plant by overexpression of glyoxalase II. Transgenic Res 17:171–180

Smita S, Lenka SK, Katiyar A, Jaiswal P, Preece J, Bansal KC (2011) QlicRice: a web interface for abiotic stress responsive QTL and loci interaction channels in rice. Database (Oxford). 2011:bar037. doi:10.1093/database/bar037

Smita S, Katiyar A, Pandey DM, Chinnusamy V, Archak S, Bansal KC (2013) Identification of conserved drought stress responsive gene-network across tissues and developmental stages in rice. Bioinformation 9:72–78

Subbaiyan GK, Waters DL, Katiyar SK, Sadananda AR, Vaddadi S, Henry RJ (2012) Genome-wide DNA polymorphisms in elite indica rice inbreds discovered by whole-genome sequencing. Plant Biotechnol J 10:623–634

Sujatha K, Natarajkumar P, Laha GS, Viraktamath BC, Reddy CS, Mishra B, Balachandran SM, Ram T, Srinivasarao K, Hari Y, Kirti PB, Sundaram RM (2010) Molecular mapping of a recessive bacterial blight resistance gene derived from Oryza. Rice Genet Newsl 25:1–2

Sujatha K, Prasad MS, Pandey M, Ramesha MS, Neeraja CN, Balachandran SM, Shobharani N, Kemparaju B, Madhanmohan K, Sama VSAK, Hajira SK, Baachiranjeevi CH, Pranathi K, Reddy GA, Madhav MS, Sundaram RM (2013) Marker-assisted introgression of bacterial blight and blast resistance into IR 58025B, an elite maintainer line of rice. Plant Breed 132:586–594

Sundaram RM, Vishnupriya MR, Biradar SK, Laha GS, Reddy GA, Shoba Rani N, Sarma NP, Sonti RV (2008) Marker assisted introgression of bacterial blight resistance in Samba Mahsuri, an elite indica rice variety. Euphytica 160:411–422

Sundaram RM, Vishnupriya MR, Laha GS, Rani NS, Rao PS, Balachandran SM, Reddy GA, Sarma NP, Sonti RV (2009) Introduction of bacterial blight resistance into Triguna, a high yielding, mid-early duration rice variety. Biotechnol J 4:400–407

Swapna L, Khurana R, Kumar SV, Tyagi AK, Rao KV (2011) Pollen-specific expression of Oryza sativa indica pollen allergen gene (OsIPA) promoter in rice and Arabidopsis transgenic systems. Mol Biotechnol 48:49–59

Thakur JK, Agarwal P, Parida S, Bajaj D, Pasrija R (2013) Sequence and expression analyses of KIX domain proteins suggest their importance in seed development and determination of seed size in rice, and genome stability in Arabidopsis. Mol Genet Genomics 288:329–346

Thakur S, Singh PK, Das A, Rathour R, Variar M, Prashanthi SK, Singh AK, Singh UD, Chand D, Singh NK, Sharma TR (2015) Extensive sequence variation in rice blast resistance gene Pi54 makes it broad spectrum in nature. Front Plant Sci 6:345

Tiwari M, Sharma D, Singh M, Tripathi RD, Trivedi PK (2014) Expression of OsMATE1 and OsMATE2 alters development, stress responses and pathogen susceptibility in Arabidopsis. Sci Rep 4:3964

Tiwari KK, Singh A, Pattnaik S, Sandhu M, Kaur S, Jain S, Tiwari S, Mehrotra S, Anumalla M, Samal R, Bhardwaj J, Dubey N, Sahu V, Kharshing GA, Zeliang PK, Sreenivasan K, Kumar P, Parida SK, Mithra SVA, Rai V, Tyagi W, Agarwal PK, Rao AR, Pattanayak A, Chandel G, Singh AK, Bisht IS, Bhat KV, Rao GJN, Khurana JP, Singh NK, Mohapatra T (2015a) Identification of a diverse mini-core panel of Indian rice germplasm based on genotyping using microsatellite markers. Plant Breed 134:164–171

Tiwari LD, Mittal D, Chandra Mishra R, Grover A (2015b) Constitutive over-expression of rice chymotrypsin protease inhibitor gene OCPI2 results in enhanced growth, salinity and osmotic stress tolerance of the transgenic Arabidopsis plants. Plant Physiol Biochem 92:48–55

Trivedi DK, Yadav S, Vaid N, Tuteja N (2012) Genome wide analysis of Cyclophilin gene family from rice and Arabidopsis and its comparison with yeast. Plant Signal Behav 7:1653–1666

Tuteja N, Sahoo RK, Garg B, Tuteja R (2013) OsSUV3 dual helicase functions in salinity stress tolerance by maintaining photosynthesis and antioxidant machinery in rice (Oryza sativa L. cv. IR64). Plant J 76:115–127

Tyagi H, Jha S, Sharma M, Giri J, Tyagi AK (2014) Rice SAPs are responsive to multiple biotic stresses and overexpression of OsSAP1, an A20/AN1 zinc-finger protein, enhances the basal resistance against pathogen infection in tobacco. Plant Sci 225:68–76

Umate P (2010) Genome-wide analysis of thioredoxin fold superfamily peroxiredoxins in Arabidopsis and rice. Plant Signal Behav 5:1543–1546

Umate P (2011) Genome-wide analysis of lipoxygenase gene family in Arabidopsis and rice. Plant Signal Behav 6:335–338

Umate P, Tuteja N (2010) microRNA access to the target helicases from rice. Plant Signal Behav 5:1171–1175

Umate P, Tuteja R, Tuteja N (2010) Genome-wide analysis of helicase gene family from rice and Arabidopsis: a comparison with yeast and human. Plant Mol Biol 73:449–465

Vaid N, Pandey PK, Tuteja N (2012) Genome-wide analysis of lectin receptor-like kinase family from *Arabidopsis* and rice. Plant Mol Biol 80:365–388

Vhora Z, Trivedi R, Chakraborty S, Ravikiran R, Sasidharan N (2013) Molecular studies of aromatic and non-aromatic rice (*Oryza sativa* L.) genotypes for quality traits using microsatellite markers. Bioscan 8:359–362

Vij S, Tyagi AK (2006) Genome-wide analysis of the stress associated protein (SAP) gene family containing A20/AN1 zinc-finger(s) in rice and their phylogenetic relationship with Arabidopsis. Mol Genet Genomics 276:565–575

Vij S, Gupta V, Kumar D, Vydianathan R, Raghuvanshi S, Khurana P, Khurana JP, Tyagi AK (2006) Decoding the rice genome. Bioessays 28:421–432

Vij S, Giri J, Dansana PK, Kapoor S, Tyagi AK (2008) The receptor-like cytoplasmic kinase (*OsRLCK*) gene family in rice: organization, phylogenetic relationship, and expression during development and stress. Mol Plant 1:732–750

Vijayan J, Devanna BN, Singh NK, Sharma TR (2015) Cloning and functional validation of early inducible *Magnaporthe oryzae* responsive CYP76M7 promoter from rice. Front Plant Sci 6:371

Vikram P, Swamy BP, Dixit S, Ahmed HU, Teresa Sta Cruz M, Singh AK, Kumar A (2011) qDTY$_{1.1}$, a major QTL for rice grain yield under reproductive-stage drought stress with a consistent effect in multiple elite genetic backgrounds. BMC Genet 12:89

Wang M, Yu Y, Haberer G, Marri PR, Fan C, Goicoechea JL, Zuccolo A, Song X, Kudrna D, Ammiraju JS, Cossu RM, Maldonado C, Chen J, Lee S, Sisneros N, de Baynast K, Golser W, Wissotski M, Kim W, Sanchez P, Ndjiondjop MN, Sanni K, Long M, Carney J, Panaud O, Wicker T, Machado CA, Chen M, Mayer KF, Rounsley S, Wing RA (2014) The genome sequence of African rice (*Oryza glaberrima*) and evidence for independent domestication. Nat Genet 46:982–988

Wankhede DP, Kumar K, Singh P, Sinha AK (2013) Involvement of mitogen activated protein kinase kinase 6 in UV induced transcripts accumulation of genes in phytoalexin biosynthesis in rice. Rice 6:35

2

High-resolution QTL mapping for grain appearance traits and co-localization of chalkiness-associated differentially expressed candidate genes in rice

Likai Chen[1], Weiwei Gao[1], Siping Chen[1], Liping Wang[1], Jiyong Zou[1,2], Yongzhu Liu[1], Hui Wang[1], Zhiqiang Chen[1*] and Tao Guo[1*] (ID)

Abstract

Background: Grain appearance quality is a main determinant of market value in rice and one of the highly important traits requiring improvement in breeding programs. The genetic basis of grain shape and endosperm chalkiness have been given significant attention because of their importance in affecting grain quality. Meanwhile, the introduction of NGS (Next Generation Sequencing) has a significant part to play in the area of genomics, and offers the possibility for high-resolution genetic map construction, population genetics analysis and systematic expression profile study.

Results: A RIL population derived from an inter-subspecific cross between indica rice PYZX and japonica rice P02428 was generated, based on the significant variations for the grain morphology and cytological structure between these two parents. Using the Genotyping-By-Sequencing (GBS) approach, 2711 recombination bin markers with an average physical length of 137.68 kb were obtained, and a high-density genetic map was constructed. Global genetic mapping of QTLs affecting grain shape and chalkiness traits was performed across four environments and the newly identified stable loci were obtained. Twelve important QTL clusters were detected, four of which were coincident with the genomic regions of cloned genes or fine mapped QTL reported. Eight novel QTL clusters (including six for grain shape, one for chalkiness, and one for both grain shape and chalkiness) were firstly obtained and highlighted the value and reliability of the QTL analysis. The important QTL cluster on chromosome 5 affects multiple traits including circularity (CS), grain width (GW), area size of grain (AS), percentage of grains with chalkiness (PGWC) and degree of endosperm chalkiness (DEC), indicating some potentially pleiotropic effects. The transcriptome analysis demonstrated an available gene expression profile responsible for the development of chalkiness, and several DEGs (differentially expressed genes) were co-located nearby the three chalkiness-related QTL regions on chromosomes 5, 7, and 8. Candidate genes were extrapolated, which were suitable for functional validation and breeding utilization.

Conclusion: QTLs affecting grain shape (grain width, grain length, length-width ratio, circularity, area size of grain, and perimeter length of grain) and chalkiness traits (percentage of grains with chalkiness and degree of endosperm chalkiness) were mapped with the high-density GBS-SNP based markers. The important differentially expressed genes (DEGs) were co-located in the QTL cluster regions on chromosomes 5, 7 and 8 affecting PGWC and DEC parameters. Our research provides a crucial insight into the genetic architecture of rice grain shape and chalkiness, and acquired potential candidate loci for molecular cloning and grain quality improvement.

Keywords: Rice, QTL mapping, GBS, Grain shape, Chalkiness, Transcriptome profiling

* Correspondence: chenlin@scau.edu.cn; guoguot@scau.edu.cn
[1]National Engineering Research Center of Plant Space Breeding, South China Agricultural University, Guangzhou 510642, China
Full list of author information is available at the end of the article

Background

The production and consumption of rice is concentrated in Asia where more than 90 % of the world's rice is grown and consumed (Muthayya et al. 2014; Kong et al. 2015; Jones and Sheats 2016). Appearance (including grain shape and endosperm chalk), cooking properties and texture time were the most important traits affecting grain quality (Fitzgerald et al. 2009). As one of the major aspects of grain quality, grain appearance affects market demand significantly (Tanabata et al. 2012). Even though preferences relating to grain quality properties vary across countries and regions (Calingacion et al. 2014; Concepcion et al. 2015), consumers typically desire rice with uniform shape and translucent endosperm, therefore the quality of appearance directly affects consumer acceptance (Zhao et al. 2015). Grain shape and chalkiness have attracted significant attention in rice genetic research, however, as a practical matter, grain appearance quality is mostly conditioned by quantitative trait locus QTL, representing a major problem for rice improvement programs and production.

Grain shape, widely accepted as a complex quantitative trait, is usually measured as grain length, width, thickness and length-to-width ratio (Bai et al. 2010). Furthermore, digital imaging technology was introduced for computational methods, which could enable us to automatically measure the grain shape parameters of circularity, seed area and perimeter length, etc. (Tanabata et al. 2012). Over the past thirty years, QTL mapping and association analysis have become widely used for analysis of grain appearance traits (Bai et al. 2010; Han and Huang 2013; Huang et al. 2013). By utilizing a variety of mapping populations, such as F_2, recombinant inbred lines (RILs), backcross and doubled haploid (DH), many QTLs associated with these traits have been identified (Huang et al. 2013). Bai et al. (2010) identified 28 QTLs related to grain shape using a RIL population derived from the cross between japonica and indica rice, and suggested that a mapping population derived from two contrasting parents in grain shape is expected to give rise to a larger number of QTLs. By using map-based cloning strategies, several valuable genes regulating grain shape have been isolated, including *GS3* (Fan et al. 2006), *GW2* (Song et al. 2007), *GS5* (Weng et al. 2008), *qSW5* (Shomura et al. 2008), *OsSPL16* (Wang et al. 2012), *qGL3.1/qGL3* (Qi et al. 2012; Zhang et al. 2012), *GS6* (Sun et al. 2013), *GW7* (Wang et al. 2015a), *SLG7* (Zhou et al. 2015b) and *GL7* (Wang et al. 2015b), which have enhanced our knowledge of the molecular regulatory mechanisms responsible for grain shape and enables breeders to develop high-yield varieties with improved grain-quality (Wang et al. 2015b).

Chalkiness is the other appearance-related trait that affects consumer acceptance of rice (Fitzgerald et al. 2009). Grain chalk is an important indicator of rice quality evaluation and a highly undesirable quality trait in marketing and consumption of the rice grain (Li et al. 2014b). As a polygenic quantitative trait with complex inheritance pattern, chalkiness is highly influenced by the environment. Thus the genetic basis of grain chalkiness is still poorly understood, even though many QTLs for chalkiness or related components have also been identified (http://www.gramene.org). Peng et al. (2014) mapped multiple QTLs associated with six chalkiness traits (chalkiness rate, white core rate, white belly rate, chalkiness area, white core area, and white belly area) using five populations and suggested that most of the QTLs clustered together and could be detected in different backgrounds. Two loci controlling PGWC were mapped by Zhou et al. (2009), and the *qPGWC-7* was narrowed to a 44-kb region. *Chalk5*, regulating grain chalkiness was isolated by Li et al. (2014b), which encodes a vacuolar H^+-translocating pyrophosphatase (V-PPase) with PPi hydrolysis and H^+-translocation activity. Elevated expression of *Chalk5* increases chalkiness of the endosperm by disturbing the pH homeostasis in the endomembrane trafficking system in developing seeds, which affects the biogenesis of protein bodies coupled with a great increase in small vesicle-like structures, thus forming air spaces among endosperm storage substances, resulting in chalky grain (Li et al. 2014b). However, the regulation pathway and interaction mechanisms of rice chalkiness associated genes remain unclear.

For traditional QTL mapping, molecular marker genotyping was time consuming and labor-intensive (Chen et al. 2014a). Low-throughput molecular markers such as simple sequence repeats (SSRs) were the most commonly used for linkage maps construction in rice QTL mapping analysis. They are mostly of low density and not able to provide precise and complete information about the numbers and locations of the QTLs controlling the interesting traits (Yu et al. 2011). Single nucleotide polymorphisms (SNPs) are currently the marker of choice due to their large numbers in virtually all populations of individuals (Kumar et al. 2012) and next-generation sequencing (NGS) has enabled the discovery of numerous SNPs for many plant species. Therefore, high-density genetic maps based on SNP markers are achievable and can be developed to improve the efficiency and accuracy of gene or QTL mapping (Li et al. 2014a). In rice, previous studies have demonstrated that the improved quality and resolution of the linkage map based on sequencing-based SNP has greatly facilitated QTL dissection (Huang et al. 2009; Yu et al. 2011; Gao et al. 2013; Zhang et al. 2015).

In this study, we demonstrate the discrepancy in the morphology and cytological structure of two contrasting genotypes and the phenotypic variance of the grain shape and chalkiness traits across a RIL population. Using NGS, a high-density genetic map was constructed based on the new developed bin markers. The QTLs associated with grain shape and chalkiness were identified under four different environments. Moreover, we performed transcriptome expression profiling and identified differentially expressed genes located in the chalkiness-related QTL regions, providing valuable information for candidate gene verification and dissection of gene regulatory networks affecting rice grain appearance quality.

Results

The grain appearance and cytological difference between PYZX and P02428

Considerable distinct variations in panicle structure, grain shape and chalkiness traits between PYZX and P02428 were observed (Fig. 1). PYZX showed much longer and slenderer grain, with limpid kernels, whereas the grain of P02428 was wide and short, along with chalky kernel. The grain appearance traits, including grain length, grain width, the ratio of grain length and width, circularity, area size of grain, and perimeter length of grain for grain shape parameters, and chalk property including percentage of grain with chalkiness and degree of endosperm chalkiness were examined under four environments (Table 1 and Additional file 1: Table S1). The GW and GL of PYZX were at an average of 2.26 mm and 12.17 mm, whereas averages were 3.75 mm and 7.16 mm respectively for P02428. Thus, the LWR of PYZX was almost 3 times the level of P02428, and the CS of the latter was about 2 times greater than the former. Extreme differences in PGWC

(0.95 and 91.67 % on avg. for PYZX and P02428 respectively) and DEC (0.09 and 54.16 at avg. for PYZX and P02428 respectively) between parental genotypes were detected continuously across all environments.

Microscopic observation with a cross-section of spikelets indicated that P02428 contained substantially higher cell numbers when compared to that of PYZX, with only an insignificant increase in cell length (Fig. 2a, b). A scanning electron microscope investigation of outer glume surfaces demonstrated a significant increase in cell numbers and decrease in cell length for P02428 compared to PYZX (Fig. 2c, d). This histological analysis established the major origins of the observed grain shape and size variation between the parental lines. Transverse sections of the endosperm bellies of mature seeds were also examined using scanning electron microscopy, and revealed that the endosperm of chalky grains of P02428 contained loosely packed starch granules with large air spaces, while those of PYZX were filled with densely packed granules (Fig. 2e).

Phenotypic variation of grain shape and chalkiness parameters in RIL population

Generally, the RIL population exhibited an extremely wide variation in rice grain shape and chalkiness traits and continuous distributions were observed for all eight investigated traits (Fig. 3), consistent with quantitative traits controlled by multi-genes. All of the grain shape related parameters were evenly distributed as single peak patterns, whereas the phenotypic values of PGWC and DEC were exhibited in the specific asymmetric distributed pattern. Furthermore, PGWC and DEC for the RIL population also showed higher standard deviations

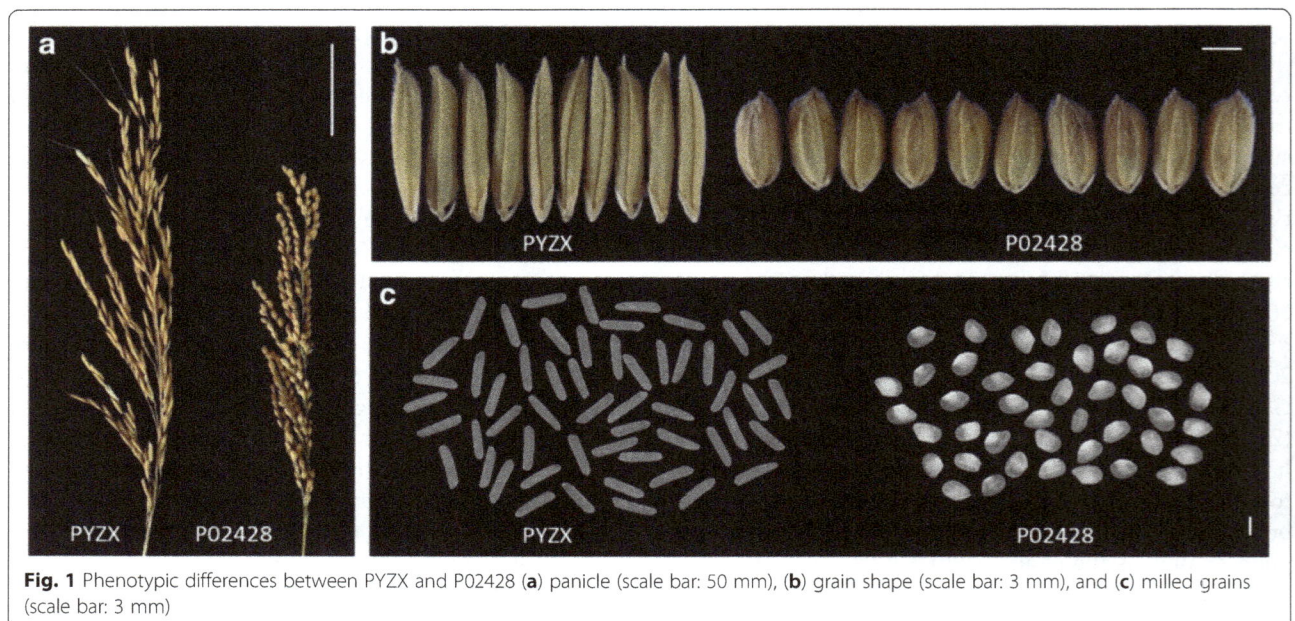

Fig. 1 Phenotypic differences between PYZX and P02428 (**a**) panicle (scale bar: 50 mm), (**b**) grain shape (scale bar: 3 mm), and (**c**) milled grains (scale bar: 3 mm)

Table 1 Phenotypic performances and correlation coefficients among the grain shape and chalkiness traits in the RIL population

Phenotypic performances

Lines		GW (mm)	GL (mm)	LWR	CS	AS (mm^2)	PL (mm)	PGWC (%)	DEC
PYZX	Mean	3.002	9.826	3.714	0.548	20.848	23.295	44.446	25.334
	S.D	0.043	0.171	0.116	0.003	0.842	0.278	0.460	0.043
P02428	Mean	3.754	7.163	1.837	0.729	19.425	18.644	91.667	54.158
	S.D	0.087	0.225	0.161	0.008	0.642	0.747	4.659	4.192
RILs	Mean	2.743	9.404	3.461	0.516	19.650	21.950	28.260	19.430
	S.D	0.028	0.161	0.026	0.002	0.588	0.359	5.936	7.989

Correlation coefficients

		GW (mm)	GL (mm)	LWR	CS	AS (mm^2)	PL (mm)	PGWC (%)	DEC
GW			−0.378[b]	−0.766[b]	0.822[b]	0.557[b]	−0.125[b]	0.556[b]	0.357[b]
GL				0.759[b]	−0.760[b]	0.608[b]	0.919[b]	−0.130[a]	−0.038
LWR					−0.886[b]	−0.047	0.669[b]	−0.462[b]	−0.248[a]
CS						0.043	−0.672[b]	0.455[b]	0.286[b]
AS							0.702[b]	0.374[b]	0.298[b]
PL								−0.049	−0.014
PGWC									0.626[b]

[a], [b] significant at the level of 0.05 and 0.01, respectively

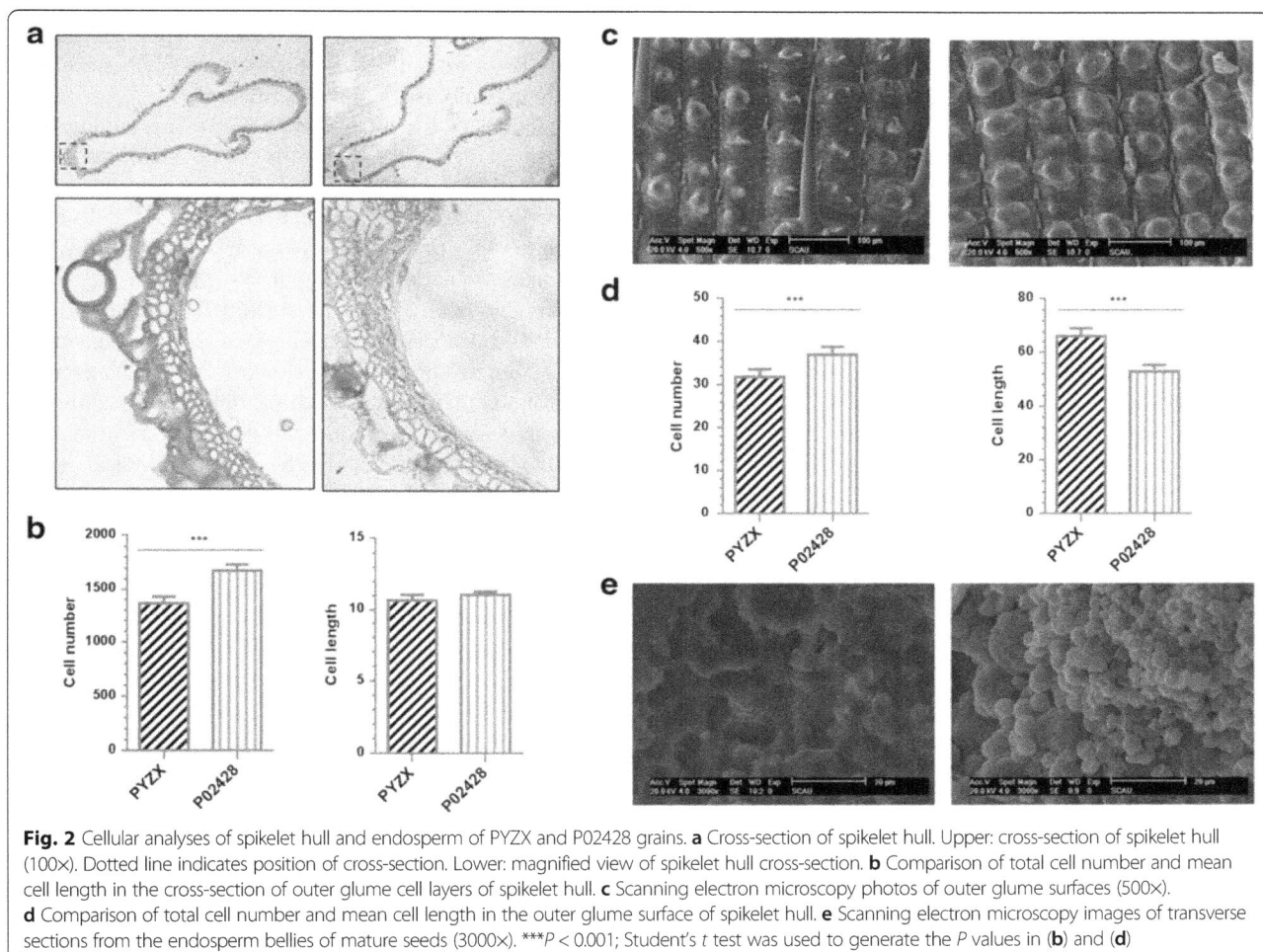

Fig. 2 Cellular analyses of spikelet hull and endosperm of PYZX and P02428 grains. **a** Cross-section of spikelet hull. Upper: cross-section of spikelet hull (100×). Dotted line indicates position of cross-section. Lower: magnified view of spikelet hull cross-section. **b** Comparison of total cell number and mean cell length in the cross-section of outer glume cell layers of spikelet hull. **c** Scanning electron microscopy photos of outer glume surfaces (500×). **d** Comparison of total cell number and mean cell length in the outer glume surface of spikelet hull. **e** Scanning electron microscopy images of transverse sections from the endosperm bellies of mature seeds (3000×). ***$P < 0.001$; Student's t test was used to generate the P values in (**b**) and (**d**)

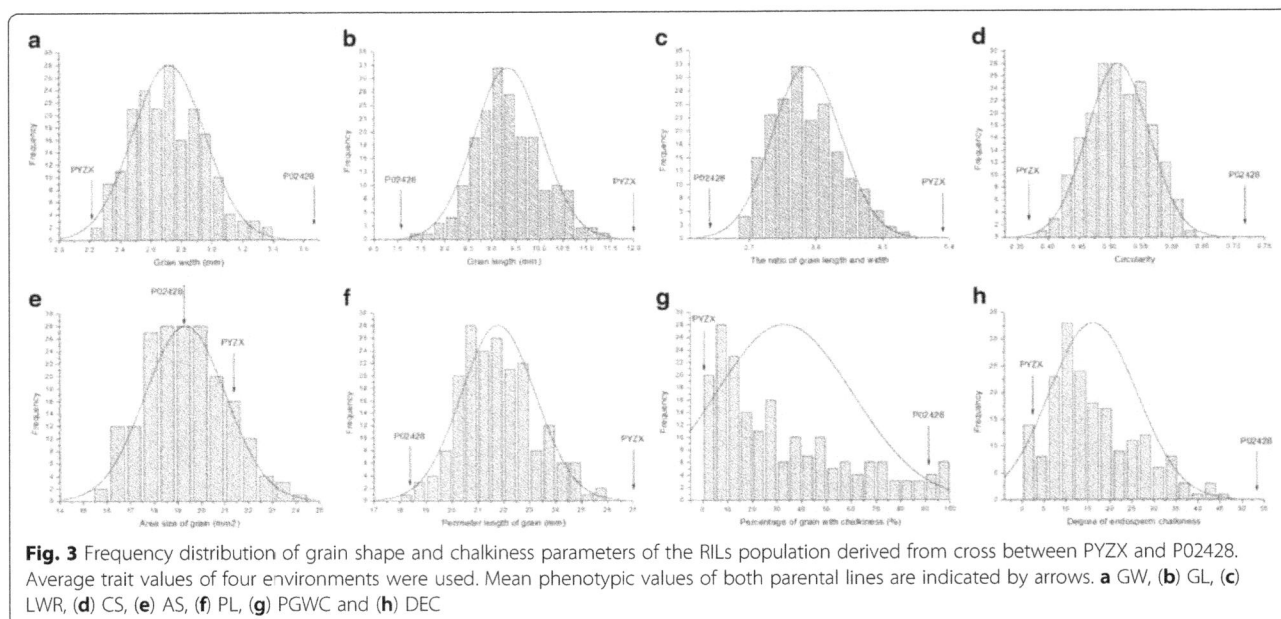

Fig. 3 Frequency distribution of grain shape and chalkiness parameters of the RILs population derived from cross between PYZX and P02428. Average trait values of four environments were used. Mean phenotypic values of both parental lines are indicated by arrows. **a** GW, (**b**) GL, (**c**) LWR, (**d**) CS, (**e**) AS, (**f**) PL, (**g**) PGWC and (**h**) DEC

(Table 1), indicating that they are significantly affected by environments. For GW, GL, LWR and CS, the average value of traits measured in the RIL population was between the two parental lines, and none of the individual lines exhibit values that surpass either P02428 or PYZX. A large amount of variation and the greatest transgressive segregation was observed for the trait of AS.

The correlation among the grain shape and chalkiness parameters in the RIL population was analyzed (Table 1). The results showed that significant correlations were detected between grain shape and chalkiness trait. PGWC and DEC were positively correlated with GW, CS, and AS, while negatively correlated with LWR. The correlation coefficient between PGWC and DEC was high ($r = 0.626$). For grain shape traits, CS, AS, and PL were significantly correlated with GW and GL concurrently. We also found a considerably high positive correlation between LWR and CS ($r = 0.886$).

Genotyping by sequencing and bin markers establishment

In this study, a total of 83.88 Gb high-quality sequence data from 559,213,384 pair-end reads was obtained and about 97.48 % of those reads were mapped to the Nipponbare reference genome. For 192 RILs individuals, total mapped regions covered by the captured fragments were about ~7.0 % of the genome sequence and with a coverage depth of ~11.76× on average for the captured regions. Initial analysis identified 1,534,036 SNPs between the two parents (see Additional file 1: Table S2 for annotation statistics of location) of which 1,334,454 were consistent with the "aa × bb" type (presence of polymorphism between the genotypes of parents and both of them were homozygous) were

selected for further analysis. Genotypes of the RIL individuals at these SNP sites were determined, and 123,982 loci with more than 4 base depth remained. After filtering for abnormality, a total of 85,742 high-quality SNPs were validated for recombinant event determination.

Consecutive SNPs were examined (with a sliding window size of 15 SNPs) and the same genotypes were lumped into recombination bins. Bins with an interval of less than 300 kb and the number of sequenced SNPs fewer than five were masked as missing data to avoid false double recombinations (Xie et al. 2010). Adjacent bins of the same genotype across the entire RIL population were merged and transition between two different genotype bins was determined as a breakpoint. After this processing, a total of 2711 recombination bin markers along the 12 chromosomes were adopted to construct a bin map for the RIL population (Fig. 4). The average physical length between the recombination bin markers was 137.68 kb, and the average annotated gene loci among the markers ranged from 15.34 for Chr04 to 22.09 for Chr01 (Table 2). Using the R/qtl package (est.map function with Lander-Green algorithm), we constructed a genetic linkage map with an average distance of 0.86 cM between adjacent bin markers and a maximum spacing between markers ranging from 9.93 cM for Chr03 to 5.84 cM for Chr10. The average genetic size of the 12 chromosomes is about 195.31 cM, and the average physical distance between markers was 137.68 Kb (Table 2).

Comprehensive QTL mapping for the grain shape and chalkiness traits

With the inclusive composite interval mapping method, a total of 136 loci affecting grain shape and chalkiness traits were detected according to the LOD threshold across four environments (47 for G-DS, 28 for Z-DS, 38 for G-WS and 23 for Z-WS respectively) (Additional file 1: Table S3 and Fig. 4). The majority of the QTLs associated with grain width and circularity traits had a negative additive effect, indicating that alleles from the parent P02428 contributed to increasing phenotype, whereas the phenotypes of GL, LWR and PL were mainly contributed by PYZX. For AS traits, 62.5 and 37.5 % of the QTLs had a negative and positive effect respectively, indicating both parents contributed favorable alleles. All of the QTLs related to chalkiness were endowed with the additive effect contributed by P02428 (Additional file 1: Table S3).

Seventeen QTLs were detected for GW and each QTL explained 3.354 ~ 12.377 % of the phenotypic variation. Among all QTLs identified for GW, five QTLs showed high PEV value of more than 10 %, including three major QTLs on Chr05 and two on Chr07. Sixteen QTLs were detected for GL, explaining 3.904 ~ 20.799 % of phenotypic variation for each QTL, and six showed major QTL with higher LOD and effect value, located on Chr03 and Chr07, remarkably similar to that of the PL and LWR parameter. QTLs associated with AS were also identified in these regions as well and a QTL with LOD of 7.044 and PEV of 10.657 % was detected on Chr10. Twenty-four QTLs for CS parameter were detected across ten chromosomes using single-

environment analysis and each QTL explained 3.403 ~ 26.139 % of phenotypic variation, and six major QTLs with higher PEV value were located on chromosomes 3, 4, 5 and 7. There were ten QTLs for DEC and each QTL explained 5.391 ~ 12.779 % of phenotypic variation, and sixteen QTLs associated with PWGC were detected explaining 4.457 ~ 12.975 % of phenotypic variation. More than 81 % of the QTLs for PGWC and DEC provided a lower PEV value, indicating that the chalkiness-related parameters were regulated with polygene and minor effects. However, four major genomic regions harboring QTLs with higher LOD and PEV values for PGWC or DEC were identified, and more importantly, six common QTL regions were detected for both PGC and DEC and distributed on chromosomes 4, 5, 7, and 8 (Table 3).

Stable QTLs and major QTL clusters

As many of the QTLs detected here overlapped, they were classified into same loci according to the genetic position. Three or more QTLs detected for the same trait within the consistent confidence interval using single environment analysis were defined as stable in this study. Notably, most of these QTLs had good reproducibility across multiple environments. For instance, 54.17 % of the grain circularity QTLs were reproducible under the various environments. The main effect QTL located on Chr03 between mk690 ~ mk698 was detected

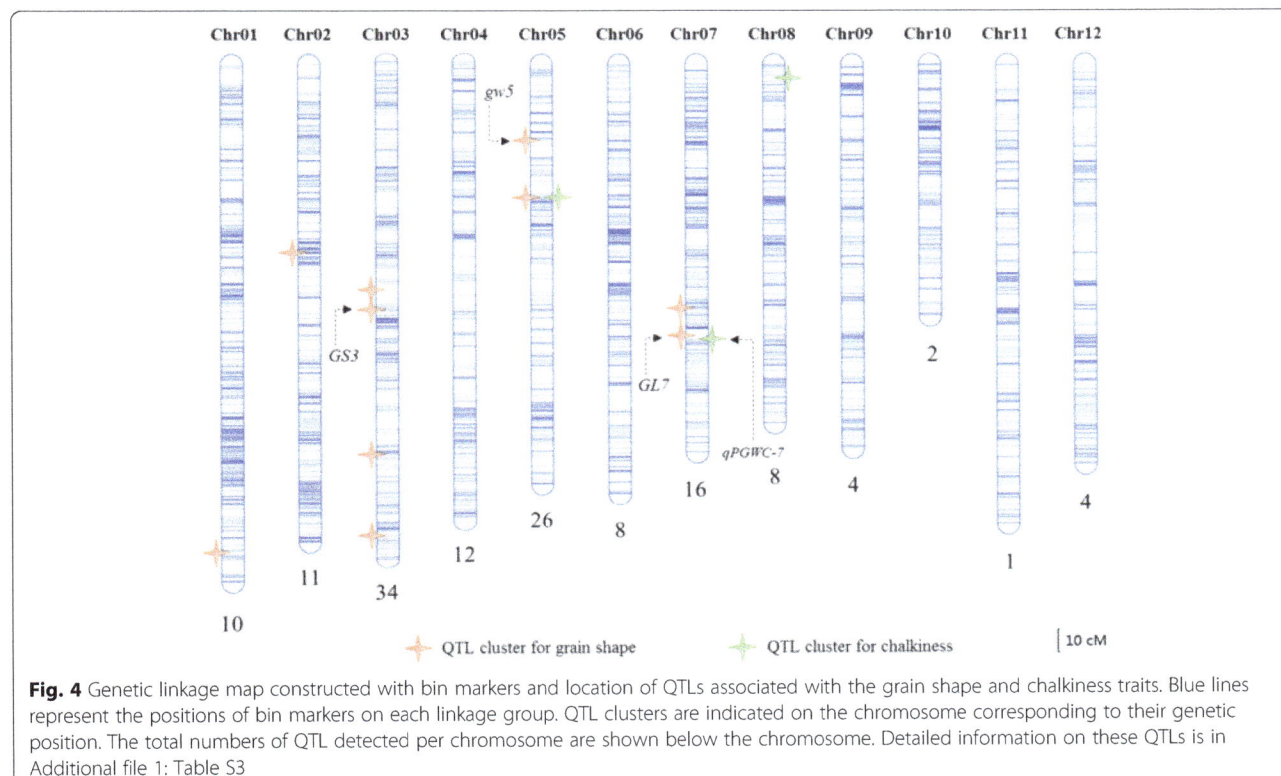

Fig. 4 Genetic linkage map constructed with bin markers and location of QTLs associated with the grain shape and chalkiness traits. Blue lines represent the positions of bin markers on each linkage group. QTL clusters are indicated on the chromosome corresponding to their genetic position. The total numbers of QTL detected per chromosome are shown below the chromosome. Detailed information on these QTLs is in Additional file 1: Table S3

Table 2 Distribution of genetic markers across the 12 chromosomes in rice

Chromosome	Number of bin markers	Length (cM)	Average genetic distance between markers (cM)	Maximum spacing between markers (cM)	Average physical distance between markers (Kb)	Average number of annotated gene loci among markers[a]
1	295	232.15	0.79	8.25	146.68	22.09
2	258	221.65	0.86	7.20	139.29	20.90
3	311	226.44	0.73	9.93	117.09	18.66
4	278	223.22	0.80	8.53	127.71	15.34
5	182	194.90	1.07	6.85	164.61	22.03
6	229	199.68	0.87	9.75	136.46	17.31
7	232	180.72	0.78	6.50	128.01	16.24
8	179	167.37	0.94	6.51	158.90	19.14
9	178	178.69	1.00	7.19	129.28	15.55
10	162	118.78	0.73	5.84	143.25	17.47
11	208	213.82	1.03	9.37	139.52	15.42
12	199	186.26	0.94	9.37	138.35	14.99
Overall	2711	2343.68	0.86	9.93	137.68	18.12

[a] annotation on Os-Nipponbare-Reference-IRGSP-1.0 (http://rapdb.dna.affrc.go.jp/)

up to four times, and the QTL located in the genetic interval between mk1743 ~ mk1747 were stably reproducible in three separate environments (Additional file 1: Table S3). QTLs associated with different traits located within the same confidence marker intervals were grouped together as major QTL clusters. As a result, we concluded that there were twelve QTL clusters distributed over six chromosomes (Table 4). Some of the traits with high inter-trait correlations appeared to cluster together, which was in accordance with our correlation analysis between the traits (Table 1), revealing the main genetic determinants of grain shape and chalkiness characteristics in rice. Among these, the QTL cluster of *qGS7.2* (mk1743-mk1745) was identified as the major grain shape QTL explaining highest phenotypic variance in our study (Table 4). The previously reported QTL of *GL7*, which overlaps with this QTL cluster region, encodes a protein homologous to *Arabidopsis thaliana* LONGIFOLIA proteins and was reported to regulate grain appearance quality mainly by affecting the grain length to width ratio and the formation of starch granules in endosperm (Wang et al. 2015b). In addition, three other QTL clusters for grain shape or chalkiness traits were mapped to relatively narrow genomic regions that coincided with QTLs in previously published reports including *GS3* (Fan et al. 2006; Mao et al. 2010), *gw5/qSW5* (Shomura et al. 2008; Weng et al. 2008), and *qPGWC-7* (Zhou et al. 2009) (Table 4), thus supporting the accuracy of our linkage map and mapping analysis.

Significantly, the other eight novel QTL clusters detected in this study (Table 4) contributed the stable effects on the phenotype across different environments which underscores the value and reliability of the QTL

analysis. The QTL cluster of *qGS5.2*, *qPGWC5* (*qDEC5*) (mk1189-mk1192) on Chr05 (Table 4 and Fig. 4) simultaneously affected the traits of CS, GW, AS, PGWC and DEC and explained a phenotypic variation of 6.388 ~ 12.378 % for each trait, suggesting some pleiotropic effect for this QTL cluster. The *qGS5.2* allele for increasing grain CS, GW and AS was contributed by the parent P02428, which also contributed the *qPGWC5* (*qDEC5*) allele for increased PGWC and DEC. These results strongly suggest that *qGS5.2* and *qPGWC5* (*qDEC5*) represent the same locus controlling both grain shape and chalkiness. Around this QTL region, Gao et al. (2015) also roughly mapped five major QTLs (the interval was large) affecting GW, LWR, GT (grain thickness), PGWC, and TGW (1000-grain weight), each explaining up to 44.30, 55.29, 62.30, 30.94, and 28.78 % of the variation. Consequently, this QTL cluster could be a novel genetic region controlling multiple grain quality traits.

The QTL clusters of *qGS3.1* and *qGS7.1* associated with grain shape on Chr03 and Chr07 harbor more than four QTL loci, showing high stability and PVE value in multiple environments. Whereas the *qGS3.3* was extrapolated to be a minor-effect QTL, and allele of P0428 contributed positively to grain shape (GL, CS, LWR and PL). *qPGWC8* (*qDEC8*), which has not been reported previously, was identified for both PGWC and DEC, explaining the phenotypic variation of 6.70 ~ 10.07 %. In addition, there were three other QTL clusters with relatively smaller effects: *qGS1* (QTLs for GW, LWR and CS), *qGS2* (QTLs for GL, LWR and PL), and *qGS3.4* (QTLs for GL and PL). Due to high correlations between the examined traits, it is highly likely that these loci have pleiotropic effects on multiple characters, rather than

Table 3 QTLs associated with chalkiness traits detected in the different environments

Trait	Chromosome	Peak Position (cM)	Interval (cM)		Left Marker	Right Marker	LOD	PVE (%)[a]	Add[b]	Environment
PGWC	2	168	167.52	169.65	mk473	mk475	3.165	5.594	−7.291	Z-DS
	4	113	111.78	113.37	mk998	mk1000	7.035	12.975	−10.997	Z-DS
	4	135	134.51	136.37	mk1005	mk1007	5.635	9.738	−9.579	G-DS
	5	13	9.76	14.4	mk1153	mk1155	2.638	4.971	−8.125	G-WS
	5	35	34.39	35.18	mk1174	mk1175	5.250	8.985	−8.999	G-DS
	5	60	56.09	61.75	mk1189	mk1191	6.347	11.890	−10.337	Z-DS
	5	60	56.09	61.75	mk1189	mk1191	4.007	8.143	−9.706	G-WS
	5	84	83.81	85.4	mk1228	mk1230	2.773	5.182	−8.332	G-WS
	6	55	54.62	56.51	mk1378	mk1379	3.002	5.819	−8.807	G-WS
	6	78	77.75	78.01	mk1408	mk1409	2.568	4.457	−6.255	Z-DS
	7	132	130.52	132.94	mk1743	mk1745	3.662	7.123	−7.934	Z-DS
	7	134	133.2	134.53	mk1746	mk1747	2.946	4.763	−6.593	G-DS
	8	1	0.53	1.85	mk1786	mk1789	4.804	9.866	−10.693	Z-WS
	8	4	1.85	3.46	mk1789	mk1791	4.076	6.701	−7.893	G-DS
	8	17	16.85	17.11	mk1803	mk1804	4.580	8.645	−10.740	G-WS
	9	108	107.39	108.72	mk2078	mk2079	2.642	5.430	−7.888	Z-WS
DEC	4	110	109.95	110.21	mk991	mk992	3.016	6.342	−3.745	Z-WS
	4	136	134.51	136.37	mk1005	mk1007	3.779	7.760	−2.188	G-DS
	5	18	17.86	21.8	mk1160	mk1161	3.589	5.831	−4.473	G-WS
	5	35	34.39	35.18	mk1174	mk1175	3.354	6.847	−2.014	G-DS
	5	60	56.09	61.75	mk1189	mk1191	3.660	8.462	−1.813	Z-DS
	5	61	56.09	61.75	mk1189	mk1191	3.516	7.464	−4.003	G-WS
	6	23	22.67	23.19	mk1339	mk1340	2.516	5.391	−3.390	Z-WS
	6	64	63.57	65.46	mk1387	mk1388	7.378	12.779	−6.611	G-WS
	7	134	133.2	134.53	mk1746	mk1747	3.610	6.532	−0.015	Z-WS
	8	6	5.87	8.33	mk1792	mk1793	6.002	10.074	−5.931	G-WS
	8	130	129.72	130.24	mk1924	mk1925	3.411	7.027	−2.046	G-DS

[a] phenotypic variation explained; [b] additive effects

closely linked loci affecting individual characters. These QTL intervals were assumed to harbor novel gene loci affecting grain shape or chalkiness traits, and therefore are worth further investigating.

Gene expression profile and identification of candidate genes associated with chalkiness

To investigate the gene regulation patterns during grain development and perform a large-scale inspection of different expressed genes (DEGs) correlated with the chalkiness traits, RNA-Seq analysis was performed on the parental lines and two bulked pools made of RILs exhibiting extreme PGWC and DEC phenotypes. The total number of genes detected in each of the bulks is shown in Fig. 5c. After a significance test, 3603 genes with increased expression and 1949 genes with decreased expression were identified for P02428 compared to those of PYZX (Additional file 2: Table S4, Fig. 5a, d).

Such a large number of differentially expressed genes could not be responsible for the variation in chalkiness traits between PYZX and P02428. Normalization to remove the background noise of DEGs not related to chalkiness resulted in the identification of 88 genes with increased expression and 623 genes with decreased expression in the L-Pool (pool with extremely low levels of PGWC and DEC) compared to the H-Pool (pool with high levels of PGWC and DEC) (Additional file 3: Table S5, Fig. 5b, d). Functional annotations of the DEGs were analyzed and the investigation of GO enrichment (Gene Ontology Enrichment) was performed (Additional file 2: Table S4, Additional file 3: Table S5, and Additional file 4: Figure S1).

We analyzed the region bound by the flanking markers and co-located these DEGs near the QTL regions for PGWC and DEC including cluster 8, 11 and 12. A total of thirty-three DEGs were co-located in these three QTL

Table 4 Major QTL clusters associated with grain shape and chalkiness traits detected in this study

QTL cluster	Chromosome	Marker interval	Physical interval (100 kb)	Involved traits	LOD	PVE (%)[a]	Overlapped QTL reported
qGS1	1	mk289-mk295	425.5–432.5	GW, LWR and CS	3.122–4.948	3.781–7.943	
qGS2	2	mk401-mk405	162.5–167.5	GL, LWR and PL	2.765–4.789	3.540–8.689	
qGS3.1	3	mk686-mk692	156.5–162.5	AS, PL, CS, GL and LWR	3.161–12.450	7.228–18.648	
qGS3.2	3	mk693-mk698	163.5–169	CS, GL, LWR and PL	3.151–14.296	5.068–21.024	GS3 (Fan et al. 2006)
qGS3.3	3	mk794-mk795	286.5–287.5	CS, GL, LWR and PL	2.648–5.080	4.176–6.470	
qGS3.4	3	mk819-mk822	311.5–314.5	GL and PL	2.921–5.690	5.600–8.828	
qGS5.1	5	mk1174-mk1175	51.5–55.5	CS, GW and LWR	3.047–8.912	3.729–11.617	gw5 (Weng et al. 2008)
							qSW5 (Shomura et al. 2008)
qGS5.2, qPGWC5 (qDEC5)	5	mk1189-mk1192	76.5–79.5	CS, GW and AS; PGWC and DEC	2.686–6.347	6.388–12.377	
qGS7.1	7	mk1734-mk1739	232.5–239	CS, LWR, GL and PL	4.297–11.568	8.141–17.573	
qGS7.2	7	mk1743-mk1745	244–248	CS, GW, LWR and GL	18.346–18.346	12.326–26.139	GL7 (Wang et al. 2015a, b)
							GW7 (Wang et al. 2015a)
							SLG7 (Zhou et al. 2015b)
qPGWC7 (qDEC7)	7	mk1743-mk1747	244–251.5	PGWC and DEC	2.946–3.662	4.763–7.123	qPGWC-7 (Zhou et al. 2009)
qPGWC8 (qDEC8)	8	mk1786-mk1793	1.5–7.5	PGWC and DEC	4.076–6.002	6.701–10.074	

[a] phenotypic variation explained

regions (Table 5). Most were expressed more highly in H-Pool than in L-Pool and were higher in P02428 than in PYZX. Three genes with significantly different expression were located at the qPGWC5 (qDEC5) locus: the bidirectional sugar transporter SWEET3a (Os05g0214300), the UDP-glucuronosyl/UDP-glucosyltransferase family protein (Os05g0215300), and the class III peroxidase 73 (Os05g0231900) (Table 5). qRT-PCR analysis indicated that the expression level of SWEET3a, a gene involved in sugar transport, was about fifteen times higher in the H-Pool than in the L-Pool. This gene was also strongly upregulated in P02428 compared to PYZX in grain tissue. Os05g0215300 was expressed 4.55 fold higher in the H-Pool than comparing to L-Pool. The remaining thirty DEGS were located at qPGWC7 (qDEC7) (17) and qPGWC8 (qDEC8) (13) (Table 5). Of particular interest, the UDP-arabinopyranose mutase 3 (Os07g0604800) gene which was up-regulated in P02428 is located in the qPGWC7 (qDEC7) region and beta-glucosidase, GBA2 type domain containing protein (Os08g0111200) and

fructose-bisphosphate aldolase (Os08g0120600) mapped to the qPGWC8 (qDEC8) region. Differential expression of these genes was also validated by qRT-PCR and the results of the two experiments were basically consistent. These genes are the most suitable candidates for molecular cloning and development of new functional gene-target markers to facilitate marker assisted breeding.

Discussion

Cell division (cell number or cell size) is considered to contribute to the development and patterning of grain shape (Zhou et al. 2015b). Our finding in this study is that outer glume epidermal cell numbers and cell length were both significantly different between PYZX and P02428. In brief, the slender grains of PYZX were produced by longitudinally increasing cell length and cell number while transversely decreasing cell number (Fig. 2). The LWR of PYZX is more than twice the value of P02428, which to the best of our knowledge is the

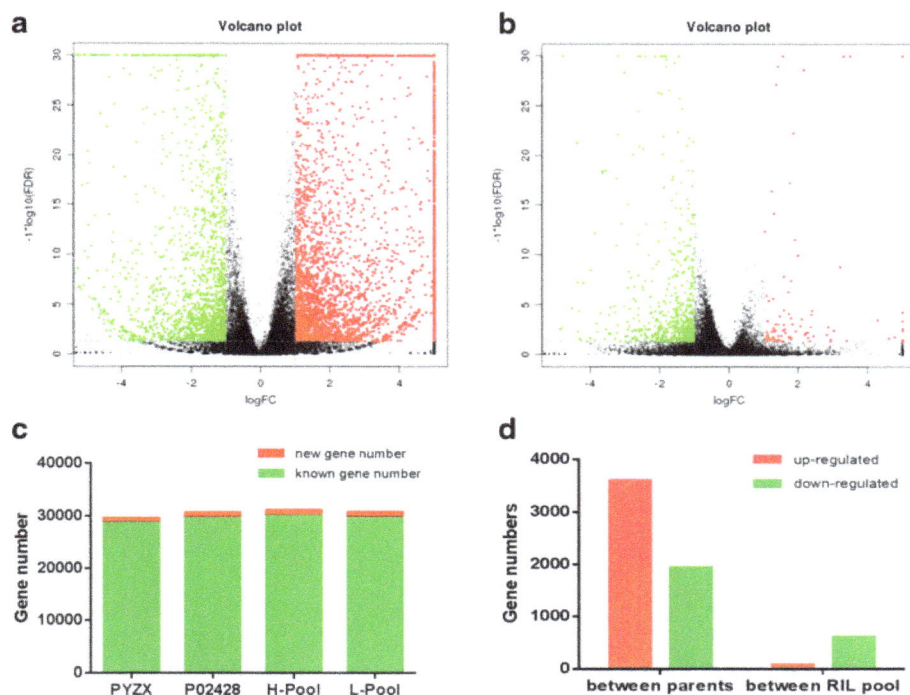

Fig. 5 Transcriptome profiling of parents and bulked RILs. **a** Differentially expressed genes between parents, (**b**) Differentially expressed genes between bulked RILs, (**c**) Gene number mapped across four sequencing library, (**d**) Up and down regulated genes detected between different samples

most extreme difference between parents of mapping populations used for QTL analysis of rice grain shape. Examination of the microstructures of rice endosperm of mature seeds demonstrated that the arrangement of the endosperm of PYZX was more compact than that of P02428, which exhibited starch granules with more spherical surfaces and uniform size. The differences in the starch granule shape and the arrangement of the granules resulted in the higher PGWC and DEC percentage in P02428, which was in agreement with previous research (Guo et al. 2011; Li et al. 2014b). The PGWC and DEC parameters were related to multiple investigated grain shape traits, and gave a maximum correlation coefficient with grain width (Table 1). These results were in accordance with Adu-Kwarteng et al. (2003); Zhou et al. (2015a), considering grain width had positive and high correlation with chalkiness. Starch granules in translucent areas of grains are bigger and more tightly packed than the small loosely packed granules in chalky areas of the grain (Lisle et al. 2000), and the hypothesis is that source-sink interactions involved in grain-filling are involved in the formation of chalk. Hence, the processes of starch synthesis were the focus of many studies about grain chalk (Fitzgerald et al. 2009). Previous studies demonstrate a complex mechanism for chalkiness formation in the rice endosperm. Although many starch-metabolic genes have been

characterized in the rice mutants, few corresponding to the QTLs for grain chalkiness have been addressed (Sun et al. 2015). Our results provide valuable background information on the structural characteristics of hull and endosperm tissues, which facilitate the understanding of molecular mechanisms determining grain shape and chalkiness.

Since the advent of molecular markers, crop researchers and breeders have dedicated huge amounts of effort on QTL mapping in biparental populations and marker-assisted selection (MAS) (Chen et al. 2014a). High-throughput SNP genotyping and estimation of recombination points based on resequencing of recombinant inbred lines were recently utilized, even though the sequencing coverage was insufficient (Huang et al. 2009; Xie et al. 2010). In this study, a total of 85,743 high-quality population SNPs with even distribution throughout the entire genome were detected using GBS strategy. Recombination breakpoints were determined by checking the positions where genotypes change. By this way, raw SNPs were converted into effective recombination bin, and these small recombination bins can be regarded as an effective type of genetic marker (Wang et al. 2011). The number of bin makers (a total of 2711) of the linkage map were increased significantly, compared to the previous study (1495 in total) using the RICE6K SNP array, which mapped QTLs of grain shape using a

Table 5 Annotated function of differentially expressed genes identified within or near the QTLs affecting chalkiness trait

No.	Gene ID	H-Pool vs. L-Pool					PYZX vs. P02428				Symbol	Description
		Fold change	$\log_2$(FC)	FDR	Sig.	Fold change (qRT-PCR)	Fold change	$\log_2$(FC)	FDR	Sig.		
qPGWC5 (qDEC5)												
1	*Os05g0214300*	(−)11.74	−3.554	0.003	yes	(−)15.33	(+)430.00	8.748	0.029	yes	*OsSWEET3a*	Similar to Bidirectional sugar transporter SWEET3a.
2	*Os05g0215300*	(−)2.07	−1.046	0.000	yes	(−)4.55	(+)22.53	4.494	0.000	yes		UDP-glucuronosyl/UDP-glucosyltransferase family protein.
3	*Os05g0231900*	(+)2.85	1.511	0.013	yes		(−)1263.33	−10.303	0.009	yes	*prx73*	Hypothetical conserved gene.
qPGWC7 (qDEC7)												
1	*Os07g0597000*	(−)2.24	−1.163	0.000	yes		(+)1.84	0.878	0.000	no		Similar to Eukaryotic translation initiation factor 5A (eIF-5A).
2	*Os07g0597050*	(−)2.78	−1.473	0.000	yes		(+)6.30	2.654	0.000	yes		Hypothetical gene.
3	*Os07g0597100*	(−)2.62	−1.392	0.000	yes		(+)5.74	2.522	0.000	yes		Similar to Saccharopine dehydrogenase.
4	*Os07g0597400*	(−)7.63	−2.931	0.000	yes		(+)168.27	7.395	0.000	yes		Conserved hypothetical protein.
5	*Os07g0599300*	(−)4.27	−2.094	0.033	yes		(+)15.84	3.986	0.002	yes		Hypothetical protein.
6	*Os07g0599500*	(−)3.73	−1.900	0.004	yes		(+)2483.33	11.278	0.000	yes		Hypothetical protein.
7	*Os07g0599600*	(−)4.39	−2.135	0.000	yes		(+)6.29	2.652	0.000	yes		Hypothetical protein.
8	*Os07g0599700*	(−)3.69	−1.885	0.000	yes		(+)3.88	1.955	0.000	yes		Similar to Surface protein PspC.
9	*Os07g0599900*	(−)6.15	−2.620	0.000	yes		(+)14.53	3.861	0.000	yes		Conserved hypothetical protein.
10	*Os07g0600300*	(−)2.67	−1.415	0.005	yes		(+)2.22	1.153	0.052	no		Protein of unknown function DUF794, plant family protein.
11	*Os07g0601100*	(−)6.04	−2.594	0.000	yes		(+)4.60	2.201	0.000	yes	*DHFR*	Similar to NADPH HC toxin reductase (Fragment).
12	*Os07g0602000*	(−)2.21	−1.142	0.001	yes		(+)26.60	4.733	0.000	yes	*DHFR*	Similar to NADPH HC toxin reductase (Fragment).
13	*Os07g0602900*	(−)4.42	−2.144	0.001	yes		(+)1.75	0.805	0.029	no		Protein of unknown function DUF1675 domain containing protein;Similar to UPF0737 protein 1.
14	*Os07g0604800*	(−)3.52	−1.815	0.000	yes	(−)2.63	(+)2.09	1.062	0.000	yes	*OsUAM3*	Similar to Alpha-1,4-glucan-protein synthase [UDP-forming] 1 (EC 2.4.1.112) (UDP- glucose:protein transglucosylase 1) (UPTG 1).
15	*Os07g0607500*	(−)2.09	−1.067	0.049	yes		(+)2.59	1.371	0.035	yes		Protein of unknown function DUF1195 family protein.
16	*Os07g0616750*	(+)2.13	1.090	0.034	yes		(−)2.34	−1.229	0.012	yes		Hypothetical gene.
17	*Os07g0617100*	(−)2.17	−1.120	0.000	yes		(+)1.01	0.017	1.000	no		Plant disease resistance response protein family protein.
qPGWC8 (qDEC8)												
1	*Os08g0101500*	(−)2.64	−1.403	0.000	yes		(+)1.22	0.281	0.538	no	*OsSultr5;2*	Similar to sulfate transporter.
2	*Os08g0101800*	(−)2.80	−1.485	0.020	yes		(+)2.68	1.420	0.000	yes		Protein of unknown function DUF821, CAP10-like family protein.

Table 5 Annotated function of differentially expressed genes identified within or near the QTLs affecting chalkiness trait *(Continued)*

#	Gene											Function
3	Os08g0104400	9.27	−3.212	0.000	yes		(+)2.94	1.554	0.000	yes		Conserved hypothetical protein.
4	Os08g0106100	(−)5.57	−2.477	0.000	yes		(+)5793.33	12.500	0.000	yes		Conserved hypothetical protein.
5	Os08g0111200	(−)3.14	−1.652	0.007	yes	(−)2.87	(+)1.69	0.760	0.362	no		Beta-glucosidase, GBA2 type domain containing protein.
6	Os08g0112300	(−)3.54	−1.824	0.006	yes		(+)2.10	1.071	0.002	yes		Transferase domain containing protein.
7	Os08g0113000	(−)3.67	−1.877	0.006	yes		(+)2.43	1.279	0.000	yes	*prx117*	Similar to Peroxidase 47 precursor (EC 1.11.1.7) (Atperox P47) (ATP32).
8	Os08g0114300	(−)3.16	−1.660	0.000	yes		(+)1.62	0.699	0.008	no		D-arabinono-1,4-lactone oxidase domain containing protein.
9	Os08g0114400	(−)4.85	−2.277	0.000	yes		(+)1.88	0.912	0.003	no		Hypothetical protein.
10	Os08g0116800	(−)3.77	−1.913	0.016	yes		(+)4.95	2.309	0.155	no		Exoribonuclease domain containing protein.
11	Os08g0120600	(−)2.33	−1.219	0.005	yes	(−)5.67	(+)1.42	0.501	0.343	no		Similar to Fructose-bisphosphate aldolase, cytoplasmic isozyme (EC 4.1.2.13).
12	Os08g0122800	(−)6.45	−2.689	0.003	yes		(+)240.50	7.910	0.000	yes		Conserved hypothetical protein. (Os08t0122800-01); Kringle, conserved site domain containing protein.
13	Os08g0124500	(−)306.67	−8.261	0.018	yes		(+)96.67	6.595	0.114	no		Similar to Resistance protein candidate (Fragment); Similar to Resistance protein candidate (Fragment).

similar population (197 RILs derived from the cross between indica variety ZS97 and japonica variety XZ2) (Hu et al. 2013). It is generally considered that the efficiency of QTL mapping largely depends on marker density and QTL mapping resolution can be improved with greater marker density to detect the locations of recombination events more precisely (Pan et al. 2012; Chen et al. 2014b). Yu et al. (2011) detected QTL using an ultra-high density SNP map based on population sequencing relative to traditional RFLP/SSR markers, and indicated that compared to RFLP/SSR and array-based SFP genotyping methods, the sequence-based method produces a map of the highest density, while the accuracy and the quality of the SNP markers was enhanced by using information of adjacent SNPs to form bins. In our study, take the example of QTLs of PGWC, the genetic intervals range from 0.26 to 5.66 cM, with an average of 2.14 cM, while the QTL clusters of PGWC were delimited into a physical region of few hundred kb, which had narrowed significantly compared to previous QTL mapping studies about grain chalkiness with SSR (Mei et al. 2013; Peng et al. 2014; Zhao et al. 2016b).

In general, it was expected that more QTLs for target traits with a mapping population derived from two contrasting parents would be detected (Bai et al. 2010). In our study, the mapping population derived from PYZX and P02428 showed an extremely wide diversity in rice grain shape and chalkiness traits, thus it was ideal for identification of main and minor effect QTL. Using single environment analysis, 109 and 27 QTLs associated with grain shape and chalkiness were detected respectively. Among these, 58 (53.2 %) of the QTLs for grain shape were detected in two or more environments whereas only 6 (22.2 %) of the chalkiness-related QTLs were observed in more than one of the environments (Additional file 1: Table S3). This confirmed that chalkiness was considerably affected by environment and exhibited a pattern of instability, whereas rice grain shape was fixed as long as the panicle was normally differentiated and mainly controlled by genotype and had higher heritability (Bai et al. 2010). *GS3*, which is reported as a major QTL for grain length and weight and a minor QTL for grain width and thickness in rice, encodes a putative transmembrane protein. It was located on the *qGS3.2* region in our QTL analysis, which was recurrent across four environments for GL, LWR, CS, PL and AS, showing the most stable expression and pleiotropic effects (Fan et al. 2006). The reported QTL of *GL7* (Wang et al. 2015b) regulates longitudinal cell elongation and results in an increase in grain length and improvement of grain appearance quality. In our study, *GL7* was detected in the QTL region of *qGS7.2* (associated with CS,

GW, LWR, PL and GL) and also displayed consistent heredity and pleiotropism. Similarly, we detected a new QTL cluster of qGS5.2 associated with GW, LWR, CS and AS with high PVE levels and stable performance across multiple environments, representing potential targets for gene cloning. In addition, there were some new-found important QTL clusters, such as qGS3.1 associated with AS, PL, CS, GL and LWR, qGS7.1 associated with CS, LWR, GL and PL, and qPGWC8 (qDEC8) associated with PGWC and DEC, which offer an opportunity for improving grain appearance in rice for the future. As the main grain quality related traits, we intend to develop backcrossing introgression lines of the novel QTLs in the background of elite high-yielding varieties which should be useful to improve the grain quality. "Consumer-Targeted Rice Breeding" is emphasized especially in the rice quality area, and the grain shape QTLs should been utilized based on the specificity of consumer preference (Calingacion et al. 2014). The stable QTLs across multiple environments gained in this study enable incorporation of favorable alleles into agronomically superior germplasm by using technology of MAS combined with backcrossing method. Moreover, grain appearance quality had a close relation with milling quality (Lou et al. 2009). And milling yield of short and medium grain rice is typically higher than long grain rice (Kepiro et al. 2008). QTLs associated to more grain shape parameters (including GW, GL, LWR, CS, PL, and AS) had been analysis in this study, and different effects of the alterable combinations of target QTLs could probably strike a balance between grain shape and milling trait. In our practice, MAS have being performing by introgression of multiple QTL alleles of low chalk and grain length (preferred by local consumers of south China) into the elite restorer lines (R) and maintainer lines (B) in our lab, with the aim of improving the grain quality characters in hybrid rice.

Another advantageous feature of our study was the integration of the QTL mapping of chalkiness-related traits and the transcriptome profiling. Obtaining gene expression information using bulked RIL pools is simple, effective and aim-focused (Kloosterman et al. 2010), making it possible to assay a large number of our subjects. Hence it became possible to identify and narrow down our search for putative candidate genes affecting chalkiness-related traits. By this analysis, several differentially expressed genes in the three QTL intervals on chromosomes 5, 7 and 8 appear to have high probability of the target genes, considering their protein function prediction information as well. For instance, bidirectional sugar transporter SWEET3a, which was identified to be a DEG located on the region of qPGWC5 (qDEC5), is involved in carbohydrate transmembrane transport process (http://www.ebi.ac.uk/QuickGO/) and predicted to mediate both low-affinity uptake and efflux of sugar across the plasma membrane (http://string-db.org/news-tring_cgi/). The major up-regulated expression observed in P02428 is highly likely to influence carbohydrate transport and result in high levels of chalkiness. In another example, Os07g0604800 (OsUAM3) which co-located with the qPGWC7 (qDEC7) QTL in our study and the previously reported qPGWC7 QTL (Zhou et al. 2009) was detected to be sharply down-regulated in the L-Pool and PYZX compared to the H-Pool and P02428. OsUAM3 gene has an annotated as having UDP-arabinopyranose mutase activity (http://www.shigen.ni-g.ac.jp/rice/oryzabase/) and was predicted to be analpha-1,4-glucan-protein synthase and UDP-glucose/protein transglucosylase 1 (http://rapdb.dna.affrc.go.jp/). Given that carbohydrate metabolism is considered to play an important role in endosperm chalkiness (Wang et al. 2008), these DEGs represent strong candidates for genes underlying these chalkiness-associated QTLs. For instance, the rice GIF1 (GRAIN INCOMPLETE FILLING 1) gene is required for carbon partitioning during early grain-filling and the gif1 mutant grains showed more chalkiness, which accumulated lower levels of glucose and fructose, as well as sucrose (Wang et al. 2008). The UGPase1 gene played a key role in seed carbohydrate metabolism and inactivation of the UGPase1 gene caused endosperm chalkiness in rice (Woo et al. 2008). The several DEGs we identified here were involved in carbohydrate synthesis or transportation, indicating their potential role in the formation of endosperm chalkiness. Our identification of 668 functionally annotated DEGs between bulked RIL pools (Additional file 3: Table S5) provides a basis for dissecting the regulatory network governing the chalkiness trait. While detailed insight into these genes will be of great importance in unraveling the complex nature of rice chalkiness and also in elucidating the role of the diverse QTLs involved. The integrated approach utilizing the identified stable QTLs and transcriptome profiling could serve as a platform for candidate gene identification for genetic dissection and provide basal tools for molecular breeding in rice.

Conclusion

Using the Genotyping-By-Sequencing approach, a genetic linkage map with an average distance of 0.865 cM between adjacent markers was constructed based on a RIL population in rice. A global mapping of quantitative trait loci affecting the grain shape and chalkiness traits were detected on four environments and the stable QTL clusters were highlighted and analyzed. The results of the transcriptome analysis demonstrated an available gene expression profile responsible for the development of chalkiness, and several important differentially expressed genes were co-located on the chalkiness-related QTL regions on chromosomes 5, 7, and 8. Critical loci were investigated and identified as candidate genes, which were suitable for functional validation and breeding utilization.

Methods

Plant material

In this study, the mapping population consisting of 192 recombinant inbred lines (RILs) derived by single-seed descent from an inter-subspecific cross of the indica PYZX and the japonica P02428, was used to perform QTL analyses for grain shape and chalkiness traits. RILs and the parental lines were field planted in Guangzhou (traditional flatland field) and Zengcheng (low hill district) in Guangdong province of China at the dry season (DS) in 2014 and the wet season (WS) in 2015, which have been named G-DS, Z-DS, G-WS, and Z-WS respectively in this study.

Morphological and cellular analyses

Paraffin-embedded sections of spikelet samples were prepared according to Kim et al. (2014), with minor modifications. The materials were fixed in FAA and stored at 4 °C for 24 h. The fixed spikelets were dehydrated in a gradient ethanol series, and were then incubated in 100 % ethanol overnight. Dehydrated spikelets were embedded in Paraplast Plus (Sigma). The transverse sections of each spikelet were stained with 0.5 % toluidine blue, and viewed using an SZX10 stereomicroscope (Olympus, Tokyo, Japan). For scanning electron microscopy analysis, spikelet hull and endosperm samples were processed according to Wang et al. (2015b) and Li et al. (2014b). Young spikelets were fixed in 4 % (w/v) paraformaldehyde and 0.25 % glutaraldehyde in 0.1 M sodium phosphate buffer, pH 7.2, at 4 °C overnight. Fixed spikelets were dehydrated in a graded ethanol series, and 100 % ethanol was replaced with 3-methylbutyl acetate (Toriba et al. 2010). Milled rice grains were transversely cut in the middle with a knife and were coated with gold under vacuum conditions. Samples were dried at their critical point, sputter coated with platinum, and observed with the XL-30-ESEM instrument at an accelerating voltage of 5 kV.

Evaluation of grain shape and chalkiness traits

Images of the mature grain were captured on a CanoScan 5600 F (Canon, Japan) scanner with the supplied software without image enhancement, and the grain shape parameters of GW, GL, LWR, CS, PL, and AS were measured precisely using SmartGrain Software (Tanabata et al. 2012). The chalkiness parameters were measured with an automatic machine JMWT 12 according to Xu et al. (2012). Two metrics were used to describe grain chalkiness as previously described (Zhao

et al. 2016a): percentage of grains with chalk (PGWC) and degree of endosperm chalkiness (DEC) which is the ratio of total chalky area to the total kernel area of all sampled grains. The statistical analysis was performed with SPSS statistics 18.0 and Microsoft Excel.

DNA extraction, genotyping by sequencing, and SNP identification

Leaf samples were collected from two parental lines and 192 RILs at F_7 generation. DNAs were extracted using the CTAB method and quantified using both a Nano-Drop ND-1000 Spectrophotometer and agarose gel electrophoresis. In our study, the genome of parental lines, PYZX and P02428, were directly sequenced to ~25× coverage, while the RILs were subjected to Genotyping By Sequencing (GBS) as described by Elshire et al. (2011). The DNA samples of the RIL population digested using *Mse*I and *Hae*III. The other basic schematic of the protocol used for performing GBS was according to Duan et al. (2013).

Sequencing was performed on the Illumina HiSeq 2500 platform to generate 150 bp paired-end reads (Novogene Bioinformatics Technology Co., Ltd, China). The original image data generated were converted into sequence data via base calling (Illumina pipeline CASAVA v1.8.2) and then subjected to the quality control (QC) procedure to remove unusable reads: 1) reads contain the Illumina library construction adapters; 2) reads contain more than 10 % unknown bases (N bases); 3) one end of a read contains more than 50 % low quality bases 4) Sequencing reads were aligned to the reference genome (http://plants.ensembl.org/Oryza_sativa/) using BWA with default parameters. Subsequent processing, including duplicate removal was performed using SAMtools and PICARD (http://picard.sourceforge.net). The raw SNP/InDel sets are called by SAMtools with the parameters as '-q 1 -C 50 −m 2 -F 0.002 -d 1000'. Then we filtered these sets using the following criteria: (1) mapping quality >20; (2) depth of the variant position >4.

Bin marker production and QTL analysis

To overcome the false positive of SNPs genotype of the population, the sliding window approach adopted by Huang et al. (2009) with some modification was used to evaluate a group of consecutive SNPs for genotyping. The genotypic maps of the RILs were aligned and split into recombination bins according to the recombination breakpoints, with the parameter of window size of 15 SNPs. Bins less than 300-kb were merged with the next bin. Genotypes of bins for regions at the transitions between two different genotype blocks were set to missing data. Segregation distortion markers showing distorted segregation ($P < 0.01$) were discarded. For this step, a total of 85,743 high-quality SNPs were used for bin map construction, containing 2711 bin markers.

SNP bin markers were used to construct the genetic linkage map using the est.map function of the R/qtl package (Broman et al. 2003) with the Kosambi map method, and the marker genetic distances were estimated. The QTLs were mapped with the inclusive composite interval mapping method using the QTL IciMapping software version 4 (Li et al. 2007; Meng et al. 2015) with single-environment phenotypic values. QTLs were calculated using the ICIM-ADD mapping method, with mapping parameters of 1 cM step and 0.001 probabilities in a stepwise regression. The threshold for logarithm of odds (LOD) scores was set as 2.5, and the QTLs in a particular genomic region with the LOD values larger than this threshold were called (Li et al. 2014a). The regional genes were annotated and analyzed via the database of RAP (http://rapdb.dna.affrc.go.jp) and Ensembl (http://plants.ensembl.org/Oryza_sativa).

Experimental design for transcriptome profiling study
To obtain an overview of the transcriptome profiling and differential gene expression pattern relating to the chalky trait, two bulked RIL pools with extreme tails of the chalky trait were developed, consisting of 13 bulks RIL individuals respectively. The L-Pool was constructed from individuals with extremely low levels at PGWC and DEC, conversely the H-Pool bears high levels of PGWC and DEC parameters (the details are available in Additional file 1: Table S6). The pools were used for RNA-Seq analysis along with the parental lines. After approximately 20 days after fertilization, grain samples of each group were collected and stored at −80 °C in preparation for RNA-Seq.

RNA isolation, sequencing and statistical analysis of gene expression profile
The total RNA of each of the above listed samples was homogenized using mortar and pestle with liquid nitrogen and purified using the Plant Total RNA Purification Kit (Dakewe Biotech Company) following the manufacturer's instructions. RNA quality was verified using Agilent 2100 Bio-analyzer (Agilent Technologies, Santa Clara, CA) and was also checked by RNase free agarose gel electrophoresis. Next, Poly (A) mRNA was isolated using oligo-dT beads (Qiagen). All mRNA was broken into short fragments by adding fragmentation buffer. First-strand cDNA was generated using random hexamer-primed reverse transcription, followed by the synthesis of the second-strand cDNA using RNase H and DNA polymerase I. The cDNA fragments were purified using a QIAquick PCR extraction kit. The cDNA library was sequenced on the Illumina sequencing platform (IlluminaHiSeq™ 2500) using the paired-end

technology by Gene Denovo Co. (Guangzhou, China). A Perl program was written to select clean reads by removing low quality sequences (there were more than 50 % bases with quality lower than 20 in one sequence), reads with more than 5 % N bases (bases unknown) and reads containing adaptor sequences. Sequencing reads in FASTQ format were mapped to the reference genome (http://plants.ensembl.org/Oryza_sativa/) and splice junctions were identified using TopHat (Kim et al. 2013). The Cufflinks package (Trapnell et al. 2012) was used for genome guided transcript assembly and the expression abundance was estimated.

After the expression level of each transcript and gene was calculated, differential expression analysis was conducted using edgeR (Robinson et al. 2010). The false discovery rate (FDR) was used to determine the threshold of the p value in multiple tests, and for the analysis, a threshold of the FDR ≤ 0.05 and an absolute value of $\log_2$Ratio ≥ 1 were used to judge the significance of the gene expression differences. The differentially expressed genes were used for GO (Gene Ontology) and KEGG (Kyoto Encyclopedia of Genes and Genomes) enrichment analyses according to a method similar to that described by Zhang et al. (2013). Both GO terms and KEGG pathways with a Q-value ≤ 0.05 are significantly enriched in DEGs. To compare the differential gene expression between PYZX versus P02428, and between bulked RIL pools, we took PYZX and the H-pool as baseline controls, respectively.

Validation of gene expression by qRT–PCR
Expression levels of five genes (*Os05t0214300-00*, *Os05t0215300-01*, *Os07t0604800-01*, *Os08t0101500-01*, and *Os08t0120600-01*) were selected for the validation of RNA-seq results using quantitative real-time PCR (qRT-PCR). The mRNA sequences of the five genes were downloaded from EnsemblPlants database (http://plants.ensembl.org/Oryza_sativa/), and were used for primers design using Primer3 software. The primer sequences are listed in Additional file 2: Table S4. First strand cDNA was prepared from 2 µg of total RNA in 50 µl of reaction volume using the high-capacity cDNA Archive kit (Applied Biosystems, USA). Two µl of the first strand cDNA reaction was used for quantitative real time PCR. qRT-PCR was conducted using the AceQ qPCR SYBR Green Master Mix Kit (Vazyme Biotech) according to standard protocol, and the expression levels of the genes were determined on the StepOnePlus System (Applied Biosystems, USA). Three biological and three technical replicates were taken for each treatment. As an endogenous control, *Actin* was used for the normalization of Ct value obtained and the relative expression values were calculated by ΔΔCt method.

Additional files

Additional file 1: Table S1. Comparison of eight traits between PYZX, P02428 and RILs in four environments in this study. Table S2. Annotation statistics of the SNPs between PYZX and P02428. Table S3.QTLs detected in the 192 RIL lines derived from the cross between PYZX and P02428 in four environments. Table S6.Details of PGWC and DEC parameters of the L-Pool and H-Pool. Table S7.Primers used for qRT-PCR.

Additional file 2: Table S4. Annotated function of differentially expressed genes identified between parents

Additional file 3: Table S5. Annotated function of differentially expressed genes identified between RIL segregant pools.

Additional file 4: Figure S1. GO (Gene Ontology) Enrichment of differentially expressed genes.

Abbreviations
AS: Area size of grain; CS: Circularity; DEC: Degree of endosperm chalkiness; DEG: Differentially expressed gene; DS: Dry season; GBS: Genotyping-by-sequencing; GL: Grain length; GW: Grain width; LWR: The ratio of grain length and width; PGWC: Percentage of grain with chalkiness; PL: Perimeter length of grain; WS: Wet season

Acknowledgments
Financial support for this research was provided in part by a grant from the National Key Technology Research and Development Program of China (No. 2016YFD0102102), Science and technology project of Guangdong Province (No. 2015B020231011) and the earmarked fund for Modern Agro-industry Technology Research System (No. CARS-01-12). We thank Dr. Kunshen Wu for the critical reading and modification of the manuscript.

Authors' contributions
ZC, TG and LC designed the project and LC performed all the experiments and wrote the manuscript. WG, SC, LW, YL and JZ assisted in conducting experiments and data analysis. HW and ZC provided the direction for the study and the correction of the manuscript. All authors read and approved the final manuscript.

Competing interests
The authors declare that they have no competing interests.

Author details
[1]National Engineering Research Center of Plant Space Breeding, South China Agricultural University, Guangzhou 510642, China. [2]Guangdong Agricultural Technology Extension, Guangzhou 510520, China.

References
Adu-Kwarteng E, Ellis WO, Oduro I, Manful JT (2003) Rice grain quality: a comparison of local varieties with new varieties under study in Ghana. Food Control 14:507–514

Bai XF, Luo LJ, Yan WH, Kovi MR, Zhan W, Xing YZ (2010) Genetic dissection of rice grain shape using a recombinant inbred line population derived from two contrasting parents and fine mapping a pleiotropic quantitative trait locus qGL7. BMC Genet 11:1–11

Broman KW, Wu H, Sen Ś, Churchill GA (2003) R/qtl: QTL mapping in experimental crosses. Bioinformatics 19:889–890

Calingacion M, Laborte A, Nelson A, Resurreccion A, Concepcion JC, Daygon VD, Mumm R, Reinke R, Dipti S, Bassinello PZ, Manful J, Sophany S, Lara KC, Bao J, Xie L, Loaiza K, El-hissewy A, Gayin J, Sharma N, Rajeswari S, Manonmani S, Rani NS, Kota S, Indrasari SD, Habibi F, Hosseini M, Tavasoli F, Suzuki K, Umemoto T, Boualaphanh C, Lee HH, Hung YP, Ramli A, Aung PP, Ahmad R, Wattoo JI, Bandonill E, Romero M, Brites CM, Hafeel R, Lur H-S, Cheaupun K, Jongdee S, Blanco P, Bryant R, Thi Lang N, Hall RD, Fitzgerald M (2014) Diversity of global rice markets and the science required for consumer-targeted rice breeding. PLoS One 9:e85106

Chen W, Chen HD, Zheng TQ, Yu RB, Terzaghi WB, Li ZK, Deng XW, Xu JL, He H (2014a) Highly efficient genotyping of rice biparental populations by

GoldenGate assays based on parental resequencing. Theor Appl Genet 127: 297–307

Chen Z, Wang B, Dong X, Liu H, Ren L, Chen J, Hauck A, Song W, Lai J (2014b) An ultra-high density bin-map for rapid QTL mapping for tassel and ear architecture in a large F(2) maize population. BMC Genomics 15:433

Concepcion JCT, Ouk M, Zhao D, Fitzgerald MA (2015) The need for new tools and investment to improve the accuracy of selecting for grain quality in rice. Field Crop Res 182:60–67

Duan M, Sun Z, Shu L, Tan Y, Yu D, Sun X, Liu R, Li Y, Gong S, Yuan D (2013) Genetic analysis of an elite super-hybrid rice parent using high-density SNP markers. Rice 6:21

Elshire RJ, Glaubitz JC, Sun Q, Poland JA, Kawamoto K, Buckler ES, Mitchell SE (2011) A robust, simple genotyping-by-sequencing (GBS) approach for high diversity species. PLoS One 6:e19379

Fan C, Xing Y, Mao H, Lu T, Han B, Xu C, Li X, Zhang Q (2006) GS3, a major QTL for grain length and weight and minor QTL for grain width and thickness in rice, encodes a putative transmembrane protein. Theor Appl Genet 112:1164–1171

Fitzgerald MA, McCouch SR, Hall RD (2009) Not just a grain of rice: the quest for quality. Trends Plant Sci 14:133–139

Gao F, Zeng L, Qiu L, Lu X, Ren J, Wu X, Su X, Gao Y, Ren G (2015) QTL mapping of grain appearance quality traits and grain weight using recombinant inbred populations in rice (Oryza sativa L.). J Integr Agr 4:1961–1968

Gao ZY, Zhao SC, He WM, Guo LB, Peng YL, Wang JJ, Guo XS, Zhang XM, Rao YC, Zhang C, Dong GJ, Zheng FY, Lu CX, Hu J, Zhou Q, Liu HJ, Wu HY, Xu J, Ni PX, Zeng DL, Liu DH, Tian P, Gong LH, Ye C, Zhang GH, Wang J, Tian FK, Xue DW, Liao Y, Zhu L, Chen MS, Li JY, Cheng SH, Zhang GY, Wang J, Qian Q (2013) Dissecting yield-associated loci in super hybrid rice by resequencing recombinant inbred lines and improving parental genome sequences. Proc Natl Acad Sci U S A 110:14492–14497

Guo T, Liu X, Wan X, Weng J, Liu S, Liu X, Chen M, Li J, Su N, Wu F, Cheng Z, Guo X, Lei C, Wang J, Jiang L, Wan J (2011) Identification of a stable quantitative trait locus for percentage grains with white chalkiness in rice (Oryza sativa). J Integr Plant Biol 53:598–607

Han B, Huang X (2013) Sequencing-based genome-wide association study in rice. Curr Opin Plant Biol 16:133–138

Hu W, Wen M, Han Z, Tan C, Xiong Y (2013) Scanning QTLs for grain shape using a whole genome SNP array in rice. J Plant Biochem Physiol 16:104

Huang R, Jiang L, Zheng J, Wang T, Wang H, Huang Y, Hong Z (2013) Genetic bases of rice grain shape: so many genes, so little known. Trends Plant Sci 18:218–226

Huang X, Feng Q, Qian Q, Zhao Q, Wang L, Wang A, Guan J, Fan D, Weng Q, Huang T, Dong G, Sang T, Han B (2009) High-throughput genotyping by whole-genome resequencing. Genome Res 19:1068–1076

Jones JM, Sheats DB (2016) Consumer trends in grain consumption, Reference module in food science

Kepiro JL, McClung AM, Chen MH, Yeater KM, Fjellstrom RG (2008) Mapping QTLs for milling yield and grain characteristics in a tropical japonica long grain cross. J Cereal Sci 48:477–485

Kim D, Pertea G, Trapnell C, Pimentel H, Kelley R, Salzberg SL (2013) TopHat2: accurate alignment of transcriptomes in the presence of insertions, deletions and gene fusions. Genome Biol 14:R36

Kim DM, Lee HS, Kwon SJ, Fabreag ME, Kang JW, Yun YT, Chung CT, Ahn SN (2014) High-density mapping of quantitative trait loci for grain-weight and spikelet number in rice. Rice 7:1–11

Kloosterman B, Oortwijn M, Uitdewilligen J, America T, de Vos R, Visser RGF, Bachem CWB (2010) From QTL to candidate gene: genetical genomics of simple and complex traits in potato using a pooling strategy. BMC Genomics 11:500

Kong X, Kasapis S, Bao J (2015) Viscoelastic properties of starches and flours from two novel rice mutants induced by gamma irradiation. LWT-Food Sci Techn 60:578–582

Kumar S, Banks TW, Cloutier S (2012) SNP discovery through next-generation sequencing and its applications. J Plant Genom 2012:831460

Li B, Tian L, Zhang JY, Huang L, Han FX, Yan SR, Wang LZ, Zheng HK, Sun JM (2014a) Construction of a high-density genetic map based on large-scale markers developed by specific length amplified fragment sequencing (SLAF-seq) and its application to QTL analysis for isoflavone content in Glycine max. BMC Genomics 15:1–16

Li HH, Ye GY, Wang JK (2007) A modified algorithm for the improvement of composite interval mapping. Genetics 175:361–374

Li Y, Fan C, Xing Y, Yun P, Luo L, Yan B, Peng B, Xie W, Wang G, Li X, Xiao J, Xu C, He Y (2014b) Chalk5 encodes a vacuolar H + –translocating pyrophosphatase influencing grain chalkiness in rice. Nat Genet 46:398–404

Lisle A, Martin M, Fitzgerald M (2000) Chalky and translucent rice grains differ in starch composition and structure and cooking properties. Cereal Chem 77:627–632

Lou J, Chen L, Yue GH, Lou QJ, Mei HW, Xiong L, Luo LJ (2009) QTL mapping of grain quality traits in rice. J Cereal Sci 50:145–151

Mao H, Sun S, Yao J, Wang C, Yu S, Xu C, Li X, Zhang Q (2010) Linking differential domain functions of the GS3 protein to natural variation of grain size in rice. Proc Natl Acad Sci U S A 107:19579–19584

Mei DY, Zhu YJ, Yu YH, Fan YY, Huang DR, Zhuang JY (2013) Quantitative trait loci for grain chalkiness and endosperm transparency detected in three recombinant inbred line populations of indica rice. J Integr Agr 12:1–11

Meng L, Li H, Zhang L, Wang J (2015) QTL IciMapping: integrated software for genetic linkage map construction and quantitative trait locus mapping in biparental populations. Crop J 3:269–283

Muthayya S, Sugimoto JD, Montgomery S, Maberly GF (2014) An overview of global rice production, supply, trade, and consumption. Ann NY Acad Sci 1324:7–14

Pan QC, Ali F, Yang XH, Li JS, Yan JB (2012) Exploring the genetic characteristics of two recombinant inbred line populations via high-density SNP markers in maize. PLoS One 7:e52777

Peng B, Wang LQ, Fan CC, Jiang GH, Luo LJ, Li YB, He YQ (2014) Comparative mapping of chalkiness components in rice using five populations across two environments. BMC Genet 15:1–14

Qi P, Lin Y-S, Song X-J, Shen J-B, Huang W, Shan J-X, Zhu M-Z, Jiang L, Gao J-P, Lin H-X (2012) The novel quantitative trait locus GL3.1 controls rice grain size and yield by regulating Cyclin-T1;3. Cell Res 22:1666–1680

Robinson MD, McCarthy DJ, Smyth GK (2010) edgeR: a bioconductor package for differential expression analysis of digital gene expression data. Bioinformatics 26:139–140

Shomura A, Izawa T, Ebana K, Ebitani T, Kanegae H, Konishi S, Yano M (2008) Deletion in a gene associated with grain size increased yields during rice domestication. Nat Genet 40:1023–1028

Song X-J, Huang W, Shi M, Zhu M-Z, Lin H-X (2007) A QTL for rice grain width and weight encodes a previously unknown RING-type E3 ubiquitin ligase. Nat Genet 39:623–630

Sun L, Li X, Fu Y, Zhu Z, Tan L, Liu F, Sun X, Sun X, Sun C (2013) GS6, a member of the GRAS gene family, negatively regulates grain size in rice. J Integr Plant Biol 55:938–949

Sun W, Zhou Q, Yao Y, Qiu X, Xie K, Yu S (2015) Identification of genomic regions and the isoamylase gene for reduced grain chalkiness in rice. PLoS One 10:e0122013

Tanabata T, Shibaya T, Hori K, Ebana K, Yano M (2012) SmartGrain: high-throughput phenotyping software for measuring seed shape through image analysis. Plant Physiol 160:1871–1880

Toriba T, Suzaki T, Yamaguchi T, Ohmori Y, Tsukaya H, Hirano HY (2010) Distinct regulation of adaxial-abaxial polarity in anther patterning in rice. Plant Cell 22:1452–1462

Trapnell C, Roberts A, Goff L, Pertea G, Kim D, Kelley DR, Pimentel H, Salzberg SL, Rinn JL, Pachter L (2012) Differential gene and transcript expression analysis of RNA-seq experiments with TopHat and Cufflinks. Nat Protoc 7:562–578

Wang E, Wang J, Zhu XD, Hao W, Wang LY, Li Q, Zhang LX, He W, Lu BR, Lin HX, Ma H, Zhang GQ, He ZH (2008) Control of rice grain-filling and yield by a gene with a potential signature of domestication. Nat Genet 40:1370–1374

Wang L, Wang A, Huang X, Zhao Q, Dong G, Qian Q, Sang T, Han B (2011) Mapping 49 quantitative trait loci at high resolution through sequencing-based genotyping of rice recombinant inbred lines. Theor Appl Genet 122:327–340

Wang S, Li S, Liu Q, Wu K, Zhang J, Wang S, Wang Y, Chen X, Zhang Y, Gao C, Wang F, Huang H, Fu X (2015a) The OsSPL16-GW7 regulatory module determines grain shape and simultaneously improves rice yield and grain quality. Nat Genet 47:949–954

Wang S, Wu K, Yuan Q, Liu X, Liu Z, Lin X, Zeng R, Zhu H, Dong G, Qian Q, Zhang G, Fu X (2012) Control of grain size, shape and quality by OsSPL16 in rice. Nat Genet 44:950–954

Wang YX, Xiong GS, Hu J, Jiang L, Yu H, Xu J, Fang YX, Zeng LJ, Xu EB, Xu J, Ye WJ, Meng XB, Liu RF, Chen HQ, Jing YH, Wang YH, Zhu XD, Li JY, Qian Q (2015b) Copy number variation at the GL7 locus contributes to grain size diversity in rice. Nat Genet 47:944–948

Weng J, Gu S, Wan X, Gao H, Guo T, Su N, Lei C, Zhang X, Cheng Z, Guo X, Wang J, Jiang L, Zhai H, Wan J (2008) Isolation and initial characterization of GW5, a major QTL associated with rice grain width and weight. Cell Res 18:1199–1209

Woo MO, Ham TH, Ji HS, Choi MS, Jiang W, Chu SH, Piao R, Chin JH, Kim JA, Park BS, Seo HS, Jwa NS, McCouch S, Koh HJ (2008) Inactivation of the UGPase1

gene causes genic male sterility and endosperm chalkiness in rice (Oryza sativa L.). Plant J 54:190–204

Xie W, Feng Q, Yu H, Huang X, Zhao Q, Xing Y, Yu S, Han B, Zhang Q (2010) Parent-independent genotyping for constructing an ultrahigh-density linkage map based on population sequencing. Proc Natl Acad Sci U S A 107:10578–10583

Xu J, Jiang J, Dong X, Ali J, Mou T (2012) Introgression of bacterial blight (BB) resistance genes Xa7 and Xa21 into popular restorer line and their hybrids by molecular marker-assisted backcross (MABC) selection scheme. Afr J Biotechnol 11:8225–8233

Yu H, Xie W, Wang J, Xing Y, Xu C, Li X, Xiao J, Zhang Q (2011) Gains in QTL detection using an ultra-high density SNP Map based on population sequencing relative to traditional RFLP/SSR markers. PLoS One 6:e17595

Zhang B, Ye WJ, Ren DY, Tian P, Peng YL, Gao Y, Ruan BP, Wang L, Zhang GH, Guo LB, Qian Q, Gao ZY (2015) Genetic analysis of flag leaf size and candidate genes determination of a major QTL for flag leaf width in rice. Rice 8:1–10

Zhang J, Wu K, Zeng S, Teixeira da Silva JA, Zhao X, Tian CE, Xia H, Duan J (2013) Transcriptome analysis of Cymbidium sinense and its application to the identification of genes associated with floral development. BMC Genomics 14:279

Zhang XJ, Wang JF, Huang J, Lan HX, Wang CL, Yin CF, Wu YY, Tang HJ, Qian Q, Li JY, Zhang HS (2012) Rare allele of OsPPKL1 associated with grain length causes extra-large grain and a significant yield increase in rice. Proc Natl Acad Sci U S A 109:21534–21539

Zhao X, Daygon VD, McNally KL, Hamilton RS, Xie F, Reinke RF, Fitzgerald MA (2016a) Identification of stable chalk in rice grains in nine environments. Theor Appl Genet 129:141–153

Zhao XQ, Daygon VD, McNally KL, Hamilton RS, Xie FM, Reinke RF, Fitzgerald MA (2016b) Identification of stable QTLs causing chalk in rice grains in nine environments. Theor Appl Genet 129:141–153

Zhao XQ, Zhou LJ, Ponce K, Ye GY (2015) The usefulness of known genes/Qtls for grain quality traits in an indica population of diverse breeding lines tested using association analysis. Rice 8:1–13

Zhou L, Chen L, Jiang L, Zhang W, Liu L, Liu X, Zhao Z, Liu S, Zhang L, Wang J, Wan J (2009) Fine mapping of the grain chalkiness QTL qPGWC-7 in rice (Oryza sativa L.). Theor Appl Genet 118:581–590

Zhou LJ, Liang SS, Ponce K, Marundon S, Ye GY, Zhao XQ (2015a) Factors affecting head rice yield and chalkiness in indica rice. Field Crop Res 172:1–10

Zhou Y, Miao J, Gu HY, Peng XR, Leburu M, Yuan FH, Gu HW, Gao Y, Tao YJ, Zhu JY, Gong ZY, Yi CD, Gu MH, Yang ZF, Liang GH (2015b) Natural variations in SLG7 regulate grain shape in rice. Genetics 201:1591–1599

3

Molecular Dissection of Seedling Salinity Tolerance in Rice (*Oryza sativa L.*) Using a High-Density GBS-Based SNP Linkage Map

Teresa B. De Leon[1], Steven Linscombe[2] and Prasanta K. Subudhi[1*]

Abstract

Background: Salinity is one of the many abiotic stresses limiting rice production worldwide. Several studies were conducted to identify quantitative trait loci (QTLs) for traits associated to salinity tolerance. However, due to large confidence interval for the position of QTLs, utility of reported QTLs and the associated markers has been limited in rice breeding programs. The main objective of this study is to construct a high-density rice genetic map for identification QTLs and candidate genes for salinity tolerance at seedling stage.

Results: We evaluated a population of 187 recombinant inbred lines (RILs) developed from a cross between Bengal and Pokkali for nine traits related to salinity tolerance. A total of 9303 SNP markers generated by genotyping-by-sequencing (GBS) were mapped to 2817 recombination points. The genetic map had a total map length of 1650 cM with an average resolution of 0.59 cM between markers. For nine traits, a total of 85 additive QTLs were identified, of which, 16 were large-effect QTLs and the rest were small-effect QTLs. The average interval size of QTL was about 132 kilo base pairs (Kb). Eleven of the 85 additive QTLs validated 14 reported QTLs for shoot potassium concentration, sodium-potassium ratio, salt injury score, plant height, and shoot dry weight. Epistatic QTL mapping identified several pairs of QTLs that significantly contributed to the variation of traits. The QTL for high shoot K^+ concentration was mapped near the *qSKC1* region. However, candidate genes within the QTL interval were a CC-NBS-LRR protein, three uncharacterized genes, and transposable elements. Additionally, many QTLs flanked small chromosomal intervals containing few candidate genes. Annotation of the genes located within QTL intervals indicated that ion transporters, osmotic regulators, transcription factors, and protein kinases may play essential role in various salt tolerance mechanisms.

Conclusion: The saturation of SNP markers in our linkage map increased the resolution of QTL mapping. Our study offers new insights on salinity tolerance and presents useful candidate genes that will help in marker-assisted gene pyramiding to develop salt tolerant rice varieties.

Keywords: *Oryza sativa*, Genotyping by sequencing, Quantitative trait locus, Salt tolerance, Candidate gene, Single nucleotide polymorphism

Background

Rice is a staple food crop for many countries in Asia, Africa, and Latin America. In spite of increased production worldwide, rice growers are faced with challenges caused by both biotic and abiotic stresses. Hence, breeding programs targeted to address those problems are implemented. Among the abiotic stresses, soil and water salinity is a problem not only in the coastal areas but also in areas where crop production heavily relies on irrigation with poor drainage system. Previous studies have indicated that rice is sensitive to salt stress during seedling stage and reproductive stage (Pearson and Bernstein 1959; Zeng et al. 2001). Rice seedlings wither and eventually die at 10dSm^{-1} salt stress (Munns et al. 2006) while yield loss can be as high as 90 % at 3dSm^{-1} salt level (Asch et al. 2000). Progress in breeding rice with salt tolerance is slow due to genetic complexity of salinity tolerance (Flowers and Flowers 2005). Some

* Correspondence: psubudhi@agcenter.lsu.edu
[1]School of Plant, Environmental, and Soil Sciences, Louisiana State University Agricultural Center, Baton Rouge, LA, USA
Full list of author information is available at the end of the article

germplasms with high salt tolerance are available. However, majority of these germplasms possess many undesirable traits. Pokkali, Nona Bokra, and Hasawi, which are highly tolerant and often used as donors in breeding for salt tolerance, are tall, photosensitive, low yielding, and have red kernel. In addition, salt tolerance screening is difficult because the phenotypic response of rice to salt stress is highly affected by other confounding environmental factors (Gregorio and Senadhira 1993; Flowers 2004). Hence, the search for QTLs and DNA markers tightly linked to traits related to salt tolerance becomes a major objective in most breeding programs. It is assumed that molecular markers will facilitate a fast and cost-effective screening of large populations (Munns and James 2003).

Since the advent of molecular markers, QTL analyses for salinity tolerance at seedling stage were conducted using RIL (Koyama et al. 2001; Gregorio et al. 2002; Wang et al. 2012), $F_{2:3}$ lines (Lin et al. 2004), and backcross populations (Thomson et al. 2010; Alam et al. 2011). QTLs for visual scoring, survival, shoot and root lengths, Na^+/K^+ ratio, Na^+ and K^+ concentrations, in root and shoot were frequently investigated at 100–120 mM salt stress. Most of the QTL mapping studies have indicated polygenic nature of salinity tolerance. Among the QTLs for traits related to salt tolerance, only qSKC1 was successfully isolated by map-based cloning (Ren et al. 2005). The SKC1 gene from Nona Bokra encodes an HKT-type transporter that regulates the Na^+/K^+ homeostasis under salt stress. In earlier reports, the QTL designated as Saltol (Gregorio 1997) and a gene 'SalT' (Causse et al. 1994) for Na^+/K^+ ratio were located on chromosome 1.

Numerous QTL mapping studies for salinity tolerance were based on linkage maps constructed using AFLP (Gregorio 1997), RFLP (Koyama et al. 2001; Bonilla et al. 2002; Lin et al. 2004), and SSR markers (Thomson et al. 2010; Wang et al. 2012). The population size was usually small and the markers were sparse due to limited polymorphism between the parents. The rapid development in the sequencing technology makes single nucleotide polymorphism (SNP) to become the marker of choice for QTL mapping. Bimpong et al. (2013) used 194 polymorphic SNP markers for mapping QTLs related to salinity tolerance. More recently, Kumar et al. (2015) applied the genome-wide association (GWAS) mapping on 220 rice varieties using a custom-designed array containing 6000 SNPs. Major association of Na^+/K^+ ratio still co-localized to the Saltol locus with additional QTLs on chromosome 4, 6, and 7. Significant SNPs were identified and some candidate genes were suggested. However, tight association of candidate genes in or around a single variant still needs enrichment with more markers at a locus to avoid false association. Moreover, complete resequencing of the locus in tolerant and non-tolerant lines or

in bi-parental population are needed to add credence to the robustness of GWAS using SNP array.

The introduction of genotyping-by-sequencing (GBS) and the availability of whole genome sequence of rice have accelerated the identification of millions of SNPs across the whole genome. To date, GBS is becoming popular for population studies, genetic diversity, QTL mapping, and genomic selection (He et al. 2014). GBS enabled the construction of high-density linkage map and QTL analysis in maize, wheat, barley (Poland et al. 2012; Chen et al. 2014), oat (Huang et al. 2014), and chickpea (Jaganathan et al. 2015). In rice, GBS has been applied in QTL mapping for leaf width and aluminum tolerance (Spindel et al. 2013), pericarp color and some agronomic traits (Arbelaez et al. 2015), and rice blast resistance (Liu et al. 2015). Several QTL mapping studies for salinity tolerance have been reported. However, QTLs and markers flanking QTLs for salinity tolerance are not being utilized in breeding programs. The main reason for this is attributed to the large chromosome intervals delimited by those QTLs. Thus, identification of candidate genes and understanding of salinity tolerance mechanism still remained a challenge.

In this study, a recombinant inbred line population at F_6 generation, developed from the cross Bengal x Pokkali, was used. Bengal is a high yielding, early maturing, semi-dwarf, medium grain cultivar developed from the cross of MARS//M201/MARS (Linscombe et al. 1993). It is sensitive to salinity stress (De Leon et al. 2015). Pokkali is a highly tolerant landrace often used as a donor for salinity tolerance. However, it is notable for many undesirable traits such as low-yield, tall, and highly susceptible to lodging. It is photoperiod-sensitive, awned, with red pericarp and poor cooking quality (Gregorio et al. 2002). We used the GBS technique to construct a high-resolution genome-wide SNP genetic map for identification of additive and epistatic QTLs for salinity tolerance. Segregation distortion loci (SDLs) and QTLs for plant height were mapped to show the quality and accuracy of the genetic map and QTL mapping. Our ultra-high density map allowed us to map QTLs with high resolution and identify candidate genes that may play important role in the mechanism of salt tolerance in rice. The candidate genes identified in this study will serve as useful targets for functional genomics, gene pyramiding, and for gene-based marker-assisted breeding for salinity tolerance.

Results

Phenotypic Characterization Under Salt Stress

The parents and RIL population were evaluated under salt stress for salt injury score (SIS), chlorophyll content (CHL), shoot length (SHL), root length (RTL), shoot length to root length ratio (SRR), dry shoot weight (DWT), shoot Na^+ and K^+ concentrations, and Na^+/K^+

ratio (NaK ratio). At $12 dSm^{-1}$ salt stress, the RILs and parents showed varying levels of tolerance. Bengal and Pokkali showed significant contrasting response in SIS, SHL, RTL, DWT, and NaK ratio (Table 1). However, the differences in CHL, SRR, Na^+ and K^+ concentrations, were not statistically significant between parents. Pokkali showed consistently lower SIS, Na^+ concentration, NaK ratio, and higher K^+ concentration than Bengal. Among the RILs, all traits showed significant genotypic differences (p <0.0001), indicating a wide range of variation. The RIL population had a mean value between the parental means for all traits except in CHL and SRR. Pokkali had an average SIS of 3; Bengal had 8.4, while the RILs had a mean SIS of 4.7. The RIL population had a mean Na^+ accumulation of 1430 $mmolkg^{-1}$ in shoot, which is much lower than Bengal (1700 $mmolkg^{-1}$), and marginally higher than Pokkali (1424 $mmolkg^{-1}$). In contrast, the mean K^+ accumulation was highest in Pokkali (591 $mmolkg^{-1}$), followed by RILs (547 $mmolkg^{-1}$) and lowest in Bengal (420 $mmolkg^{-1}$). The RIL population had mean chlorophyll content greater than either parent. As indicated in the frequency distribution (Fig. 1) and the range of RIL values for each trait (Table 1), several lines were phenotypically superior to the parents. There were many transgressive segregants with much lower Na^+ than Bengal, lower NaK ratio and SHL and higher CHL, DWT, RTL and SRR than Bengal or Pokkali. Similarly, some lines accumulated twice the K^+ concentration of Pokkali. But there was no line that showed higher tolerance than Pokkali as judged by SIS (Fig. 1). There was wide variation for heritability values for traits. Heritabilities for Na^+, K^+ concentrations, and SHL were 0.98, 0.95, and 90, respectively. In contrast, NaK ratio, SIS, CHL, RTL, and SRR had moderate heritability of 0.24–0.63 while DWT has very low heritability.

Correlation of Traits

Correlations among all traits (Table 2) revealed that SIS was highly significant and positively correlated to Na^+ concentration and NaK ratio. The SIS was highly significant and negatively correlated to CHL, SHL, RTL, DWT and SRR, indicating the negative effect of salt stress on the overall growth and photosynthetic capability of plants. On the other hand, K^+ concentration was positively correlated to Na^+ concentration, SHL, CHL, DWT, and SRR but negatively correlated to NaK ratio, SIS, and RTL. The relationships between traits in RIL population were consistent to the correlation of traits observed among the 30 US rice genotypes (De Leon et al. 2015), thus indicating reliability and reproducibility of our salt tolerance screening.

Linkage Mapping

GBS generated a total of 33,987 SNP markers which were furtherly filtered for polymorphic markers and for markers with less than 10 % missing data across the population. A total of 9303 SNPs markers were retained and used in the linkage map construction (Fig. 2, Additional file 1: Table S1). On the average, about 775 SNP markers were placed per chromosome (Table 3). The final linkage map had a total length of 1650 cM with 2817 recombination sites. The average distance between adjacent markers was 0.59 cM or 39,798 bp, with maximum resolution of 0.27 cM. The average marker density was 5.6 SNP markers per cM or 3.3 SNP markers per recombination point. The map was saturated with SNP markers across all chromosomes. However, twenty large gaps were observed on chromosomes 1, 2, 3, 4, 6, 7, 8, 10, 11, and 12 that ranged between 5 cM to 13 cM. With 9303 SNP

Table 1 Phenotypic response of parents and F_6 RIL population for traits related to salt tolerance at seedling stage

Trait Name	Bengal Mean	Pokkali Mean[β]	RIL Mean	Std. Dev.	RIL Range	RIL Pr > F[§]	Heritability[¥]
Na^+ ($mmolkg^{-1}$)	1700.00	1424.3[ns]	1430.7	246.24	861.97–2733.35	<0.0001	0.98
K^+ ($mmolkg^{-1}$)	420.00	591[ns]	547.3	107.59	335.99–884.18	<0.0001	0.95
NaK (ratio)	4.07	2.38**	2.8	0.56	1.25–5.32	<0.0001	0.24
SIS	8.40	3.00***	4.7	0.72	3.00–8.73	<0.0001	0.44
CHL (SPAD unit)	20.56	19.54[ns]	24.2	4.25	13.72–43.67	<0.0001	0.45
SHL (cm)	32.07	44.52***	40.7	3.21	22.60–59.73	<0.0001	0.90
RTL (cm)	6.73	10.08**	7.4	0.64	4.67–11.27	<0.0001	0.61
DWT (g)	0.06	0.11*	0.1	0.01	0.04–0.16	<0.0001	0.01
SRR (ratio)	4.98	4.53[ns]	5.6	0.53	3.08–9.79	<0.0001	0.63

Na^+: shoot sodium concentration, K^+: shoot potassium concentration, NaK: ratio of the shoot sodium and shoot potassium content, SIS: salt injury score, CHL: chlorophyll content, SHL: shoot length, RTL: root length, DWT: shoot dry weight, SRR: shoot length to root length ratio

[β]Significant differences between Bengal and Pokkali, [ns]no significant differences, *significant at 0.05 probability level, **significant at 0.01 probability level, ***significant at 0.001 probability level

[§]Genotypic differences among RIL

[¥]Broad sense heritability computed on family mean basis

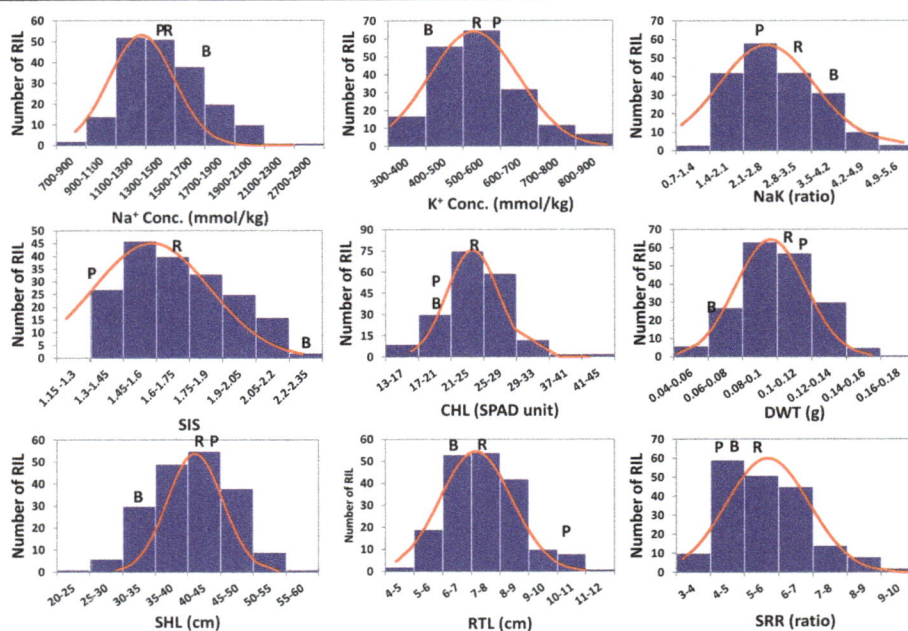

Fig. 1 Frequency distribution of Bengal/Pokkali F$_6$ RIL population for traits related to seedling salinity tolerance. Na$^+$ Conc., Na$^+$ concentration; K$^+$ conc., K$^+$ concentration; NaK, Na$^+$/K$^+$ ratio; SIS, log transformed salt injury score; CHL, chlorophyll content measured by SPAD-502 unit; DWT, dry weight; SHL, shoot length; RT_, root length; SRR, Shoot length to root length ratio

markers, the linkage map had a physical to genetic map length ratio of 225 Kb/cM.

Identification of Additive and Di-genic Epistatic QTLs for Traits Related to Salinity Tolerance

To detect novel additive and epistatic QTLs for traits related to salinity tolerance, the phenotype and GBS data were used in interval mapping (IM) and inclusive composite interval mapping (ICIM) methods.

QTLs for Shoot Na$^+$ Concentration

The IM and ICIM methods consistently detected three additive QTLs for shoot Na$^+$ concentration (Table 4). The QTLs were located on chromosomes 2, 6, and 12. Each additive QTL explained at least 5.5 % of the phenotypic variation. Pokkali alleles of qNa2.7 and qNa12.18 had increasing effect while for qNa6.5 Bengal allele had the increasing effect. Interval mapping of epistatic QTLs detected seven pairs of QTLs with significant contribution to the variation in Na$^+$ concentration

Table 2 Pearson correlation matrix of traits measured in response to salt stress at 12dSm^{-1} in Bengal/Pokkali F$_6$ RIL population at seedling stage

	Na$^+$	K$^-$	NaK	SIS	CHL	SHL	RTL	DWT	SRR
Na$^+$	1								
K$^+$	0.1271**	1							
NaK	0.594***	-C.649***	1						
SIS	0.337***	-C.129**	0.337***	1					
CHL	-0.128**	0.092*	-0.157***	-0.214***	1				
SHL	0.039	0.253***	-0.151***	-0.236***	0.221***	1			
RTL	0.057	-0.105*	0.095*	-0.109**	0.059	0.204***	1		
DWT	0.006	0.144***	-0.099*	-0.475***	0.177***	0.539***	0.279***	1	
SRR	-0.024	0.277***	-0.195***	-0.099*	0.111**	0.593***	-0.638***	0.173*	1

Na$^+$: shoot sodium concentration, K$^+$: shoot potassium concentration, NaK: ratio of the shoot sodium and shoot potassium content, SIS: salt injury score, CHL: chlorophyll content, SHL: shoot length, RTL: root length, DWT: dry weight, SRR: shoot length to root length ratio
*significant at 0.05 probability level, **significant at 0.01 probability level, ***significant at 0.001 probability level

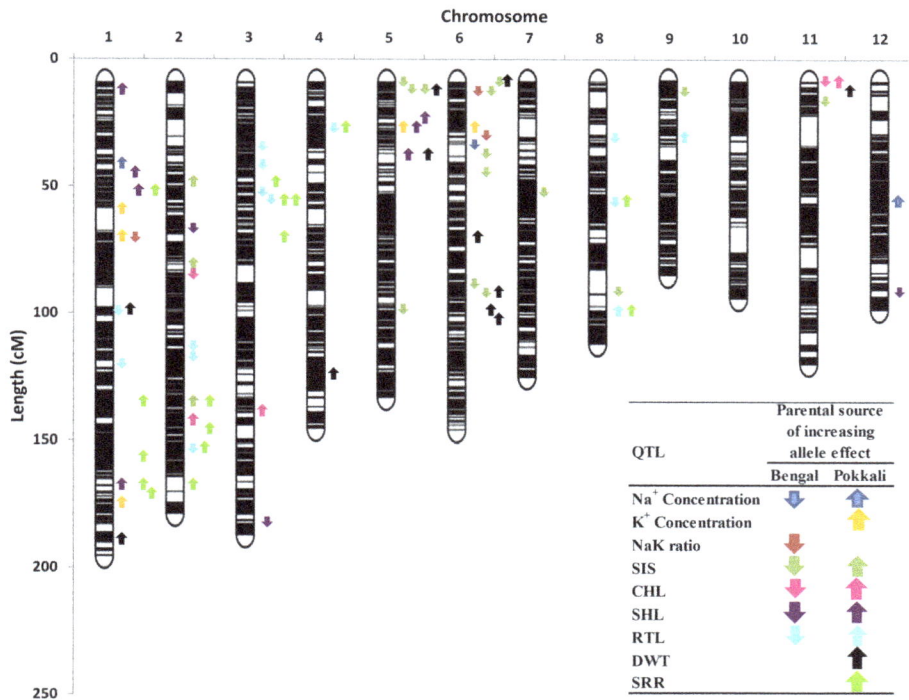

Fig. 2 Molecular genetic map showing the positions of QTLs for nine traits investigated under salt stress. Linkage and QTL mapping were implemented in ICIM QTL Mapping 4.0 using 9303 GBS-SNP markers in 187 Bengal/Pokkali F_6 RILs. Chromosome regions that are dark indicate the saturation of markers while regions that are white indicate the absence of marker placed in those segments. Genetic distance in centimorgan was determined by Kosambi map function. Each arrow represents a single QTL for a particular trait

Table 3 Summary distribution, coverage, and intervals of SNP markers in the Bengal/Pokkali RIL linkage map

Chromosome	No. of SNP markers used	Chromosome length coverage (Mb)	Genetic length (cM)	No. of recombination points	No. of SNP markers/cM	No. of SNP markers/unique position	Minimum interval (cM)	Maximum interval (cM)	Average Interval (cM)	No. of Gaps >5 cM
1	1245	43,237,333	199.8	363	6.2	3.4	0.27	9.98	0.55	2
2	1001	35,875,736	182.9	324	5.5	3.1	0.27	7.19	0.56	2
3	1068	36,405,799	191.2	320	5.6	3.3	0.28	8.01	0.60	2
4	822	35,501,387	148.1	244	5.6	3.4	0.27	6.88	0.60	2
5	780	29,507,277	135.6	243	5.8	3.2	0.27	4.35	0.56	0
6	842	30,869,147	148.6	258	5.7	3.3	0.27	5.33	0.57	1
7	736	29,582,943	127.4	225	5.8	3.3	0.27	8.7	0.57	1
8	471	28,399,689	113.9	162	4.1	2.9	0.27	10.57	0.70	3
9	584	22,779,506	85.9	164	6.8	3.6	0.27	6.49	0.52	1
10	517	23,117,196	95.2	149	5.4	3.5	0.28	11.55	0.64	1
11	622	28,973,227	121.8	187	5.1	3.3	0.28	13.09	0.65	3
12	615	27,488,377	99.9	178	6.2	3.5	0.27	6.75	0.56	2
Total	9303	371,737,617	1650.2	2817	67.7	39.7	3.27	98.89	7.08	20
Average[a]	775.3	30,978,134.75	137.5	234.8	5.6	3.3	0.27	8.24	0.59	1.7

[a]Average value per chromosome

Table 4 Additive QTLs for traits related to seedling-stage salt tolerance in Bengal/Pokkali F_6 RIL population identified by IM and ICIM methods

Phenotype	QTL	Chr[a]	Position (cM)	Left Marker	Right Marker	QTL Interval Size (bp)	LOD	PVE (%)	Additive Effect	Parental Source of Increasing Allele Effect[b]	No. of genes in QTL interval
Na+ concentration-IM	qNa2.7	2	48	S2_7769844	S2_7939496	169,652	2.2969	5.55	-66.59	P	24
	qNa6.5	6	34	S6_5269698	S6_5533752	264,054	2.399	5.97	69.15	B	34
	qNa12.18	12	60	S12_18687038	S12_18741493	54,455	2.2496	5.51	-66.36	P	5
Na+ concentration-ICIM	qNa2.7	2	48	S2_7769844	S2_7939496	169,652	2.2969	5.55	-66.59	P	21
	qNa6.5	6	34	S6_5269698	S6_5533752	264,054	2.399	5.97	69.15	B	34
	qNa12.18	12	60	S12_18687038	S12_18741493	54,455	2.2496	5.51	-66.36	P	5
K+ concentration-IM	qK1.8	1	63	S1_8656025	S1_8901503	245,478	5.7933	13.65	-46.34	P	33
	qK1.11	1	71	S1_11529325	S1_11581799	52,474	5.9263	13.66	-45.13	P	6
	qK1.38	1	173	S1_38794029	S1_39047133	253,104	3.5096	8.30	-33.32	P	40
	qK5.4	5	31	S5_4699921	S5_5326365	626,444	2.2496	5.51	-27.00	P	86
	qK6.4	6	31	S6_4890290	S6_5269698	379,408	3.333	8.21	-32.92	P	61
K+ concentration-ICIM	qK1.11	1	71	S1_11529325	S1_11581799	52,474	7.7414	16.08	-48.95	P	6
	qK1.38	1	173	S1_38794029	S1_39047133	253,104	5.3793	10.71	-37.86	P	40
NaK ratio-IM	qNaK1.11	1	71	S1_11529325	S1_11581799	52,474	4.154	9.83	0.29	B	6
	qNaK6.2	6	15	S6_2927160	S6_2962502	35,342	3.5777	8.46	0.26	B	7
	qNaK6.5	6	33	S6_5269698	S6_5533752	264,054	5.1164	13.21	0.32	B	34
NaK ratio-ICIM	qNaK1.11	1	71	S1_11529325	S1_11581799	52,474	2.6375	5.66	0.22	B	6
	qNaK6.5	6	33	S6_5269698	S6_5533752	264,054	3.7097	8.85	0.26	B	34
Salt injury score-IM	qSIS2.8	2	50	S2_8730258	S2_8927908	197,650	3.5375	8.58	-0.06	P	25
	qSIS2.19	2	81	S2_19331684	S2_19454952	123,268	3.2133	7.66	-0.06	P	14
	qSIS2.28	2	131	S2_28239596	S2_28274467	34,871	2.6449	6.37	-0.05	P	8
	qSIS5.03	5	1	S5_312457	S5_329699	17,242	2.8257	6.74	0.06	B	4
	qSIS5.1a	5	12	S5_1686924	S5_1707475	20,551	2.8266	6.76	0.06	B	5
	qSIS5.24	5	106	S5_24057323	S5_24281632	224,309	3.1266	7.51	0.06	B	39
	qSIS6.2	6	15	S6_2927160	S6_2962502	35,342	2.0824	5.04	0.05	B	7
	qSIS6.5	6	37	S6_5848568	S6_5905669	57,101	3.0401	7.23	0.06	B	11
	qSIS6.7	6	48	S6_7646442	S6_7661883	15,441	3.1238	7.41	0.06	B	3
	qSIS6.20	6	90	S6_20929261	S6_20929283	22	3.9605	9.44	0.07	B	1
	qSIS11.2	11	18	S11_2838776	S11_3716306	877,530	2.664	8.36	0.06	B	136
Salt injury score-ICIM	qSIS5.1b	5	11	S5_1441967	S5_1454837	12,870	9.7068	13.33	0.08	B	2
	qSIS6.2b	6	9	S6_2123411	S6_2242943	119,532	3.5933	4.46	0.05	B	23
	qSIS6.21	6	92	S6_21253244	S6_21256132	2888	6.9204	9.11	0.07	B	1

Table 4 Additive QTLs for traits related to seedling-stage salt tolerance in Bengal/Pokkali F_6 RIL population identified by IM and ICIM methods (*Continued*)

Trait	QTL			Marker 1	Marker 2						
	qSIS7.14	7	57	S7_14598897	S7_14625841	26,944	3.6209	4.50	0.05	B	7
	qSIS8.24	8	93	S8_24763939	S8_25110888	346,949	2.6235	3.28	0.04	B	47
	qSIS9.8	9	13	S9_8608506	S9_9070610	462,104	7.09	9.19	0.07	B	51
	qSIS11.2	11	21	S11_2838776	S11_3716306	877,530	2.336	3.53	0.04	B	136
Chlorophyll content-IM	qCHL11.1	11	5	S11_1086712	S11_1293020	206,308	2.1922	5.41	-1.00	P	34
	qCHL11.2	11	14	S11_2666525	S11_2724222	57,697	2.0172	4.86	-0.95	P	7
Chlorophyll content-ICIM	qCHL2.20	2	86	S2_20258450	S2_20346560	88,110	3.6938	7.44	1.18	B	7
	qCHL2.30	2	143	S2_30355435	S2_30402468	49,033	2.3418	4.69	-0.94	P	7
	qCHL3.26	3	136	S3_26705619	S3_26709038	3419	3.2263	6.42	-1.10	P	1
Shoot length-IM	qSHL1.1	1	11	S1_1708228	S1_1747144	38,916	2.0367	5.03	-1.42	P	7
	qSHL1.7a	1	48	S1_7259818	S1_7296346	36,528	3.9307	9.26	-1.95	P	7
	qSHL1.38	1	168	S1_38286772	S1_38611845	325,073	25.3529	48.03	-4.43	P	52
	qSHL3.34	3	185	S3_34720589	S3_35060080	339,491	2.361	5.65	1.54	B	69
	qSHL5.4	5	29	S5_4565557	S5_4699921	134,364	2.3239	5.64	-1.52	P	23
	qSHL5.6	5	44	S5_6356744	S5_6433933	77,189	2.0324	4.96	-1.41	P	13
Shoot length-ICIM	qSHL1.7b	1	50	S1_7520182	S1_7569628	49,446	6.2678	5.86	-1.57	P	5
	qSHL1.38	1	168	S1_38286772	S1_38611845	325,073	36.9075	51.64	-4.59	P	52
	qSHL2.18	2	77	S2_18806154	S2_18937362	131,208	3.0049	2.71	1.04	B	25
	qSHL3.34	3	185	S3_34720589	S3_35060080	339,491	4.3994	3.96	1.29	B	69
	qSHL5.3	5	25	S5_3353753	S5_3506138	152,385	7.0797	6.79	-1.66	P	21
	qSHL12.25	12	93	S12_25709174	S12_25887173	177,999	2.246	2.05	0.91	B	30
Root length-IM	qRTL1.26	1	121	S1_26421289	S1_26447134	25,845	2.7346	6.52	0.32	B	6
	qRTL2.24	2	114	S2_24961302	S2_24961342	40	4.1447	9.72	0.39	B	0
	qRTL2.26	2	120	S2_26028043	S2_26070191	42,148	4.2053	9.91	0.39	B	9
	qRTL2.33	2	160	S2_33573567	S2_33614297	40,730	3.9359	9.50	0.39	B	7
	qRTL3.6	3	36	S3_6011601	S3_6027452	15,851	3.4683	8.23	0.36	B	2
	qRTL3.7	3	44	S3_7130220	S3_7209963	79,743	4.4685	10.70	0.41	B	15
	qRTL3.10	3	57	S3_10116591	S3_10132745	16,154	5.0358	11.99	0.43	B	2
	qRTL4.10	4	24	S4_10625625	S4_10726368	100,743	2.0131	4.88	0.28	B	14
	qRTL8.4	8	37	S8_4558562	S8_4858127	299,565	2.121	5.34	0.36	B	41
	qRTL8.19	8	59	S8_19884635	S8_19898432	13,797	3.272	7.75	0.41	B	2
	qRTL8.27	8	109	S8_27238050	S8_27304101	66,051	2.1036	5.13	-0.28	P	9
	qRTL9.14	9	39	S9_14960521	S9_14976723	16,202	2.6572	6.45	-0.36	P	3
Root length-ICIM	qRTL1.22	1	102	S1_22666852	S1_22677418	10,566	2.2657	3.54	0.23	B	2
	qRTL1.26	1	121	S1_26421289	S1_26447134	25,845	2.1764	3.41	0.23	B	6

Table 4 Additive QTLs for traits related to seedling-stage salt tolerance in Bengal/Pokkali F_6 RIL population identified by IM and ICIM methods (*Continued*)

Trait	QTL	Chr[a]	Pos	Left marker	Right marker	Interval	LOD	PVE	Add	Source[b]	
	qRTL3.9	3	56	S3_9853159	S3_9891061	37,902	4.2907	7.59	0.34	B	7
Dry weight-IM	qDWT1.21	1	97	S1_21707357	S1_21733437	26,080	2.3413	5.60	-0.01	P	6
	qDWT4.32	4	126	S4_32367131	S4_32367159	28	2.3915	5.73	-0.01	P	1
	qDWT5.2	5	15	S5_2116055	S5_2167880	51,825	4.6334	10.80	-0.01	P	4
	qDWT5.4	5	29	S5_4565557	S5_4699921	134,364	6.5783	15.04	-0.01	P	23
	qDWT5.5	5	42	S5_5997340	S5_6196044	198,704	6.5368	15.47	-0.01	P	32
	qDWT6.13	6	72	S6_13046472	S6_13097774	51,302	2.0119	5.16	0.00	P	10
	qDWT6.20	6	90	S6_20929261	S6_20929283	22	3.7538	8.95	-0.01	P	1
	qDWT6.23	6	102	S6_23812023	S6_24039384	227,361	3.7054	8.91	-0.01	P	32
	qDWT11.2	11	10	S11_2379158	S11_2402109	22,951	2.4389	6.03	-0.01	P	3
Dry weight-ICIM	qDWT1.40	1	185	S1_40372283	S1_40412316	40,033	2.0722	3.13	0.00	P	6
	qDWT4.32	4	126	S4_32367131	S4_32367159	28	3.6566	5.93	-0.01	P	1
	qDWT5.4	5	29	S5_4565557	S5_4699921	134,364	7.5727	12.98	-0.01	P	23
	qDWT6.06	6	3	S6_692773	S6_782975	90,202	3.7128	6.02	-0.01	P	13
	qDWT6.24	6	104	S6_24107596	S6_24228831	121,235	4.4604	7.46	-0.01	P	19
Shoot-root ratio-IM	qSRR1.7	1	50	S1_7520182	S1_7569628	49,446	3.7944	9.09	-0.38	P	5
	qSRR1.29	1	135	S1_29561423	S1_29568978	7555	3.1151	7.42	-0.33	P	2
	qSRR1.36	1	159	S1_36158467	S1_36189206	30,739	5.8447	13.42	-0.45	P	5
	qSRR1.382	1	170	S1_38286772	S1_38611845	325,073	10.3107	23.01	-0.59	P	52
	qSRR2.28	2	133	S2_28317911	S2_28375704	57,793	4.7052	10.96	-0.41	P	7
	qSRR2.31	2	146	S2_31037977	S2_31043939	5962	3.2045	7.62	-0.34	P	1
	qSRR2.33	2	160	S2_33573567	S2_33614297	40,730	4.1813	9.90	-0.39	P	7
	qSRR2.34	2	168	S2_34660774	S2_35085922	425,148	2.9367	7.37	-0.33	P	68
	qSRR3.8	3	49	S3_8327882	S3_8353264	25,382	2.6453	6.32	-0.31	P	6
	qSRR3.10	3	57	S3_10116591	S3_10132745	16,154	2.6902	6.58	-0.31	P	2
	qSRR3.11	3	70	S3_11848358	S3_11865689	17,331	2.4751	5.93	-0.30	P	1
	qSRR4.10	4	24	S4_10625625	S4_10726368	100,743	2.438	5.91	-0.30	P	14
	qSRR8.19	8	59	S8_19884635	S8_19898432	13,797	2.3793	5.70	-0.35	P	2
Shoot root ratio-ICIM	qSRR1.7	1	50	S1_7520182	S1_7569628	49,446	6.9282	8.73	-0.37	P	5
	qSRR1.386	1	171	S1_38636497	S1_38768787	132,290	15.6449	22.43	-0.59	P	22
	qSRR2.33	2	160	S2_33573567	S2_33614297	40,730	8.5304	10.92	-0.41	P	7
	qSRR3.9	3	56	S3_9853159	S3_9891061	37,902	4.3278	5.25	-0.28	P	7
	qSRR8.26	8	107	S8_26716230	S8_26744324	28,094	2.533	3.01	0.21	B	5

[a] Chromosome where the QTL was located. [b] Parental source of increasing allele effect was either Pokkali (P) or Bengal (B)

(Table 5). Four of the seven pairs of epistatic QTLs had large effects (PVE = 11–16 %) while the other three pairs had small effects (PVE = 8–9 %). Nine interacting QTLs with increasing effect were from Bengal and five were from Pokkali. None of the additive QTLs co-localized with epistatic QTLs.

QTLs for Shoot K+ Concentration

The IM method detected five additive QTLs (*qK1.8, qK1.11, qK1.38, qK5.4,* and *qK6.4*) for shoot K+ concentration. The *qK1.8* and *qK1.11* were large-effect QTLs, each accounting for at least 13 % of the variation for shoot K+. The other three QTLs had small effects (5–8 % PVE) and were located on chromosomes 1, 5, and 6. The *qK1.11 and qK1.38* were also detected by ICIM with LOD values of 7.7 and 5.4, respectively. Both *qK1.11* and *qK1.38* were large-effect QTLs in ICIM method with PVE of 16 and 10 %. In contrast, *qK1.8, qK5.4,* and *qK6.4* were not detected in ICIM. All additive QTLs for K+ concentration had increasing effects that originated from Pokkali, indicating the importance of Pokkali alleles for increased uptake of K+ in the leaves. Five pairs of epistatic QTLs were detected for K+ concentration. The *qK1.7* and the *qK2.3* pair had a PVE of 21 % and LOD score of 3.5, with Pokkali allele contributing for increased K+ accumulation. The *qK1.7* also interacted with *qK12.17* and accounted for 9 % of the variation in K+ accumulation. Additionally, *qK11.19* and *qK12.18* pair had a PVE of 10 % while the remaining two pairs accounted for 9 % of the phenotypic variation. Six and four interacting QTLs with increasing effects involved Pokkali and Bengal alleles, respectively. All additive QTL positions were independent of epistatic QTLs.

QTLs for NaK Ratio

For NaK ratio, three additive QTLs (*qNaK1.11, qNaK6.2, qNaK6.5*) were significant in IM method but only two of the additive QTLs (*qNaK1.11, qNaK6.5*) were detected in ICIM. The *qNaK 6.5* explained 13 % of the phenotypic variation while *qNaK6.2* and *qNaK1.11* were small-effect QTLs. All NaK ratio QTLs had increasing effects due to Bengal alleles. Of the seven pairs of epistatic QTLs, two pairs were large-effect QTLs (PVE = 11 and 18 %) and five pairs were minor QTLs with PVEs lower than 9 %. There was no epistatic QTL found in the same chromosome intervals of additive QTLs for NaK ratio, K+, or Na+ concentrations. Most of the QTL alleles with increasing effects were from Bengal. But four epistatic QTLs with increasing effects were from Pokkali.

QTLs for SIS

A total of 11 chromosomal regions with significant additive effect were detected on chromosomes 2, 5, 6, and 11 by IM. All QTLs are having small effects of at least 5 % but not more than 9 % of the phenotypic variation. Three QTLs were mapped on chromosome 2 (*qSIS2.8, qSIS2.29,* and *qSIS2.28*) with increasing effects from Pokkali alleles. In contrast, ICIM detected seven QTLs. The additive QTLs were distributed on chromosomes 5, 6, 7, 8, 9, and 11. The *qSIS5.1b* was a major QTL, explaining about 13 % of the phenotypic variation. However, *qSIS5.1b* had increasing salt sensitivity effect from Bengal allele. Except for QTLs on chromosome 2, all other additive QTLs had increasing effects due to Bengal alleles. Between the two mapping methods, all QTLs were different except for *qSIS11.2*. For epistatic QTLs, five pairs of interacting QTLs were significant of which four pairs explained 11–15 % of the SIS variation. Among the additive QTLs, *qSIS6.2* was significantly interacting with *qSIS6.30* and increased the PVE from 5 to 15 % (Table 5). All interacting QTLs had increasing effects from Bengal alleles except the *qSIS2.20*.

QTLs for Chlorophyll Content

A total of five chromosome regions with additive effects were detected for chlorophyll content under salt stress. Two QTLs were detected on chromosome 11 by IM while ICIM detected two QTLs on chromosome 2 and one QTL on chromosome 3. All additive QTLs were minor-effect QTLs, with increasing CHL effects from Pokkali alleles except *qCHL2.20*. In contrast, epistatic QTL mapping detected ten significant pairs of interacting QTLs. Eight QTL pairs had large effects with PVE as high as 36 %. All additive QTLs were independent of epistatic QTLs for CHL.

QTLs for Shoot Length

Six additive QTLs were detected by IM and another six QTLs were detected by ICIM. The *qSHL1.38* and *qSHL3.34* were significant QTLs in both methods. The *qSHL1.38* was a major QTL with LOD value of 37 and accounted for 48–52 % of the phenotypic variation. The additive effect of *qSHL1.38* had increasing effect from Pokkali allele. Other SHL QTLs were located on chromosome 2, 3, 5, and 12 with small effects. Seven pairs of QTLs were significant in epistatic QTL mapping. Five pairs had 11 % PVE and the other two pairs had 9 % PVE. There was no epistatic QTL that co-localized with additive QTL.

QTLs for Root Length

Twelve additive QTLs were detected for root length by IM. In contrast, ICIM detected only three QTLs, with *qRTL1.26* common in both methods. Two large-effect QTLs in chromosome 3 (*qRTL3.7* and *qRTL3.10*) were highly significant and accounted for 10 and 12 % of the phenotypic variation, respectively. Both QTLs had increasing effects from Bengal alleles. All other QTLs were minor-effect QTLs, with increasing effects

Table 5 Di-genic epistatic QTLs for traits related to salt tolerance at seedling stage in Bengal/Pokkali F6 RIL population identified by interval mapping

Phenotype	QTL1	Chr.1	Position1	LeftMarker1	RightMarker1	QTL2	Chr.2	Position2 (cM)	LeftMarker2	RightMarker2	LOD	PVE(%)	Add1	Add2	Add x Add
Na+ concentration	qNa4.25	4	90	S4_25549517	S4_26622324	qNa4.29	4	110	S4_29966056	S4_29968457	3.08	11.02	57.92	-21.10	-102.91
	qNa3.26	3	135	S3_26536286	S3_26542118	qNa5.008	5	0	S5_87749	S5_96410	3.24	7.90	-13.88	-0.72	78.59
	qNa1.12	1	75	S1_12583448	S1_12685974	qNa6.2	6	10	S6_2266152	S6_2272501	3.34	9.09	43.52	-14.46	90.61
	qNa6.17	6	80	S6_17631626	S6_17780076	qNa6.19	6	85	S6_19446057	S6_19585327	3.10	15.64	106.90	-89.69	-208.67
	qNa6.4a	6	25	S6_4631489	S6_4771954	qNa10.21	10	85	S10_21364298	S10_21407693	3.83	13.99	86.22	46.66	88.20
	qNa6.4b	6	30	S6_4890290	S6_5269698	qNa11.1	11	5	S11_1086712	S11_1293020	3.26	13.91	58.04	39.35	80.06
	qNa3.2	3	15	S3_2171559	S3_2250307	qNa11.23	11	100	S11_23611942	S11_23708208	3.16	8.69	5.27	10.52	82.87
K+ concentration	qK1.7	1	60	S1_7778029	S1_8656025	qK2.3	2	15	S2_3207423	S2_3207477	3.50	21.42	-44.62	-22.75	36.16
	qK2.25	2	115	S2_25166702	S2_25192275	qK3.22	3	115	S3_22976923	S3_23020366	3.04	8.02	5.08	-5.45	31.47
	qK1.40	1	190	S1_40584495	S1_40894634	qK7.19	7	70	S7_19334046	S7_19406235	3.14	9.07	-12.92	1.97	-32.60
	qK1.7	1	50	S1_7520182	S1_7569628	qK12.17	12	55	S12_17065005	S12_17195754	3.23	9.03	-13.81	8.04	-32.71
	qK11.19	11	70	S11_19222100	S11_19245359	qK12.18	12	60	S12_18687038	S12_18741493	3.63	10.31	16.55	-3.15	-34.00
NaK ratio	qNaK1.42	1	195	S1_42138516	S1_42310908	qNaK3.21	3	110	S3_21445493	S3_21628785	3.08	8.71	0.10	0.02	-0.24
	qNaK6.30	6	145	S6_30296317	S6_30370989	qNaK8.2	8	20	S8_2341829	S8_2949528	3.09	8.68	0.07	-0.13	0.26
	qNaK6.4a	6	25	S6_4631489	S6_4771954	qNaK10.213	10	85	S10_21364298	S10_21407693	3.58	17.65	0.34	0.12	0.26
	qNaK7.22	7	90	S7_22936622	S7_22936634	qNaK10.217	10	90	S10_21749293	S10_21786307	3.47	8.73	-0.03	0.03	-0.26
	qNaK5.16	5	65	S5_16290294	S5_16307102	qNaK11.2	11	15	S11_2838776	S11_3716306	3.55	10.80	-0.05	0.09	-0.26
	qNaK3.2	3	20	S3_2776106	S3_2780171	qNaK11.24	11	105	S11_24319577	S11_24335733	3.46	9.89	0.00	-0.12	0.26
	qNaK1.5	1	35	S1_5501756	S1_5792183	qNaK12.19	12	65	S12_19926993	S12_20016304	3.01	8.66	0.06	-0.08	0.24
Salt injury score	qSIS6.2a	6	15	S6_2927160	S6_2962502	qSIS6.30	6	145	S6_30296317	S6_30370989	3.82	15.04	0.04	0.04	0.07
	qSIS5.18	5	80	S5_18942631	S5_18997491	qSIS9.9	9	15	S9_9351804	S9_9857266	3.16	12.06	0.05	0.05	0.06
	qSIS3.10	3	65	S3_10992290	S3_11053944	qSIS10.2	10	5	S10_2799960	S10_2837737	4.11	11.59	0.01	0.04	0.07
	qSIS2.20	2	85	S2_20153436	S2_20182321	qSIS10.11	10	25	S10_11045261	S10_11244588	3.07	14.67	-0.07	0.02	-0.06
	qSIS3.11	3	70	S3_11848358	S3_11865689	qSIS12.2	12	15	S12_2315570	S12_2397199	3.33	7.96	0.00	0.00	0.06
Chlorophyll content	qCHL1.20	1	90	S1_20242882	S1_21276489	qCHL1.21	1	95	S1_21276489	S1_21352851	7.27	29.81	-4.09	4.17	-5.07
	qCHL3.17	3	105	S3_17083355	S3_17143997	qCHL3.21	3	110	S3_21445493	S3_21628785	4.59	28.05	4.16	-4.42	-4.73
	qCHL3.21	3	110	S3_21445493	S3_21628785	qCHL7.7	7	50	S7_7781645	S7_7839200	3.31	8.45	-0.29	-0.20	-1.23
	qCHL8.23	8	90	S8_23657286	S8_24738259	qCHL8.24	8	95	S8_24763939	S8_25110888	5.38	35.72	-4.73	5.06	-3.76
	qCHL9.12	9	25	S9_12217170	S9_12366675	qCHL9.12	9	30	S9_12915373	S9_14359383	3.89	34.51	-4.77	4.51	-4.09
	qCHL2.5	2	40	S2_5800279	S2_5848583	qCHL9.18	9	60	S9_18667894	S9_18669560	3.04	7.74	-0.17	-0.44	-1.20
	qCHL10.18	10	65	S10_18819950	S10_19941928	qCHL10.18	10	70	S10_18819950	S10_19941928	5.68	34.16	-4.44	4.32	-4.33
	qCHL7.27	7	110	S7_27772814	S7_27803479	qCHL10.21	10	90	S10_21749293	S10_21786307	3.52	13.31	1.14	0.91	1.28
	qCHL11.4	11	35	S11_4854309	S11_4863888	qCHL11.6	11	40	S11_6970703	S11_7012013	3.97	11.76	-1.18	1.25	-2.15

Table 5 Di-genic epistatic QTLs for traits related to salt tolerance at seedling stage in Bengal/Pokkali F6 RIL population identified by interval mapping (*Continued*)

Trait	QTL_i	Chr	Pos	Marker interval i	Marker interval i	QTL_j	Chr	Pos	Marker interval j	Marker interval j	LOD				
	qCHL3.4	3	25	S3_4116916	S3_4311471	qCHL11.24	11	105	S11_24319577	S11_24335733	3.68	11.82	0.28	-0.53	-1.30
Shoot length	qSHL2.1	2	10	S2_1653448	S2_2064517	qSHL2.5	2	40	S2_5800279	S2_5848583	3.84	9.77	-0.12	-0.48	-2.02
	qSHL4.25	4	95	S4_25549517	S4_26622324	qSHL5.008	5	0	S5_87749	S5_96410	4.32	10.89	-0.16	-0.22	-2.07
	qSHL4.27	4	100	S4_27678052	S4_27715999	qSHL8.1	8	15	S8_1995144	S8_2005542	3.81	10.15	-0.23	0.09	1.97
	qSHL2.32	2	155	S2_32339457	S2_32429009	qSHL9.12	9	25	S9_12217170	S9_12366675	3.67	11.31	0.82	-1.54	2.23
	qSHL1.28	1	130	S1_28157998	S1_28247178	qSHL9.19	9	65	S9_19628929	S9_19696641	3.09	11.19	-0.72	-0.54	1.75
	qSHL2.34	2	165	S2_34519074	S2_34545438	qSHL10.20	10	80	S10_20682624	S10_20733813	3.34	9.36	-0.15	0.46	1.83
	qSHL4.32	4	130	S4_32867449	S4_33074444	qSHL10.21	10	90	S10_21749293	S10_21786307	3.47	10.80	-1.40	-0.05	-1.86
Root length	qRTL1.32	1	145	S1_32327040	S1_32418346	qRTL3.10	3	65	S3_10992290	S3_11053944	4.80	17.47	-0.03	0.30	0.41
	qRTL4.16	4	35	S4_16669714	S4_16706375	qRTL6.25	6	115	S6_25296416	S6_25363541	3.79	12.10	0.19	-0.12	0.37
	qRTL3.28	3	145	S3_28513488	S3_29240341	qRTL8.23	8	90	S8_23657286	S8_24738259	3.26	9.34	-0.11	-0.05	-0.38
	qRTL6.15	6	75	S6_15734275	S6_15881397	qRTL9.16	9	50	S9_16775205	S9_16882286	3.02	11.22	0.15	-0.29	0.36
	qRTL4.33	4	135	S4_33557881	S4_33861248	qRTL10.19	10	75	S10_19941928	S10_20082337	4.15	10.32	-0.01	0.02	-0.41
Dry weight	qDWT3.17	3	105	S3_17083355	S3_17143997	qDWT6.7	6	50	S6_7662391	S6_7749349	3.20	10.64	0.00	0.00	-0.01
	qDWT6.4	6	30	S6_4890290	S6_5269698	qDWT6.30	6	145	S6_30296317	S6_30370989	3.06	12.61	0.00	-0.01	-0.01
	qDWT7.1	7	5	S7_1021298	S7_1051320	qDWT7.27	7	110	S7_27772814	S7_27803479	3.43	10.08	0.00	0.00	0.01
	qDWT5.2	5	20	S5_2483311	S5_2495045	qDWT10.16	10	50	S10_16848745	S10_16898283	3.30	16.39	-0.01	0.00	-0.01
	qDWT4.16	4	35	S4_16669714	S4_16706375	qDWT10.19	10	75	S10_19941928	S10_20082337	3.34	9.61	0.00	0.00	0.01
	qDWT3.5	3	35	S3_5859095	S3_5904925	qDWT12.09	12	10	S12_977852	S12_1386213	3.80	12.95	0.00	0.00	-0.01
Shoot-root ratio	qSRR2.1	2	5	S2_1103758	S2_1653448	qSRR4.27	4	100	S4_27678052	S4_27715999	3.47	11.08	0.00	-0.19	0.35
	qSRR5.2	5	15	S5_2116055	S5_2167880	qSRR5.5	5	40	S5_5798670	S5_5909747	3.85	9.93	0.12	-0.07	-0.39
	qSRR2.3	2	25	S2_3978527	S2_4234638	qSRR9.14	9	40	S9_14976723	S9_15092089	3.20	10.65	-0.14	0.24	-0.39
	qSRR1.20	1	90	S1_20242882	S1_21276489	qSRR9.21	9	75	S9_21030508	S9_21085376	3.09	10.97	-0.11	-0.13	0.37
	qSRR4.18	4	45	S4_18779374	S4_18826971	qSRR10.001	10	0	S10_103050	S10_160013	3.18	9.97	-0.13	-0.16	0.34

contributed by Bengal allele. Five significant pairs of interacting QTLs with PVE ranging between 9 and 17 % were detected. None of the interacting QTLs were found similar or co-localizing to additive QTLs.

QTLs for dry Weight

For shoot dry weight, nine additive QTLs were significant by IM. Three QTLs located on chromosome 5 (*qDWT5.2*, *qDWT5.4* and *qDWT5.5*) were large-effect QTLs that accounted for 11, 15 and 15 % of the phenotypic variation, respectively. Other QTLs were distributed on chromosomes 1, 4, 6, and 11, with PVE of at least 5 %. In contrast, ICIM detected five significant QTLs for DWT. Two QTLs (*qDWT4.32* and *qDWT5.4*) were common in both methods. Among the five QTLs by ICIM, *qDWT5.4* had the largest effect (PVE = 13 %) with LOD score of 7.6. All DWT additive QTLs had increasing effects coming from Pokkali alleles. Analysis of epistatic QTLs detected six pairs of interacting QTLs. All pairs of interacting QTLs except *qDWT4.16* and *qDWT10.19* had large effects of at least 10 % PVE. Intervals of all epistatic QTLs were independent of additive QTLs.

QTLs for Shoot-to-Root Ratio

Additive QTL mapping by IM detected three large-effect and two small-effect QTLs located on chromosomes 1 and 2. The *qSRR1.382*, *qSRR1.36* and *qSRR2.28* were highly significant and had PVE of 23, 13 and 11 %, respectively. Conversely, ICIM method identified five significant additive QTLs. Among the QTLs, two were

large-effect QTLs (*qSRR1.386* and *qSRR2.33*) with PVE of 22 and 11 %, respectively. Pokkali alleles had increasing effects in all additive QTLs for SRR. For interacting QTLs, five large-effect QTL pairs of Bengal and Pokkali origin were detected. All interacting QTLs were mapped to chromosomal regions different from additive QTLs.

Quality and Accuracy of QTL Mapping

Segregation distortion is commonly observed in populations developed from crosses between *indica* and *japonica* rice varieties. We mapped the regions of segregation distortion to determine if significant SDLs co-localized to the QTLs detected in this study. Interval mapping for SDLs detected 16 significant intervals that were skewed toward either parent (Table 6). For each chromosome, at least one SDL was mapped, except on chromosomes 2, 4, and 12. In most of the SDLs, Pokkali allele transmission was favored. In chromosome 11 alone, four significant intervals showed segregation distortion favoring inheritance of Pokkali alleles. The average interval size of SDLs was about 198Kb, with the smallest and largest interval size of 600 bp (*sdl11.26*) and 1.4 Mb (*sdl9.12*), respectively. By comparing the positions of QTLs against the positions of SDLs, the additive QTL *qK1.8* and epistatic QTL *qCHL9.12* overlapped exactly with *sdl1.8* and *sdl9.12* intervals. Therefore, these two QTLs should be considered with caution as they deviate from the expected 1:1 segregation ratio in the RIL population. The Bengal allele was transmitted to progeny lines more frequently than the Pokkali allele in *sdl1.8*. In contrast, Pokkali allele was favorably inherited in *sdl9.12*. Overall,

Table 6 Interval mapping of segregation distortion loci (SDLs) in Bengal/Pokkali F$_6$ RIL population

SDL	Chromosome	Position (cM)	Left Marker	Right Marker	Interval size (bp)	LOD	Segregation ratio	
							Bengal	Pokkali
sdl1.8	1	63	S1_8656025	S1_8901503	245,478	7.0103	1	0.419
sdl1.12	1	74	S1_12394007	S1_12414777	20,770	6.6211	1	0.4304
sdl3.29	3	153	S3_29855008	S3_30045852	190,844	4.0197	0.5244	1
sdl3.34	3	181	S3_34487907	S3_34521908	34,001	3.4639	0.5504	1
sdl5.22	5	96	S5_22077219	S5_22142421	65,202	3.2006	0.5639	1
sdl6.4	6	23	S6_4269744	S6_4327404	57,660	3.2649	0.5605	1
sdl6.9	6	57	S6_9246940	S6_9317830	70,890	3.0751	1	0.5706
sdl7.26	7	109	S7_26680214	S7_26796826	116,612	2.5927	0.5983	1
sdl8.7	8	43	S8_7488739	S8_7668333	179,594	29.5389	0.1136	1
sdl8.16	8	52	S8_16619372	S8_16941109	321,737	22.0004	0.1761	1
sdl9.12	9	29	S9_12915373	S9_14359383	1,444,010	13.3385	0.2847	1
sdl10.12	10	31	S10_12765359	S10_12968073	202,714	2.5777	0.5992	1
sdl11.17	11	61	S11_17286328	S11_17316420	30,092	3.5648	0.5455	1
sdl11.22	11	91	S11_22242895	S11_22274274	31,379	2.8801	0.5814	1
sdl11.23	11	101	S11_23708208	S11_23866022	157,814	2.9439	0.5778	1
sdl11.26	11	115	S11_26254930	S11_26255530	600	5.3304	0.4724	1

most additive and epistatic QTLs mapped in this study were in chromosomal regions not affected by segregation distortion.

Plant height is one of most frequently studied traits in QTL mapping. Several studies showed that plant height has high heritability and stable at different growth stages at different environments (Yan et al. 1998). In rice, 1011 QTLs were reported for plant height (gramene.org). Among these QTLs, *sd1* is the main QTL that played a major role in the development of semi-dwarf varieties in rice (Khush 1999). To assess the quality of our phenotypic data and the accuracy of our QTL mapping, we surveyed plant height QTLs in rice under normal or stress conditions and compared the positions of our SHL QTLs to see if we can detect any of the previously reported plant height QTLs. In both mapping methods, the green revolution gene *sd1* gene, LOC_Os01g66100 (Spielmeyer et al. 2002) was located within our major QTL designated as *qSHL1.38*, with LOD value as high as 36 and PVE of 51 %. The *sd1* gene is about 95 Kb away from the left SNP marker and 226 Kb from the right SNP marker of *qSHL1.38*. Moreover, *qSHL12.25* was found within the region of *qPHT12-1* on chromosome 12 located between 23,603,156-26,017,884 bp region (Hemamalini et al. 2000). Also, *qSHL3.34* was covered within the interval of *QPh3c* located between 32,945,649-36,396,286 bp of chromosome 3 (Li et al. 2003). The minor QTL *qSHL1.7* was flanked within *ph1.2* located in 5,941,464-7,445,919 bp region on chromosome 1 (Marri et al. 2005); while *qSHL2.18* was found within the reported QTL on chromosome 2 at 17,484,665-33,939,159 bp region (Huang et al. 1996). Additionally, *qSHL5.6* was confirmed within the QTL region of chromosome 5 located in between 5255, 880-

6,700,408 bp region (Mei et al. 2003) and in *ph5* located between 6,132,767-18,875,558 bp region on chromosome 5 (Zhuang et al. 1997). In summary, the locations of six *SHL* QTLs matched with previously reported plant height QTLs. In addition, four new minor QTLs were mapped in this study, each contributing at least 5 % of the plant height variation. Together with other QTLs for other traits, a total of eleven QTLs in this study were validated (Table 7). Therefore, our QTL mapping by IM and ICIM methods using ultra-high density genetic map is robust and informative.

Identification of Candidate Genes Within QTLs

The saturation of SNP markers in our linkage map allowed us to detect QTLs at an interval size much shorter than previously reported QTLs. In this study, the average interval size of a QTL was 132 Kb, with minimum and maximum interval size of 22 bp and 877 Kb, respectively (Table 4). For nine traits, IM and ICIM mapped 64 and 36 additive QTLs. Fifteen QTLs were commonly detected in both methods with a total of 85 QTLs. To identify candidate genes underlying fitness of rice under salt stress, we looked at all genes in the QTL region using flanking markers. For 36 additive QTLs by ICIM, a total of 704 genes were present within QTLs (Additional file 2: Table S3), of which, 110 were annotated while the 594 genes were identified as expressed proteins, hypothetical proteins, transposon, and retrotransposon proteins. Similarly, for 64 additive QTLs identified by IM method, only 111 of 1046 genes were annotated. For the 1344 gene models in the 85 QTLs for nine traits, 79 genes were classified in 7 biological processes, 50 genes were classified into 7 molecular functions, and 49 genes were classified into 16 protein classes (Fig. 3).

Table 7 Summary of additive QTLs co-localizing to previously reported QTLs

Trait	QTL in this study	Previous QTL	Reference
K+ concentration	qK1.11	qSKC1	Thomson et al. (2010)
	qK6.4	QTL on chr. 6, at 30 cM	Koyama et al. (2001)
NaK ratio	qNaK1.11	qSNK1	Thomson et al. (2010)
		QTL on chr. 1, at 74 cM	Koyama et al. (2001)
Salt injury score	qSIS9.8	qSES9	Thomson et al. (2010)
Plant height	qSHL1.38	sd1	Spielmeyer et al. (2002)
	qSHL1.7	ph1.2	Marri et al. (2005)
		qPH1.2	Bimpong et al. (2013)
	qSHL2.18	QTL on chr. 2 at 17-33 Mb	Huang et al. (1996)
	qSHL3.34	QPh3c	Li et al. (2003)
	qSHL5.6	ph5	Zhuang et al. (1997)
		QTL on chr. 5 at 5.2- 6.7 Mb	Mei et al. (2003)
	qSHL12.25	qPHT12-1	Hemamalini et al. (2000)
Shoot dry weight	qDWT6.24	qDWT6.1	Bimpong et al. (2013)

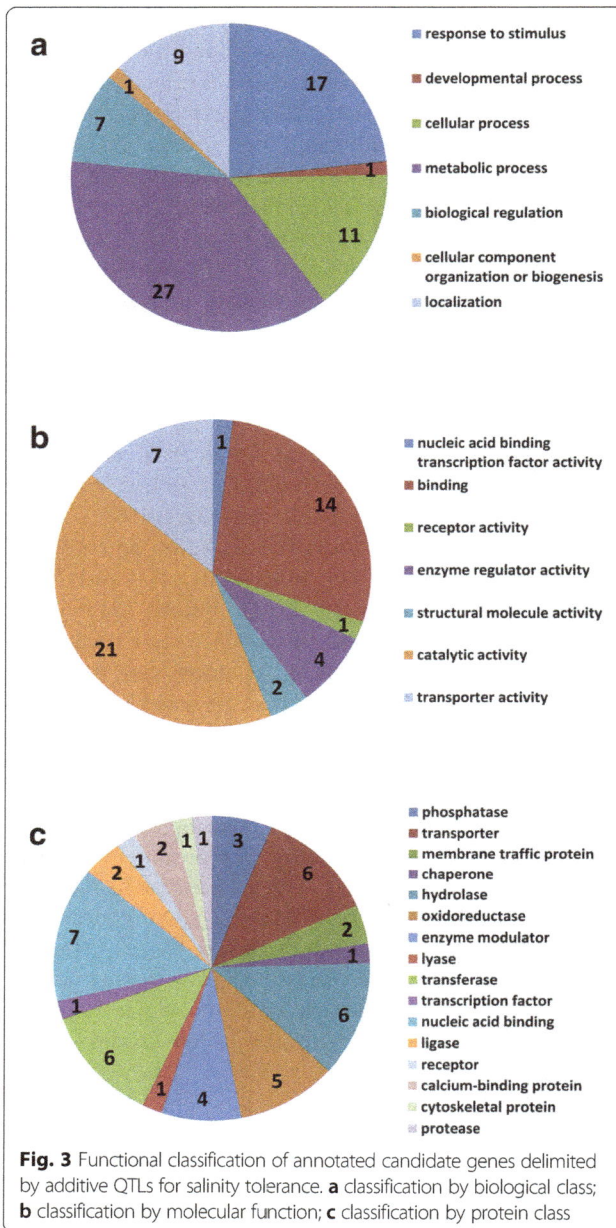

Fig. 3 Functional classification of annotated candidate genes delimited by additive QTLs for salinity tolerance. **a** classification by biological class; **b** classification by molecular function; **c** classification by protein class

A large portion of the candidate genes was involved in metabolic processes and responses to stimuli. Candidate genes classified in biological regulation and localization (six transporters) were found within QTLs.

Discussion

QTL mapping has been implemented in many breeding programs to discover genes underlying quantitative traits. However, many of these reported QTLs covered large chromosome intervals, thus, limiting the application of flanking markers in predicting the phenotype of the plant. A major constraint to previous QTL mapping studies is the number of available polymorphic markers.

However, with reduction in DNA sequencing cost, high resolution QTL mapping is now possible using SNP markers. In this study, we utilized the GBS approach to develop an ultra-high-density genetic linkage map of rice for identification of QTLs for traits related to salinity tolerance. Thirty-eight SNP calls segregating in the RIL population were validated by re-sequencing the target region in both parents. Out of 38 SNP markers, only one SNP call in Bengal was not in agreement (Additional file 3: Table S2). Therefore, the GBS data have high quality SNP calls for linkage and QTL mapping. In spite of the large number of SNP markers placed on the linkage map, there were 20 gaps of about 5 cM intervals. These gaps could be due to removal of SNP markers during filtering process. Due to multiplexing of large number of DNA samples in the GBS, representation of a SNP in all samples was greatly reduced resulting in the removal of more than two-thirds of the GBS data. The linkage map closely resembled the rice genetic map of Harushima et al. (1998). Mapping of segregation distortion loci using this map indicated 16 intervals showing segregation distortion (Table 6). Two SDLs co-localized to QTLs for salinity tolerance (qK1.8 and qCHL9.12). Therefore, genetic variances contributed by these QTLs may not be accurate due to segregation distortion. In addition to availability of numerous SNP markers for linkage map construction, the quality of phenotypic estimates is equally important for QTL mapping. We assessed this by comparing our shoot length QTLs with reported plant height QTLs. Ten QTLs for SHL were detected (Table 4), of which, six QTLs for plant height including the major sd1 (qSHL1.38) co-localized to previously reported plant height QTLs. Validation of those QTLs suggests that our phenotypic and genotypic data for QTL mapping are of high quality (Table 7). With five to six markers per cM, the average QTL interval size was 132Kb. The maximum resolution of QTL was about 22 bp interval (qSIS6.20) and the largest QTL interval size was about 877Kb (qSIS11.2) (Table 4).

Previous QTL mapping studies for salinity tolerance mainly focused on detecting additive QTLs despite the complex nature of salinity tolerance. In this study, we also mapped interacting QTLs significantly contributing to the phenotypic variation of each trait under salt stress (Table 5). Di-genic interval mapping for epistatic QTLs revealed interaction of alleles from Pokkali and Bengal. In general, interacting QTLs were located in chromosome intervals independent of additive QTLs. Likewise, the variance explained by epistatic QTL pair was higher than the variance explained by individual additive QTL. For example, additive QTLs for Na^+ concentration and CHL revealed very few small-effect QTLs. In contrast, many of the epistatic QTL pairs for Na^+ and CHLs had larger PVE as high as 35 %. Therefore, these findings indicated the importance of epistatic QTLs in salt stress response in rice. Many of the QTLs flanked small intervals with few candidate genes. Overall, the ultra-high density genetic map and the high-quality phenotypic data facilitated a high resolution QTL mapping for salinity tolerance. In addition, the genetic map will be useful in discovery of novel QTLs for other contrasting agronomic traits between Bengal and Pokkali.

Since the beginning of the search for QTLs underlying salinity tolerance, Na$^+$ concentration, K$^+$ concentration, NaK ratio, and salt injury score were often investigated. Similar to previous reports, Na$^+$ concentration was highly correlated to SIS or standard evaluation score (SES) and survival of rice plants under salt stress (Yeo et al. 1990; Platten et al. 2013). The shoot Na$^+$ concentration also had significant positive correlation to NaK ratio and shoot K$^+$ concentration (Table 2). The Na$^+$ and K$^+$ relationship implies that as shoot Na$^+$ concentration increases, shoot K$^+$ concentration also increases. It is likely that during salt stress, many lines do not discriminate these cations, thus, suggesting possible accumulation of Na$^+$ and K$^+$ in the shoot through non-selective cation channels (Demidchik and Maathuis 2007). This is evident in the high heritability of Na$^+$ and K$^+$ concentrations in the population (Table 1). In previous studies of QTLs for shoot Na$^+$ concentration, QTLs were mapped on chromosomes 1 (Thomson et al. 2010), 3, 9, 11, (Wang et al. 2012), 4 (Koyama et al. 2001), and 7 (Lin et al. 2004). None of our additive QTLs for Na$^+$ concentration co-localized to previous QTLs, but, the epistatic QTL qNa6.4 is possibly the same additive QTL in chromosome 6 at 24 cM (Koyama et al. 2001). The effects of additive QTLs for Na$^+$ concentration were all minor. Surprisingly, four pairs of interacting intervals had significant larger effects (11–15 % PVE), suggesting that interactions among Na$^+$ QTLs were important in the accumulation of Na$^+$ in shoot. Alleles of Na$^+$ QTLs from both parents contributed to shoot Na$^+$ accumulation. In contrast, all alleles of additive QTLs for shoot K$^+$ concentration were from Pokkali (Table 4). Therefore, it is interesting to know the underlying genes for K$^+$ accumulation and their role in accumulation of other cations like Na$^+$. The presence of transgressive segregants exhibiting higher concentration of shoot K$^+$ and lower NaK ratio than Pokkali suggests the presence of positive alleles in both parents for selective cation transport during salt stress (Fig. 1). In case of Pokkali, salt tolerance response could be due to maintenance of high K$^+$ concentration or low NaK ratio (Ren et al. 2005) and by compartmentalization of Na$^+$ ions into the shoot vacuoles (Kader and Lindberg 2005).

The strong relationship among Na$^+$, K$^+$, and SIS prompted us to look for the co-location of QTLs underlying these traits. Our result showed that qNa6.5 and qNaK6.5, qK1.11 and qNaK1.11, and qSIS6.2 and qNaK6.2 co-localized in the same intervals (Table 4). Therefore, it is possible that these traits shared the same underlying causal genes. The co-location of qNa6.5 and qNaK6.5 is more likely not coincidental because both alleles of the two QTLs came from Bengal and had increasing effect in the concentration of Na$^+$ ions. On the other hand, the co-location of qK1.11 and qNaK1.11 is consistent with co-location of shoot K$^+$ concentration, SKC1 and shoot Na$^+$/K$^+$ ratio, SNK1 (Thomson et al. 2010, Wang et al. 2012). Allele substitution of Bengal with Pokkali at qK1.11 had increasing effect in the shoot K$^+$ concentration. In contrast, Bengal allele of qNaK1.11 had increasing effect on NaK ratio, thus, corroborating the desirability of Pokkali allele at the locus for salt tolerance. In previous studies, SKC1 was responsible for 10–40 % of the variation in shoot K$^+$ concentration (Koyama et al. 2001; Lin et al. 2004; Thomson et al. 2010; Wang et al. 2012). Here, the qK1.11 accounts for only 16 % of the variation. The discrepancy in the estimation of PVE is likely attributed to differences in population size and number of markers used in different studies. The qK1.11 is covering a 52Kb interval between 11.52–11.58 Mb region on chromosome 1 with six genes. This interval is within the reported SKC1 by Thomson et al. (2010), but, downstream of 11.46 Mb region of the cloned HKT1;5 (Ren et al. 2005). While Thomson et al. (2010) assumed HKT1;5 (LOC_Os01g20160) as the underlying gene for qSKC1 or Saltol, it is also possible that other genes contributing toward salt tolerance might be present in the SKC1 region. This possibility is supported by the findings from a genome-wide association mapping study (Kumar et al. 2015), where 12 significant SNPs were located between 9.6 and 14.5 Mb region of chromosome 1. One of the 12 SNPs with high linkage disequilibrium (LD) at 11.6 Mb region (1:11608731) is 26Kb away from the right marker of qK1.11. Furthermore, HKT1;5 allele mining in several rice cultivars showed a weak association of HKT1;5 allele to low Na$^+$ concentration to account for salinity tolerance. The HKT1;5 allele in aromatic group that included Pokkali showed low Na$^+$ concentration. However, several cultivars having different HKT1;5 alleles (Aus, FL478, Hasawi, Daw, Japonica lines, and O. glaberrima) also showed low Na$^+$ concentration and high salt tolerance (Platten et al. 2013). Additionally, our genetic map data showed the availability of markers that flanked HKT1;5 gene (Additional file 1: Table S1, at 70.2 cM) and the absence of segregation distortions in these regions (Table 6), but the IM and ICIM methods both detected QTL for high shoot K$^+$ concentration downstream of HKT1;5. Interestingly, the qK1.11 interval contained two transposons, three uncharacterized expressed proteins, and a CC-NBS-LRR-encoding gene (LOC_Os01g20720). NBS-LRR genes are the largest class of resistance genes implicated in the recognition of pathogen-derived avirulence protein. In rice, a gene encoding a CC-NBS-LRR, Pb1, provided a durable panicle blast resistance by interacting with WRKY45 transcription factor for the activation of signal transduction pathway (Inoue et al. 2013). On the other hand, overexpression of *ADR1* gene encoding a CC-NBS-LRR in *A. thaliana* showed enhanced drought tolerance (Chini et al. 2004). Therefore, the role of LOC_Os01g20720 gene in signal transduction pathway and shoot K^+ ion accumulation should be investigated. Other QTLs for shoot K^+ concentration such as *qK1.8*, *qK1.38*, *qK5.4*, *qK6.4*, and *qK6.5* covered at least 250 kb intervals containing 33, 40, 86, 61, and 34 gene models, respectively. Candidate genes present in these QTL intervals include protein kinases, transcription factors, ethylene, auxin-responsive proteins, flavin-containing monooxygenases, and several expressed proteins of unknown function. In contrast, *qNa2.7* is saturated with transposons and retrotransposons except for a putative membrane lipid channel, scramblase protein (LOC_Os02g14290). The *qNa12.18* flanked four transposons and a hypothetical protein.

For NaK ratio QTLs, the co-location of *qSIS6.2* and *qNaK6.2* confirmed the significant correlation of NaK ratio to SIS. For both QTLs, Bengal alleles were undesirable. Only seven genes including a WRKY113 transcription factor (LOC_Os06g06360) were present in this QTL interval. Whether WRKY113 is interacting with the CC-NBS-LRR in *qK1.11* or *qNaK1.11* like the *Pb1*, presents an interesting perspective to study gene interactions and salt tolerance. In contrast, the large-effect *qNaK6.5* (or *qNa6.5*) still covered a 264Kb interval and contained 34 gene models. Candidate genes in this interval are MYB transcription factor (LOC_Os06g10350), cyclic nucleotide-gated ion channel (LOC_Os06g10580), transcription elongation factor SPT5 (LOC_Os06g10620), and leaf senescence-related protein (LOC_Os06g10560). Among the NaK QTLs, *qNaK1.11* is likely the same QTL as *qSNK1* (Koyama et al. 2001; Thomson et al. 2010).

SIS reflects the overall plant's response to salt stress. Hence, we are particularly curious in finding QTLs to identify underlying genes for this trait. Among the additive QTLs, *qSIS5.1b* had PVE of 13 % with increasing effect from Bengal allele. Therefore, in breeding for low SIS, the corresponding Pokkali allele at *qSIS5.1b* is desirable. The variance explained by *qSIS6.2* alone was only 5 %, but, interaction to *qSIS6.30* increased the PVE to 15 % (Table 5). This result indicated the additive and epistatic effect of a locus and emphasized the importance of QTL interactions in understanding the complexity of SIS or salt tolerance. Among previously mapped QTLs for salt evaluation score (SES) or salt tolerance rating (STR), the *qSIS9.8* is located within the interval of *qSES9* (Thomson et al. 2010). The *qSIS2.8* interval contained 25 genes, one of which encoded a cyclic nucleotide-gated ion channel (LOC_Os02g15580). In contrast, *qSIS5.1b* and *qSIS6.20* contained two and one gene, respectively. Both QTLs delimited a lectin protein

kinase (LOC_Os06g35870, LOC_Os05g03450). In *A. thaliana*, lectin protein kinases were involved in the protein-protein interactions for structural stability of plasma membrane and plant cell wall (Gouget et al. 2006). Therefore, it will be interesting to see if plasma membrane stability conferred by lectin protein kinase enhances salinity tolerance. Similarly, the *qSIS6.21* interval confined a single candidate gene that encodes a receptor-like protein kinase 5 precursor (LOC_Os06g36270). In *qSIS5.03*, a vacuolar ATP synthase (LOC_Os05g01560) is one of the four genes in the interval while a trehalose phosphatase is one of the five candidate genes in *qSIS5.1a*. In rice, transcript expression of a mitochondrial ATP synthase (RMtATP6) was induced in leaves by NaCl and NaHCO$_3$ treatments and overexpression of RMtATP6 in tobacco plants showed enhanced seedling salt tolerance (Zhang et al. 2006). On the other hand, overexpression of trehalose-6-phosphate phosphatase and trehalose-6-phosphate synthase increased tolerance to drought, salt, and cold in rice (Jang et al. 2003). Also, of great interest is the *qSIS6.7* interval that delimited only three genes including a pyrophosphate fructose-6-phosphate 1-phosphotransferase (LOC_Os06g13810) and a flavin monooxygenase in *qSIS7.14*. Pyrophosphate: fructose-6-phosphate 1-phosphotransferase was associated to seedling salt tolerance (Lim et al. 2014) while overexpression of a flavin monooxygenase designated as YUCCA enhanced drought tolerance of *A. thaliana* (Cha et al. 2015). Additionally, *qSIS8.24*, *qSIS9.8*, and *qSIS11.2* delimited genes involved in signal transduction pathway.

Plant vigor under salt stress is a good predictor of tolerance. In addition to common traits investigated under salt stress, CHL, and growth parameters such SHL, RTL, SRR, and DWT were also examined. In soybean, salinity tolerance was determined by a major QTL for chlorophyll content (Patil et al. 2016). In contrast, additive QTLs for CHL were all minor-effect QTLs while several pairs of epistatic QTLs had PVE as high as 35 % (Tables 4 and 5). Comparison of CHL QTLs with earlier reported QTLs co-localized *qCHL2.20* and *qCHL3.26* within the intervals of *qCHL2* and *qCHL3* (Thomson et al. 2010). All other CHL QTLs are novel, thus, offering new targets for further analysis. The *qCHL3.26* interval flanked a single unknown expressed protein (LOC_Os03g47190) while *qCHL2.20* contained six retrotransposons and one expressed protein. Aldehyde dehydrogenase (LOC_Os02g49720) and zinc-knuckle family protein (LOC_Os02g49670) were found in *qCHL2.30* interval. *Arabidopsis* plants overexpressing aldehyde dehydrogenase improved salinity tolerance of plants by reducing the accumulation of reactive oxygen species (Sunkar et al. 2003). Among the 44 genes in the *qCHL11.1*, a NAC transcription factor and a glutathione S transferase are promising candidate genes. In rice, overexpression of a NAC transcription factor showed increased tolerance to drought and salt stress (Zheng et al. 2009). Conversely, glutathione S-transferase had negative effect to drought and salt tolerance in *Arabidopsis* plants (Chen et al. 2012). On the other hand, *qCHL11.2* interval contained seven genes, one of which encodes an HVA22. In barley and *Arabidopsis*, aleurone cells transformed with HVA22 inhibited the formation of GA-induced formation of vacuoles and programmed cell death (Gou and Ho 2008). Since vacuoles are important storage of Na^+ for salt tolerance, HVA22 is a promising candidate gene for salt tolerance.

Among the SHL QTLs, *qSHL1.38* and *qSHL2.18* were validating the *qPH1.2* (Bimpong et al. 2013) and *qPH2* (Thomson et al. 2010), respectively, for plant height QTLs investigated under salt stress. The SHL QTLs contained many candidate genes. In addition to the major *sd1* gene within *qSHL1.38*, other candidate genes were AP2 domain containing protein (LOC_Os01g04020) in *qSHL1.1*, KH domain containing protein (LOC_Os01g13100) in *qSHL1.7a*, auxin response factor1 in *qSHL1.7b*, potassium transporter (LOC_Os01g13520) in *qSHL2.18*, gibberellin 2-oxidase (LOC_Os05g06670) in *qSHL5.3*, gibberellin 3-beta-dioxygenase (LOC_Os05g08540), cytokinin-O-glucosyltransferase (LOC_Os05g08480) and auxin OsIAA15 (LOC_Os05g08570) in *qSHL5.4*, OsMAD66 transcription factor (LOC_Os05g11380) in *qSHL5.6*, and OsSAUR57 in *qSHL12.25*. A putative RNA-binding protein containing a KH domain was reported to be important in *Arabidopsis* plants for heat stress tolerance (Guan et al. 2013). In other plants, AP2/ERF transcription factors were implicated in the control of metabolism, growth, and development, and in responses to environmental stress (Licausi et al. 2013).

The relationship of Na^+ concentration with SHL, RTL, DWT, and SRR were not significant. However, correlation of these traits to SIS, indicated growth inhibition with increasing sensitivity to salt stress (Table 2). For RTL, large-effect additive QTLs were detected on chromosome 3 (*qRTL3.7*, *qRTL3.10*) while the rest were minor-effect QTLs located on chromosomes 1, 2, 3, 4, 8, and 9. The majority of root length variation was explained in the epistatic QTLs. Similarly, QTLs for DWT detected only three large-effect QTLs on chromosome 5 (*qDWT5.2*, *qDWT5.4*, *qDWT5.5*) and all epistatic QTL pairs had PVE not lower than 10 %. In contrast, five large-effect additive QTLs were mapped on chromosomes 1 and 2 for SRR. The *qSRR1.382* was located on the same interval of *qSHL1.38* and so, the same *sd1* gene determined the increased SRR. The fact that all DWT and SRR additive QTLs were contributed by Pokkali suggested the growth-increasing effect of Pokkali alleles under salt stress. On the other hand, the significant epistatic QTLs identified in all traits emphasized the importance of additive and epistatic effects for salinity tolerance.

The growth of roots during seedling stage under salt stress was not investigated before. All RTL QTLs in this study were new QTLs. A total of 117 genes models was delimited by 14 QTLs. In *qRTL1.22*, only two gene models were present, a retrotransposon and an uncharacterized expressed protein. Of particular interest is the VQ domain containing protein (LOC_Os01g46440) within *qRTL1.26*. In *Arabidopsis*, VQ-containing proteins interact with WRKY transcription factors and negatively regulate plant resistance to pathogen infection (Wang et al. 2015). Other candidate genes within RTL QTLs are aldehyde dehydrogenase (LOC_Os02g43194) and polyamine oxidase (LOC_Os02g43220) in *qRTL2.26*, ankyrin repeat-reach protein (LOC_Os02g54860) and trehalose-6-phosphate (LOC_Os02g54820) among seven genes contained in *qRTL2.33*, an integral membrane protein (LOC_Os03g11590) in *qRTL3.6*, MYB transcription factor (LOC_Os03g13310) and transporters (LOC_Os03g13240, LOC_Os03g13250, LOC_Os03g17740) in *qRTL3.7* and *qRTL3*. An asparagine synthetase (LOC_Os03g18130) is within *qRTL3.10*, while a vacuolar protein sorting-associated protein 18 (LOC_Os08g08060), transporter (LOC_Os08g08070), and an RLK gene (LOC_Os08g08140) are delimited in *qRTL8.4*. The *qRTL9.14* contains only three genes, one of which is a WRKY gene (LOC_Os09g25060). The *qRTL8.27* contains a PDR ABC transporter gene (LOC_Os08g43120).

Koyama et al. (2001) detected one QTL for dry mass on chromosome 6 at 34 cM. A total of six DWT QTLs were mapped on chromosome 6 by IM and ICIM. However, none of our QTLs are localized at 34 cM region. The *qDWT6.24*, however, validated the *qDWT6.1* detected by Bimpong et al. (2013). Notable candidate genes within DWT QTLs are transporters (LOC_Os01g38670, LOC_Os01g38680, LOC_Os05g04600, LOC_Os05g08430) in the intervals of *qDWT1.2*, *qDWT5.2*, and *qDWT5.4*, calmodulin-binding transcription factors (LOC_Os01g69910, LOC_Os05g10840) in *qDWT1.40*, a REX1 DNA repair gene (LOC_Os05g10980) in *qDWT5.5*, a MYB transcription factor (LOC_Os06g02250) in *qDWT6.06*, and a lectin protein kinase (LOC_Os06g35870) in *qDWT6.20*. In addition, a calcium-binding mitochondrial carrier (LOC_Os06g40200) is within *qDWT6.23* while an ABC-type transporter gene (LOC_Os06g40550) is in *qDWT6.24*.

SRR QTLs under salt stress were not investigated in previous QTL mapping studies. All QTLs for SRR are new QTLs for further understanding of plant's fitness under salinity stress. The large effect QTL *qSRR1.36* spanned five genes including a WRKY119 gene (LOC_Os01g62510). The *qSRR1.382* and *qSRR1.386* contained an amino acid transporter (LOC_Os01g66010) and several receptor-like protein kinases. In contrast, *qSRR2.31* delimited a single expressed protein. Again, a

trehalose-6-phosphate (LOC_Os02g54820) and ankyrin repeat rich protein (LOC_Os02g54860) are two of the seven
genes found in the *qSRR2.33* interval while a HEAT repeat
protein is within *qSRR2.34* interval and another transporter
is located in *qSRR3.9*. In addition to few candidate genes
with known functions present within small-effect QTLs
qSRR3.10, *qSRR3.11*, *qSRR4.10*, *qSRR8.19*, and *qSRR8.26*,
there were several uncharacterized expressed proteins.

Taken together, at least six transporter genes were
located within six QTLs, of which, three transporter genes
were found in QTLs for root length (LOC_Os3g11590 in
qRTL3.6; LOC_Os3g17770 in *qRTL3.9*, and LOC_Os3g11
590 in *qRTL3.7*), while one transporter gene was
contained in *qSIS11.2* (LOC_Os11g06810), *qCHL11.2*
(LOC_Os11g05800), and *qSHL3.34* (LOC_Os03g61290).
In addition to transporters and genes for detoxification or
osmotic adjustment (flavin monooxygenase, trahalose-6-
phosphate), the prevalence of protein kinases suggest the
role of signal transduction pathway and possible regulation of biological and cellular processes by transcription
factors (Fig. 3).

Conclusion

The availability of ultra-high density genetic map and robust phenotypic data enabled us to identify additive
QTLs with high resolution and facilitated identification
of candidate genes. Detection of significant epistatic
QTLs in addition to additive QTLs validated the complex architecture of salinity tolerance, which is possibly
determined by concerted interactions of several genes.
While *Saltol* or *SKC1* may provide salinity tolerance and
already being introgressed to several rice varieties in
Asia, it may not provide adequate tolerance to salt
stress. Our result suggested the use of multiple QTLs,
especially the genes for low salt injury score to enhance
salinity tolerance. The candidate genes identified in this
study will be useful targets for functional genomics,
gene-pyramiding, and gene-based marker-assisted breeding. Our study demonstrated the power and application
of GBS for QTL mapping of a complex genetic trait like
salinity tolerance.

Methods
Plant Materials and Population Development

A mapping population was developed by crossing Bengal
and Pokkali as female and male parent, respectively. The
resulting F_1 lines were selfed and advanced by single
seed descent method to generate 230 recombinant inbred lines (RILs) in F_6 generation. RILs grown in unsali-
nized condition were extracted for DNA and were
genotyped by the Cornell Genomic Diversity Facility
using the GBS method.

Phenotypic Characterization and Tissue Collection

The phenotypic evaluation was conducted in the greenhouse with day time and night time temperature settings
at 26–29 °C. The hydroponics system was used in the
screening for seedling salt tolerance following the IRRI
standard evaluation technique (Gregorio et al. 1997).
The parental lines and 230 RILs were pre-germinated in
a paper towel for 2 days and then transplanted to hydroponic set up containing 1 g/L of Jack's Professional (20-
20-20) (J.R. Peters, Inc.), supplemented with 300 mg/L
of ferrous sulfate. The pH of the solution was maintained at 5.0–5.1 and plants were allowed to grow for
2 weeks. The whole experiment was conducted in randomized complete block design replicated three times,
with ten plants per line per replicate.

At 14th day after planting, the plants were subjected to
$6dSm^{-1}$ for 2 days and then into $12dSm^{-1}$ salt stress.
After 6 days of salt stress, the amount of chlorophyll
content was measured on the mid-length of the second
youngest leaf using a SPAD-502 chlorophyll meter
(Spectrum Technologies, Inc.). Five plants per line of
uniform growth were evaluated for traits related to salinity tolerance. On the 9th or 11th day, when the susceptible check plants were dead, lines were phenotyped for
salt injury score, shoot length, and root length. A score
of 1 was given to unaffected plants, score of 3 to healthy
plants but stunted, score of 5 to plants showing green
leaves and stem with some tip burning and leaf rolling,
score of 7 to plants with green stem but all leaves are
dead, and a score of 9 to completely dead plants. Shoot
length and root length were measured in centimeter.
Shoot length was measured from the base of the culm to
the tip of the tallest leaf. Root length was measured from
the base of the culm to the tip of the longest root. Shoot
length to root length ratio was derived by dividing the
shoot length by the root length. For dry weight, five
plants per line were collected and dried at 65 °C oven
for 5 days prior to weighing.

Measurement of Na^+ and K^+ Concentration in Shoot

The amount of Na^+ and K^+ in the shoot was measured
from 100 mg ground tissue taken from a pool of five
plants per line. Briefly, the shoots of the plants were collected, rinsed with water, oven dried for 5 days and
ground to fine powder. The tissue was digested with
5 ml of nitric acid and 3 ml of 30 % hydrogen peroxide
at 152–155 °C heating block for 3 h (Jones and Case
1990). The digested tissue was diluted to a final volume
of 125 ml. Flame photometer (model PFP7, Bibby Scientific Ltd, Staffordshire, UK) was used to quantify the Na^+
and K^+ concentrations in each sample. The final concentrations were computed from the derived standard curve
of different dilutions of Na^+ and K^+ and the ratio of Na^+

and K^+ concentration (NaK) was calculated from these values.

Statistical Analyses

The phenotypic data for each trait were analyzed by ANOVA and LS mean of each line was extracted using the GLIMMIX procedure. The RIL line was entered as a fixed effect and replication as a random effect. Broad sense heritability for each trait was computed by family mean basis following Holland et al. (2003). CORR procedure was implemented to determine the relationship among traits. All data analysis was conducted using Statistical Analysis System (SAS) software version 9.4 for Windows (SAS Institute Inc 2012). Frequency distribution for each trait was constructed in Microsoft Excel 2010.

Genotyping-by-Sequencing of Bengal, Pokkali, and RIL Population

Leaf tissues were collected from each of the parental lines and RIL. The DNA was extracted using the Qiagen DNeasy Plant Mini Kit following the manufacturer's protocol (Qiagen Inc., Valencia, CA, USA). Genomic DNA libraries were prepared as described by Elshire et al. (2011). Each DNA was cut by *ApeKI* enzyme and the adapters were ligated to barcode the DNA of each line. Pooled DNA from parents, 189 RILs, 94 other lines and 3 blanks was sequenced in one lane with the Illumina HiSeq sequencer at Genomic Diversity Facility, Cornell University Institute of Biotechnology (http://www.biotech.cornell.edu/brc/genomic-diversity-facility). The Tassel GBS pipeline was used to process the data and SNP calling was based on the Nipponbare reference genome MSU release 7 (Kawahara et al. 2013).

Construction of Linkage Map and QTL Analysis

Sequence alignment and SNP calling were done by the Genomic Diversity Facility, Cornell University. A total of 1,593,692 tags were sequenced, of which, 1,215,287 (76.3 %) were aligned to unique positions, 134,210 (8.4 %) had multiple alignments and 244,195 (15.3 %) were not aligned. Upon processing and filtering of SNPs, the resulting SNPs markers were reduced to a total of 33,987, with an average individual depth of 5.5 or site depth of 4.6 and individual mean missingness of 0.28. Pokkali and two RILs were declared as failed samples for having less than 10 % of the mean reads per sample. They were removed before further analysis, resulting to a total of 187 RILs for final analysis. The hapmap data file containing the filtered SNP calls were further analyzed prior to linkage map construction and QTL analysis. The Bengal parent was successfully sequenced, thus providing data for differentiation of alleles among

RILs. To validate the GBS SNP calling, we amplify and re-sequenced 38 positions of GBS SNP calls in Bengal and Pokkali. Allele differentiation and allele origin among RILs were confirmed with Bengal and Pokkali re-sequenced data available in our laboratory. With the breeding scheme of the mapping population, only three possible genotypes may exist at polymorphic loci with bi-allelic SNP calling. The 2, 0, -1 coding numbers were then used to code for different alleles in the genotype data. SNP call for each marker across the population was coded as 2 if the allele was the same as Bengal. A code of 0 was given to the alternative allele and was assumed as the allele from Pokkali. Since our materials are F_6 RIL, most of the loci were homozygous and should be segregating into 1:1. However, with low read depth due to highly multiplexed nature of GBS, all heterozygous SNPs (Y = T|C, M = A|C, W = T|A, R = A|G, S = C|G, K = G|T) and missing SNP (N) calls were coded as -1. All SNP markers monomorphic across the 187 RILs were removed. Likewise, all SNP markers with more than 10 % missing SNP calls were purged before further analysis. As a result, only 9303 SNP markers were retained and used for linkage and QTL mapping. The order of SNP markers along the chromosome was fixed based on the physical position of SNPs in the MSU Rice Genome Annotation (Osa1) Release 7. Genetic distances of SNP markers based on recombination rates were converted using the Kosambi mapping function. To see if segregation distortion of markers occurs in the QTLs detected in this study, interval mapping of segregation distortion locus (SDL) was also conducted. Significant SDLs were declared for loci exceeding the 2.0 LOD threshold level.

Nine traits were used for QTL mapping. The mean of three replications was used as phenotypic score for each trait. Except for salt injury score, Na^+ concentration, K^+ concentration, Na^+/K^+ ratio, chlorophyll content, shoot length, root length, shoot dry weight and shoot length to root length ratio showed normal distribution. Hence, the data were directly used for QTL mapping. For SIS, data were log transformed to improve the normality of RIL distribution prior to QTL mapping. Analysis of additive QTLs for traits related to salinity tolerance was performed by interval mapping (IM-ADD), and inclusive composite interval mapping (ICIM-ADD) methods. By interval mapping method, parameters for QTL detection were set to a scanning window size of every 1 cM with LOD threshold value set at 2.0 to declare significant QTLs. In ICIM-ADD, the parameters were set as follows: missing phenotype by mean replacement, stepwise regression method every 1 cM window size with the probability levels of entering and removing variables set at 0.001, and a second step scanning by interval mapping for significant QTL detection at LOD threshold of

2.0. Epistatic QTLs were identified by interval mapping every 5 cM window with LOD threshold set at 3.0. The phenotypic variation explained by QTLs and their genetic effect were estimated. Confidence interval of each QTL was delimited by the flanking markers within the 1-LOD drop from the estimated QTL position. QTL interval size is computed from the distance between the physical positions of left and right flanking markers. Significant QTL for each trait was named with the trait, followed by numbers indicating the chromosome location and megabase (Mb) position of the QTL. For example, *qK1.8* indicates the presence of a QTL for shoot K^+ concentration in chromosome 1 located at 8 Mb region. All linkage, SDL and QTL analyses were implemented in QTL IciMapping software version 4.0.6.0 (Meng et al. 2015).

Candidate Gene Prediction

To identify potential candidate genes within QTL intervals, the physical positions of SNP markers flanking the QTLs were searched in MSU Rice Genome Annotation (Osa1) Release 7. Genes contained in each QTLs were listed (Additional file 2: Table S3). To understand the roles of candidate genes in the mechanism of salinity tolerance, classification and annotation of candidate genes were inquired using the Panther Classification System (Mi et al. 2016).

Abbreviations

cM: centi-Morgan; GBS: Genotyping-by-sequencing; PVE: Phenotypic variance explained; QTL: Quantitative trait locus; RILs: Recombinant inbred lines; SDL: Segregation distortion loci; SIS: Salt injury score; SNP: Single nucleotide polymorphism

Acknowledgments

We thank Dennis Alcalde and Anna Borjas for helping in greenhouse experiments. This research was supported by the National Institute of Food and Agriculture, U.S. Department of Agriculture (Grant No. 2013-67013-21238) and the Louisiana Rice Research Board. This manuscript is approved for publication by the Director of Louisiana Agricultural Experiment Station, USA as manuscript number 2016-306-27533.

Authors' contributions

PKS conceived and designed the experiment. TBD conducted the experiment, analyzed the data, and wrote the manuscript. TBD and PKS critically revised the manuscript. All authors read and approved the final manuscript.

Authors' information

TBD: Graduate student, School of Plant, Environmental, and Soil Sciences, Louisiana State University, USA. SL: Director and Rice Breeder, Rice Research Station, Louisiana State University Agricultural Center. PKS: Professor of Plant Genetics, School of Plant, Environmental, and Soil Sciences, Louisiana State University.

Competing interests

The authors declare that they have no competing interests.

Author details

[1]School of Plant, Environmental, and Soil Sciences, Louisiana State University Agricultural Center, Baton Rouge, LA, USA. [2]Rice Research Station, Louisiana State University Agricultural Center, Rayne, LA, USA.

References

Alam R, Rahman MS, Seraj ZI, Thomson MJ, Ismail AM, Tumimbang-Raiz E, Gregorio GB (2011) Investigation of seedling-stage salinity tolerance QTLs using backcross lines derived from *Oryza sativa* L. Pokkali. Plant Breed 130:430–437

Arbelaez JD, Moreno LT, Singh N, Tung CW, Maron LG, Ospina Y, Martinez CP, Grenier C, Lorieux M, McCouch S (2015) Development and GBS-genotyping of introgression lines (ILs) using two wild species of rice, *O. meridionalis* and *O. rufipogon*, in a common recurrent parent, *O. sativa* cv. Curinga. Mol Breeding 35(2):81

Asch F, Dingkuhn M, Dörffling K, Miezan K (2000) Leaf K/Na ratio predicts salinity induced yield loss in irrigated rice. Euphytica 113:109–118

Bimpong IK, Manneh B, El-Namaky R, Diaw F, Amoah NKA, Sanneh B, Ghislain K, Sow A, Singh RK, Gregorio G, Bizimana JB, Wopereis M (2013) Mapping QTLs related to salt tolerance in rice at the young seedling stage using 384-plex single nucleotide polymorphism SNP marker sets. Mol Plant Breed 5(9):47–62

Bonilla P, Dvorak J, Mackill D, Deal K, Gregorio G (2002) RFLP and SSLP mapping of salinity tolerance genes in chromosome 1 of rice (*Oryza sativa* L.) using recombinant inbred lines. Philip Agric Sci 85:68–76

Causse MA, Fulton TM, Cho YG, Ahn SN, Chunwongse J, Wu K, Xiao J, Yu Z, Ronald PC, Harrington SE, Second G, McCouch SR (1994) Saturated molecular map of the rice genome based on an interspecific backcross population. Genetics 138:1251–1274

Cha JY, Kim WY, Kang S, Kim J, Baek D, Jung I, Kim MR, Li N, Kim HJ, Nakajima M, Asami T, Sabir JS, Park HC, Lee SY, Bohnert HJ, Bressan RA, Pardo JM, Yun DJ (2015) A novel thiol-reductase activity of Arabidopsis YUC6 confers drought tolerance independently of auxin biosynthesis. Nat Commun 6:8041. doi:10.1038/ncomms9041

Chen JH, Jiang HW, Hsieh EJ, Chen HY, Chien CT, Hsieh HL, Lin TP (2012) Drought and salt stress tolerance of an Arabidopsis glutathione S-transferase U17 knockout mutant are attributed to the combined effect of glutathione and abscisic acid. Plant Physiol 158(1):340–351

Chen Z, Wang B, Dong X, Liu H, Ren L, Chen J, Hauck A, Song W, Lai J (2014) An ultra-high density bin-map for rapid QTL mapping for tassel and ear architecture in a large F$_2$ maize population. BMC Genomics 15:433. doi:10.1186/1471-2164-15-433

Chini A, Grant J, Seki M, Shinozaki K, Loake G (2004) Drought tolerance established by enhanced expression of CC-NBS-LRR gene ADR1 requires salicylic acid, EDS1, and ABI1. Plant J 38:810–822

De Leon TB, Linscombe S, Gregorio G, Subudhi PK (2015) Genetic variation in Southern rice genotypes for seedling salinity tolerance. Front Plant Sci 6:374. doi:10.3389/fpls.2015.00374

Demidchik V, Maathuis F (2007) Physiological roles of nonselective cation channels in plants:from salt stress to signalling and development. New Phytol 175:387–404

Elshire RJ, Glaubitz JC, Sun Q, Poland JA, Kawamoto K, Buckler ES, Mitchell SE (2011) A robust, simple genotyping-by-sequencing (GBS) approach for high diversity species. PLoS One 6(5):e19379

Flowers TJ (2004) Improving crop salt tolerance. J Exp Bot 55:307–319

Flowers T, Flowers S (2005) Why does salinity pose such a difficult problemfor plant breeders? Agricul Water Manage 78:15–24

Gou JW, Ho TH (2008) An abscisic acid-induced protein, HVA22, inhibits gibberellin-mediated programmed cell death in cereal aleurone cells. Plant Physiol 147:1710–1722

Gouget A, Senchou V, Govers F, Sanson A, Barre A, Rougé P, Pont-Lezica R, Canut H (2006) Lectin receptor kinases participate in protein-protein interactions to mediate plasma membrane-cell wall adhesions in Arabidopsis. Plant Physiol 140:81–90

Gregorio GB (1997) Tagging salinity tolerance genes in rice using amplified fragment length polymorphism (AFLP). Dissertation, University of the Philippines, Los Baños

Gregorio GB, Senadhira D, Mendoza RD (1997) Screening rice for salinity tolerance. IRRI Disc. Paper Series No.22:1-30. International Rice Research Institute, Los Baños, Philippines.

Gregorio GB, Senadhira D (1993) Genetic analysis of salinity tolerance in rice (Oryza sativa L.). Theor Appl Genet 86:333–338

Gregorio GB, Senadhira D, Mendoza RD, Manigbas NL, Roxas JP, Cuerta CQ (2002) Progress in breeding for salinity tolerance and associated abiotic stresses in rice. Field Crops Res 76:91–101

Guan Q, Wen C, Zeng H, Zhu J (2013) A KH domain-containing putative RNA-binding protein is critical for heat stress-responsive gene regulation and thermotolerance in Arabidopsis. Mol Plant 6:386–395

Harushima Y, Yano M, Shomura A, Sato M, Shimano T, Kuboki Y, Yamamoto T, Lin SY, Antonio BA, Parco A, Kajiya H, Huang N, Yamamoto K, Nagamura Y, Kurata N, Khush GS, Sasaki T (1998) A high-density rice genetic linkage map with 2275 markers using a single F_2 population. Genetics 148:479–494

He J, Zhao X, Laroche A, Lu ZX, Liu HK, Li Z (2014) Genotyping-by-sequencing (GBS), an ultimate marker-assisted selection (MAS) tool to accelerate plant breeding. Front Plant Sci 5:1–8

Hemamalini G, Shashidhar H, Hittalmani S (2000) Molecular marker assisted tagging of morphological and physiological traits under two contrasting moisture regimes at peak vegetative stage in rice (Oryza sativa). Euphytica 112:69–78

Holland JB, Nyquist WE, Cervantes-Martínez CT (2003) Estimating and interpreting heritability for plant breeding: an update. Plant Breed Rev 22:9–111

Huang N, Courtois B, Khush GS, Lin H, Wang G, Wu P, Zheng K (1996) Association of quantitative traits loci for plant height with major dwarfing genes in rice. Heredity 77:130–137

Huang YF, Poland JA, Wight CP, Jackson EW, Tinker NA (2014) Using genotyping-by-sequencing (GBS) for genomic discovery in cultivated oat. PLoS One 9(7): e102448. doi:10.1371/journal.pone.0102448

Inoue H, Hayashi N, Matsushita A, Xinqiong L, Nakayama A, Sugano S, Jiang CJ, Takatsuji H (2013) Blast resistance of CC-NBS-LRR protein Pb1 is mediated by WRKY45 through protein-protein interaction. Proc Natl Acad Sci U S A 110: 9577–9582

Jaganathan D, Thudi M, Kale S, Azam S, Roorkiwal M, Gaur PM, Kishor PB, Nguyen H, Sutton T, Varshney RK (2015) Genotyping-by-sequencing based intra-specific genetic map refines a "QTL-hotspot" region for drought tolerance in chickpea. Mol Genet Genomics 290:559–571

Jang IC, Oh SJ, Seo JS, Choi WB, Song SI, Kim CH, Kim YS, Seo HS, Choi YD, Nahm BH, Kim JK (2003) Expression of a bifunctional fusion of the Escherichia coli genes for trehalose-6-phosphate synthase and trehalose-6-phosphate phosphatase in transgenic rice plants increases trehalose accumulation and abiotic stress tolerance without stunting growth. Plant Physiol 131:516–524

Jones JB, Case VW (1990) Sampling, handling, and analyzing plant tissue samples In: Westerman RL (Ed), Soil Testing and Plant Analysis. third ed., Soil Science Society of America, Book Series No. 3, Madison, Wisconsin, pp 389–427.

Kader M, Lindberg S (2005) Uptake of sodium in protoplasts of salt-sensitive and salt-tolerant cultivars of rice, Oryza sativa L. determined by fluorescent dye SBFI. J Exp Bot 56:3149–3158

Kawahara Y, de la Bastide M, Hamilton JP, Kanamori H, McCombie WR, Ouyang S, Schwartz DC, Tanaka T, Wu J, Zhou S, Childs KL, Davidson RM, Lin H, Quesada-Ocampo L, Vaillancourt B, Sakai H, Lee SS, Kim J, Numa H, Itoh T, Buell CR, Matsumoto T (2013) Improvement of the Oryza sativa Nipponbare reference genome using next generation sequence and optical map data. Rice 6:1–10. doi:10.1186/1939-8433-6-1

Khush G (1999) Green Revolution: preparing for the 21st century. Genome 42(4):646–655

Koyama ML, Levesley A, Koebner R, Flowers TJ, Yeo AR (2001) Quantitative trait loci for component physiological traits determining salt tolerance in rice. Plant Physiol 125:406–422

Kumar V, Singh A, Mithra SV, Krishnamurthy SL, Parida SK, Jain S, Tiwari KK, Kumar P, Rao AR, Sharma SK, Khurana JP, Singh NK, Mohapatra T (2015) Genome-wide association mapping of salinity tolerance in rice (Oryza sativa). DNA Res 22(2):133–145. doi:10.1093/dnares/dsu046

Li ZK, Yu SB, Lafitte HR, Huang N, Courtois B, Hittalmani S, Vijayakumar CH, Liu GF, Wang GC, Shashidhar HE, Zhuang JY, Zheng KL, Singh VP, Sidhu JS, Srivantaneeyakul S, Khush GS (2003) QTL x environment interactions in rice. I. heading date and plant height. Theor Appl Genet 108:141–153

Licausi F, Ohme-Takagi M, Perata P (2013) APETALA2/ethylene responsive factor (AP2/ERF) transcription factors: mediators of stress responses and developmental programs. New Phytol 199:639–649

Lim H, Cho MH, Bhoo S, Hahn TR (2014) Pyrophosphate: fructose-6-phosphate 1-phosphotransferase is involved in the tolerance of Arabidopsis seedlings to salt and osmotic stresses. In Vitro Cell Dev Biol 50:84–91

Lin HX, Zhu MZ, Yano M, Gao JP, Liang ZW, Su WA, Ren ZH, Chao DY (2004) QTLs for Na^+ and K^+ uptake of the shoots and roots controlling rice salt tolerance. Theor Appl Genet 108:253–260

Linscombe SD, Jodari F, Mckenzie KS, Bollich PK, White LM, Groth DE, Dunand RT (1993) Registration of Bengal rice. Crop Sci 33:645–646

Liu Y, Qi X, Young ND, Olsen KM, Caicedo AL, Jia Y (2015) Characterization of resistance genes to rice blast fungus Magnaporthe oryzae in a "Green Revolution" rice variety. Mol Breed 35:52

Marri P, Sarla N, Reddy L, Siddiq E (2005) Identification and mapping of yield and yield related QTLs from an Indian accession of Oryza rufipogon. BMC Genet 6(33):1–14

Mei HW, Luo LJ, Ying CS, Wang YP, Yu XQ, Guo LB, Paterson AH, Li ZK (2003) Gene actions of QTLs affecting several agronomic traits resolved in a recombinant inbred rice population and two testcross populations. Theor Appl Genet 107:89–101

Meng L, Li H, Zhang L, Wang J (2015) QTL IciMapping: Integrated software for genetic linkage map construction and quantitative trait locus mapping in biparental populations. Crop J 3(3):269–283

Mi H, Poudel S, Muruganujan A, Casagrande J, Thomas P (2016) PANTHER version 10: expanded protein families and functions, and analysis tools. Nucleic Acids Res 44(D1):D336–D342

Munns R, James RA (2003) Screening methods for salinity tolerance: a case study with tetraploid wheat. Plant Soil 253:201–218

Munns R, James RA, Lauchli A (2006) Approaches to increasing the salt tolerance of wheat and other cereals. J Exp Bot 57:1025–1043

Patil G, Do T, Vuong TD, Valliyodan B, Lee JD, Chaudhary J, Shannon JG, Nguyen HT (2016) Genomic-assisted haplotype analysis and the development of high-throughput SNP markers for salinity tolerance in soybean. Sci Rep 6: 19199. doi:10.1038/srep19199

Pearson GA, Bernstein L (1959) Salinity effects at several growth stages of rice. Agron J 51:654–657

Platten JD, Egdane JA, Ismail AM (2013) Salinity tolerance, Na^+ exclusion and allele mining of HKT1;5 in Oryza sativa and O. glaberrima: many sources, many genes, one mechanism? BMC Plant Biol 13:32. doi:10.1186/1471-2229-13-32

Poland JA, Brown PJ, Sorrells ME, Jannink JL (2012) Development of high-density genetic maps for barley and wheat using a novel two-enzyme genotyping-by-sequencing approach. PLoS One 7(2):e32253. doi:10.1371/journal.pone.0032253

Ren ZH, Gao JP, Li LG, Cai XL, Huang W, Chao DY, Zhu MZ, Wang ZY, Luan S, Lin HX (2005) A rice quantitative trait locus for salt tolerance encodes a sodium transporter. Nat Genet 37:1141–1146

SAS Institute Inc (2012) SAS® 9.4 System Options: Reference, 2nd edn. SAS Institute Inc., Cary, NC

Spielmeyer W, Ellis MH, Chandler PM (2002) Semidwarf (sd-1), "green revolution" rice, contains a defective gibberellin 20-oxidase gene. Proc Natl Acad Sci U S A 99(13):9043–9048

Spindel J, Wright M, Chen C, Cobb J, Gage J, Harrington S, Lorieux M, Ahmadi N, McCouch S (2013) Bridging the genotyping gap: using genotyping by sequencing (GBS) to add high-density SNP markers and new value to traditional bi-parental mapping and breeding populations. Theor Appl Genet 126:2699–2716

Sunkar R, Bartels D, Kirch HH (2003) Overexpression of a stress-inducible aldehyde dehydrogenase gene from Arabidopsis thaliana in transgenic plants improves stress tolerance. Plant J 35:452–464

Thomson MJ, de Ocampo M, Egdane J, Rahman MA, Sajise AG, Adorada DL, Tumimbang-Raiz E, Blumwald E, Seraj ZI, Singh RK, Gregorio GB, Ismail A (2010) Characterizing the Saltol quantitative trait locus for salinity tolerance in rice. Rice 3(2):148–160

Wang Z, Chen Z, Cheng J, Lai Y, Wang J, Bao Y, Huang J, Zhang H (2012) QTL analysis of Na^+ and K^+ concentrations in roots and shoots under different levels of NaCl stress in rice (Oryza sativa L.). PLoS One 7(12):e51202. doi:10.1371/journal.pone.0051202

Wang H, Hu Y, Pan Y, Yu D (2015) Arabidopsis VQ motif-containing proteins VQ12 and VQ29 negatively modulate basal defense against Botrytis cinerea. Sci Rep 5:14185. doi:10.1038/srep14185

Yan J, Zhu J, He C, Benmoussa M, Wu P (1998) Molecular dissection of developmental behavior of plant height in rice (Oryza sativa). Genetics 150:1257–1265

Yeo AR, Yeo ME, Flowers SA, Flowers TJ (1990) Screening of rice (Oryza sativa) genotypes for physiological characters contributing to salinity resistance, and their relationship to overall performance. Theor Appl Genet 79:377–384

Zeng L, Shannon MC, Lesch SM (2001) Timing of salinity stress affects rice growth and yield components. Agric Water Manage 48:191–206

Zhang X, Takano T, Liu S (2006) Identification of a mitochondiral ATP synthase small subunit gene (RMtATP6) expressed in response to salt and osmotic stresses in rice (*Oryza sativa L.*). J Exp Bot 57:193–200

Zheng X, Chen B, Lu G, Han B (2009) Overexpression of NAC transcription factor enhances rice drought and salt tolerance. Biochem Biophys Res Commun 379:985–989

Zhuang J, Lin H, Lu J, Qian H, Hittalmani S, Huang N, Zheng K (1997) Analysis of QTL x environment interaction for yield components and plant height in rice. Theor Appl Genet 95:799–808

4

Rice Chloroplast Genome Variation Architecture and Phylogenetic Dissection in Diverse *Oryza* Species Assessed by Whole-Genome Resequencing

Wei Tong[1], Tae-Sung Kim[1,2] and Yong-Jin Park[1,3]*

Abstract

Background: Chloroplast genome variations have been detected, despite its overall conserved structure, which has been valuable for plant population genetics and evolutionary studies. Here, we described chloroplast variation architecture of 383 rice accessions from diverse regions and different ecotypes, in order to mine the rice chloroplast genome variation architecture and phylogenetic.

Results: A total of 3677 variations across the chloroplast genome were identified with an average density of 27.33 per kb, in which wild rice showing a higher variation density than cultivated groups. Chloroplast genome nucleotide diversity investigation indicated a high degree of diversity in wild rice than in cultivated rice. Genetic distance estimation revealed that African rice showed a low level of breeding and connectivity with the Asian rice, suggesting the big distinction of them. Population structure and principal component analysis revealed the existence of clear clustering of African and Asian rice, as well as the *indica* and *japonica* in Asian cultivated rice. Phylogenetic analysis based on maximum likelihood and Bayesian inference methods and the population splits test suggested and supported the independent origins of *indica* and *japonica* within Asian cultivated rice. In addition, the African cultivated rice was thought to be domesticated differently from Asian cultivated rice.

Conclusions: The chloroplast genome variation architecture in Asian and African rice are different, as well as within Asian or African rice. Wild rice and cultivated rice also have distinct nucleotide diversity or genetic distance. In chloroplast level, the independent origins of *indica* and *japonica* within Asian cultivated rice were suggested and the African cultivated rice was thought to be domesticated differently from Asian cultivated rice. These results will provide more candidate evidence for the further rice chloroplast genomic and evolution studies.

Keywords: Chloroplast, African rice, Phylogenetic, Asian rice, Resequencing, Variation

Background

The chloroplast is maternally inherited in most angiosperms and possesses its own genome encoding many chloroplast-specific components (Hagemann 2010; Palmer et al. 1988; Sugiura 1989). The chloroplast has a circular genome, ranging in size from 39.4 to 200.8 kb among photosynthetic plant species (Kohler et al. 1997; Turmel et al. 1999). More than 800 eukaryotic *viridiplantae* chloroplast genomes have been described to date (http://www.ncbi.nlm.nih.gov/genomes/GenomesGroup.cgi?taxid=2759&opt=plastid). The chloroplast genome sequence of rice Nipponbare (*O. sativa* L. ssp. *japonica*) was reported to have a length of 134,525 bp (Hiratsuka et al. 1989). Chloroplasts contain both highly conserved genes fundamental to plant life and more variable regions, which have been informative over broad time scales. Comparative studies of the genomic architecture showed that the order of genes and the contents of essential genes are highly conserved among most chloroplast genomes (De Las Rivas et al. 2002; Kato et al. 2000). Nevertheless,

---
* Correspondence: yjpark@kongju.ac.kr
[1]Department of Plant Resources, College of Industrial Sciences, Kongju National University, Yesan 32439, Republic of Korea
[3]Center for Crop Genetic Resource and Breeding (CCGRB), Kongju National University, Cheonan 31080, Republic of Korea
Full list of author information is available at the end of the article

variations among different and closely related genomes have occurred during evolution (Provan et al. 1997; Tang et al. 2004).

The availability of rice nuclear (Goff et al. 2002; Yu et al. 2002) and chloroplast (Hiratsuka et al. 1989) reference genomes has enabled detailed studies of the origin, domestication, and phylogenetic relationships within this group. In particular, whole chloroplast genome analysis provides high-resolution plant phylogenies (Parks et al. 2009). Due to the high level of conservation, analysis of the chloroplast genome has become a valuable tool for plant phylogenetic studies (Waters et al. 2012; Yang et al. 2013). Previously, only a few chloroplast markers have been applied in studies of plant diversity and evolution (Ishii et al. 2001; King and Ferris 2000; Schroeder et al. 2011; Soejima and Wen 2006). From the conventional sequencing of plant chloroplast genomes to next-generation sequencing (NGS), it has become increasingly feasible to investigate the entire genome of the chloroplast, rather than targeting individual regions (McPherson et al. 2013; Nock et al. 2011; Straub et al. 2012). Whole chloroplast genome sequencing for phylogenetic analysis without prior isolation or amplification is now relatively straightforward for plant species (Nock et al. 2011). However, the chloroplast genome only represents the maternal evolutionary history. In addition, it also cannot be fully applied to rapidly diverging taxa, as the chloroplast has a slow rate of evolution (Moore et al. 2010; Parks et al. 2009). Therefore, chloroplast-based evolutionary studies must sometimes be complemented by nuclear genomic information.

Asian cultivated rice (*O. sativa* L.) is generally considered to have been domesticated from *Oryza rufipogon* several thousands of years ago (Cheng et al. 2003; Huang et al. 2012; Khush 1997; Oka 1988). However, there has been some debate regarding the origin of cultivated rice over the past several years, which centered on whether the two major rice cultivars, *O. sativa* L. ssp. *indica* and *japonica*, were derived from a single ancestor or were domesticated independently at different locations (Jin et al. 2008; Kawakami et al. 2007; Li et al. 2006; Molina et al. 2011; Zhang et al. 2009; Zhu et al. 2011; Huang et al. 2012; Xu et al. 2012; Zhu and Ge 2005). While, African cultivated rice (*O. glaberrima*), which was thought to be domesticated from the wild progenitor *O. barthii* ~3000 years ago, had been demonstrated to be domesticated in a single region along the Niger river with independent and distinct process in regard to Asian cultivated rice (Wang et al. 2014). A closer evolutionary relationship between *indica* and *aus* strains were observed using both nuclear and chloroplast genome data, as well as among the *tropical japonica*, *temperate japonica*, and *aromatic* groups (Garris et al. 2005). The *indica* subpopulation was shown to contain the highest degree of chloroplast diversity (Garris et al. 2005). Kim et al. (2014) evaluated 67 improved varieties and 13 landraces from the Democratic People's Republic of Korea (DPRK) at both nuclear and chloroplast levels, and they found a *temperate japonica* subgroup that was less diverse than the *indica* ancestor group at the nuclear level but more diverse at the chloroplast level (Kim et al. 2014). Whole chloroplast genome phylogenetic analysis revealed that the *Oryza nivara* is closed to *O. sativa* L. spp. *indica* and the *O. sativa* L. spp. *japonica* is closed to *Oryza rufipogon* in Asian cultivated and wild rice (Brozynska et al. 2014) and the African rice (*Oryza glaberrima* and *Oryza barthii*) were cluster together but in separate group with the Asian rice (Wambugu et al. 2015). Our previous studies indicated that the use of chloroplast genome variation to study diversity, population genetics, and phylogenetic analysis was quite convincing and also supported some previous outcomes (Tong et al. 2015). Despite these chloroplast-related studies, a large number of accessions must be applied to extend these studies from limited collections and specific varieties. In addition to rice, chloroplast genome-dependent phylogenetic analyses have also been performed in apple, tangerine, and other species. (Nikiforova et al. 2013; Carbonell-Caballero et al. 2015).

In the present study, a collection of 383 rice accessions with diverse ecotypes, including Asian cultivated and wild rice (*O. sativa* L. spp. *indica* and *japonica*, *Oryza rufipogon*, *Oryza nivara*) and African cultivated and wild rice (*Oryza glaberrima*, *Oryza barthii*) were selected to investigate the variation, diversity, and phylogenetic of rice chloroplast genome. The chloroplast genome of *O. rufipogo*n [Genbank: NC_017835], which is thought to be the immediate ancestral progenitor of cultivated rice, was chosen as the reference. Chloroplast variations in the collection were mined and subjected to comparative analysis among different groups. Diversity, population structure, and principal component analysis were also performed in the current collection. Phylogenetic analysis that conducted using the maximum likelihood (ML) and Bayesian inference (BI) methods and population splits evaluation were investigated, which could provide evidence to illustrate the phylogenetic relationships among rice subgroups, with a focus on Asian cultivated rice, as well as African rice (*Oryza glaberrima* and *Oryza barthii*). This report provides a further case study for the rice chloroplast genome, and the data generated here could be applied to further analyses of rice chloroplast evolution and genetics.

Results

Re-Sequencing and Variation Architecture Across the Chloroplast Genome

In this study, we re-sequenced 295 accessions of Asian cultivated rice with a high mean coverage (~7.34×), generating ~920Gbp raw sequence base with ~9.18 billion reads. After removing the low quality bases, a total of ~8.89 billion clean reads (with a clean read rate of 96.96 %) and ~860Gbp clean bases (with a clean base rate of 93.73 %) were obtained (Additional file 1: Table S1). Then this data was carried out for rice chloroplast genome variations detecting and phylogenetic analysis together with other 88 rice accessions.

Variations in 383 rice accessions, including 335 Asian cultivated rice (*O. sativa* L.), 10 Asian wild rice (*O. rufipogon, O. nivara*), 19 African cultivated rice (*Oryza glaberrima*), and 19 African wild rice (*Oryza barthii*), were characterized based on whole-genome resequencing data using the chloroplast genome of *O. rufipogon* as a reference. A total of 3677 variations, including 3592 SNPs and 85 indels (insertions/deletions), were identified in the whole collection (Table 1). A variation density of 27.33 per 1kb were observed through the total SNPs/indels. However, after excluding missing genotypes with MAF (Minor Allele Frequency) ≥ 0.01, high-quality (HQ) variations were dramatically decreased to 242, including 227 SNPs (93.8 % of the total HQ variations) and 15 indels (6.2 % of the total HQ variations) with a variation density of 1.8 per 1 kb (Table 1). The overall variations across the genome and groups specific variations were also extracted, suggesting that the African wild rice hold about 82.9 % of the total variations on its own (Fig. 1, Table 1). What's more, the distribution of the variations across the chloroplast genome is uneven (Fig. 1). Except the African wild rice, which harbored 2982 HQ variations (97.8 % of all variations), the Asian wild rice possessed the most number of HQ variations, even with only 10 accessions. Interestingly, a greatest abundance of

variations in the African wild rice were observed both in all and HQ variations among all of the groups, however, the African cultivated rice had minimal variations.

After determination of the location of variations across the genome, 2156 SNPs/indels were found within the genic region scattered over 87 genes in whole variations, including those encoding tRNAs and rRNAs (Table 1, Additional file 2: Table S2). Only 141 variations were retained by HQ selection in the genic region, involved 27 genes. In the four different groups, maximum variations in African wild rice were found both in all and HQ variations, as expected, which including 86 and 81 genes, respectively. In HQ variations, the Asian wild rice held the most number of genic variations and involved genes except the African wild rice.

Different allele types were also investigated, which indicated that T/C and A/T have the most number in all variations, while A/G and C/T are the major types in HQ variations. The overall Ts/Tv (Transition/Transversion) ratio in chloroplast genome of whole collection was 0.7328, which indicates that the mutations within the same type of nucleotide were less than those from a pyrimidine to a purine or vice versa (Additional file 3: Figure S1). In the four groups, the Asian wild rice holds the highest Ts/Tv ratio (1.047), while the African wild rice holds the lowest (0.7093).

Genetic Diversity Evaluation of Rice Chloroplast Genome

The nucleotide diversity (*pi*) of the whole collection and different groups (Asian cultivated and wild rice, African cultivated and wild rice) was calculated with a mean *pi* of 0.000918 in whole collection. While among the subgroups, the African wild rice has the highest diversity

Table 1 Summary of the total variations (SNPs/indels) detected in the germplasm and subgroups and the location distribution of the variations

Group	All variations				HQ variations[a]			
	Total	SNPs	Indels	Density/kb	Total	SNPs	Indels	Density/kb
Whole collection (383)[b]	3677	3592	85	27.33	242	227	15	1.8
Asian cultivated (332)[c]	723	671	52	5.37	308	281	27	2.29
Asian wild (10)	413	374	39	3.07	354	321	33	2.63
African cultivated (19)	418	385	33	3.11	280	255	25	2.08
African wild (19)	3049	3016	33	22.66	2982	2958	24	22.16
	Genic[d]	Intergenic		Genes[e]	Genic	Intergenic		Genes
Whole collection (383)	2156	1521		87	141	101		27
Asian cultivated (332)	312	411		45	149	159		28
Asian wild (10)	187	226		37	167	187		32
African cultivated (19)	171	247		35	108	172		29
African wild (19)	1948	1101		86	1915	1067		81

[a]HQ variation: High-quality variation, referring to variations excluding missing genotypes and MAF ≥ 0.01
[b]Subgroups in the whole collection. The numbers in brackets indicate the number of accessions
[c]Three mixed accessions belong to Group III, IV and V in the 50 cultivated and wild rice group were excluded here
[d]The genic region also includes tRNAs and rRNAs. [e]The total number of genes that the variations harbored

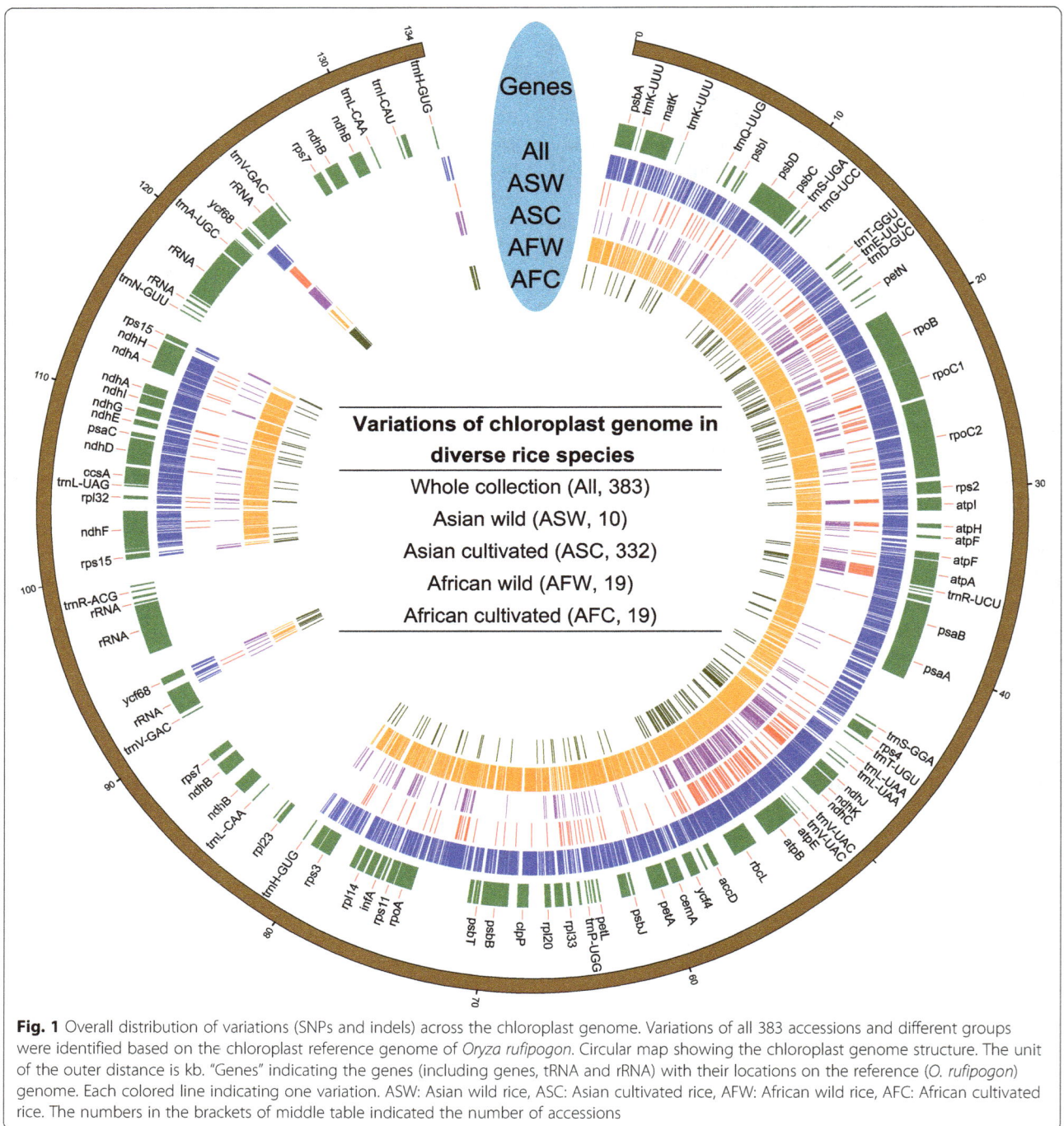

Fig. 1 Overall distribution of variations (SNPs and indels) across the chloroplast genome. Variations of all 383 accessions and different groups were identified based on the chloroplast reference genome of *Oryza rufipogon*. Circular map showing the chloroplast genome structure. The unit of the outer distance is kb. "Genes" indicating the genes (including genes, tRNA and rRNA) with their locations on the reference (*O. rufipogon*) genome. Each colored line indicating one variation. ASW: Asian wild rice, ASC: Asian cultivated rice, AFW: African wild rice, AFC: African cultivated rice. The numbers in the brackets of middle table indicated the number of accessions

(0.001959), and the African cultivated rice has the lowest (0.000548) (Fig. 2a, b, Additional file 4: Table S3). The Asian wild rice also holds a high *pi* (0.001665), and the Asian cultivated rice has the similar *pi* with whole collection (0.000987).

A long genetic distance (*Fst*) between Asian rice and African rice was observed (~0.43), which indicates the low levels of breeding and low connectivity between them (Fig. 2a). The African cultivated rice (*O. glaberrima*) has

very high breeding level with the African wild rice (*Oryza barthii*). These results may suggest and support the distinct domestication between African and Asian rice. Tajima's *D* value of the chloroplast genome was also examined for detection of balancing selection (Fig. 2c, Additional file 5: Table S4). The negative value indicated population size expansion and/or purifying selection, while a positive value indicated a decrease in population size and/or balancing selection. Values closer to 0 indicate

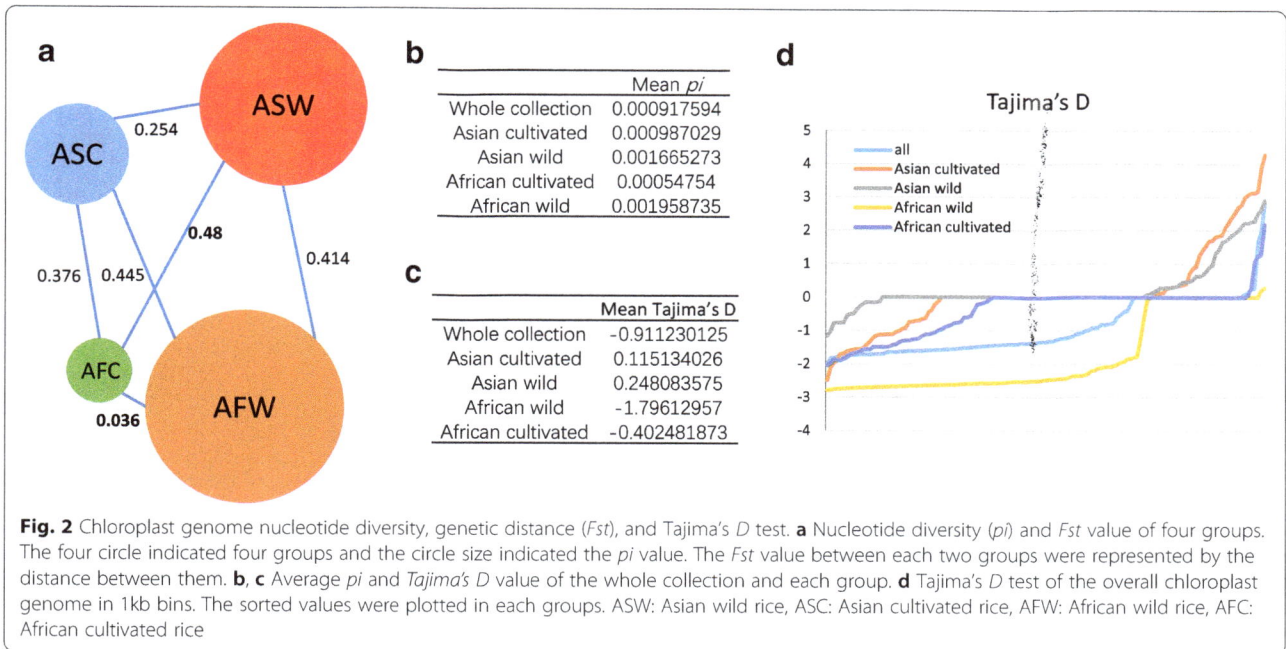

Fig. 2 Chloroplast genome nucleotide diversity, genetic distance (*Fst*), and Tajima's *D* test. **a** Nucleotide diversity (*pi*) and *Fst* value of four groups. The four circle indicated four groups and the circle size indicated the *pi* value. The *Fst* value between each two groups were represented by the distance between them. **b, c** Average *pi* and *Tajima's D* value of the whole collection and each group. **d** Tajima's *D* test of the overall chloroplast genome in 1kb bins. The sorted values were plotted in each groups. ASW: Asian wild rice, ASC: Asian cultivated rice, AFW: African wild rice, AFC: African cultivated rice

less evidence for the occurrence of selection. According to the distribution (Fig. 2d), Tajima's *D* value of all groups showed a location fluctuation in 1kb bins with positive, negative value and also 0. Excepting the African wild rice and whole collection, which showing more negative positions, other groups showing a relative even distribution of positive, negative and 0. The mean Tajima's D value of whole collection and different groups was shown in Fig. 2c, the whole collection (because of diverse rice accessions) and African wild rice showing a relative high divergence. While closer Tajima's *D* value to 0 indicated rare selection in the chloroplast genome.

Population Structure and Principal Component Analysis Based on Chloroplast Genome

The population structure of the whole collection was investigated based on the HQ variations using SRUCTURE, which estimates individual ancestry and admixture proportions assuming *K* populations. With increasing *K* (number of populations) values from 1 to 10 with 10 iterations each, we analyzed the population structure for each *K* value (Fig. 3a, from *K* = 2 to 4). We distinguished the major substructure groups using an optimal *K* value of 4 (highest Δ*K*, Additional file 6: Figure S2a). All the collected accessions formed four subpopulations, denoted as *indica* type, *japonica* type, Admixture, and African rice (wild and cultivated). In addition, a validation of population structure was conducted using ADMIXTURE from *K* = 1 to 10. With a cross-validation procedure, a good *K* value of 8 was adopted, which exhibited a lowest cross-validation error in all *K* values (Additional file 6: Figure S2b). The population structure form *K* = 4 to 8

was illustrated (Results in *K* = 2 and 3 were almost same using SREUCTURE and ADMIXTURE), which clustered the population into four subgroups (African cultivated and wild, Asian wild, and Asian cultivated with *indica* and *japonica* type) tightly (Fig. 3b, from *K* = 4 to 8). The results were consistent using two software, which indicates the clear separation of African and Asian rice. A similar clustering within the Asian cultivated rice (*indica* and *japonica*) was also observed, which actually also consistent with the clusters in nuclear genome test.

PCA using the whole variation data was conducted in TASSEL, with the first two PCs explaining more than 81.9 % of the proportional variance; therefore, we constructed the PCA using PC1 and PC2 (Fig. 3c). Four main groups were inferred, *indica* type, *japonica* type, African wild and cultivated rice, as well as several scattered accessions (Asian wild rice) and admixed among them. Multidimensional scaling (MDS) analysis was also conducted with TASSEL, which reveals four major groups that were almost same with PCA result (Fig. 3d). Even though no perfect clustering was found according to nuclear genome structure, these variations and the present case study also suggested that chloroplast genome-based analyses can be applied in population genetics studies.

Rice Phylogeny Based on the Chloroplast Genome

Phylogenetic analysis of the whole rice collection was performed using a ML iterative model-based method with a bootstrap of 1000 replicates to assess the reliability of the phylogeny reconstructed using PhyML. In parallel, phylogenetic analysis was also inferred using a

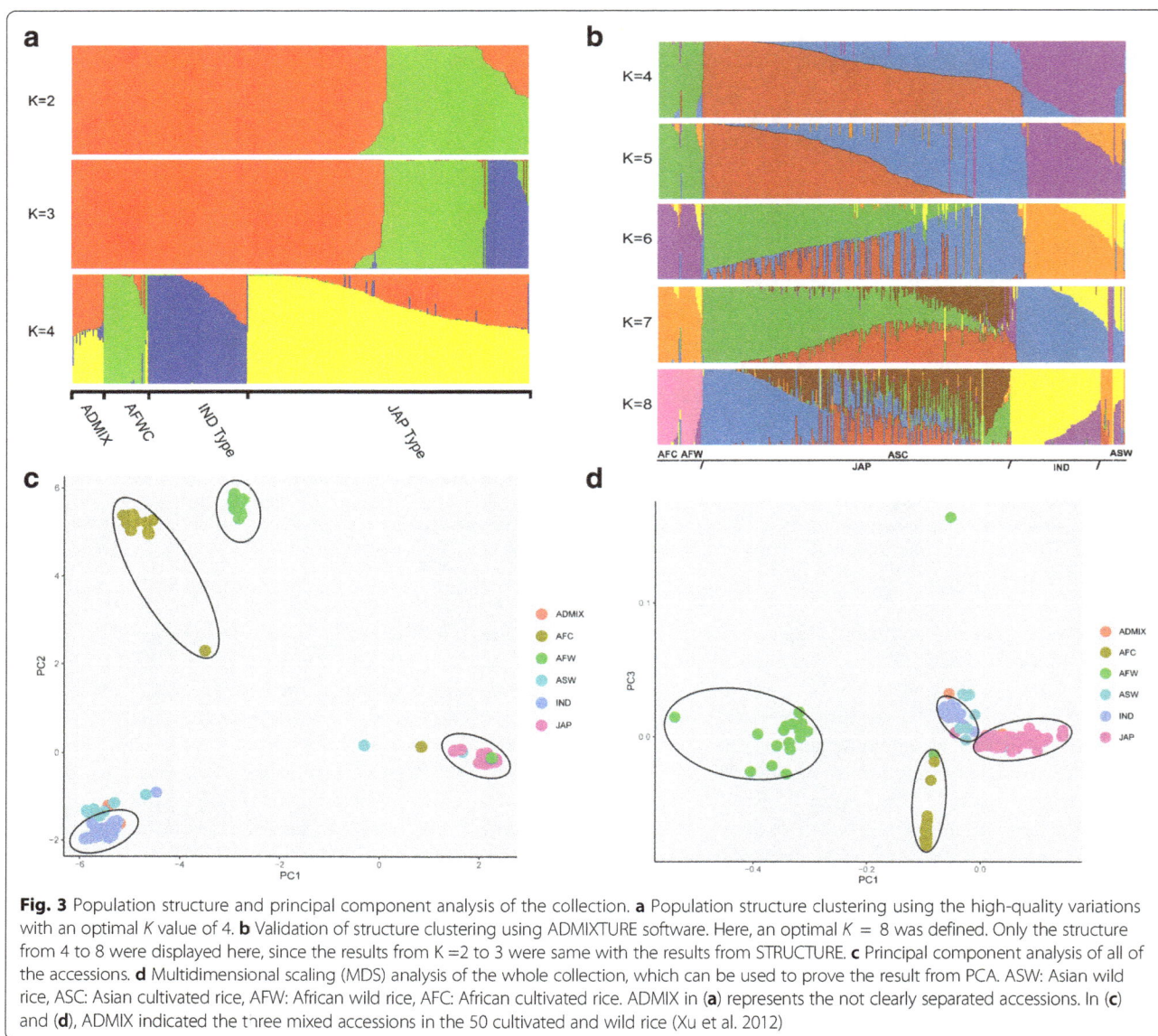

Fig. 3 Population structure and principal component analysis of the collection. **a** Population structure clustering using the high-quality variations with an optimal *K* value of 4. **b** Validation of structure clustering using ADMIXTURE software. Here, an optimal *K* = 8 was defined. Only the structure from 4 to 8 were displayed here, since the results from K =2 to 3 were same with the results from STRUCTURE. **c** Principal component analysis of all of the accessions. **d** Multidimensional scaling (MDS) analysis of the whole collection, which can be used to prove the result from PCA. ASW: Asian wild rice, ASC: Asian cultivated rice, AFW: African wild rice, AFC: African cultivated rice. ADMIX in (**a**) represents the not clearly separated accessions. In (**c**) and (**d**), ADMIX indicated the three mixed accessions in the 50 cultivated and wild rice (Xu et al. 2012)

Bayesian MCMC search method. The ML method suggested three clear groups (*indica* type, *japonica* type, and African rice), with the Asian wild rice scattered between *indica* and *japonica* (Fig. 4a). Most of the accessions showed clear separation into the *japonica* group, *indica* type, or African rice group, indicated by the clustering of the 50 cultivated and wild rice accessions. Similar phylogenetic results were also obtained using the BI method displayed in Fig. 4b. A comparison of the trees from two methods was implemented in a tanglegram, which reveals that the overall phylogenetic structure and clustering of the accessions in the two trees are nearly same (the same accession in two trees can connect with each other at the same location in the clusters), even the outward shape of the two trees are not well unified. The *indica*, *japonica*, and African rice groups showed almost

the same clustering in two methods, but the Asian wild rice showed closer with *indica* group in the BI method. From the results, we inferred that *indica* and *japonica* may have an independent domestication, as the Asian wild rice was clustered between them (4 of the wild rice are mixed inside the *indica* and *japonica* group). Meanwhile, it was obvious that African rice, including cultivated (*O. glaberrima*) and wild (*O. barthii*) were in an independent group, even the wild and cultivated are not well separated. It also can be inferred that *O. glaberrima* was from *O.barthii* and have an independent domestication process distinct with *O. sativa* L.

In the *TreeMix* test, the subpopulation relationships were evaluated among two subsets with four and six subpopulations, which revealed that the Asian cultivated rice (*indica* and *jaoponica*) may have different origin (Fig. 5a), since the

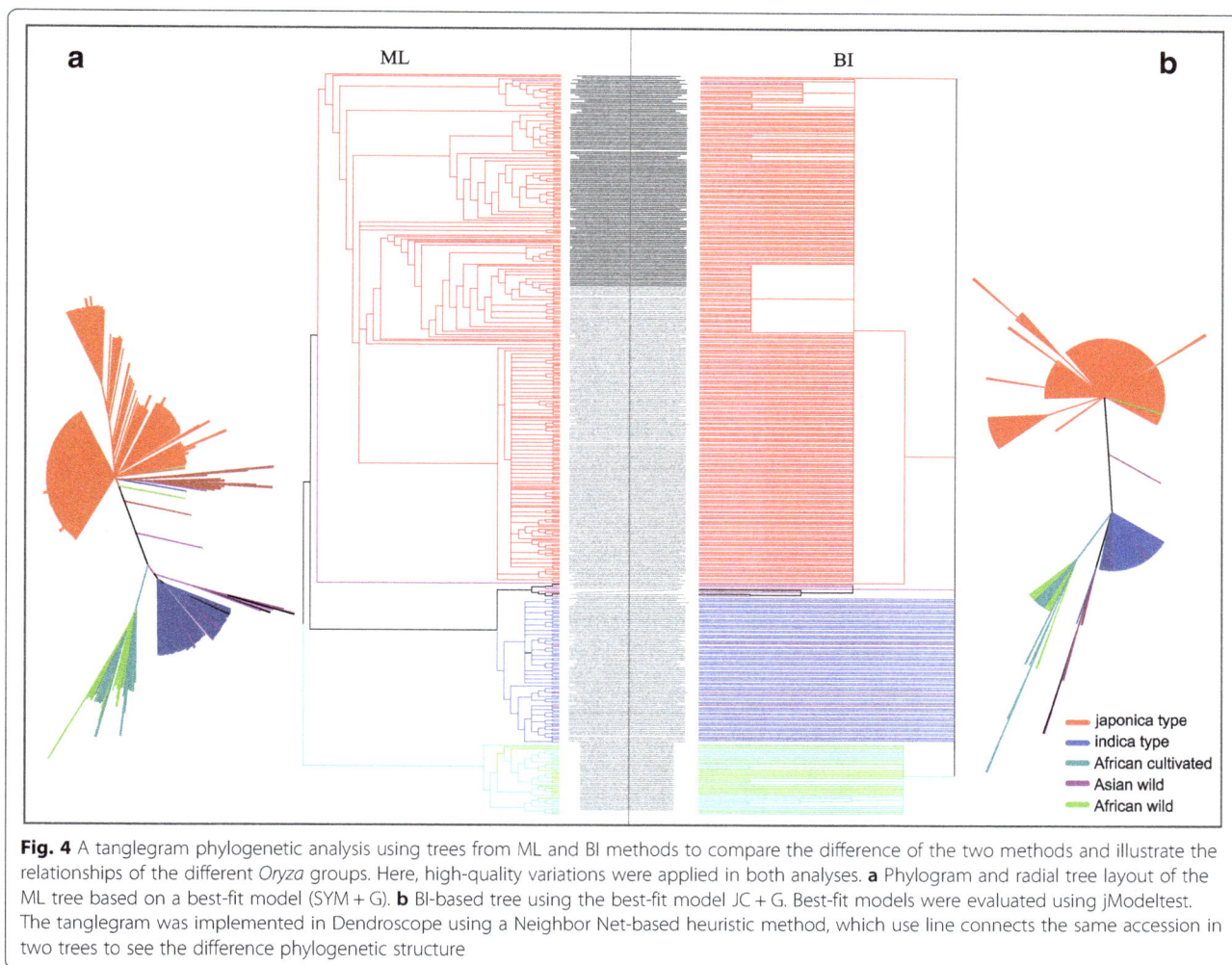

Fig. 4 A tanglegram phylogenetic analysis using trees from ML and BI methods to compare the difference of the two methods and illustrate the relationships of the different *Oryza* groups. Here, high-quality variations were applied in both analyses. **a** Phylogram and radial tree layout of the ML tree based on a best-fit model (SYM + G). **b** BI-based tree using the best-fit model JC + G. Best-fit models were evaluated using jModeltest. The tanglegram was implemented in Dendroscope using a Neighbor Net-based heuristic method, which use line connects the same accession in two trees to see the difference phylogenetic structure

two subgroups located on different side of the Asian wild rice. By evaluating the population splits between Asian and African rice, different domestication process can be inferred since very distinct clustering was observed (Fig. 5b). When six groups were applied, similar results were obtained, and in addition, the *indica* is closer to the *O. nivara* and the *japonica* is closer to *O.rufipogon* (Fig. 5c).

Together with the results of previous studies regarding the origins of rice, we concluded that *O. sativa* L. spp. *indica* may have evolved from *O. nivara*, and that *O. sativa* L. spp. *japonica* may be domesticated from *O. rufipogon*. Simultaneously, African cultivated rice may have a different and separated domestication process with Asian cultivated rice.

Discussion

Genetic Variation and Population Structure in Chloroplast Genome Level

Chloroplast DNA shows a much lower substitution rate than does nuclear DNA, which is significantly reduced

even in the inverted repeat regions (Wolfe et al. 1987). The overall sequence differences among rice subspecies varieties is ~130-fold higher in the nuclear than chloroplast genomes (0.12 %) (Yu et al. 2002). Therefore, in practice, detecting useful polymorphisms at the population level is difficult, due to the low substitution rates in plant chloroplast genomes. Highly accurate whole-genome sequencing and reference genome based assembly of chloroplast genome become a more economical approach and can be used for the further genomic studies (Wu et al. 2012). In this case, investigating the variations of chloroplast genome based on higher genome coverage sequencing could decrease the number of missing values and heterozygotes, and thus obtain more accurate results. In this report, we evaluated the chloroplast genome variations in a diverse collection of 383 rice accessions with relative high coverage re-sequencing, as well as the variation distribution in different groups (Table 1, Fig. 1). Intersection of variations in different groups was characterized, and only 130 variations were overlapped in four

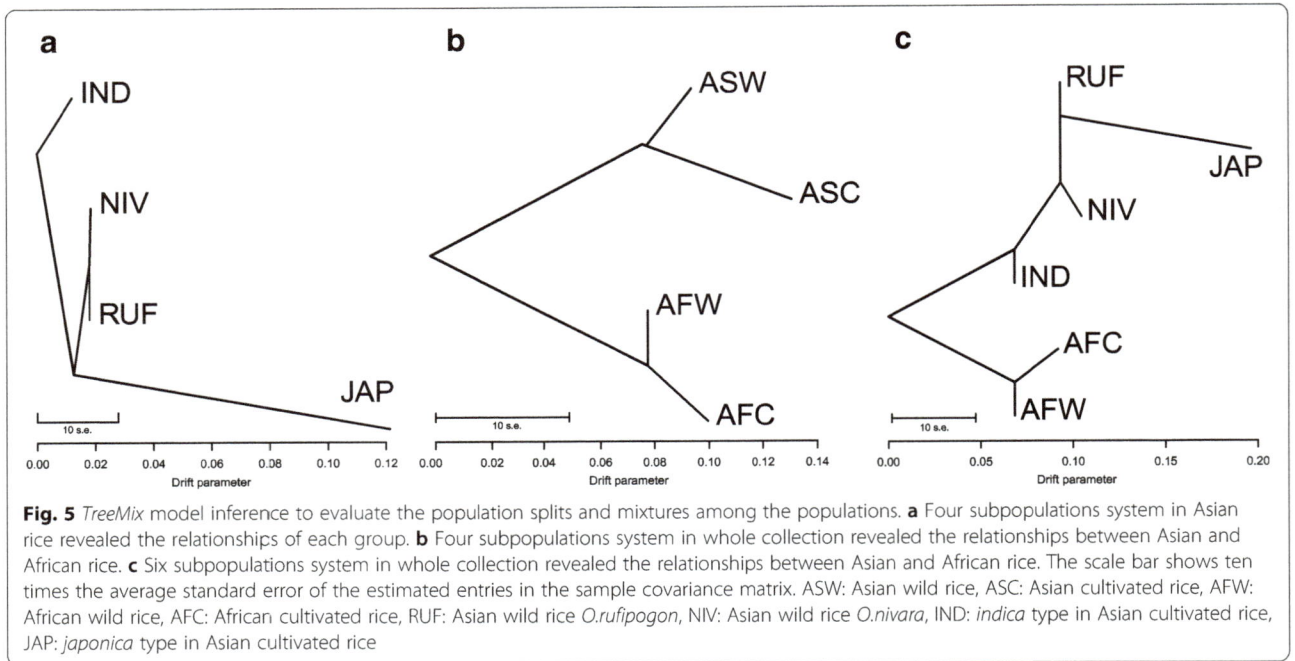

Fig. 5 *TreeMix* model inference to evaluate the population splits and mixtures among the populations. **a** Four subpopulations system in Asian rice revealed the relationships of each group. **b** Four subpopulations system in whole collection revealed the relationships between Asian and African rice. **c** Six subpopulations system in whole collection revealed the relationships between Asian and African rice. The scale bar shows ten times the average standard error of the estimated entries in the sample covariance matrix. ASW: Asian wild rice, ASC: Asian cultivated rice, AFW: African wild rice, AFC: African cultivated rice, RUF: Asian wild rice *O.rufipogon*, NIV: Asian wild rice *O.nivara*, IND: *indica* type in Asian cultivated rice, JAP: *japonica* type in Asian cultivated rice

groups. While the African wild rice shown much more total and unique variations than other groups, which may indicate the huge difference between African wild and Asian rice (Fig. 6). And very few overlaps were found that only in African cultivated &Asian wild &African wild, Asian wild &African wild, African cultivated &Asian wild. Besides, considering the HQ variations, the Asian wild rice has the most variations except the African wild rice, inferring that wild type has much higher diversity than cultivated type (Table 1). Moreover, the variations showed a

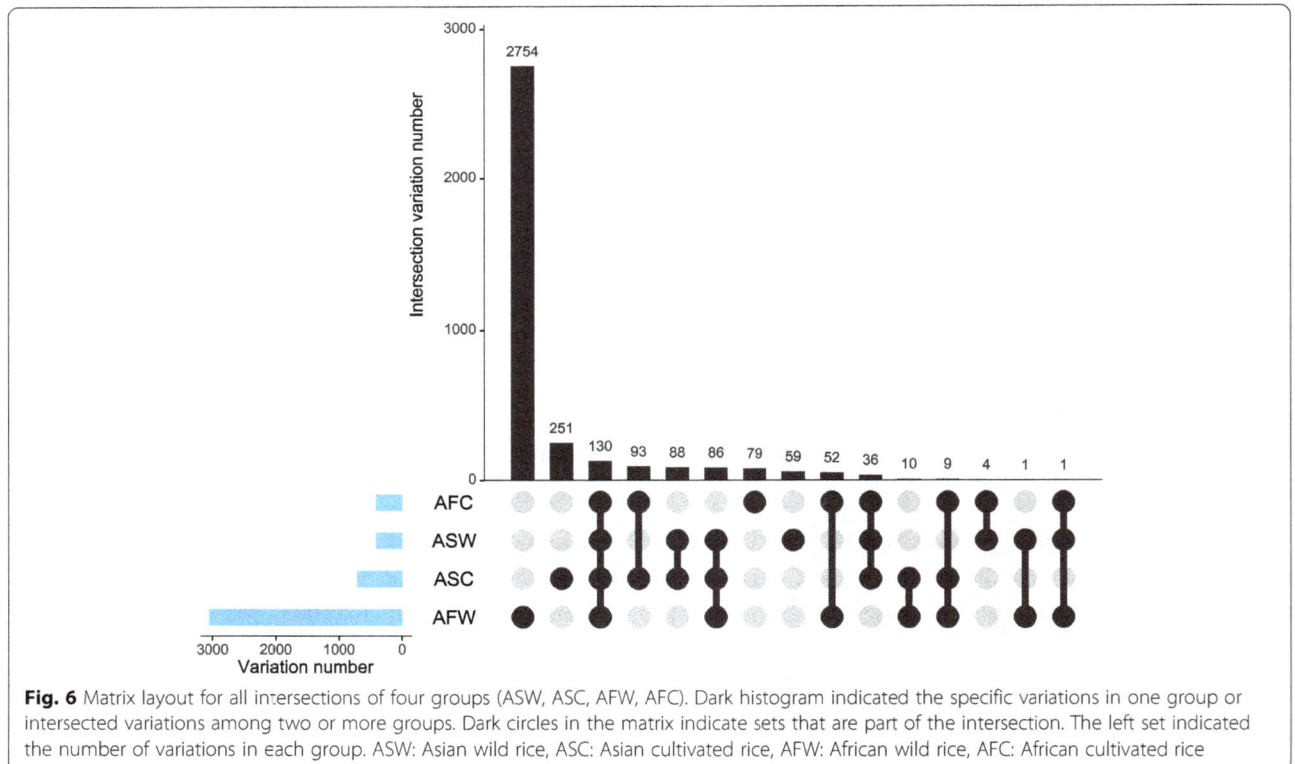

Fig. 6 Matrix layout for all intersections of four groups (ASW, ASC, AFW, AFC). Dark histogram indicated the specific variations in one group or intersected variations among two or more groups. Dark circles in the matrix indicate sets that are part of the intersection. The left set indicated the number of variations in each group. ASW: Asian wild rice, ASC: Asian cultivated rice, AFW: African wild rice, AFC: African cultivated rice

heterogeneity across the chloroplast genome, which leads to no variations in some specific regions (Fig. 1). The average *pi* of the overall genome was low (~0.0009), as were those in other groups, while the wild rice showed higher diversity than their cultivated type. A high *Fst* value (>0.37) was observed between the Asian and African rice, indicating their far genetics distance. Tajima's *D* test in chloroplast level of African rice showed a negative value, which may indicate some purifying selection or a signature of a recent population expansion. Whereas, the Asian rice that have a positive value may indicate an overdominant selection or population bottleneck.

The results of the population structure analysis indicated that population clustering based on chloroplast genomes was consistent with the results based on nuclear genomes in most accessions. Besides 2 admixed accessions from the 50 cultivated and wild rice group (Group III and IV), we also found 1 *indica* accession, 2 African accessions, are clustered into or close to *japonica* and several Asian wild accessions scattered between *indica* and *japonica* were observed to be closer to *indica* (Fig. 3c, d). Fortunately, African rice and Asian rice can be well grouped in most accessions at the chloroplast genome level, with well clustering of African wild and cultivated rice. We can infer that the African cultivated rice has distinct genetic background with Asian rice.

Chloroplast Genome Indicates Independent Origin of *Indica* and *Japonica*

In rice, the evolutionary rate of chloroplast DNA is three-fold higher than that of mitochondrial DNA (Tian et al. 2006). Therefore, its maternal inheritance and relatively high mutation rate are useful for elucidating the phylogeny of the species. The advent of NGS (next-generation sequencing) has allowed detection of substitutions in large populations both easily and accurately, leading to a better understanding in evolutionary studies. It may not be necessary to assemble whole chloroplast genomes for molecular ecology studies by exploring chloroplast variation (McPherson et al. 2013). Chloroplast DNA provides the advantage of a high copy number without recombination, which is a critical issue in nuclear genome-based phylogenetic studies (Poke et al. 2006; Takahashi et al. 2008). Interspecific hybridization can lead to chloroplast capture, whereby the plastome of one species introgresses into another, and this has been used to explain the inconsistencies between chloroplast and nuclear gene trees.

In spite of the debate of the domestication of Asian cultivated rice (*O. sativa* L.), which focus on whether the two major subgroups were in single (Huang et al. 2012; Molina et al. 2011) or independent origins (Londo et al. 2006; Ma and Bennetzen 2004; Vitte et al. 2004; Yang et al. 2012; Zhu and Ge 2005), new opinion about three

geographically separate domestications of Asian rice had been proposed recently (Civáň et al. 2015). From these different data and analyzing methods, different or entirely opposite result have obtained. According to Civáň et al. (2015), they got different results using the same data from Huang et al. (2012), which suggested that extreme complicated issues might happen during the long history of rice domestication in nuclear genome level complemented with the response points from Huang and Han (2015) against the new analyzing results. While the chloroplast genome can narrow down this problem for its non-recombination and high level of conservation.

Londo et al. (2006) detected the haplotype network of chloroplast atpB-rbcL region, they concluded that *japonica* rice is less diverse than *indica* rice and also demonstrated that *O. sativa* L. was domesticated from *O. rufipogon* at least twice (Khush 1997; Londo et al. 2006). A recent study of the wild and domesticated rice AA genome species using whole chloroplast genome sequences indicated that the *O. rufipogon* (Asian) and *O. nivara* are always separately clustered with *japonica* and *indica*, respectively (Wambugu et al. 2015). By applied a chloroplast genome-wide variation analysis in current report, we found the evidence support the independent domestication of Asian cultivated rice, *O. sativa* L. spp. *indica* and *japonica*, which were thought to be originated from *O. nivara* and *O. rufipogon*, respectively. According to a chloroplast whole genome sequence investigation from several references, we previously found that the *indica* and *japonica* were closer with *O. nivara* and *O. rufipogon*, respectively, which also indicated the independent origin of Asian cultivated rice (Tong et al. 2015). Population structure of a KRICE_CORE set, which hold 137 accessions in current collection, also supported the hypothesis of the independent origin of *indica* and *japonica* in nuclear genome (Kim et al. 2016). However, as we mentioned previously, the chloroplast genome only represents the maternal evolutionary history, which cannot be fully applied to rapidly diverging taxa. Whereas, in nuclear genome level, different dataset applied with different method sometimes generated different results. Therefore, in some cases, chloroplast genome based evolutionary studies should be complemented with nuclear genome data, and vice versa, to obtain more reliable results.

Evidence for Distinct Domestication of African Cultivated Rice

O. glaberrima was thought to be independently domesticated from the wild progenitor, *Oryza barthii*, ~3000 years ago (Sweeney and McCouch 2007), which is 6000–7000 years after the domestication of Asian rice (*O. sativa* L.) (Vaughan et al. 2008). *O. glaberrima* was domesticated in a single region along the Niger River, as opposed to noncentric domestication events across Africa, which has experienced geographically and culturally distinct domestication processes (Wang et al. 2014). Here, we presented evidence supporting the domestication of *O. glaberrima*, as well as insights into the genetic distance and population structure analyses of the chloroplast genome. In chloroplast *Fst* analyses, the genetic distance value (*Fst*) of African rice, especially African cultivated rice (*O. glaberrima*) with Asian rice was much higher than the distances between the other groups (Fig. 2a), indicating a low level of breeding with the Asian rice. While, a very low value was observed between African cultivated and wild rice. What observed in the population structure and principal component analyses were that African rice always classified as a separated group (Fig. 3c, d), which also can be clearly seen in the phylogenetic trees using both ML and BI methods and in the TreeMix test (Figs. 4 and 5). One more thing we need to note is that in African rice, the cultivated and wild type are not well grouped into only two clusters but intersect, which was also observed in the nuclear genome analysis conducted by Wang et al. (2014). Even though, same conclusion can be inferred according the present result. These observations indicated that *O. glaberrima* was distant from Asian rice and had a distinct domestication process at chloroplast genome level.

Conclusions

In current report, we described chloroplast variation architecture of 383 rice accessions from diverse regions and different ecotypes. A total of 3677 variations across the chloroplast genome were identified. The chloroplast genome variation architecture in Asian and African rice are different, as well as within Asian or African rice. Wild rice and cultivated rice also have distinct nucleotide diversity or genetic distance. Chloroplast genome nucleotide diversity and genetic distance were investigated, indicated a high degree of diversity in wild rice than in cultivated rice. African rice showed a low level of breeding and connectivity with the Asian rice, suggesting the big distinction of them. Population structure and principal component analysis revealed the existence of clear clustering of African and Asian rice, as well as the *indica* and *japonica* in Asian cultivated rice. Phylogenetic analysis and the population splits test suggested and supported the independent origins of *indica* and *japonica* within Asian cultivated rice. In addition, the African cultivated rice was thought to be domesticated differently from Asian cultivated rice. We hope these results could provide more candidate evidence for the further rice chloroplast genomic and evolution studies.

Methods
Samples and Whole-Genome Resequencing
A core set containing 137 rice accessions with diverse types (landrace, weedy, cultivated) previously generated

from worldwide varieties collected from the National Genebank of the Rural Development Administration (RDA-Genebank, Republic of Korea) using the program PowerCore (Kim et al. 2007; Zhao et al. 2010; Kim et al. 2016) and 158 bred accessions were selected and sequenced for chloroplast genomic evaluation (Additional file 7: Table S5). In addition, 50 accessions of cultivated and wild rice developed by Xu et al. (Xu et al. 2012) and 19 accessions of African cultivated rice (*O. glaberrima*) and 19 accessions of African wild rice (*O. barthii*) (Wang et al. 2014) were also combined in the present study (Additional file 8: Table S6). Raw data from the 50 cultivated and wild rice, 19 African cultivated rice, and 19 African wild rice accessions were downloaded from the European Nucleotide Archive (http://www.ebi.ac.uk/ena) under accession numbers [SRA023116, SRP038750, and SRP037996] respectively.

For our germplasm (295 accessions with diverse origin), young leaves from a single plant were sampled and stored at −80°C prior to genomic DNA extraction using the DNeasy Plant Mini Kit (Qiagen). Qualified DNA was used for whole-genome resequencing of the collected rice varieties (295 accessions), with an average coverage of approximately 7.34× on the Illumina HiSeq 2000 Sequencing Systems Platform (Illumina Inc.).

Data preparation, Identification of Variation, and Statistics
Resequencing raw data (Fastq format) of all the accessions were trimmed using Sickle v1.2 (Joshi and Fass 2011) to remove low-quality reads. BWA v0.6.2 (Li and Durbin 2009) was used to align the raw data to the *O. rufipogon* chloroplast genome sequence. A Sequence Alignment/Map (SAM) file was created during the mapping and converted to a binary SAM (BAM) file with sorting. Removal of duplicates and addition of read group IDs were performed using Picard Tools v1.88 (https://broadinstitute.github.io/picard/). Final realignment and identification of variation were performed using GATK software v3 (McKenna et al. 2010). The variant call format file describing the variation result was processed by two python scripts, generating a HapMap (Haplotype Map) file.

Statistical analyses were performed to summarize the number and distribution of single nucleotide polymorphisms (SNPs) and indels (insertions and deletions) based on the HapMap file. The positions of high-quality (HQ, sites without missing and MAF ≥ 0.01, determined by the smallest group number 5 both in *O. nivara* and *O. rufipogon*) SNPs and indels in this population and subgroups were established according to the reference genome of *O. rufipogon*. For the Asian cultivated group, three admixed accessions in the 50 cultivated and wild rice were excluded for the further subgroup comparative analyses.

Chloroplast Genome Diversity Architecture

Analyses of chloroplast genome nucleotide diversity (pi), population divergence (Fst value), Ts/Tv (Transition/ Transversion ratio) and Tajima's D value were conducted using VCFtools (Danecek et al. 2011). Assessments of these calculations in whole collection and different sub-groups (Asian cultivated and wild, African cultivated and wild) were performed using VCFtools with a sliding window 1000 bp in length and a 500-bp step size.

Population Structure and Principal Component Analysis

The population structures of the collection were investigated using the model-based program STRUCTURE v2.3.4 (Pritchard et al. 2000) with a burn-in period length of 100,000 and a Markov chain Monte Carlo (MCMC) rep number of 200,000, which implements a Bayesian approach to identify subpopulations with distinct allelic frequencies and places individuals into K clusters. The distribution of L (K) revealed a continuously increasing curve without a clear maximum for true K. To overcome these difficulties in identifying the true K value, an ad hoc quantity (ΔK) was calculated based on the second-order rate of change of likelihood (ΔK) using the software Structure Harvest (Evanno et al. 2005; Earl 2012). Besides, the population structure was also validated using another model-based software ADMIXTURE (Alexander et al. 2009). By using ADMIXTURE's cross-validation procedure, a good value of K can be obtained, which will exhibit a low cross-validation error compared to other K values. Principal component analysis (PCA) and multidimensional scaling (MDS) was conducted using TASSEL 5 (Bradbury et al. 2007), which could provide more evidence and complement the population structure analyses. MDS produces results that are similar to PCA but starts with a distance matrix and results in coordinate axes that are scaled differently.

Chloroplast-Based Phylogenetic and Population Splits

ML and BI methods were applied to construct a phylogenetic tree for all 383 accessions. Briefly, appropriate nucleotide substitution models were assessed using jModeltest 2.1.7 (Darriba et al. 2012). A phylogenetic tree was conducted using PhyML 3.0 (Guindon et al. 2010) complemented by the best nucleotide substitution model SYM + G (symmetrical model + gamma distribution) selected by the hierarchical LRT (Hierarchical Likelihood Ratio Test) (Felsenstein 1988) and the Akaike Information Criterion (AIC) (Akaike 1974) with 1000 bootstrap replicates. A Bayesian tree was constructed using MrBayes 3.2.5 (Ronquist et al. 2012) implemented with a Bayesian MCMC search, with two parallel runs of 2 million generations and four chains each. Best-fit model JC + G (Jukes-Cantor + gamma distribution) were selected according to the Bayesian Information Criterion (BIC) (Schwarz 1978).

The phylogenetic tree was displayed and modified using Figtree v1.4.2 (http://tree.bio.ed.ac.uk/software/figtree/). The consensus tree of the bootstrap in the ML method was integrated using Phylip software (Phylogeny Inference Package v3.695, http://evolution.genetics.washington.edu/phylip.html). A tanglegram for two trees was implemented in Dendroscope (Huson and Scornavacca 2012) using a Neighbor Net-based heuristic, which is one good way to visualize similarities and differences between two phylogenetic trees side by side connected with lines between taxa that correspond to each other.

Additionally, a *TreeMix* model for inferring the set of population splits and mixtures in the history of a set of populations was performed using genome-wide allele frequency data in TreeMix (Pickrell and Pritchard 2012). In the collection, four (African cultivated and wild, Asian cultivated and wild, as well as the four groups of Asian rice) and six subpopulations (African wild and cultivated rice, Asian wild and cultivated rice, and the *indica* and *japonica* groups in Asian cultivated rice) were implemented to identify the relationships among the populations.

Additional Files

Additional file 1: Table S1. Summary of the 295 rice whole genome re-sequencing.

Additional file 2: Table S2. Location of all the SNPs and Indels and their gene region in the reference detected in this study.

Additional file 3: Figure S1. Overall Ts/Tv (Transition/ Transversion ratio) in 1kb bins of the whole collection and different groups. ASW: Asian wild rice, ASC: Asian cultivated rice, AFW: African wild rice, AFC: African cultivated rice.

Additional file 4: Table S3. Nucleotide diversity of the overall chloroplast genome with a 1000bp sliding window and 500bp step size.

Additional file 5: Table S4. Overall Tajima's D testing of the chloroplast genome in a 1kb bin.

Additional file 6: Figure S2. Magnitude of ΔK as a function of K and cross-validation error estimation to find the optimal K value for the population structure in STRUCTURE and ADMIXTURE. In this case, the maximum value of ΔK for all of the accessions was identified as $K = 4$ in STRUCTURE. While a lowest error value in K = 8 was identified in ADMIXTURE. But the values were similar from K = 5 to 10.

Additional file 7: Table S5. The 295 accessions information sequenced by ourselves and subpopulation designations used in this study.

Additional file 8: Table S6. Fifty cultivated and wild rice accessions and 38 African rice (including 19 African cultivated rice and 19 African wild rice) accessions used in the chloroplast genome study.

Acknowledgements

This work was supported by a grant from the BioGreen 21 Program (No. PJ01116101), Rural Development Administration, Republic of Korea. We also thank the anonymous reviewers for their helpful comments.

Authors' Contributions

YP led and conceived the manuscript. YP, TK developed the idea, editing and revised the manuscript. WT did the data analysis and wrote the text of the manuscript. All authors have read and approved the final manuscript.

Competing Interests
The authors declare that they have no competing interests.

Author details
[1]Department of Plant Resources, College of Industrial Sciences, Kongju National University, Yesan 32439, Republic of Korea. [2]Department of Agricultural Sciences, College of Natural Sciences, Korea National Open University, Seoul 03087, Republic of Korea. [3]Center for Crop Genetic Resource and Breeding (CCGRB), Kongju National University, Cheonan 31080, Republic of Korea.

References
Akaike H (1974) A new look at the statistical model identification. IEEE Trans Automatic Control 19(6):716–723

Alexander DH, Novembre J, Lange K (2009) Fast model-based estimation of ancestry in unrelated individuals. Genome Res 19:1655–1664

Bradbury PJ, Zhang Z, Kroon DE, Casstevens TM, Ramdoss Y, Buckler ES (2007) TASSEL: software for association mapping of complex traits in diverse samples. Bioinformatics 23(19) 2633–2635. doi:10.1093/bioinformatics/btm308

Brozynska M, Omar ES, Furtado A, Crayn D, Simon B, Ishikawa R, Henry RJ (2014) Chloroplast genome of novel rice germplasm identified in Northern Australia. Trop Plant Biol 7(3-4):111–120. doi:10.1007/s12042-014-9142-8

Carbonell-Caballero J, Alonso R, Ibanez V, Terol J, Talon M, Dopazo J (2015) A phylogenetic analysis of 34 chloroplast genomes elucidates the relationships between wild and domestic species within the genus citrus. Mol Biol Evol 32(8):2015–2035. doi:10.1093/molbev/msv082

Cheng C, Motohashi R, Tsuchimoto S, Fukuta Y, Ohtsubo H, Ohtsubo E (2003) Polyphyletic origin of cultivated rice: based on the interspersion pattern of SINEs. Mol Biol Evol 20(1):67–75

Civáň P, Craig H, Cox CJ, Brown TA (2015) Three geographically separate domestications of Asian rice. Nat Plants 1:15164

Danecek P, Auton A, Abecasis G, Albers CA, Banks E, DePristo MA, Handsaker RE, Lunter G, Marth GT, Sherry ST, McVean G, Durbin R, Genomes Project Analysis G (2011) The variant call format and VCFtools. Bioinformatics 27(15): 2156–2158. doi:10.1093/bioinformatics/btr330

Darriba D, Taboada GL, Doallo R, Posada D (2012) jModelTest 2: more models, new heuristics and parallel computing. Nat Methods 9(8):772. doi:10.1038/nmeth.2109

De Las Rivas J, Lozano JJ, Ortiz AR (2002) Comparative analysis of chloroplast genomes: functional annotation, genome-based phylogeny, and deduced evolutionary patterns. Genome Res 12(4):567–583. doi:10.1101/gr.209402

Earl DA (2012) STRUCTURE HARVESTER: a website and program for visualizing STRUCTURE output and implementing the Evanno method. Conserv Genet Resour 4(2):359–361

Evanno G, Regnaut S, Goudet J (2005) Detecting the number of clusters of individuals using the software STRUCTURE: a simulation study. Mol Ecol 14(8): 2611–2620. doi:10.1111/j.1365-294X.2005.02553.x

Felsenstein J (1988) Phylogenies from molecular sequences: inference and reliability. Annu Rev Genet 22:521–565. doi:10.1146/annurev.ge.22.120188.002513

Garris AJ, Tai TH, Coburn J, Kresovich S, McCouch S (2005) Genetic structure and diversity in Oryza sativa L. Genetics 169(3):1631–1638. doi:10.1534/genetics.104.035642

Goff SA, Ricke D, Lan TH, Presting G, Wang R, Dunn M, Glazebrook J, Sessions A, Oeller P, Varma H, Hadley D, Hutchison D, Martin C, Katagiri F, Lange BM, Moughamer T, Xia Y, Budworth P, Zhong J, Miguel T, Paszkowski U, Zhang S, Colbert M, Sun WL, Chen L, Cooper B, Park S, Wood TC, Mao L, Quail P, Wing R, Dean R, Yu Y, Zharkikh A, Shen R, Sahasrabudhe S, Thomas A, Cannings R, Gutin A, Pruss D, Reid J, Tavtigian S, Mitchell J, Eldredge G, Scholl T, Miller RM, Bhatnagar S, Adey N, Rubano T, Tusneem N, Robinson R, Feldhaus J, Macalma T, Oliphant A, Briggs S (2002) A draft sequence of the rice genome (Oryza sativa L. ssp. japonica). Science 296(5565):92–100. doi:10.1126/science.1068275

Guindon S, Dufayard JF, Lefort V, Anisimova M, Hordijk W, Gascuel O (2010) New algorithms and methods to estimate maximum-likelihood phylogenies: assessing the performance of PhyML 3.0. Syst Biol 59(3): 307–321. doi:10.1093/sysbio/syq010

Hagemann R (2010) The foundation of extranuclear inheritance: plastid and mitochondrial genetics. Mol Genet Genomics 283(3):199–209. doi:10.1007/s00438-010-0521-z

Hiratsuka J, Shimada H, Whittier R, Ishibashi T, Sakamoto M, Mori M, Kondo C, Honji Y, Sun CR, Meng BY et al (1989) The complete sequence of the rice

(Oryza sativa) chloroplast genome: intermolecular recombination between distinct tRNA genes accounts for a major plastid DNA inversion during the evolution of the cereals. Mol Gen Genet 217(2-3):185–194

Huang X, Han B (2015) Rice domestication occurred through single origin and multiple introgressions. Nat Plants 1:15207

Huang X, Kurata N, Wei X, Wang ZX, Wang A, Zhao Q, Zhao Y, Liu K, Lu H, Li W, Guo Y, Lu Y, Zhou C, Fan D, Weng Q, Zhu C, Huang T, Zhang L, Wang Y, Feng L, Furuumi H, Kubo T, Miyabayashi T, Yuan X, Xu Q, Dong G, Zhan Q, Li C, Fujiyama A, Toyoda A, Lu T, Feng Q, Qian Q, Li J, Han B (2012) A map of rice genome variation reveals the origin of cultivated rice. Nature 490(7421): 497–501. doi:10.1038/nature11532

Huson DH, Scornavacca C (2012) Dendroscope 3: an interactive tool for rooted phylogenetic trees and networks. Syst Biol 61(6):1061–1067. doi:10.1093/sysbio/sys062

Ishii T, Xu Y, McCouch S (2001) Nuclear-and chloroplast-microsatellite variation in A-genome species of rice. Genome 44(4):658–666

Jin J, Huang W, Gao JP, Yang J, Shi M, Zhu MZ, Luo D, Lin HX (2008) Genetic control of rice plant architecture under domestication. Nat Genet 40(11): 1365–1369. doi:10.1038/ng.247

Joshi N, Fass J (2011) Sickle: A sliding-window, adaptive, quality-based trimming tool for FastQ files (Version 1.33) [Software]. Available at https://github.com/najoshi/sickle

Kato T, Kaneko T, Sato S, Nakamura Y, Tabata S (2000) Complete structure of the chloroplast genome of a legume, Lotus japonicus. DNA Res 7(6):323–330

Kawakami S, Ebana K, Nishikawa T, Sato Y, Vaughan DA, Kadowaki K (2007) Genetic variation in the chloroplast genome suggests multiple domestication of cultivated Asian rice (Oryza sativa L.). Genome 50(2):180–187. doi:10.1139/g06-139

Khush GS (1997) Origin, dispersal, cultivation and variation of rice. Plant Mol Biol 35(1-2):25–34

Kim H, Jeong EG, Ahn S-N, Doyle J, Singh N, Greenberg AJ, Won YJ, McCouch SR (2014) Nuclear and chloroplast diversity and phenotypic distribution of rice (Oryza sativa L.) germplasm from the democratic people's republic of Korea (DPRK; North Korea). Rice 7(1):1–15

Kim KW, Chung HK, Cho GT, Ma KH, Chandrabalan D, Gwag JG, Kim TS, Cho EG, Park YJ (2007) PowerCore: a program applying the advanced M strategy with a heuristic search for establishing core sets. Bioinformatics 23(16):2155–2162. doi:10.1093/bioinformatics/btm313

Kim T-S, He Q, Kim K-W, Yoon M-Y, Ra W-H, Li FP, Tong W, Yu J, Oo WH, Choi B (2016) Genome-wide resequencing of KRICE_CORE reveals their potential for future breeding, as well as functional and evolutionary studies in the post-genomic era. BMC Genomics 17(1):1

King RA, Ferris C (2000) Chloroplast DNA and nuclear DNA variation in the sympatric alder species, Alnus cordata (Lois.) Duby and A. glutinosa (L.) Gaertn. Biol J Linn Soc 70(1):147–160

Kohler S, Delwiche CF, Denny PW, Tilney LG, Webster P, Wilson RJ, Palmer JD, Roos DS (1997) A plastid of probable green algal origin in Apicomplexan parasites. Science 275(5305):1485–1489

Li C, Zhou A, Sang T (2006) Rice domestication by reducing shattering. Science 311(5769):1936–1939. doi:10.1126/science.1123604

Li H, Durbin R (2009) Fast and accurate short read alignment with Burrows-Wheeler transform. Bioinformatics 25(14):1754–1760. doi:10.1093/bioinformatics/btp324

Londo JP, Chiang YC, Hung KH, Chiang TY, Schaal BA (2006) Phylogeography of Asian wild rice, Oryza rufipogon, reveals multiple independent domestications of cultivated rice, Oryza sativa. Proc Natl Acad Sci U S A 103(25):9578–9583. doi:10.1073/pnas.0603152103

Ma J, Bennetzen JL (2004) Rapid recent growth and divergence of rice nuclear genomes. Proc Natl Acad Sci U S A 101(34):12404–12410. doi:10.1073/pnas.0403715101

McKenna A, Hanna M, Banks E, Sivachenko A, Cibulskis K, Kernytsky A, Garimella K, Altshuler D, Gabriel S, Daly M, DePristo MA (2010) The Genome Analysis Toolkit: a MapReduce framework for analyzing next-generation DNA sequencing data. Genome Res 20(9):1297–1303. doi:10.1101/gr.107524.110

McPherson H, van der Merwe M, Delaney SK, Edwards MA, Henry RJ, McIntosh E, Rymer PD, Milner ML, Siow J, Rossetto M (2013) Capturing chloroplast variation for molecular ecology studies: a simple next generation sequencing approach applied to a rainforest tree. BMC Ecol 13:8. doi:10.1186/1472-6785-13-8

Molina J, Sikora M, Garud N, Flowers JM, Rubinstein S, Reynolds A, Huang P, Jackson S, Schaal BA, Bustamante CD, Boyko AR, Purugganan MD (2011) Molecular evidence for a single evolutionary origin of domesticated rice.

Proc Natl Acad Sci U S A 108(20):8351–8356. doi:10.1073/pnas.1104686108

Moore MJ, Soltis PS, Bell CD, Burleigh JG, Soltis DE (2010) Phylogenetic analysis of 83 plastid genes further resolves the early diversification of eudicots. Proc Natl Acad Sci 107(10):4623–4628

Nikiforova SV, Cavalieri D, Velasco R, Goremykin V (2013) Phylogenetic analysis of 47 chloroplast genomes clarifies the contribution of wild species to the domesticated apple maternal line. Mol Biol Evol 30(8):1751–1760. doi:10.1093/molbev/mst092

Nock CJ, Waters DL, Edwards MA, Bowen SG, Rice N, Cordeiro GM, Henry RJ (2011) Chloroplast genome sequences from total DNA for plant identification. Plant Biotechnol J 9(3):328–333. doi:10.1111/j.1467-7652.2010.00558.x

Oka HI (1988) Origin of cultivated rice. Developments in crop science, vol 14. Japan Scientific Societies Press; Elsevier; Exclusive sales rights for the U.S.A. and Canada, Elsevier Science Pub. Co., Tokyo Amsterdam Netherlands; New York, N.Y.

Palmer JD, Jansen RK, Michaels HJ, Chase MW, Manhart JR (1988) Chloroplast DNA variation and plant phylogeny. Ann Mo Bot Gard 75(4):1180–1206

Parks M, Cronn R, Liston A (2009) Increasing phylogenetic resolution at low taxonomic levels using massively parallel sequencing of chloroplast genomes. BMC Biol 7(1):84

Pickrell JK, Pritchard JK (2012) Inference of population splits and mixtures from genome-wide allele frequency data. PLoS Genet 8(11):e1002967

Poke FS, Martin DP, Steane DA, Vaillancourt RE, Reid JB (2006) The impact of intragenic recombination on phylogenetic reconstruction at the sectional level in Eucalyptus when using a single copy nuclear gene (cinnamoyl CoA reductase). Mol Phylogenet Evol 39(1):160–170. doi:10.1016/j.ympev.2005.11.016

Pritchard JK, Stephens M, Donnelly P (2000) Inference of population structure using multilocus genotype data. Genetics 155(2):945–959

Provan J, Corbett G, Powell W, McNicol J (1997) Chloroplast DNA variability in wild and cultivated rice (Oryza spp.) revealed by polymorphic chloroplast simple sequence repeats. Genome 40(1):104–110

Ronquist F, Teslenko M, van der Mark P, Ayres DL, Darling A, Hohna S, Larget B, Liu L, Suchard MA, Huelsenbeck JP (2012) MrBayes 3.2: efficient Bayesian phylogenetic inference and model choice across a large model space. Syst Biol 61(3):539–542. doi:10.1093/sysbio/sys029

Schroeder H, Höltken A, Fladung M (2011) Chloroplast SNP-marker as powerful tool for differentiation of Populus species in reliable poplar breeding and barcoding approaches. BMC Proc 5(Suppl 7):56, BioMed Central Ltd

Schwarz G (1978) Estimating the dimension of a model. Ann Stat 6(2):461–464

Soejima A, Wen J (2006) Phylogenetic analysis of the grape family (Vitaceae) based on three chloroplast markers. Am J Bot 93(2):278–287. doi:10.3732/ajb.93.2.278

Straub SC, Parks M, Weitemier K, Fishbein M, Cronn RC, Liston A (2012) Navigating the tip of the genomic iceberg: Next-generation sequencing for plant systematics. Am J Bot 99(2):349–364. doi:10.3732/ajb.1100335

Sugiura M (1989) The chloroplast chromosomes in land plants. Annu Rev Cell Biol 5:51–70. doi:10.1146/annurev.cb.05.110189.000411

Sweeney M, McCouch S (2007) The complex history of the domestication of rice. Ann Bot 100(5):951–957. doi:10.1093/aob/mcm128

Takahashi H, Y-i S, Nakamura I (2008) Evolutionary analysis of two plastid DNA sequences in cultivated and wild species of Oryza. Breed Sci 58(3):225–233. doi:10.1270/jsbbs.58.225

Tang J, Xia H, Cao M, Zhang X, Zeng W, Hu S, Tong W, Wang J, Wang J, Yu J, Yang H, Zhu L (2004) A comparison of rice chloroplast genomes. Plant Physiol 135(1):412–420. doi:10.1104/pp.103.031245

Tian X, Zheng J, Hu S, Yu J (2006) The rice mitochondrial genomes and their variations. Plant Physiol 140(2):401–410. doi:10.1104/pp.105.070060

Tong W, He Q, Wang XQ, Yoon MY, Ra WH, Li F, Yu J, Oo WH, Min SK, Choi BW (2015) A chloroplast variation map generated using whole genome re-sequencing of Korean landrace rice reveals phylogenetic relationships among Oryza sativa subspecies. Biol J Linn Soc 115(4):940–952

Turmel M, Otis C, Lemieux C (1999) The complete chloroplast DNA sequence of the green alga Nephroselmis olivacea: insights into the architecture of ancestral chloroplast genomes. Proc Natl Acad Sci U S A 96(18):10248–10253

Vaughan DA, Lu B-R, Tomooka N (2008) The evolving story of rice evolution. Plant Sci 174(4):394–408

Vitte C, Ishii T, Lamy F, Brar D, Panaud O (2004) Genomic paleontology provides evidence for two distinct origins of Asian rice (Oryza sativa L.). Mol Genet Genomics 272(5):504–511. doi:10.1007/s00438-004-1069-6

Wambugu PW, Brozynska M, Furtado A, Waters DL, Henry RJ (2015) Relationships of wild and domesticated rices (Oryza AA genome species) based upon whole chloroplast genome sequences. Sci Rep 5:13957

Wang M, Yu Y, Haberer G, Marri PR, Fan C, Goicoechea JL, Zuccolo A, Song X, Kudrna D, Ammiraju JS, Cossu RM, Maldonado C, Chen J, Lee S, Sisneros N, de Baynast K, Golser W, Wissotski M, Kim W, Sanchez P, Ndjiondjop MN, Sanni K, Long M, Carney J, Panaud O, Wicker T, Machado CA, Chen M, Mayer KF, Rounsley S, Wing RA (2014) The genome sequence of African rice (Oryza glaberrima) and evidence for independent domestication. Nat Genet 46(9):982–988. doi:10.1038/ng.3044

Waters DL, Nock CJ, Ishikawa R, Rice N, Henry RJ (2012) Chloroplast genome sequence confirms distinctness of Australian and Asian wild rice. Ecol Evol 2(1):211–217. doi:10.1002/ece3.66

Wolfe KH, Li WH, Sharp PM (1987) Rates of nucleotide substitution vary greatly among plant mitochondrial, chloroplast, and nuclear DNAs. Proc Natl Acad Sci U S A 84(24):9054–9058

Wu J, Liu B, Cheng F, Ramchiary N, Choi SR, Lim YP, Wang XW (2012) Sequencing of chloroplast genome using whole cellular DNA and solexa sequencing technology. Front Plant Sci 3:243. doi:10.3389/fpls.2012.00243

Xu X, Liu X, Ge S, Jensen JD, Hu F, Li X, Dong Y, Gutenkunst RN, Fang L, Huang L, Li J, He W, Zhang G, Zheng X, Zhang F, Li Y, Yu C, Kristiansen K, Zhang X, Wang J, Wright M, McCouch S, Nielsen R, Wang J, Wang W (2012) Resequencing 50 accessions of cultivated and wild rice yields markers for identifying agronomically important genes. Nat Biotechnol 30(1):105–111. doi:10.1038/nbt.2050

Yang CC, Kawahara Y, Mizuno H, Wu J, Matsumoto T, Itoh T (2012) Independent domestication of Asian rice followed by gene flow from japonica to indica. Mol Biol Evol 29(5):1471–1479. doi:10.1093/molbev/msr315

Yang JB, Tang M, Li HT, Zhang ZR, Li DZ (2013) Complete chloroplast genome of the genus Cymbidium: lights into the species identification, phylogenetic implications and population genetic analyses. BMC Evol Biol 13:84. doi:10.1186/1471-2148-13-84

Yu J, Hu S, Wang J, Wong GK, Li S, Liu B, Deng Y, Dai L, Zhou Y, Zhang X, Cao M, Liu J, Sun J, Tang J, Chen Y, Huang X, Lin W, Ye C, Tong W, Cong L, Geng J, Han Y, Li L, Li W, Hu G, Huang X, Li W, Li J, Liu Z, Li L, Liu J, Qi Q, Liu J, Li L, Li T, Wang X, Lu H, Wu T, Zhu M, Ni P, Han H, Dong W, Ren X, Feng X, Cui P, Li X, Wang H, Xu X, Zhai W, Xu Z, Zhang J, He S, Zhang J, Xu J, Zhang K, Zheng X, Dong J, Zeng W, Tao L, Ye J, Tan J, Ren X, Chen X, He J, Liu D, Tian W, Tian C, Xia H, Bao Q, Li G, Gao H, Cao T, Wang J, Zhao W, Li P, Chen W, Wang X, Zhang Y, Hu J, Wang J, Liu S, Yang J, Zhang G, Xiong Y, Li Z, Mao L, Zhou C, Zhu Z, Chen R, Hao B, Zheng W, Chen S, Guo W, Li G, Liu S, Tao M, Wang J, Zhu L, Yuan L, Yang H (2002) A draft sequence of the rice genome (Oryza sativa L. ssp. indica). Science 296(5565):79–92. doi:10.1126/science.1068037

Zhang LB, Zhu Q, Wu ZQ, Ross-Ibarra J, Gaut BS, Ge S, Sang T (2009) Selection on grain shattering genes and rates of rice domestication. New Phytol 184(3):708–720. doi:10.1111/j.1469-8137.2009.02984.x

Zhao W, Cho G-T, Ma K-H, Chung J-W, Gwag J-G, Park Y-J (2010) Development of an allele-mining set in rice using a heuristic algorithm and SSR genotype data with least redundancy for the post-genomic era. Mol Breed 26(4):639–651

Zhu BF, Si L, Wang Z, Zhou Y, Zhu J, Shangguan Y, Lu D, Fan D, Li C, Lin H, Qian Q, Sang T, Zhou B, Minobe Y, Han B (2011) Genetic control of a transition from black to straw-white seed hull in rice domestication. Plant Physiol 155(3):1301–1311. doi:10.1104/pp.110.168500

Zhu Q, Ge S (2005) Phylogenetic relationships among A-genome species of the genus Oryza revealed by intron sequences of four nuclear genes. New Phytol 167(1):249–265

5

Genetic Diversity and Population Structure of Rice Varieties Cultivated in Temperate Regions

Juan L. Reig-Valiente[1], Juan Viruel[2,3], Ester Sales[4], Luis Marqués[5], Javier Terol[1], Marta Gut[6,7], Sophia Derdak[6,7], Manuel Talón[1] and Concha Domingo[1]* (iD)

Abstract

Background: After its domestication, rice cultivation expanded from tropical regions towards northern latitudes with temperate climate in a progressive process to overcome limiting photoperiod and temperature conditions. This process has originated a wide range of diversity that can be regarded as a valuable resource for crop improvement. In general, current rice breeding programs have to deal with a lack of both germplasm accessions specifically adapted to local agro-environmental conditions and adapted donors carrying desired agronomical traits. Comprehensive maps of genome variability and population structure would facilitate genome-wide association studies of complex traits, functional gene investigations and the selection of appropriate donors for breeding purposes.

Results: A collection of 217 rice varieties mainly cultivated in temperate regions was generated. The collection encompasses modern elite and old cultivars, as well as traditional landraces covering a wide genetic diversity available for rice breeders. Whole Genome Sequencing was performed on 14 cultivars representative of the collection and the genomic profiles of all cultivars were constructed using a panel of 2697 SNPs with wide coverage throughout the rice genome, obtained from the sequencing data. The population structure and genetic relationship analyses showed a strong substructure in the temperate rice population, predominantly based on grain type and the origin of the cultivars. Dendrogram also agrees population structure results.

Conclusions: Based on SNP markers, we have elucidated the genetic relationship and the degree of genetic diversity among a collection of 217 temperate rice varieties possessing an enormous variety of agromorphological and physiological characters. Taken together, the data indicated the occurrence of relatively high gene flow and elevated rates of admixture between cultivars grown in remote regions, probably favoured by local breeding activities. The results of this study significantly expand the current genetic resources available for temperate varieties of rice, providing a valuable tool for future association mapping studies.

Keywords: SNPs, *Oryza sativa*, Infinium SNP genotyping array

Background

Rice is a major crop with an enormous economic impact worldwide and it is widely cultivated throughout both tropical and temperate regions (Lu and Chang 1980). Modern rice (*Oryza sativa* L.) domestication occurred in southern China and, concomitant with human migrations, expanded to a wide range of geographical regions with diverse climates (Gross and Zhao 2014). As a consequence, it was generated an extensive and vast array of genetic diversity that in principle can be predominantly structured in two main subgroups (Childs 2004), including the *indica* and *japonica* varietal groups. These genetic groups are characterized by adaptations to specific climates, according to the agro-ecological conditions where they were cultivated. *Indica* genotypes are grown exclusively in tropical latitudes, whereas *japonica* genotypes can be found either in tropical or temperate climates (Mackill and Lei 1997).

* Correspondence: domingo_concar@gva.es
[1]Centro de Genómica, Instituto Valenciano de Investigaciones Agrarias, Carretera CV 315 Km 10,7 (Carretera Moncada – Náquera Km 4.5), 46113 Moncada, Spain
Full list of author information is available at the end of the article

Rice yield is highly influenced by cultivation practices in addition to climatic conditions. The adaptation process to new climates involved the selection of plants carrying genomic features that conferred advantages against adverse upcoming growth conditions that were transmitted through generations. During the northwards expansion of rice until the boundary limited by cold temperatures, crop adapted to new photoperiod conditions: the permissive summer temperatures with long days and short nights. While low temperature stress remains as the pivotal limitation in rice production in temperate regions (Andaya and Tai 2006), the acclimation to long day conditions of northern rice cultivars represents one of the main constrains during their expansion, and the most evident difference with the cultivars that remained in tropical latitudes (Izawa, 2007). Cultivation and breeding for centuries in diverse agro-ecological conditions gave rise to a myriad of different rice varieties that show the highest performance in the specific region where they were developed. The adaptations involved in this process are modifications in the regulation of metabolic and physiological processes that decrease plant yield and performance when grown out of their appropriated growth conditions. In this context, genes involved in flowering regulation should show allelic differentiation across environmental gradients, and their allele frequencies may reflect the mechanism of adaptation of plants to new conditions of day length through a geographic correlation pattern (Naranjo et al. 2014). In this regard, a significant intersubspecific variation concerning tolerance to cold stress could also be observed as the growing area approaches the northern limit (Baruah et al. 2009) and recently Ma et al. (2015) have described the COLD1 locus which is involved in the acclimation to cold of *japonica* rice. As a consequence of this intense and long-term breeding process, rice cultivars became adapted to specific regions around the world narrowing in this way its genetic pool, since many traits were forsaken due to the lack of interest in a certain moment. Nowadays it is difficult to reincorporate these characters because of the genetic distance raised between cultivars from tropical regions and those cultivated in temperate regions. The use of non-adapted varieties in breeding programs is challenging, as the incorporation of a new interesting trait is generally accompanied by many undesirable characters that do not meet climate adaptation requirements and consumers' preferences.

In contrast with this generalized narrow genetic pool, the region where temperate *japonica* varieties are cultivated is wide enough to hold relevant natural diversity uncovering a wide spectrum of morphological and physiological variations. The characterization of this diversity, especially that concerning agronomic traits,

constitutes the basis for genetic association analyses. Identifying loci that underlie this phenotypic variation is crucial for breeders, since it will offer opportunities to incorporate new traits of interest into local cultivars while conserving in unison those characters responsible for photoperiod adaptation.

The characterization of genome diversity can be efficiently performed by using next-generation sequencing (NGS) technologies that enable the massive identification of single nucleotide polymorphism (SNP) markers. These markers have been successfully applied for this purpose previously in many major crops (e.g. Myles et al. 2010) including rice (e.g. Huang et al. 2012; Xu et al. 2011). Databases of SNPs have been developed from the sequencing of numerous accessions of cultivated rice and wild rice (Duitama et al. 2015; McCouch et al.; 2016 Xu et al. 2011). The large-scale SNP database available at IRIC portal (http://oryzasnp.org/iric-portal/), for instance, has been generated as a result of re-sequencing 3000 rice genomes (3K RPG 2014). This huge amount of information constitutes an invaluable tool for breeders. However, breeding activities concern local varieties adapted to specific agro-climatic conditions, and therefore they should be complemented by studying the genetic variability within these local varieties to identify specific alleles that may introduced improvements by combination. In this study, we aimed the identification and characterization of high density SNP markers using NGS techniques in a collection of 217 rice varieties encompassing modern elite and old cultivars, and traditional landraces mainly cultivated under long-day photoperiod conditions. This collection attempts to represent the genetic diversity available for rice breeders, including traits that have been abandoned because of the continuous selection pressure and adaptation to changing agronomic environments and consumer demand conditions.

Results
Selection of 14 Cultivars Representative of the Genotypic Diversity of the Collection
A collection of 217 selected varieties was generated in order to analyse the population structure of rice grown under long photoperiod conditions. The collection was composed of modern and old cultivars as well as some landraces to cover a wider genetic diversity (Additional file 1: Table S1, Additional file 2: Figure S1). In this compilation (26 countries, 52.5 % arising from Europe; Additional file 1: Table S1), the core collection was composed of cultivars considered as *japonica* type from different geographical origins and cultivated in temperate climates in northern latitudes. A set of *indica* cultivars was also included in the collection as a reference of genetic divergence. The heading period of the varieties

included in the collection ranged from 48 to 107 days (Fig. 1) and the longer periods corresponded to *indica* cultivars. Furthermore, one *indica* variety, Nona Bokra, failed at flowering in our growing conditions.

From this collection, 14 cultivars which covered the neutral genotypic diversity useful for breeding were selected (Table 1). Cultivars were selected according to their known genealogical data, temperature and photoperiod requirements and grain type.

Genome Sequencing and Identification of Polymorphisms
SNP panel for diversity of rice cultivated in temperate regions

Genome sequencing of the subset of 14 varieties generated a mean of 75×10^6 short reads per cultivar that were mapped onto the Nipponbare reference genome (IRGSP-1.0). Approximately 78 % of them corresponded to unique reads (Table 1) while the mean coverage of these was x36. Comparison of the sequences with the reference genome provided a relatively high number of polymorphisms between the genomes analysed. Data were filtered according to different criteria as prediction significance, type and number of alleles, and absence of repetitive sequences. An average of 143,107 SNPs per genome were identified in the 14 genomes (Table 1), evenly distributed across the 12 chromosomes (Additional file 3: Table S2), and the number of nonredundant SNPS was 763,021.

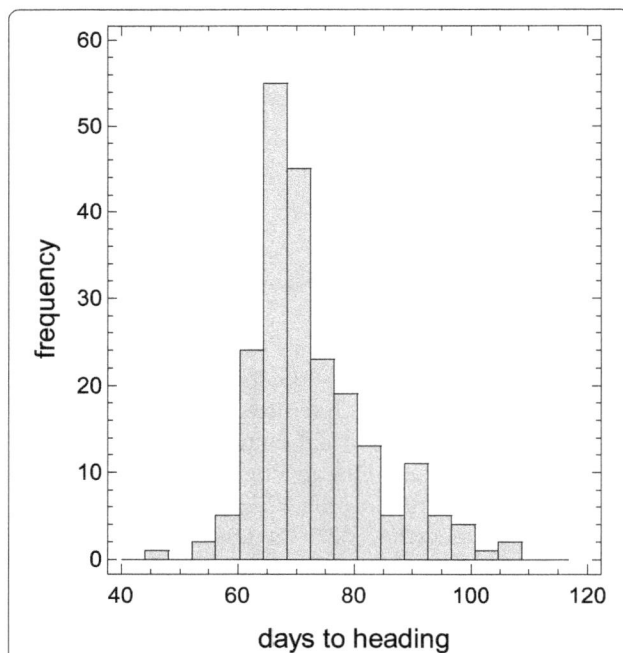

Fig. 1 Distribution of the mean flowering time in the *japonica* rice core collection grown under natural long day conditions

SNP Panel for Analysing Genetic Diversity of Japonica Rice Cultivated in Temperate Regions

A custom Infinium SNP genotyping array (Illumina) compiling 2697 SNPs selected out of the polymorphisms identified in the 14 rice cultivars was designed. The SNP panel was generated by selecting bi-allelic polymorphisms with a uniform distribution along the 12 chromosomes, but avoiding centromere regions where gene occurrence is scarce. SNPs that were detected in more than one cultivar were prioritized. These selected SNPs were manually curated using Integrative Genomics Viewer (IGV, Broad Institute) software. The mean interval distance between adjacent SNPs was 137,525 pb. The number of SNPs per chromosome ranged from 163 in chromosome 9 (the shortest one, 23.0 Mb) to 321 SNPs in chromosome 1 (the largest, 43.2 Mb). The distribution of these SNPs reflected their non-redundancy. The number of selected SNPs varied among cultivars, ranging from 761 SNPs in L-202 to 502 in Bahia, with an average of 607 SNPs per genome (Table 2).

The proportion of SNPs located within intergenic regions (54.8 %) was higher than those within genic regions (22.1 %) (Fig. 2a). Genic SNPs were classified as exonic (6.4 %), intronic (11.2 %) or UTR (4.5 %), while the proportion of SNPs distributed within promoter regions was similar to that of genic SNPs (21.9 %). Among coding SNPs, nonsynonymous substitutions (57.3 %) were higher than synonymous substitutions (41.8 %) (Fig. 2b). Meanwhile, SNPs causing large effect thus affecting the integrity of encoded proteins were not frequent, and only 0.9 % of SNPs causing disruption of the stop signal were detected.

Then, 217 rice accessions were genotyped with this array of 2697 SNP and a genetic profile was obtained for each cultivar. Comparison between the genetic profiles allowed the identification of low frequency alleles and those with a minimum allele frequency (MAF) below 5 % were removed, then originating a panel of 1713 SNPs evenly distributed along the genome at a mean distance of 215,223 pb. Using this panel we estimated the extent of linkage disequilibrium (LD) in the *japonica* subpopulation in the collection, and compared it with that estimated for the temperate *japonica* subpopulation in the 3KRPG (http://oryzasnp.org/iric-portal/). The LD decay rate was calculated as the chromosomal distance where the correlation coefficient (r^2) between SNP pairs dropped to half its maximum estimated value. LD estimation reflects the strong population structure of the *japonica* subpopulation in our collection, since r^2 drops to 0.23 when chromosomal distance extends to approximately 368 kb while in the 3KRPG subpopulation we estimated a more rapid decay that extends to 174 kb for the same value of correlation coefficient (Fig. 3).

Table 1 Country of origin, main traits interesting for breeders and sequencing statistics of 14 rice cultivars selected for genome sequencing in this study. Reads were mapped on the sequences of Nipponbare Os-Nipponbare-Reference-IRGSP-1.0. Number of reads, % of unique and duplicated reads and resulted mean coverages are reported. Number of total SNPs and number of total SNPs after filtering according to prediction significance, type and number of alleles and absence of repetitive sequences

Accession	Origin	Characteristic	Number of reads (× 10³)	% unique reads	% duplicated reads	Mean coverage	SNPs vcf	SNPs filtered
Arroz da Terra	Portugal	Cold tolerance, earliness	80,492	78.9	0.14	39.9	465,830	117,170
Bahia	Spain	Parental, grain quality	78,446	78.2	0.20	38.3	341,451	84,321
Bomba	Spain	Landrace, grain quality	80,038	77.0	0.12	38.7	771,460	200,619
Gigante Vercelli	Italy	Blast resistance	69,031	78.1	0.09	33.8	592,763	111,351
Gleva	Spain	Grain type	78,505	78.6	0.12	38.7	564,829	141,363
Italica Livorno	Italy	Earliness, cold tolerance	72,919	79.7	0.20	36.3	404,125	91,806
Kalao	France	Blast resistance	75,052	76.9	0.13	36.4	1,025,081	253,553
L202	USA	Parental, grain type	78,854	77.1	0.14	38.3	989,104	269,476
Loto	Italy	earliness	71,612	78.2	0.19	35.1	527,474	106,014
LTH	China	Cold tolerance	77,667	77.7	0.10	37.8	573,944	149,950
M202	USA	Parental	69,364	79.1	0.14	34.5	669,310	124,931
Pavlovski	Russia	Earliness, grain type	66,962	80.1	0.16	33.6	396,196	69,894
Puntal	Spain	Grain quality	69,581	77.6	0.17	34.0	928,794	182,272
Senia	Spain	Grain type	81,376	79.7	0.16	40.6	358,900	100,783

To further explore the adequacy of the SNPs array for the analysis of the population structure, we examined 2D site frequency spectra between *japonica* and *indica* varietal groups in our collection and in the temperate *japonica* 3KRPG subpopulation (Alexandrov et al. 2015). We found that the majority of SNPs that are at high frequency in *japonica* cultivars in the collection are found at markedly lower frequencies in *indica* cultivars (Fig. 4)

meanwhile the majority of SNPs that are at low frequency in *japonica* cultivars in our collection are also found at low frequency in *japonica* 3 K RPG subpopulation (Fig. 4).

Genetic Structure of the Collection

To determine the population structure of the collection, a 948 SNP panel based on the above LD decay results

Table 2 Number of SNPs per chromosome selected in each rice cultivar and number of non-redundant SNPs. The Nipponbare genome was used as the reference genome

	Arroz da Terra	Bahia	Bomba	Gigante Vercelli	Gleva	Italica Livorno	Kalao	L-202	Loto	LTH	M-202	Pavlovski	Puntal	Senia	Total non-redundant SNPs
chr01	81	96	61	87	76	89	101	105	65	56	95	58	108	77	321
chr02	71	54	36	61	53	62	57	55	42	45	50	60	61	55	259
chr03	60	49	56	56	34	54	74	74	63	50	38	44	65	48	281
chr04	53	34	63	56	35	42	65	71	56	64	44	45	56	41	248
chr05	42	45	53	49	54	50	71	74	63	53	60	50	66	59	218
chr06	52	37	52	52	52	69	52	55	57	55	37	48	62	53	225
chr07	61	27	54	39	25	48	64	57	59	53	50	46	34	23	199
chr08	54	28	60	53	59	49	63	69	45	46	61	28	62	30	211
chr09	39	16	41	39	19	38	49	52	17	37	19	27	44	16	163
chr10	34	46	58	42	43	31	41	41	48	36	45	35	52	44	179
chr11	35	32	44	40	27	31	45	47	27	39	39	38	45	24	200
chr12	38	38	52	59	55	49	59	61	34	45	50	24	64	41	194
Total	620	502	630	633	532	612	741	761	576	579	588	503	719	511	2.698

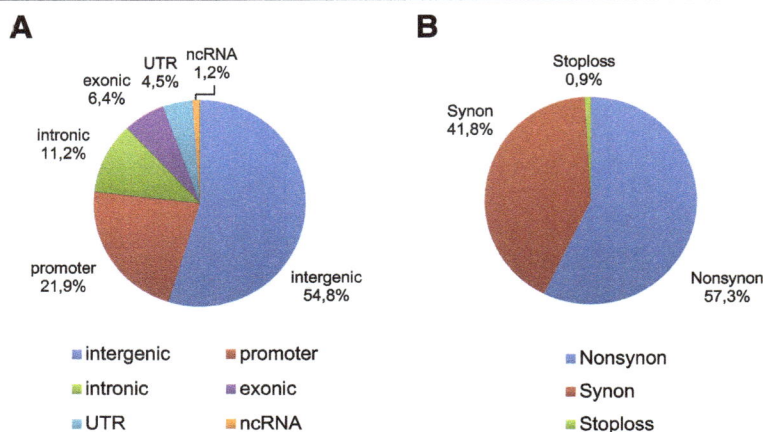

Fig. 2 a Distribution of SNPs according to the TIGR gene annotation. Genic SNPs were classified as exonic, intronic and UTR SNPs. SNPs distributed within 2.0 Kb upstream of the coding region were identified as promoter SNPs. Nuclear RNAs (ncRNA) are also represented. **b** Distribution of exonic SNPs causing synonymous (synon) or nonsynonymous (nonsynon) substitutions or producing changes in the stop codon (lossstop)

and with a mean interval distance of 390,228 pb was designed. The most probable number of subpopulations, and cultivars included in each, was estimated using STRUCTURE software. According to theses analyses, ΔK showed maximum values for K = 4 (ΔK = 101.2) indicating that the optimum number of subpopulations was 4 (Fig. 5, Additional file 1: Table S1). Long grain type accessions conformed subpopulation 2 and displayed different geographical origins from Europe, America and Australia. Medium grain type cultivars were distributed in different subpopulations in accordance with their geographical origin: cultivars in group 4 were mainly from Italy, while cultivars in group 3 originated in America, Spain and Australia. Most of the Spanish cultivars included in group 3 were of recent release, while some old accessions from Spain and Italy were included in group

1, in which Asian *japonica* cultivars were clustered, thus revealing the putative Asian ancestry of these European old varieties. Differentiation of these four subpopulations was corroborated by the high Fst values obtained for each group although the fourth genetic group, composed mainly by Italian accessions, showed the lowest value of genetic differentiation (Additional file 4: Table S3, Additional file 5: Table S4).

A further maximum of ΔK was found for K = 5 (ΔK = 1.8), which separated a new subgroup that included Spanish cultivars. Accessions showing higher than 80 % of membership to this cluster were released in the mid-twentieth century. A total of 12 accessions were classified as admixed and showed approximately 50 % of membership to the Spanish and American clusters. This admixed group includes accessions originated in Spain that were released approximately at the same time, during the second half of the twentieth century. The proportion of the genetic content belonging to these two clusters clearly reflects their breeding history, since these cultivars were obtained when American germplasm was introduced into the Spanish breeding programs in the early 20th century.

A subsequent division of long grain cluster was observed at K = 6 (ΔK = 3.5), which separated in a new subgroup a set of long grain varieties, as Moroberekan, Agami and Honduras, cultivated in tropical regions.

To deepen in the patterns of population structure we performed principal components analysis (PCA) with the set of 1713 SNPs selected from the Infinium analysis. The first PC, which accounted for 22.1 % of the variance, separated cultivars according to the grain size, which agrees with the four most probable groups described in the population structure analysis, since long-grain type cultivars included in cluster 2 are separated from the other groups (Fig. 6).

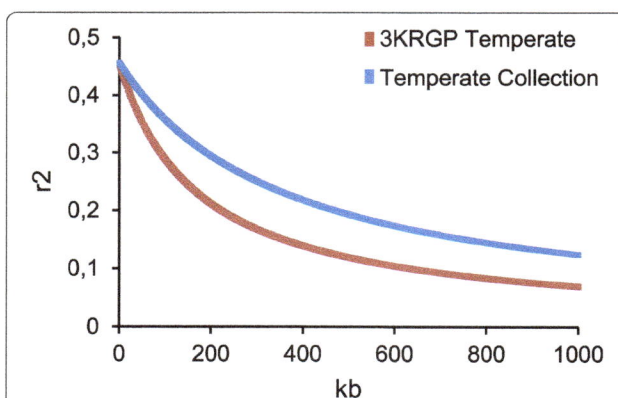

Fig. 3 Estimated LD decay from the *japonica* subpopulation in the collection and the 3KRPG *japonica* subpopulation, expressed as decay of r^2

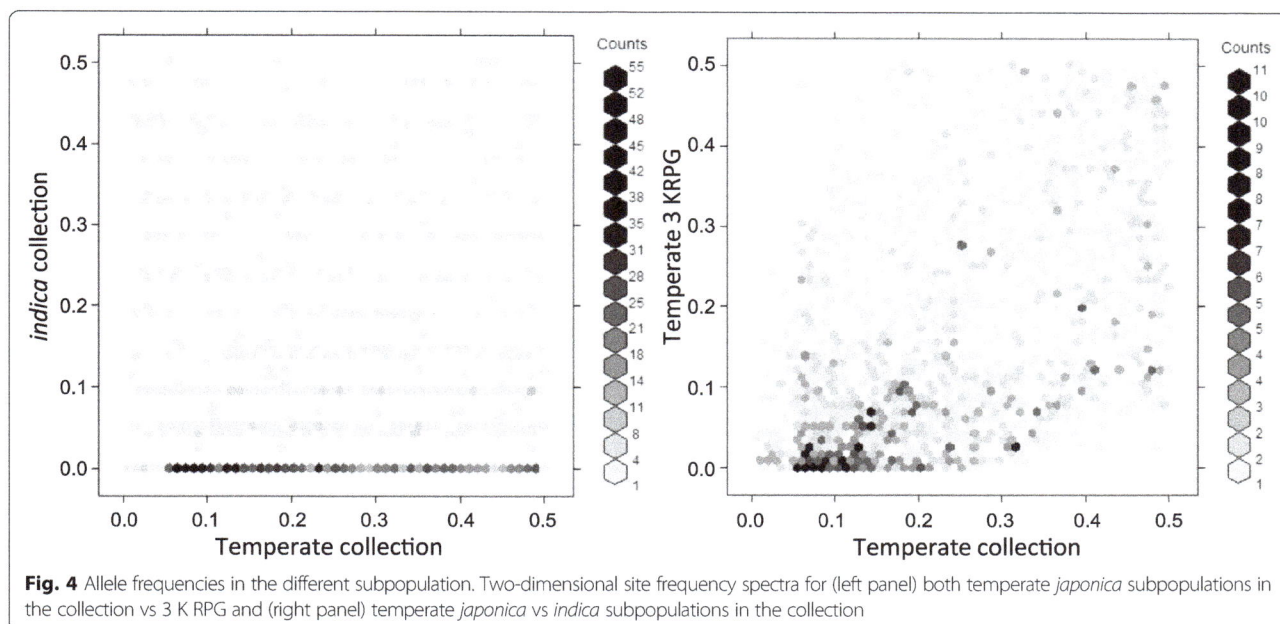

Fig. 4 Allele frequencies in the different subpopulation. Two-dimensional site frequency spectra for (left panel) both temperate *japonica* subpopulations in the collection vs 3 K RPG and (right panel) temperate *japonica* vs *indica* subpopulations in the collection

Genetic Relationships Among the Temperate Rice Collection

The 1713 SNPs panel was used to determine the relationship of cultivars in our collection and to estimate genetic distances among them. Basically, the 217 cultivars were grouped into a dendrogram with clusters arranged similarly to the distribution obtained in the STRUCTURE analysis (Fig. 7, Additional file 6: Figure S2).

The distribution of cultivars obtained was roughly in accordance with their origin or grain type. On one side of the tree, *indica* varieties appear in a highly supported cluster. Closely related to *indica* cluster, long grain type varieties diverge into a wider cluster that includes most of the varieties grouped in cluster 2 in the population structure analysis, such as *japonica* accessions from tropical regions (Azucena or Moroberekan) and temperate

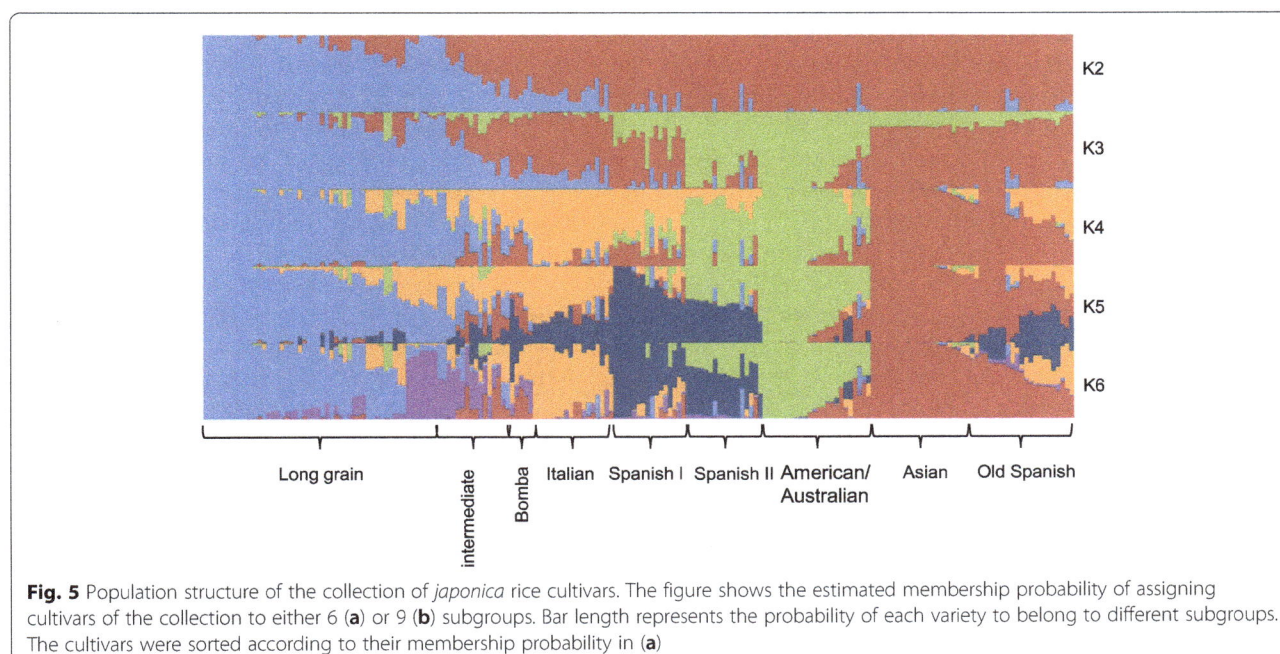

Fig. 5 Population structure of the collection of *japonica* rice cultivars. The figure shows the estimated membership probability of assigning cultivars of the collection to either 6 (**a**) or 9 (**b**) subgroups. Bar length represents the probability of each variety to belong to different subgroups. The cultivars were sorted according to their membership probability in (**a**)

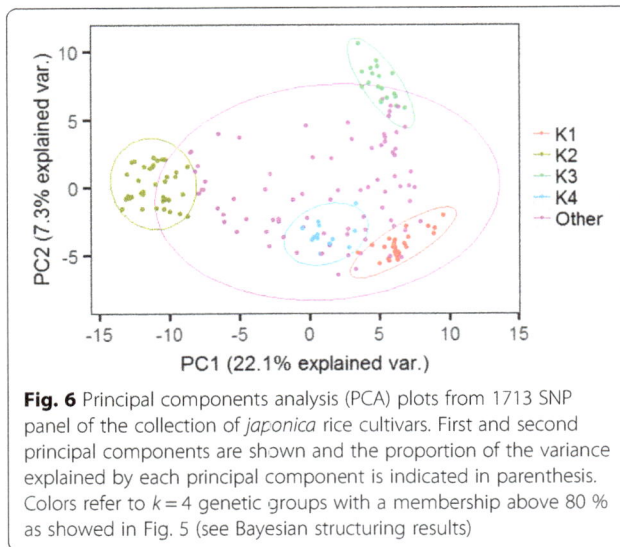

Fig. 6 Principal components analysis (PCA) plots from 1713 SNP panel of the collection of *japonica* rice cultivars. First and second principal components are shown and the proportion of the variance explained by each principal component is indicated in parenthesis. Colors refer to $k = 4$ genetic groups with a membership above 80 % as showed in Fig. 5 (see Bayesian structuring results)

regions (Cormoran or Apolo). Aromatic rice cultivars, for instance Fragance or Giglio, can also be distinguished among them.

All the remaining varieties, mostly medium-grain type, are genetically distant from the long-grain and *indica* clusters. The medium-grain varieties were grouped in 7 different clusters with a remarkable influence of the geographic origin (Fig. 7). Several European landraces were clustered together within a wider cluster that includes Asian representatives. The use of a higher number of SNPs to construct the dendrogram allowed the fragmentation of European cultivars into several small subgroups, with a distribution that clearly matches its origin and, even from some occasions, the release date of the cultivars. Asian cultivars are clearly related to a cluster formed by ancient Spanish varieties, revealing the influence of Asian germplasm in the early reported history of cultivars grown in Spain. Another two clusters are constituted by Spanish varieties, which are closely related to the American/Australian group, indicating the relevance of the introduction of American germplasm in the Spanish breeding programs. This genetic relationship was also observed in the population genetic analysis. Finally, a fourth cluster of Spanish landraces grouped together the five accessions with Bomba type grain, which possesses specific culinary qualities, emerge as an independent cluster.

Discussion

The expansion of rice to a wide geographical area after its domestication generated enormous diversification due to the progressive breeding activity through the selection of plants that showed superior performance under the different local climate conditions leading to a broad range of novel phenotypes. Genetic diversity is a natural resource for rice breeding to meet current food demands. Understanding population structure and genome variations are crucial achievements to facilitate genome-wide association studies of complex traits and functional gene investigations. In this regard, our study provides a useful pool of SNP markers with deep coverage throughout the rice genome. We have characterized 1713 SNPs that were genotyped in 217 rice varieties. Cultivars were representative of all genetic groups found across geographical regions in temperate climate as revealed by our population structure analysis and in accordance with previous studies (Courtois et al. 2012). Our genomic diversity and structure analyses have included old and modern *japonica* cultivars attempting to uncover a maximum spectrum of variability. This allowed us to reconstruct the genetic relationships and genetic diversity among several temperate rice varieties from distant geographic origins (Additional file 1: Table S1) which possess an enormous variability in agro-morphological and physiological traits.

The use of a high number of SNP markers, provided by deep sequencing of a selected group of cultivars, enabled us to analyse in detail the temperate *japonica* group, and in particular, the genetic relationship of several Spanish cultivars with other groups from different countries. This allowed us to revisit the history of rice breeding in Spain. As previous genetic studies in rice (Huang et al. 2010; Xu et al. 2011; Zhao et al. 2010), our analysis showed strong genetic substructuring within *japonica* accessions. This is in accordance with the enlarged LD decay observed in the collection compared to other reported small subpopulations (Xu et al. 2011), that can be an effect of the sampling composition and size. One of the relevant observations in our analysis is the influence of the grain type to the genetic classification of cultivars, suggesting that this trait is a pivotal factor in the morphological structure of current and old rice varieties. Accordingly, the dendrogram based on genetic distances establishes clear-cut differences among long and medium type grain cultivars with independence of their origin (Fig. 7). This pattern was also observed in the PCA analyses, which separated both types of cultivars in the first axis (Fig. 6). The origin of long grain varieties does not appear to be a discriminating factor among this group that certainly contains many varieties from several continents. Thus, this cluster includes cultivars grown in tropical regions (e.g. Philippines or Honduras), as well as in countries with temperate climate, as European countries. Furthermore, cultivars in the collection considered as tropical *japonica*, such as Azucena, Honduras, Katy and Lemont (Zhao et al. 2010), are located within the long grain cultivars.

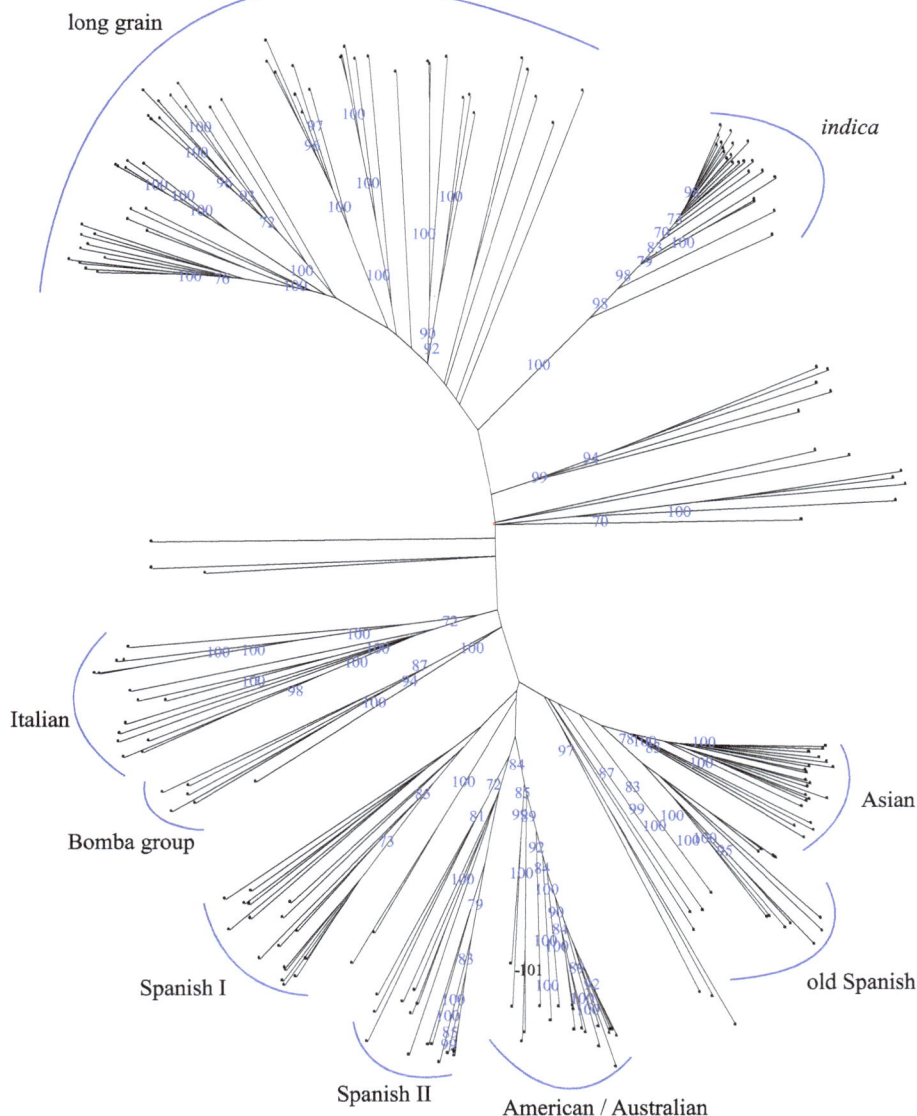

Fig. 7 Neighbour-Joining tree of 217 rice accessions based in Sokal & Michener distance as implemented in Darwin 6.0. Values in branches indicate the bootstrap (10,000 replicates) support values

On the other hand, several subpopulations of *japonica* varieties with medium grain size can be distinguished in different secondary branches. In this case, accessions from nearby locations were more closely related than samples from distant locations, showing geographic patterns of genetic variation probably produced by breeding and selection associated with local food preferences.

Spanish varieties, the most numerous group in our population, are grouped into four clusters that are predominantly related to the time they were cultivated and that clearly reflect the history of rice breeding in Spain. The Bomba family, consisting of the most ancient Spanish landraces and whose origins precede the available records, appeared in a distinct and isolated cluster in the dendrogram. Some of these cultivars, well appreciated by consumers, are still cultivated because of the peculiar characteristics of their grains and because of their agronomic characteristics that are suitable for organic farming. A second group of Spanish cultivars derived from the earliest breeding activities in Spain, remains in a distinct group close to the Asian cluster, indicating the origin for the first cultivars used as donors in initial breeding programs. In a third separate cluster we found varieties that were grown in Spain during the first half of the 20th century. This cluster appears very close to the node that separates the American-Asian clusters. Modern Spanish cultivars are located in a secondary cluster nested to the America-Australian variety group. These results, together with the genetic admixture suggested by the Bayesian structuring analyses with a similar membership probability to the American and Spanish subgroups when $K = 5$, suggest the introduction of American germplasm varieties in the Spanish breeding program.

Previous genetic studies pointed to a narrower genetic diversity in *japonica* than in *indica* rice, probably due to a more severe domestication bottleneck (Garris et al. 2005). Additionally, breeding activities during the last century also constricted the genetic pool of cultivars selected for cultivation in different agro-ecological environments. In that sense, searching for new potential donors for breeding activities has become a difficult task since the use of non-adapted donors from other latitudes to introduce particular traits entails accompaniment of characters that may not be agronomically appropriate in the region. This is the case of the hybridization between temperate and tropical *japonica* cultivars. In addition, high sterility has been observed in *indica-japonica* hybrids due to strong reproductive barriers which hampered gene flow between them (Jeung et al. 2005). On the contrary, a gene flow occurred between cultivars grown in temperate regions as revealed in the Spanish breeding history deduced from our genetic analysis, showing the influence of various genetic groups in the Spanish cultivars. During the beginning of breeding activities, genetic influence from Asia was high in medium- grain rice and some links with Italian cultivars were evident. This influence diminished as new improved varieties were generated, showing increasing membership to clusters composed by American cultivars, in detriment of Italian accessions.

Natural diversity distributed across different geographical regions during the expansion of rice constitutes a source of genetic variability highly valuable for the breeding programs to generate new varieties that are adapted to local climate conditions. Breeders need the natural genetic resources to recover agronomic traits of interest that have been left behind during the selection process, but knowledge of the chromosomal regions responsible for many phenotypic variations remains undiscovered. In this context, association mapping and genomic selection are recently developed methodologies that have been proved to be effective approaches to connect structural genomics and phenotype and, thereby, to mine elite genes in germplasm resources (e.g. Zhang et al. 2014; Begum et al. 2015). Our highly variable SNP set among and between several temperate rice varieties constitutes a short-term tool for exploring candidate genes

involved in several traits related to the adaptation to temperate climate as heading time or cold tolerance. Rice cultivation in temperate regions deals with low temperature stress as a main yield limitation (Andaya and Tai 2006) since rice varieties grown under temperatures below 15 °C usually suffer from low germination rates, yellowing or withering, reduced tillering, delayed heading or sterility (Kaneda and Beachell 1974; Mackill and Lei 1997; Yoshida et al. 1996). Some *japonica* accessions, for instance, have been characterized and selected for cultivation in temperate regions because of their high tolerance to low temperature while producing good grain quality and high yield (Fujino et al. 2004; Jiang et al. 2008). Therefore, identifying cultivars closely related to cold tolerant ones constitutes the basis to carry out breeding programs based on recombinant inbred activities.

Conclusions

The analysis of the genomic profile of 217 rice varieties revealed the genetic distances and diversity among rice cultivars grown in temperate climate regions. Our observations indicated a contribution of the grain type to the population genetic structure stronger than that of the geographic origin of temperate *japonica* cultivars. The use of a high number of SNP markers in this study also revealed gene flow and higher rates of admixture between cultivars grown in remote regions, probably as a consequence of local breeding activities. We could also observe the influence of Asian and American cultivars at the early and the most recent steps respectively, of rice breeding in Spain.

Methods

Plant Material and Growing Conditions

The rice core pool was obtained from different germplasm collections: the International Rice Research Institute (IRRI, Philippines), U.S. National Plant Germplasm System (NPGS, USA), Rice Genome Resource Center (RGRC, Japan) and Instituto Valenciano de Investigaciones Agrarias (IVIA, Spain). Seeds of different cultivars were germinated and grown in pots in greenhouses (39° 28' N) under controlled temperature (25 °C) and relative humidity (50 % RH), and in natural daylight conditions during summer in 2013 and 2014. Plantlets were manually transplanted in rows of 20 plants in May and harvested in September. Fields were irrigated by flooding. Heading time was measured as 50 % of panicle emergence in each row.

Whole Genome Sequencing

Fourteen varieties representative of *japonica* cultivars were previously selected to carry out the SNPs characterization (Table 1). Seven days-old seedlings were grown in the dark for 2 days and then nuclear genomic DNA was extracted from leaves by using a modified CTAB protocol (Schneeberger et al. 2009). Genome sequencing was performed at the Centro Nacional de Análisis Genómico (Barcelona, Spain) as follows: the short-insert paired-end libraries were prepared with NO-PCR protocol. TruSeq™DNA Sample Preparation Kit v2 (Illumina Inc.) and the KAPA Library Preparation kit (Kapa Biosystems) were used. In short, 2.0 micrograms of sheared genomic DNA was end-repaired, adenylated and ligated to Illumina specific indexed paired-end adaptors. The DNA was size selected with AMPure XP beads (Agencourt, Beckman Coulter) in order to reach the fragment size of 220-550 bp. The final libraries were quantified by Library Quantification Kit (Kapa Biosystems).

The libraries were sequenced using TruSeq SBS Kit v3-HS (Illumina Inc.), in paired end mode, 2x101bp, in ½ of a sequencing lane of HiSeq2000 flowcell v3 (Illumina Inc.) according to standard Illumina operation procedures with the yield of >14Gb and median coverage of 33-41x. Primary data analysis, the image analysis, base calling and quality scoring of the run, was processed using the manufacturer's software Real Time Analysis (RTA 1.13.48) and followed by generation of FASTQ sequence files by CASAVA. Data have been deposited at the European Nucleotide Archive (ENA) in the European Bioinformatics Institute (EBI) with the accession number PRJEB13328 (http://www.ebi.ac.uk/ena/data/view/PRJEB13328).

Sequencing Data Analysis

Reads were hard trimmed from the end up to the first base with a Phred quality of at least 10. Reads with a length of at least 40 nt were mapped to the Nipponbare rice variety reference genome (the unified-build release Os-Nipponbare-Reference-IRGSP-1.0 (IRGSP-1.0)) using the GEM toolkit (version 2) (Marco-Sola et al., 2012) allowing up to eight mismatches per read. Only uniquely mapping non-duplicate read pairs were used for subsequent analyses. The SAMtools suite (version 0.1.18) (Li et al. 2009) with default settings was used to call SNVs and short INDELS per variety. Genome annotations from http://rapdb.dna.affrc.go.jp/download/irgsp1.html such as genes and CDS were added to the resulting VCF using vcftools (Danecek et al. 2011). Variants identified in regions with low mappability, with a pronounced strand bias with a p-value of <0.001 or a pronounced tail distance bias with a p-value of < 0.05 were filtered out.

SNP Panel and Genotyping

A panel of 2697 SNPs representative of *japonica* rice cultivated under long day conditions was generated by selecting polymorphisms identified in the 14 cultivars. Bi-allelic polymorphisms with a uniform distribution along the 12 chromosomes SNPs were selected, avoiding centromeric and telomeric regions and maintaining an average distance of 137.6 kb among them. SNP polymorphisms present in more than one cultivar were prioritized. SNPs were visually inspected using Integrative Genomics Viewer software (IGV, Broad Institute). The adaptability to the Illumina iSelect detection system of the SNPs and their neighbouring sequences was scored and SNPs with a score higher than 0.4 were selected for the multiplexing Infinium Assay (Illumina Inc.).

The 217 accessions were genotyped with the 2697 SNP array. Comparison of the called genotypes was performed using GenomeStudio software (Illumina). Low frequency alleles, considered when they were present in less than 5 % of the cultivars, were removed. Genotypes were missing data if they failed in more than 50 % of the cultivars.

Genomic variation analysis and classification of SNP according to the TIGR rice gene models was implemented in platform CARMO (Comprehensive Annotation of Rice Multi-Omics data) (Wang et al. 2015, http://bioinfo.sibs.ac.cn/carmo/).

Linkage Disequilibrium (LD) Estimation

Raw data was filtered to select 2066 markers with a minimum of genotype call of 75 % individuals and to eliminate monomorphic markers showing a MAF lower than 5 %. Linkage disequilibrium was calculated using Plink 1.07 (Purcell et al. 2007) according to Zhao et al. (2011). After filtering, the 1713 remaining SNPs were used to calculate LD using "−r2 −ld-window 99999 −ld-window-r2 0" commands. The decay of linkage disequilibrium by distance was fitted using Hill and Weir (Hill and Weir 1988) expectation of r^2 between adjacent sites. Calculations were performed according to the equation by Remington et al. (2001). Nonlinear least squares, implemented in the 'nls' R package, were used to fit equations to our data as suggested by Marroni et al. (2011).

Site Frequency Spectra

Minimum allele frequency (MAF) for the 1713 SNPs panel in *japonica* and *indica* groups of our collection and for those from 3KRGP were calculated using Tassel (Tassel 5.2.26, Bradbury et al. 2007). In case of 3KRGP varieties, MAF was set after removing third state of SNPs and imputing by fillin. MAF bins values for each group of varieties were calculated using hexbin R package. The number of bins was set to 50. Since we used MAF, the values of allele frequency did not exceed 0.5.

Principal Component Analysis

A Principal Component Analysis (PCA) was performed using the *prcomp* command in R version 3.2.3. We called genotypes for 1713 SNPs from the panel used in the LD analysis. Data were analyzed using *ggbiplot* (version 0.55).

Estimation of Genetic Structure

Population genetic structure of the cultivars was estimated using the Bayesian clustering method implemented in STRUCTURE 2.3.4 (Pritchard et al. 2000). This approach estimated the optimal number of genetic clusters (K) and calculated the membership proportion of cultivars to them. Analyses were based on the admixture ancestral model for a range of K values from 1 to 15. We performed 20 runs for each K, and removed those with extreme values of L(K) that were tagged as outliners according to Evanno et al. (2005). Each run was implemented with a burn-in period of 100,000 steps followed by 1,000,000 Monte Carlo Markov Chain iterations. The optimal number of K clusters was estimated with the ad hoc parameter (ΔK) of Evanno et al. (2005) in Structure Harvester (Earl and vonHoldt 2012). We estimated the optimal alignment for the 20 replicates in CLUMPP (Jakobsson and Rosenberg 2007), employing greedy algorithm with 10,000 permutations. Cultivars were subdivided into different subgroups according to their maximum membership probability among the subgroups and the membership probabilities threshold of 0.80.

Pairwise genetic distances between rice cultivars were calculated as implemented in Darwin 6.0 (Perrier et al. 2003), by using the Sokal & Michener index and 10,000 bootstrap replicates. Then we obtained a neighbour-joining (NJ) tree, which robustness was assessed again by 10,000 replicate bootstrap analyses.

We also used the software STRUCTURE to estimate Fst values for the genetic clusters established for each group, the mean value of 20 runs was calculated. Finally, Fst pairwise values between the subpopulations defined by STRUCTURE analysis were calculated in Arlequin (Excoffier and Lischer 2010) with 10,000 permutations and including only those varieties with membership values > 80 %.

Additional files

Additional file 1: Table S1. List of cultivars included in the analysis with their country of origin, days to heading and release date (a, before 1900; b, 1900–1965; c, 1966–2000; d, after 2000). Membership (>80 %) to K = 4, K = 5 and K = 6 groups defined by STRUCTURE is also indicated. Accessions displaying approximately 50 % of membership to two subpopulations are also indicated.

Additional file 2: Figure S1. Geographic origins of cultivars used in the structure population analysis.

Additional file 3: Table S2. Number of SNPs per chromosome present in the 14 selected cultivars referred to Nipponbare genome and percentage of SNP present in each chromosome in each cultivar. The number of non-redundant SNPs in each chromosome is indicated.

Additional file 4: Table S3. Mean and standard deviation values of proportion of membership, Avdistance (expected heterozygosity) and Fst for the four genetic groups established by STRUCTURE.

Additional file 5: Table S4. Arlequin software estimates of Population pairwise differences (Fst) among the four genetic groups established by STRUCTURE.

Additional file 6: Figure S2. Neighbour joining tree of 217 rice accessions. Accessions names are shown.

Acknowledgements

This work was supported by the INNPACTO program of the Ministerio de Economía y Competitividad, (grant IPT-2011-1244-010000). We thank Spanish National Cancer Centre (CNIO-CEGEN unit) for Infinium genotyping assistance. All rice seeds used here were obtained with MTA agreements.

Authors' Contributions

CD conceived the project, designed research, selected SNP, data interpretation, generated the cultivar collection and phenotyping and wrote the manuscript. JLR performed SNP panel analysis, LD calculation, analysis of the population structure and phenotyping. JV and ES participated in analysis of the population structure and dendrogram construction and participated in manuscript preparation. JT and SD performed bioinformatics analysis of the genomic sequences. LM contributed to generate the cultivar collection and phenotyping. MG carried out genome sequencing. MT designed research, participated in manuscript preparation and revision. All authors read and approved the final manuscript.

Competing Interests

The authors declare that they have no competing interests.

Author details

[1]Centro de Genómica, Instituto Valenciano de Investigaciones Agrarias, Carretera CV 315 Km 10,7 (Carretera Moncada – Náquera Km 4.5), 46113 Moncada, Spain. [2]Dpto. Biología Vegetal y Ecología, SGI Herbario – Universidad de Sevilla, Edif. Celestino Mutis, Av. Reina Mercedes s/n, 41012 Sevilla, Spain. [3]Institut Méditerranéen de Biodiversité et d'Ecologie Marine et Continentale (IMBE), Aix Marseille Université, Chemin de la Batterie des Lions, 13007 Marseille, France. [4]Dpto. Ciencias Agrarias y del Medio Natural, Escuela Politécnica Superior, Universidad de Zaragoza, Ctra. Cuarte s/n, 22071 Huesca, Spain. [5]Cooperativa de Productores de Semillas de Arroz, Avenida del Mar 1, 46410 Sueca, Spain. [6]Centre Nacional d'Anàlisi Genòmica - Centre for Genomic Regulation (CNAG-CRG), Barcelona Institute of Science and Technology (BIST), Baldiri Reixac, 4, 08028 Barcelona, Spain. [7]Universitat Pompeu Fabra (UPF), Barcelona, Spain.

References

Alexandrov N, Tai S, Wang W, Mansueto L, Palis K, Fuentes RR, Ulat VJ, Chebotarov D, Zhang G, Li Z, Mauleon R, Hamilton RS, McNally KL (2015) SNP-Seek database of SNPs derived from 3000 rice genomes. Nucleic Acids Res 43(Database issue):D1023–D1027

Andaya VC, Tai TH (2006) Fine mapping of the qCTS12 locus, a major QTL for seedling cold tolerance in rice. Theor Appl Genet 113:467–475

Baruah AR, Ishigo-Oka N, Adachi M, Oguma Y, Tokizono Y, Onishi K, Sano Y (2009) Cold tolerance at the early growth stage in wild and cultivated rice. Euphytica 165:459–470

Begum H, Spindel JE, Lalusin A, Borromeo T, Gregorio G, Hernandez J, Virk P, Collard B, McCouch SR (2015) Genome-wide association mapping for yield and other agronomic traits in an elite breeding population of tropical rice (*Oryza sativa*). PLoS One 10:e0119873

Bradbury PJ, Zhang Z, Kroon DE, Casstevens TM, Ramdoss Y, Buckler ES (2007) TASSEL: Software for association mapping of complex traits in diverse samples. Bioinformatics 23:2633–2635

Childs N (2004) Production and utilization of rice. In: Champagne E (ed) RICE: chemistry and Technology, Ed third edition Vol Chapter 1. U.S. Department of Agriculture, Agricultural Research Service, Southern Regional Research Center, New Orleans, pp 1–23

Courtois B, Frouin J, Greco R, Bruschi G, Droc G, Hamelin C, Ruiz M, Clément G, Evrard J, van Coppenole S, Katsantonis D, Oliveira M, Negrão S, Matos C, Cavigiolo S, Lupotto E, Piffanelli P, Ahmadi N (2012) Genetic Diversity and Population Structure in a European Collection of Rice. Crop Sci 52:1663–1675

Danecek P, Auton A, Abecasis G, Albers CA, Banks E, DePristo MA, Handsaker RE, Lunter G, Marth GT, Sherry ST, McVean G, Durbin R, 1000 Genomes Project Analysis Group (2011) The variant call format and VCFtools. Bioinformatics 27: 2156–2158

Duitama J, Silva A, Sanabria Y, Cruz DF, Quintero C, Ballen C, Lorieux M, Scheffler B, Farmer A, Torres E, Oard J, Tohme J (2015) Whole genome sequencing of elite rice cultivars as a comprehensive information resource for marker assisted selection. PLoS One 10(4):e0124617

Earl D, vonHoldt B (2012) STRUCTURE HARVESTER: a website and program for visualizing STRUCTURE output and implementing the Evanno method. Conserv Genet Resour 4:359–361

Evanno G, Regnaut S, Goudet J (2005) Detecting the number of clusters of individuals using the software STRUCTURE: a simulation study. Mol Ecol 14:2611–2620

Excoffier L, Lischer HEL (2010) Arlequin suite ver 3.5: A new series of programs to perform population genetics analyses under Linux and Windows. Mol Ecol Resour 10:564–567

Fujino K, Sekiguchi H, Sato T, Kiuchi H, Nonoue Y, Takeuchi Y, Ando T, Lin SY, Yano M (2004) Mapping of quantitative trait loci controlling low-temperature germinability in rice (Oryza sativa L.). Theor Appl Genet 108:794–799

Garris AJ, Tai TH, Coburn J, Kresovich S, McCouch S (2005) Genetic structure and diversity in Oryza sativa L. Genetics 169:1631–1638

Gross BL, Zhao Z (2014) Archaeological and genetic insights into the origins of domesticated rice. Proc Natl Acad Sci U S A 111:6190–6197

Hill WG, Weir BS (1988) Variances and covariances of squared linkage disequilibria in finite populations. Theor Popul Biol 33:54–78

Huang X, Wei X, Sang T, Zhao Q, Feng Q, Zhao Y, Li C, Zhu C, Lu T, Zhang Z, Li M, Fan D, Guo Y, Wang A, Wang L, Deng L, Li W, Lu Y, Weng Q, Liu K, Huang T, Zhou T, Jing Y, Li W, Lin Z, Buckler ES, Qian Q, Zhang QF, Li J, Han B (2010) Genome-wide association studies of 14 agronomic traits in rice landraces. Nat Genet 42:961–967

Huang X, Zhao Y, Wei X, Li C, Wang A, Zhao Q, Li W, Guo Y, Deng L, Zhu C, Fan D, Lu Y, Weng Q, Liu K, Zhou T, Jing Y, Si L, Dong G, Huang T, Lu T, Feng Q, Qian Q, Li J, Han B (2012) Genome-wide association study of flowering time and grain yield traits in a worldwide collection of rice germplasm. Nat Genet 44:32–39

Izawa T (2007) Adaptation of flowering-time by natural and artificial selection in Arabidopsis and rice. J Exp Bot 58:3091–3097

Jakobsson M, Rosenberg NA (2007) CLUMPP: a cluster matching and permutation program for dealing with label switching and multimodality in analysis of population structure. Bioinformatics 23:1801–1806

Jeung JU, Hwang HG, Moon HP, Jena KK (2005) Fingerprinting temperate japonica and tropical indica rice genotypes by comparative analysis of DNA markers. Euphytica 146:239–251

Jiang L, Xun M, Wang J, Wan J (2008) QTL analysis of cold tolerance at seedling stage in rice (Oryza sativa L.) using recombination inbred lines. J Cereal Sci 48:173–179

Kaneda C, Beachell HM (1974) Response of indica-japonica rice hybrids to low temperatures. SABRAO J 6:17–32

Li H, Handsaker B, Wysoker A, Fennell T, Ruan J, Homer N, Marth G, Abecasis G, Durbin R, 1000 Genome Project Data Processing Subgroup (2009) The Sequence Alignment/Map format and SAMtools. Bioinformatics 25: 2078–2079

Lu JJ, Chang TT (1980) Rice in its temporal and spatial perspective. In: Luh BS (ed) Rice: production and utilization Westport, CT, USA., pp 1–74

Ma Y, Dai X, Xu Y, Luo W, Zheng X, Zhen D, Pan Y, Lin X, Liu H, Zhang D, Xiao J, Guo X, Xu S, Niu Y, Jin J, Zhang H, Xu X, Li L, Wang W, Qian Q, Ge S, Chong K (2015) COLD1 confers chilling tolerance in rice. Cell 160:1209–1221

Mackill DJ, Lei XM (1997) Genetic variation for traits related to temperate adaptation of rice cultivars. Crop Sci 37:1340–1346

Marco-Sola S, Sammeth M, Guigó R, Ribeca P (2012) The GEM mapper: fast, accurate and versatile alignment by filtration. Nat Methods 12:1185–1188

Marroni F, Pinosio S, Zaina G, Fogolari F, Felice N, Cattonaro F, Morgante M (2011) Nucleotide diversity and linkage disequilibrium in Populus nigra cinnamyl alcohol dehydrogenase (CAD4) gene. Tree Genet Genomes 7: 1011–1023

McCouch SR, Wright MH, Tung CW, Maron LG, McNally KL, Fitzgerald M, Singh N, DeClerck G, Agosto-Perez F, Korniliev P, Greenberg AJ, Naredo ME, Mercado SM, Harrington SE, Shi Y, Branchini DA, Kuser-Falcao PR, Leung H, Ebana K, Yano M, Eizenga G, McClung A, Mezey J (2016) Open access resources for genome-wide association mapping in rice. Nat Commun 7:10532

Myles S, Chia JM, Hurwitz B, Simon C, Zhong GY, Buckler E, Ware D (2010) Rapid genomic characterization of the genus vitis. PLoS One 5:e8219

Naranjo L, Talon M, Domingo C (2014) Diversity of floral regulatory genes of japonica rice cultivated at northern latitudes. BMC Genomics 15:101

Perrier X, Flori A, Bonnot F (2003) Data analysis method. In: Hamon P, Seguin M, Perrier X, Glaszmann JC (eds) Genetic diversity of cultivated tropical plants. Enfield, Science Publishers, Montpellier, pp 43–76

Pritchard JK, Stephens M, Donnelly P (2000) Inference of population structure using multilocus genotype data. Genetics 155:945–959

Purcell S, Neale B, Todd-Brown K, Thomas L, Ferreira MAR, Bender D, Maller J, Sklar P, de Bakker PIW, Daly MJ, Sham PC (2007) PLINK: a toolset for whole-genome association and population-based linkage analysis. Am J Hum Genet 81:559–575

Remington DL, Thornsberry JM, Matsuoka Y, Wilson LM, Whitt SR, Doebley J, Kresovich S, Goodman MM, Buckler ES IV (2001) Structure of linkage disequilibrium and phenotypic associations in the maize genome. Proc Natl Acad Sci U S A 98:11479–11484

Schneeberger K, Ossowski S, Lanz C, Juul T, Petersen AH, Nielsen KL, Jorgensen JE, Weigel D, Andersen SU (2009) SHOREmap: simultaneous mapping and mutation identification by deep sequencing. Nat Methods 6:550–551

The 3,000 rice genomes project (2014) The 3,000 rice genomes project. GigaScience 3:7. doi:10.1186/2047-217X-3-7

Wang J, Qi M, Liu J, Zhang Y (2015) CARMO: a comprehensive annotation platform for functional exploration of rice multi-omics data. Plant J 83:359–374

Xu X, Liu X, Ge S, Jensen JD, Hu F, Li X, Dong Y, Gutenkunst RN, Fang L, Huang L, Li J, He W, Zhang G, Zheng X, Zhang F, Li Y, Yu C, Kristiansen K, Zhang X, Wang J, Wright M, McCouch S, Nielsen R, Wang J, Wang W (2011) Resequencing 50 accessions of cultivated and wild rice yields markers for identifying agronomically important genes. Nat Biotechnol 30:105–111

Yoshida R, Kanno A, Sato T, Kameya T (1996) Cool-temperature-induced chlorosis in rice plants. Plant Physiol 110:997–1005

Zhang P, Liu X, Tong H, Lu Y, Li J (2014) Association mapping for important agronomic traits in core collection of rice (Oryza sativa L.) with SSR markers. PLoS One 9:e111508

Zhao K, Wright M, Kimball J, Eizenga G, McClung A, Kovach M, Tyagi W, Ali ML, Tung CW, Reynolds A, Bustamante CD, McCouch SR (2010) Genomic diversity and introgression in O. sativa reveal the impact of domestication and breeding on the rice genome. PLoS One 5:e10780

Zhao K, Tung C, Eizenga G, Wright M, Ali M, Price A, Norton G, Islam M, Reynolds A, Mezey J, McClung A, Bustamante C, McCouch S (2011) Genome-wide association mapping reveals a rich genetic architecture of complex traits in Oryza sativa. Nature Comm 2:467

6

Ability of *Rf5* and *Rf6* to Restore Fertility of Chinsurah Boro II-type Cytoplasmic Male Sterile *Oryza Sativa* (ssp. *Japonica*) Lines

Honggen Zhang, Jianlan Che, Yongshen Ge, Yan Pei, Lijia Zhang, Qiaoquan Liu, Minghong Gu and Shuzhu Tang[*]

Abstract

Background: Three-line *Oryza sativa* (ssp. *japonica*) hybrids have been developed mainly using Chinsurah Boro II (BT)-type cytoplasmic male sterility (CMS). The *Rf1* gene restores the fertility of BT-type CMS lines, and is the only fertility restorer gene (*Rf*) that has been used to produce three-line *japonica* hybrids. Using more *Rf* genes to breed BT-type restorer lines may broaden the genetic diversity of the restorer lines, and represents a viable approach to improve the heterosis level of BT-type *japonica* hybrids.

Results: We identified two major *Rf* genes from '93-11' that are involved in restoring the fertility of BT-type CMS plants. These genes were identified from resequenced chromosome segment substitution lines derived from a cross between the *japonica* variety 'Nipponbare' and the *indica* variety '93-11'. Molecular mapping results revealed that these genes were *Rf5* and *Rf6*, which are the *Rf* genes that restore fertility to Honglian-type CMS lines. The BT-type F_1 hybrids with either *Rf5* or *Rf6* exhibited normal seed setting rates, but F_1 plants carrying *Rf6* showed more stable seed setting rates than those of plants carrying *Rf5* under heat-stress conditions. Furthermore, the seed setting rates of F_1 hybrids carrying both *Rf5* and *Rf6* were more stable than that of F_1 plants carrying only one *Rf* gene.

Conclusion: *Rf6* is an important genetic resource for the breeding of BT-type *japonica* restorer lines. Our findings may be useful for breeders interested in developing BT-type *japonica* hybrids.

Keywords: *Japonica*, BT-type CMS, Fertility restorer gene (*Rf*), Gene mapping, Restoration ability

Background

Cytoplasmic male sterility (CMS) is caused by chimeric open reading frames in the mitochondrial genome, and is common in higher plants. It is a maternally inherited trait that results in an inability to produce functional pollen. Male sterility can be restored by the fertility restorer gene (*Rf*) in the nuclear genome (Hanson and Bentolila 2004). The CMS/*Rf* system has been widely used for hybrid seed production, and has helped to clarify the interactions between mitochondrial and nuclear genomes in plants (Chase 2007; Havey 2004).

The grain yields from three-line hybrid rice developed using the CMS/*Rf* system are 15–30% higher than those

of inbred rice varieties (Fujimura et al. 1996; Yuan 1994). A CMS line, maintainer line, and a restorer line that carries the *Rf* gene are combined to develop three-line hybrids. For three-line hybrid rice, wild abortive (WA) and Honglian (HL) are the representative CMS types used for commercial *indica* hybrid seed production, and Chinsurah Boro II (BT) is the representative CMS type used for generating *japonica* hybrids (Chen and Liu 2014; Huang et al. 2014; Li et al. 2007; Yuan 1994). The BT-type cytoplasm has been identified from *Oryza sativa* ssp. *indica* Chinsurah Boro II and the BT-type CMS/*Rf* system is gametophytic. The BT-type CMS is caused by *orf79* in mitochondria (Wang et al. 2006), and the recovery of pollen fertility is regulated by *Rf1*, a fertility restorer gene on chromosome 10 (Shinjyo 1975; Wang et al. 2006). The *Rf1* gene has been cloned, and two alleles, *Rf1a* and *Rf1b*, have been identified at the *Rf1* locus (Kazama and Toriyama 2003; Akagi et al.

* Correspondence: sztang@yzu.edu.cn
Jiangsu Key Laboratory of Crop Genetics and Physiology/Co-Innovation Center for Modern Production Technology of Grain Crops, Key Laboratory of Plant Functional Genomics of the Ministry of Education, College of Agriculture, Yangzhou University, Yangzhou 225009, China

2004; Komori et al. 2004; Wang et al. 2006). *Rf1a* and *Rf1b* both encode pentatricopeptide-repeat containing proteins. RF1A can promote endonucleolytic cleavage of the *atp6–orf79* mRNA, whereas RF1B promotes degradation of *atp6–orf79* mRNA (Wang et al. 2006). Although progress has been made toward characterizing the mechanisms underlying the development of BT-type CMS and the restoration of fertility by *Rf* genes, only *Rf1* has been used to breed BT-type restorers. This has resulted in limited genetic diversity among BT-type restorers, and this is a major obstacle for the further development of BT-type *japonica* hybrids (Chen and Liu 2014). Thus, three-line *japonica* hybrids are not extensively cultivated in China (Deng et al. 2006). Also, F_1 hybrids developed using the gametophytic CMS/*Rf* systems with multiple restorer gene loci can produce more than 50% normal pollen grains and exhibit good seed setting rates (Huang et al. 2012; Komori and Imaseki 2005), which may be useful for breeding BT-type *japonica* hybrids.

Several genetic loci related to the restoration of fertility in different types of CMS lines have been mapped in rice. *Rf3* and *Rf4* are two major fertility restorer genes for WA-type CMS, and have been mapped on chromosomes 1 and 10, respectively (Ahmadikhah and Karlov 2006; Tang et al. 2014; Zhang et al. 1997). *Rf5* and *Rf6*, which have been cloned and mapped on chromosomes 10 and 8, respectively, are the two major fertility restorer genes for HL-type CMS (Hu et al. 2012; Huang et al. 2000, 2012, 2015). *Rf17* (chromosome 4) and *Rf2* (chromosome 2) are the fertility restorer genes for Chinese wild rice-type CMS (Fujii and Toriyama 2009) and Lead rice-type CMS (Itabashi et al. 2011), respectively. Identifying new *Rf* genes and/or using previously reported *Rf* genes (except *Rf1*) to breed BT-type *japonica* fertility restorers may accelerate the development and application of BT-type *japonica* hybrids. This potentially represents an effective way to increase rice yields in China.

In this study, we used 57 chromosome segment substitution lines (CSSLs) derived from a cross between the *japonica* variety 'Nipponbare' (NIP; recipient parent) and the *indica* restorer '93-11' (donor parent) as males to detect *Rf* genes for BT-type CMS. To identify possible *Rf* genes for BT-type CMS in '93-11', a population of 127 CSSLs from this cross were reconstructed and resequenced. We then mapped the detected *Rf* genes and evaluated their ability to restore fertility to BT-type *japonica* CMS lines. Our results will be useful for breeding BT-type *japonica* restorer lines and for the development of three-line *japonica* hybrids.

Results
Construction of CSSLs and whole-genome resequencing
Between 2004 and 2010, a set of CSSLs was developed via crossing and backcrossing with the aid of 140 molecular markers, and the genotypes of 56 CSSLs (L1–L56) were determined using a high-throughput resequencing strategy. Based on the resequencing data, 152 substituted segments derived from '93-11' were identified. These segments covered about 89.27% of the rice genome (Zhang et al. 2011). We observed that most lines contained five or more introgressed chromosome segments with many gaps. To obtain additional lines with a single segment and a high coverage rate, we reconstructed the CSSLs from the previously generated lines using backcrosses and marker-assisted selection (MAS) with 357 polymorphic molecular markers distributed relatively evenly across the 12 rice chromosomes. We ultimately obtained 351 CSSLs, and 127 lines were selected for resequencing. According to the resequencing results, the 127 CSSLs harbored 362 substituted segments derived from line '93-11'. The segments were 0.02–24.60 Mb long (average length: 5.39 Mb). For each CSSL, the number of substituted chromosomal segments varied from one to seven. Ninety-seven CSSLs contained two or fewer introgressed chromosomal segments, with 53 lines carrying only one segment. The total length of these substituted segments was 2023.8 Mb, which was 5.08 times the length of the rice genome. Furthermore, the CSSLs covered 94.93% of the '93-11' genome (Table 1).

Two major genes restore fertility of BT-type CMS lines
In the summer of 2011, the seed setting rate of 52 test-cross F_1 lines was analyzed. Six F_1 hybrids derived from crosses between NIPA, a BT-type *japonica* CMS line with the nuclear background of NIP and L1, L18–20, L30, and L47 exhibited bagged and natural seed setting rates of 9.16–63.46% and 45.74–92.67%, respectively (Table 2). The other F_1 hybrids showed bagged seed setting rates of zero and extremely low natural seed setting rates (i.e., < 5%). Therefore, we considered that these lines were all sterile and the natural seed setting rates of these lines were caused by outcrossing.

To detect quantitative trait loci (QTLs) associated with the fertility restoration of NIPA, we constructed a bin map based on 56 introgression lines, two parents, and the high-throughput resequencing data. A total of 366 bins were generated, with sizes of 0.01–11.92 Mb. The QTL analysis was conducted using the SARS program, and three QTLs for natural spikelet fertility were detected (Table 3). Among these QTLs, *qSF8-1* was located in bin 222 on chromosome 8, *qSF10-1* was located in bin 314 on chromosome 10, and *qSF12-1* was located in bin 349 on chromosome 12. Bin 222 was present in L1, L18, L19, and L20, while L30 and L47 harbored bin 314, and L47 harbored bin 349 (Table 3). The QTL mapping results revealed that bins 349 and 314 were present in L47. The chromosome interval related to bin 314

Table 1 Distribution of substituted segments from 127 CSSLs

Chromosome	Number of donor segment	Size of donor segment (Mb)	Density (%)
1	24	93.63	92.02
2	35	277.11	100
3	45	224.13	100
4	25	149.87	86.19
5	19	94.7	97.27
6	37	268.85	100
7	42	231.5	100
8	31	198.48	98.53
9	31	153.28	100
10	16	99.95	100
11	22	128.06	91.46
12	35	104.24	81.54
Total	362	2023.8	94.93

Table 3 Mapping QTLs related to the fertility restoration of BT-type CMS lines

QTLs	Bin	Chr.	Interval	Size	Partial R-square	F
qSF8-1	222	8	1–688,039	688,038	0.6416	76.99
qSF10-1	314	10	16,949,055–18,486,880	153,7825	0.194	49.6
qSF12-1	349	12	4,816,707–6,620,947	1,804,240	0.2255	36.59

covered the *Rf1* locus (Akagi et al. 1996; Komori et al. 2004; Wang et al. 2006). Thus, we considered the possibility that *qSF12-1* does not exist.

Because of the gametophytic manner in which *Rf* genes restore fertility in the BT-type CMS/*Rf* system, plants with the *rfrf* genotype were not present in the F_2 populations. Therefore, most plants likely carried homozygous '93-11' or heterozygous genotypes at the linkage marker loci if there was only one *Rf* gene that restored the fertility of BT-type CMS. Plants carrying the homozygous NIPA genotype were considered to be recombinants. In the spring of 2012, six F_2 populations derived from the fertile F_1 hybrids were planted, with each population consisting of 100 individuals. We developed five simple sequence repeat markers on chromosome 8 (1–688,039 bp) and two insertion/deletion markers on chromosome 10 (16,949,055–18,486,880 bp). The markers were used to detect polymorphisms between NIP and '93-11'. RM1019 and RM407 on chromosome 8 and STS10-16 and STS10-27 on chromosome 10 were determined to be polymorphic. RM407 was subsequently used to detect the genotypes of plants in the F_2 populations derived from the following crosses: NIPA/L1, NIPA/L18, NIPA/L19, and NIPA/L20. All 400 analyzed

Table 2 Spikelet fertility of fertile F_1 lines in testcross populations

Combination	Bagged seec setting rate (%) (mean ± SD)	Natural seed setting rate (%) (mean ± SD)
NIPA/L1	63.46 ± 19.24	82.05 ± 3.95
NIPA/L18	46.21 ± 11.08	92.67 ± 3.46
NIPA/L19	38.92 ± 15.15	85.18 ± 2.72
NIPA/L20	35.46 ± 11.91	92.12 ± 2.70
NIPA/L30	9.16 ± 4.73	45.74 ± 6.17
NIPA/L47	48.13 ± 7.95	89.42 ± 10.74

plants carried homozygous '93-11' or heterozygous genotypes. STS10-16 and STS10-27 were used to detect the genotypes of plants in the F_2 populations derived from the NIPA/L30 and NIPA/L47 crosses, respectively. Almost all of the 200 examined plants exhibited homozygous '93-11' or heterozygous genotypes. The exception was one plant that carried a homozygous NIPA genotype at STS10-27. These results indicated that all of L1, L18–20, L30, and L47 carried only one *Rf* gene, and that *qSF8-1* and *qSF10-1* were two major QTLs conferring the ability to restore fertility to BT-type CMS lines.

In the summer of 2014, natural spikelet fertility was assessed in 123 F_1 lines derived from the cross between NIPA and the reconstructed CSSLs. Among the F_1 lines, the populations generated from crosses between NIPA and four CSSLs (i.e., N91, N93, N116, and N119) exhibited natural seed setting rates of 90.95–94.51%, and the other lines were sterile. According to the high-throughput resequencing data, N91 and N93 carried a substituted segment covering *qSF8-1*, whereas N116 and N119 carried a substituted segment spanning *qSF10-1*. These results were confirmed by molecular detection with RM407, STS10-16, and STS10-27. Based on these results, we concluded that there were two non-allelic nuclear genes in '93-11' that restored the fertility of BT-type CMS *japonica* lines.

Molecular mapping of *qSF10-1* and *qSF8-1*

qSF10-1 is located in a region that includes the *Rf1* locus, which contains *Rf1a*, or is tightly linked with *Rf1a* and *Rf1b* on chromosome 10 in BT-type *japonica* restorers (Akagi et al. 1996; Komori et al. 2004; Wang et al. 2006). *Rf5*, which is a dominant *Rf* gene associated with HL-type CMS in *indica* rice, is the same gene as *Rf1a* (Hu et al. 2012). To determine whether *qSF10-1* was *Rf1a* (*Rf5*), we sequenced the *Rf1a* allele in '93-11', L37, and L47. Comparison of the nucleotide sequences of *Rf1a* and *Rf5* with those previously cloned from different restorers revealed that the sequence of the *Rf1a* allele from '93-11'is the same as the sequences of the corresponding genes in IR8 (Akagi et al. 2004) and Miyang 23 (Hu et al. 2012). This result confirmed that *qSF10-1* is *Rf1a* (*Rf5*).

To clarify the precise genomic position of *qSF8-1*, we developed 51 markers in the region containing RM407, and 14 polymorphic markers were obtained for mapping (Additional file 1: Table S1). In the summer of 2012, two markers (i.e., RM1019 and RM22271) were used to screen recombinant individuals from more than 4000 plants in the F_2 and F_3 populations. We identified 23 and 17 recombinants using RM1019 and RM22271, respectively. Another four markers (i.e., STS8-4, STS8-23, RM407, and STS8-32) were used to genotype the 40 recombinants. Among these recombined plants, two were detected by RM407 and one was detected by STS8-32. Thus, based on the reference sequence of the NIP genome, *qSF8-1* was localized to an approximately 35.5-kb region between the markers RM407 and STS8-32 on the short arm of chromosome 8, within which an open reading frame (ORF) (Os08g0110200) encoding PPR-containing protein might be the candidate gene for *qSF8-1*. We sequenced this ORF from 93-11 to L18, respectively. The nucleotide sequences were identical between 93-11 and L18, and 93-11 contains 16 nucleotide substitutions and one insertions of 2 bp at position +2315 compared with that of NIP. During our mapping work, Huang et al. (2012) reported the mapping results of the *Rf6* gene, which is also from '93-11', and restores the fertility of HL-type *indica* CMS lines. Our mapping study essentially produced the same results. Furthermore, the cloning of *Rf6* indicated that *Rf6* is the ORF of Os08g0110200 and is also capable of restoring the fertility of BT-type CMS plants (Huang et al. 2015). Therefore, we concluded that *qSF8-1* is *Rf6*.

Ability of *Rf6* to restore fertility is stable and two non-allelic restorer genes enhance heat stress tolerance in F_1 hybrids

Among the samples sown in the field on May 10 (S1) in Yangzhou, most of the testcross F_1 lines from the CSSLs headed on August 6–9, similar to the heading dates of NIP plants. In 2011, of the six fertile testcross F_1 lines, the NIPA/L30 F_1 plants headed on August 24. The daily maximum temperature during the flowering period was 24.0–28.0 °C (Additional file 2: Figure S1a), which was lower than the temperature during the flowering periods for the five other testcross F_1 lines. The NIPA/L30 F_1 plants exhibited a low seed setting rate (Table 2). These results indicated that exposure to low temperatures may influence the seed setting rates of F_1 plants. Considering the temperature fluctuations during the flowering period of the testcross F_1 lines, two sowing dates were used in 2012 [May 10 (S1) and May 20 (S2)], and 2013 [May 10 (S1) and June 5 (S3)]. In 2012, the testcross F_1 plants and the F_2 plants with different genotypes headed on August 6 and August 9, respectively. They also exhibited normal spikelet fertility (i.e., > 80%), indicating that *Rf5* and *Rf6* were able to restore fertility to BT-type CMS lines (Table 4). In 2013, the F_1 plants and F_3 plants carrying different genotypes headed on August 6 and

August 20, respectively. The highest temperature (i.e., 37.0 °C) was first recorded on August 6, and high daytime temperatures continued for more than 1 week (Additional file 2: Figure S1b). Among the plants that headed on August 6, the seed setting rates of plants carrying heterozygous genotypes were relatively low, while the plants carrying homozygous genotypes exhibited normal seed setting rates (Table 4). In contrast, the seed setting rate was normal for all plants that headed on August 20. These results implied that heat stress affects the ability of *Rf5* and *Rf6* to restore fertility to BT-type CMS plants.

To further evaluate the ability of F_1 hybrids with one or two *Rf* genes to adapt to heat stress, we developed the following three F_1 populations: NIPA/L18 (*Rf6rf6*), NIPA/L47 (*Rf5rf5*), and (NIPA/L18) F_2//L47 (*Rf5rf5Rf6rf6*). We also selected the plants harboring homozygous genotypes in the NIPA/L18 and NIPA/L47 F_4 populations, which were exposed to high temperatures (25 °C nights, and maximum daytime temperatures of 37 °C) in 2014. These plants headed on August 6 when grown in the field, and on August 1 when grown in the greenhouse. Compared with the untreated plants, the F_1 hybrids with only one *Rf* gene exhibited poorer seed setting rates, and the seed setting rate of plants with *Rf6* was higher than that of plants with *Rf5* (Fig. 1). In contrast, F_1 plants carrying two restorer genes and plants with homozygous genotypes at *Rf* loci exhibited normal seed setting rates (Fig. 1). These results revealed that the presence of two non-allelic restorer genes in F_1 hybrids increased the stability of their seed setting rate during exposure to environmental stresses such as high temperature.

Discussion

Three-line *japonica* hybrids developed mainly with BT-type CMS lines have contributed greatly to rice production in China (Huang et al. 2014; Li et al. 2007). The BT-type CMS lines can be developed by nuclear substitution using *japonica* varieties from Japan and China as the recurrent paternal parent in backcrosses. Most BT-type *japonica* restorers contain only the *Rf1* locus (Akagi et al. 1996; Chen and Liu 2014; Huang et al. 2014; Komori et al. 2004; Wang et al. 2006), which was first transferred from *indica* varieties cultivated in Southeast Asia (Li et al. 2007). The F_1 hybrids derived from BT-type *japonica* restorers exhibit a normal seed setting rate, with 50% of the produced pollen grains being fertile. Because BT-type *japonica* restorers contain only *Rf1*, the genetic diversity of restorers is relatively low. Additionally, the systematic use of heterosis for breeding BT-type *japonica* hybrid rice has been limited. Therefore, exploiting other *Rf* genes may be an effective way to increase the genetic diversity of BT-type restorers, which will likely enhance the development of *japonica* hybrids in China.

Table 4 Fertility levels of plants carrying different genotypes in 2012 and 2013

Combination	Genotype	Seed setting rate (%) (mean ± SD)			
		2012S1	2012S2	2013S1	2013S3
NIPA/L18 F_1	Rf6rf6	87.41 ± 2.30	86.63 ± 5.95	67.22 ± 10.02	–
NIPA/L47 F_1	Rf5rf5	–	–	36.85 ± 13.12	86.31 ± 6.56
(NIPA/L18) F_2//L47 F_1	Rf6rf6Rf5rf5	–	–	75.10 ± 12.91	–
NIPA/L18 F_{2-4}	Rf6Rf6	90.12 ± 4.24	88.46 ± 3.64	82.46 ± 3.87	89.77 ± 5.17
	Rf6rf6	87.78 ± 6.89	87.32 ± 7.16	–	85.36 ± 5.39
NIPA/L47 F_{2-4}	Rf5Rf5	86.89 ± 4.40	88.12 ± 3.52	82.77 ± 5.37	–
	Rf5rf5	85.45 ± 3.02	85.01 ± 5.88	41.37 ± 8.67	–

–: missed genotype
S1, S2, and S3 means the sowing dates of May 10, May 20 and June 5, respectively

In this study, two *Rf* genes (i.e., *qSF8-1* and *qSF10-1*) involved in restoring the fertility of NIPA plants were identified from '93-11', which is an elite *indica* fertility restorer for HL-type CMS. *Rf5* and *Rf6*, the two non-allelic fertility restorer genes in '93-11' associated with HL-type CMS, have been mapped and cloned (Huang et al. 2012, 2015). *Rf5* and *Rf1a* are the same gene, and *Rf6* is also capable of restoring the fertility of BT-type CMS plants (Huang et al. 2015). We mapped *qSF10-1* to a region on chromosome 10 containing *Rf5* (*Rf1a*) and mapped *qSF8-1* to a region on chromosome 8 containing *Rf6*. The '93-11' *Rf5* and *Rf6* allele were sequenced. The sequencing results confirmed that *qSF10-1* is *Rf5* (*Rf1a*) and *qSF8-1* is *Rf6*. Therefore, the two '93-11' *Rf* genes associated with BT-type CMS are *Rf5* and *Rf6*. Indeed, rice breeding experiments have revealed similarities in the restoration and maintenance relationship between BT-type CMS and HL-type CMS (Li et al. 2007; Zhang et al. 2016), consistent with the mapping results of this study.

The *Rf5* and *Rf6* genes exhibit similar abilities to restore the fertility of HL-type *indica* CMS lines (Huang et al. 2012). We observed that only a major gene (i.e., *Rf5* or *Rf6*) can restore normal fertility to BT-type CMS lines. However, this study revealed for the first time that the ability to restore fertility to BT-type *japonica* CMS lines is more stable with *Rf6* than with *Rf5* under heat-stress conditions. In the HL-type CMS/*Rf* system, *Rf6* and *Rf5* function via distinct mechanisms to rescue the sterility of HL-type CMS (Huang et al. 2012, 2015). Therefore, we hypothesize that the molecular mechanism underlying the fertility restoration of BT-type CMS plants differs between *Rf5* and *Rf6*, leading to the observed differences in the restorer activities of *Rf5* and *Rf6*. Additional studies are required to test this hypothesis.

In this study, F_2 plants harboring *Rf5Rf5* or *Rf6Rf6* and plants harboring *Rf5rf5Rf6rf6* in the F_1 population of a three-way cross exhibited the most stable fertility levels. These observations imply that breeding BT-type *japonica* restorers with multiple dominant *Rf* genes may

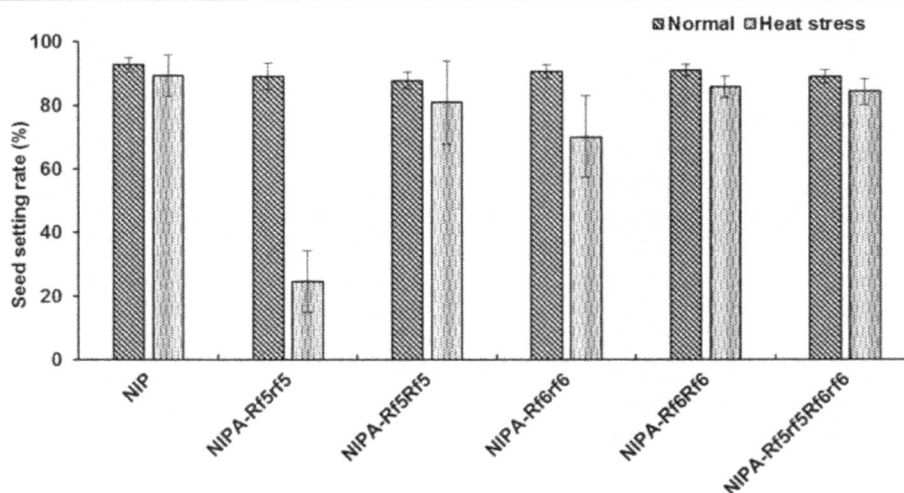

Fig. 1 Effects of high temperatureson seed setting rates of plants harboring different genotypes. *Error bars* of seed setting rate is the mean value ± SD ($n = 10$). The heat stress is treated with the maximum temperatures of 37 ℃

increase the stability of seed setting in BT-type hybrid *japonica* rice. Therefore, using *Rf6* to restore fertility to BT-type CMS plants may facilitate the exploitation of heterosis in *japonica* breeding.

Conclusion

We determined that *Rf5* and *Rf6* are the major *Rf* genes associated with BT-type CMS, and that the restorer activity of *Rf6* is more stable than that of *Rf5* under heat-stress conditions. The *Rf6* gene in '93-11' may be an important resource for breeding BT-type *japonica* restorers to broaden the genetic diversity of BT-type rice. Also, combining *Rf6* with *Rf5* (*Rf1a*) in most pre-existing BT-type restorers may be an effective way to breed new BT-type restorers that exhibit stable seed setting abilities.

Methods

Plant materials

The donor parent used in this study was '93-11', which is a typical *indica* cultivar and fertility restorer for BT-type and HL-type CMS. The recipient parent was NIP, which is a *japonica* cultivar and maintainer for BT-type CMS. The genomes of both parents have been sequenced. Between 2004 and 2010, a set of CSSLs derived from the cross between NIP and '93-11' was developed by crossing and backcrossing with MAS, of which 56 lines were genotyped in 2010 using a resequencing method (Zhang et al. 2011). Based on the resequencing results, another 127 CSSLs from the same cross were further developed by backcrossing with MAS and then genotyped by resequencing (Additional file 3: Figure S2). In the spring of 2011 and 2014, these CSSLs were used as males for crosses with 'NipponbareA' (NIPA), which has the BT-type sterile cytoplasm with the nuclear background of NIP. We generated 52,123 F_1 hybrids.

In 2011 and 2012, several F_2 and F_3 populations were generated from plants harboring the heterozygous genotype for *Rf* genes in the NIPA/CSSL testcross population. The corresponding progenies were used for the fine mapping of genes. To evaluate the ability of identified *Rf* genes to restore the fertility of BT-type CMS lines, we crossed NIPA with the CSSLs carrying various *Rf* genes. Plants harboring the homozygous or heterozygous genotypes for *Rf* genes in the F_2–F_4 populations were identified by MAS between 2012 and 2014. Plants harboring the homozygous genotype for *Rf* genes were crossed with each other in 2013 and 2014.

Field experiment

All plant materials were sown on three dates [i.e., May 10 (S1), May 20 (S2), and June 5 (S3)] between 2011 and 2014 at the experimental field of Yangzhou University in Yangzhou, Jiangsu, China. Additionally, in the summer of 2014, plants carrying different genotypes for *Rf* genes were exposed to heat stress (25 °C nights, and maximum daytime temperatures of 37 °C) in a greenhouse for 10 days during the flowering period. Each field plot consisted of four rows separated by 25 cm, with each row consisting of five plants separated by 20 cm. Plants were grown using normal rice cultivation practices.

Fertility scoring

In 2011, we analyzed the spikelet fertility in natural and bagged panicles of five plants from the CMS lines and each line in the testcross population. Between 2012 and 2014, only the spikelet fertility in natural panicles of five plants was assessed in the testcross F_1 population, F_2–F_4 populations, and the three-way-cross F_1 population. We counted the filled and unfilled grains of two panicles from one plant harvested at 20 days after flowering, and the spikelet fertility of one plant was measured as the average seed setting rate.

Genotyping CSSLs by whole-genome resequencing

We used the high-throughput genotyping method developed by Huang et al. (2009). The CSSLs were genotyped based on single nucleotide polymorphisms (SNPs) generated from whole-genome resequencing as previously described (Xu et al. 2010). Briefly, at least 5 µg genomic DNA from each sample was randomly fragmented by sonication and electrophoretically size-fractionated. The DNA fragments approximately 500 bp long were then purified. Adapters were ligated to the purified fragments, which were then clustered and sequenced using the Illumina HiSeq 2000 system according to the manufacturer's instructions (Illumina, San Diego, CA, USA). According to the published NIP genome sequence (http://rgp.dna.affrc.go.jp/IRGSP/Build4/build4.html), the detected SNPs were arranged along the chromosomes according to their physical locations. The genotypes of these CSSLs were evaluated by a group of consecutive SNPs using a sliding window approach, and different window sizes were used according to the SNP density of the CSSLs. The physical length of the substituted segments in each CSSL was estimated based on the resequencing results for each CSSL. The overlapping donor segments between or among different lines were used to divide the fragments into smaller segments (i.e., bins).

Quantitative trait locus mapping and data analysis

Statistical analyses of the fertility of the plant materials and populations were conducted using the analysis of variance package in MATLAB (version 7.0). Commonly used bin mapping schemes were constructed and the contributions of the target bins to phenotypic variability were determined as described by Xu et al. (2010). The nomenclature of QTLs for natural spikelet fertility was according to the standard procedure described elsewhere (McCouch et al. 2002).

DNA extraction, polymerase chain reaction, and sequencing

Genomic DNA was isolated from fresh leaves using cetyltrimethylammonium bromide (Rogers and Bendich 1985). Simple sequence repeat markers were identified using the Gramene database (http://www.gramene.org/). New insertion/deletion markers for the NIP (*japonica*) and '93-11' (*indica*) genome sequences (http://www.ncbi.nlm.nih.gov/) were identified using the BLAST online tool. Primers were synthesized by Shanghai Sangon Inc. (Shanghai, China). Molecular marker analyses were conducted in 1× reaction buffer containing 0.1 mM dNTP, 1.0 U *Taq* polymerase, 0.2 μM primer, and 20 ng template DNA. The final volume was adjusted to 20 μL with ultra-pure water. The polymerase chain reaction (PCR) program was as follows: 94 °C for 4 min; 30 cycles of 94 °C for 45 s, 55 °C for 45 s, and 72 °C for 50 s; with final extension at 72 °C for 5 min. The amplification products were separated by 3% (*w/v*) agarose gel electrophoresis, detected with ethidium bromide, and visualized with a Gel Doc 1000 system (Bio-Rad Laboratories, Hercules, CA, USA).

According to the mapping results, the *Rf1a* allele and *Rf6* allele in '93-11' were sequenced. The gene fragments were amplified by PCR using PrimeSTAR® GXL DNA polymerase (Takara, Dalian, China). The PCR products were purified using a TIANGEN PCR purification kit, and then ligated into the pEASY-Blunt cloning vector. The plasmid DNA was sequenced by GENEWIZ (Suzhou, China), and the correct DNA sequence was identified by comparing five plasmids. Sequence alignment was conducted with BLAST tools provided by the National Center for Biotechnology Information. The primers used to sequence the *Rf1a* and *Rf6* allele are listed in Additional file 1: Table S1.

Accession number

Sequence data generated in this study have been deposited in the GenBank/EMBL database (accession numbers [KX517796, KY387609]).

Abbreviations

BT: Chinsurah Boro II; CMS: Cytoplasmic male sterility; CSSL: Chromosome segment substitution lines; HL: Honglian; Nipponbare: NIP; NipponbareA: NIPA; PCR: Polymerase chain reaction; QTLs: Quantitative trait loci; *Rf*: Fertility restorer gene

Acknowledgments

This study was financially supported by the National Key Research and Development Program (2016YFD0101107), the National Natural Science Foundation (31071384), and the Priority Academic Program Development of Jiangsu Higher Education Institutions.

Authors' contributions

HZ analyzed the data and drafted the manuscript. JC and YG completed the phenotypic evaluations and data analyses. YP and LZ helped construct the populations. QL and MG were involved in designing the study. ST designed the study and revised the manuscript. All authors have read and approved the final manuscript.

Authors' information

HZ is a lecturer at Yangzhou University (China), JC, YG and LZ are Masters students at Yangzhou University (China), and QL, MG, and ST are professors at Yangzhou University (China).

Competing interests

The authors declare that they have no competing interests.

References

Ahmadikhah A, Karlov GI (2006) Molecular mapping of the fertility-restoration gene *Rf4* for WA-cytoplasmic male sterility in rice. Plant Breed 125:363–367

Akagi H, Yokozeki Y, Inagaki A, Nakamura A, Fujimura T (1996) A codominant DNA marker closely linked to the rice nuclear restorer gene, *Rf1*, identified with inter-SSR fingerprinting. Genome 39:1205–1209

Akagi H, Nakamura A, Yokozeki-Misono Y, Inagaki A, Takahashi H, Mori K, Fujimura T (2004) Positional cloning of the rice *Rf-1* gene, a restorer of BT-type cytoplasmic male sterility that encodes a mitochondria-targeting PPR protein. Theor Appl Genet 108:1449–1457

Chase CD (2007) Cytoplasmic male sterility: a window to the world of plant mitochondrial–nuclear interactions. Trends Genet 23:81–90

Chen LT, Liu YG (2014) Male sterility and fertility restoration in crops. Annu Rev Plant Biol 65:579–606

Deng HF, He Q, Shu F, Zhang WH, Yang F, Jing YH, Dong L, Xie H (2006) Status and technical strategy on development of japonica hybrid rice in china. Hybrid Rice 21:1–6

Fujii S, Toriyama K (2009) Suppressed expression of retrograde-regulated male sterility restores pollen fertility in cytoplasmic male sterile rice plants. Proc Natl Acad Sci U S A 106:9513–9518

Fujimura T, Akagi H, Oka M, Nakamura A, Sawada R (1996) Establishment of a rice protoplast culture and application of an asymmetric protoplast fusion technique to hybrid rice breeding. Plant Tissue Cult Lett 13:243–247

Hanson MR, Bentolila S (2004) Interactions of mitochondrial and nuclear genes that affect male gametophytic development. Plant Cell 16:S154–S169

Havey M (2004) The use of cytoplasmic male sterility for hybrid seed production. In: Daniell H, Chase C (eds) Molecular biology and biotechnology of plant organelles. Springer Publishers, Berlin, pp 617–628

Hu J, Wang K, Huang WC, Liu G, Gao Y, Wang JM, Huang Q, Ji YX, Qin XJ, Wan L, Zhu RS, Li SQ, Yang DC, Zhu YG (2012) The rice pentatricopeptide repeat protein RF5 restores fertility in Hong-Lian cytoplasmic male-sterile lines via a complex with the glycine-rich protein GRP162. Plant Cell 24:109–122

Huang QY, He YQ, Jing RC, Zhu RS, Zhu YG (2000) Mapping of the nuclear fertility restorer gene for HL cytoplasmic male sterility in rice using microsatellite markers. Chin Sci Bull 45:430–432

Huang XF, Feng Q, Qian Q, Zhao Q, Wang L, Wang A, Guan JP, Fan DL, Weng QJ, Huang T, Dong GJ, Sang T, Han B (2009) High-throughput genotyping by whole-genome resequencing. Genome Res 19:1068–1076

Huang WC, Hu J, Yu CC, Huang Q, Wan L, Wang LL, Qin XJ, Ji YX, Zhu RS, Li SQ, Zhu YG (2012) Two non-allelic nuclear genes restore fertility in a gametophytic pattern and enhance abiotic stress tolerance in the hybrid rice plant. Theor Appl Genet 124:799–807

Huang JZ, E ZG, Zhang HL, Shu QY (2014) Workable male sterility systems for hybrid rice: Genetics, biochemistry, molecular biology, and utilization. Rice 7:1–14

Huang WC, Yu CC, Hu J, Wang LL, Dan ZW, Zhou W, He CL, Zeng YF, Yao GX, Qi JZ, Zhang ZH, Zhu RS, Chen XF, Zhu YG (2015) Pentatricopeptide-repeat family protein RF6 functions with hexokinase 6 to rescue rice cytoplasmic male sterility. Proc Natl Acad Sci U S A 112:14984–14989

Itabashi E, Iwata N, Fujii S, Kazama T, Toriyama K (2011) The fertility restorer gene, *Rf2*, for Lead Rice-type cytoplasmic male sterility of rice encodes a mitochondrial glycine-rich protein. Plant J 65:359–367

Kazama T, Toriyama K (2003) A pentatricopeptide repeat-containing gene that promotes the processing of aberrant *atp6* RNA of cytoplasmic male-sterile rice. FEBS Lett 544:99–102

Komori T, Imaseki H (2005) Transgenic rice hybrids that carry the *Rf-1* gene at multiple loci show improved fertility at low temperature. Plant Cell Environ 28:425–431

Komori T, Ohta S, Murai NY, Kuraya Y, Suzuki S, Hiei Y (2004) Map-based cloning of a fertility restorer gene, *Rf-1*, in rice (*Oryza sativa* L.). Plant J 37:315–325

Li SQ, Yang DC, Zhu YG (2007) Characterization and use of male sterility in hybrid rice breeding. J Integr Plant Biol 49:791–804

McCouch S, Teytelman L, Xu Y, Lobos K, Clare K, Walton M, Fu B, Maghirang R, Li Z, Xing Y, Zhang Q, Kono I, Yano M, Fjellstrom R, DeClerck G, Schneider D, Cartinhour S, Ware D, Stein L (2002) Development and mapping of 2240 new SSR markers for ricee (*Oryza sativa* L.). DNA Res 9:199–207

Rogers SO, Bendich AJ (1985) Extraction of DNA from milligram amounts of fresh, herbarium and mummified plant tissues. Plant Mol Biol 5:69–76

Shinjyo C (1975) Genetical studies of cytoplasmic male sterility and fertility restoration in rice (*Oryza sativa* L). Sci Bull Coll Agric Univ Ryukyus 22:1–57

Tang HW, Luo DP, Zhou DG, Zhang QY, Tian DS, Zheng XM, Chen LT, Liu YG (2014) The rice restorer *Rf4* for wild-abortive cytoplasmic male sterility encodes a mitochondrial-localized PPR protein that functions in reduction of *WA352* transcripts. Mol Plant 7:1497–1500

Wang ZH, Zou YJ, Li XY, Zhang QY, Chen LT, Wu H, Su DH, Chen YL, Guo JX, Luo D, Long YM, Zhong Y, Liu YG (2006) Cytoplasmic male sterility of rice with boro II cytoplasm is caused by a cytotoxic peptide and is restored by two related PPR motif genes via distinct modes of mRNA silencing. Plant Cell 18:676–687

Xu JJ, Zhao Q, Du PN, Xu CW, Wang BH, Feng Q, Liu QQ, Tang SZ, Gu MH, Han B, Liang GH (2010) Developing high throughput genotyped chromosome segment substitution lines based on population whole-genome re-sequencing in rice (*Oryza sativa* L.). BMC Genomics 11:656

Yuan LP (1994) Increasing yield potential in rice by exploitation of heterosis. In: Virmanni SS (ed) Hybrid rice technology. New developments and future prospects. IRRI, Manila, pp 1–6

Zhang G, Lu Y, Bharaj TS, Virmani SS, Huang N (1997) Mapping of the *Rf-3* nuclear fertility-restoring gene for WA cytoplasmic male sterility in rice using RAPD and RFLP markers. Theor Appl Genet 94:27–33

Zhang H, Zhao Q, Sun ZZ, Zhang CQ, Feng Q, Tang SZ, Liang GH, Gu MH, Han B, Liu QQ (2011) Development and high-throughput genotyping of substitution lines carring the chromosome segments of *indica* 9311 in the background of *japonica* Nipponbare. J Genet Genomics 38:603–611

Zhang HG, Zhang LJ, Si H, Ge YS, Liang GH, Gu MH, Tang SZ (2016) *Rf5* is able to partially restore fertility to Honglian-type cytoplasmic male sterile japonica rice (*Oryza sativa*) lines. Mol Breed 36:1–10

7

Development and Genetic Characterization of A Novel Herbicide (Imazethapyr) Tolerant Mutant in Rice (*Oryza sativa* L.)

D. Shoba[1], M. Raveendran[1], S. Manonmani[1], S. Utharasu[1], D. Dhivyapriya[1], G. Subhasini[1], S. Ramchandar[1], R. Valarmathi[1], Nitasha Grover[2], S. Gopala Krishnan[2], A. K. Singh[2], Pawan Jayaswal[3], Prashant Kale[3], M. K. Ramkumar[3], S. V. Amitha Mithra[3], T. Mohapatra[4], Kuldeep Singh[5,10], N. K. Singh[3], N. Sarla[6], M. S. Sheshshayee[7], M. K. Kar[8], S. Robin[1]* ⓘ and R. P. Sharma[9]

Abstract

Background: Increased water and labour scarcity in major rice growing areas warrants a shift towards direct seeded rice cultivation under which management of weeds is a major issue. Use of broad spectrum non-selective herbicides is an efficient means to manage weeds. Availability of rice genotypes with complete tolerance against broad-spectrum non-selective herbicides is a pre-requisite for advocating use of such herbicides. In the present study, we developed an EMS induced rice mutant, 'HTM-N22', exhibiting tolerance to a broad spectrum herbicide, 'Imazethapyr', and identified the mutations imparting tolerance to the herbicide.

Results: We identified a stable and true breeding rice mutant, HTM-N22 (HTM), tolerant to herbicide, Imazethapyr, from an EMS-mutagenized population of approximately 100,000 M_2 plants of an upland rice variety, Nagina 22 (N22). Analysis of inheritance of herbicide tolerance in a cross between Pusa 1656-10-61/HTM showed that this trait is governed by a single dominant gene. To identify the causal gene for Imazethapyr tolerance, bulked segregant analysis (BSA) was followed using microsatellite markers flanking the three putative candidate genes viz., an Acetolactate Synthase (ALS) on chromosome 6 and two Acetohydroxy Acid Synthase (AHAS) genes, one on chromosomes 2 and another on chromosome 4. RM 6844 on chromosome 2 located 0.16 Mbp upstream of AHAS (LOC_Os02g30630) was found to co-segregate with herbicide tolerance. Cloning and sequencing of AHAS (LOC_Os02g30630) from the wild type, N22 and the mutant HTM and their comparison with reference Nipponbare sequence revealed several Single Nucleotide Polymorphisms (SNPs) in the mutant, of which eight resulted in non-synonymous mutations. Three of the eight amino acid substitutions were identical to Nipponbare and hence were not considered as causal changes. Of the five putative candidate SNPs, four were novel (at positions 30, 50, 81 and 152) while the remaining one, S627D was a previously reported mutant, known to result in Imidazolinone tolerance in rice. Of the novel ones, G152E was found to alter the hydrophobicty and abolish an N myristoylation site in the HTM compared to the WT, from reference based modeling and motif prediction studies.

Conclusions: A novel mutant tolerant to the herbicide "Imazethapyr" was developed and characterized for genetic, sequence and protein level variations. This is a HTM in rice without any IPR (Intellectual Property Rights) infringements and hence can be used in rice breeding as a novel genetic stock by the public funded organizations in the country and elsewhere.

Keywords: Rice, EMS Mutagenesis, Herbicide tolerance, Imazethapyr, AHAS

* Correspondence: robin.tnau@gmail.com
[1]Tamil Nadu Agricultural University, Coimbatore 641003, India
Full list of author information is available at the end of the article

Background

Weed management is labour intensive in irrigated rice, the most prevalent rice ecosystem. Owing to labour and water shortage, many South East Asian countries including Malaysia, Sri Lanka and Vietnam have shifted to direct seeded rice (DSR) from transplanted rice (Rao et al. 2007). In American and European countries, DSR accounts for 80–90% of total rice cultivated area (Hassan and Rao 1996). Due to predictions on increased frequency in the occurrence of water deficit/drought in Asia, DSR is expected to become popular in Bangladesh, Pakistan and India. In DSR, weed(s) are one of the major factors affecting rice production to an extent of 18–48% due to rice-weed competition for resources (Rao et al. 2007). This not only increases the cost of production due to increased labour cost for their management, but also affects productivity under situations of labour shortage. In India, the cost towards controlling the weeds accounts up to 30% of the total cost of cultivation (Rao et al. 2015). Moreover, the problem of weedy rice is being reported widely in DSR areas in India, for which herbicide tolerant rice varieties is one of the feasible and practical long term solutions (Rathore et al. 2013).

Herbicides primarily act by disrupting key enzymes/proteins involved in essential metabolic or physiological processes associated with growth and development of plants. Glyphosate, glufosinate, synthetic auxins, sulfonylurea, imidazolinones, triketones, isoxazoles, callistemone, cyclohexanediones, aryloxyphenoxypropionates and phenylpyrazolines are common herbicides for which herbicide tolerance (HT) mechanisms are well known and exploited for development of herbicide tolerant crops (Endo and Toki, 2013). Some of the herbicide tolerant crops have been developed by introducing mutations in the target site of herbicide action, whereas others have introduced genes detoxifying the herbicide molecule (Endo and Toki, 2013). Both the above mechanisms have been exploited in developing HT transgenic crops while the former approach has been achieved through mutagenesis (non-GM approach) as well (Green and Owen, 2010). Non-GM HT crops have the advantage of easier registration/release for commercial cultivation as well as wider public acceptance.

Of the various herbicides, imidazolinones are the most widely targeted ones for developing herbicide tolerant crops through non-GM approach. Imidazolinones act by disrupting Acetolactate Synthase (ALS), and/or Acetohydroxy Acid Synthase (AHAS), enzymes involved in branched chain amino acid (valine, leucine and isolecine) biosynthesis (Singh and Shaner, 1995). Nucleotide and amino acid sequences of AHAS and mutations causing tolerance against imidazolinones are well characterized in *Arabidopsis* (Sathasivan et al. 1990 and 1991). Mutations in AHAS gene is reported to confer resistance to five groups of herbicides, namely, imidazolinones (Shaner et al. 1984), sulfonylurea (Chaleff and Mauvais, 1984), pyrimidinyl thiobenzoates (Stidham, 1991), triazolopyrimidine and sulfonylaminocarbonyltriazolinones (Gerwick et al. 1990). Although ALS/AHAS mutants tolerant to imidazolinone compounds have been developed in rice, wheat, sunflower, canola and maize (Tan et al. 2005), they are all protected by patents (Croughan 1998, 2002; Livore 2003; Livore et al. 2011). The present investigation was therefore undertaken with the objective of identifying novel Imazethapyr resistant rice lines through Ethyl Methane Sulphonate (EMS) induced mutagenesis which can be used without restriction in public funded rice breeding. We report here the identification and detailed characterization of one such stable herbicide tolerant mutant in rice for the first time in India.

Methods

Mutagenesis

One kilogram seeds of a drought tolerant upland rice variety, Nagina 22 (N22) was treated with 0.8% EMS to raise M_1 (Mohapatra et al. 2014). N22 was chosen for mutagenesis for two reasons: 1. it is an upland variety sensitive to Imazethapyr spray; 2. EMS mutants in N22 have been developed in a national level consortium on rice functional genomics owing to its positive attributes such as drought tolerance and heat tolerance, etc. (Mohapatra et al. 2014; Amitha Mithra et al. 2016). The produce of all the M_1 plants was bulked to generate a large M_2 population of about 100,000 M_2 plants, which was used for screening for herbicide tolerance.

Screening for Herbicide Tolerance

All the M_2 seeds, giving rise to approximately 100,000 M_2 plants were grown under field conditions by following recommended agronomic practices. Non-selective broad spectrum herbicide, Imazethapyr, available as Pursuit™ was sprayed @2.5 ml/lit on all plants on 30 days after sowing (DAS). Observation on survival of the mutants was made on 15th day after spraying. Putative resistant plants were identified based on retention of greenness of leaves and absence of drying symptoms. Particularly, green plants surrounded by dried plants were selected to ensure that the plants in the specific area had received adequate quantity of herbicide, thus minimizing the chances of escape (Fig. 1).

Generation Advancement

Putative resistant plants were rescued and replanted in pots, harvested separately and forwarded to M_3 generation. M_3 generation was raised under field conditions in progeny to row design along with the suitable controls viz., wild type (WT) N22 and IR 64 and, sprayed

Fig. 1 Overall view of imazethapyr treated field and identification of resistant plant (inset) in M_2 generation

with Imazethapyr on 30th DAS for reconfirming the HT reaction. The HT plants were again harvested individually (single plants) and raised as progeny rows in M_4 generation along with N22 and IR 64. Herbicide spray was applied as in the M_2 and M_3 generation (Fig. 2) and individual progenies were harvested in bulk. The progeny rows were validated again for herbicide tolerance in M_5 generation under field conditions. One of the mutants, consistently breeding true over sucessive generations was finally selected and named "HTM-N22". DUS (Distinctness, Uniformity and Stability) characterization

of the mutant HTM-N22 (HTM) vis-à-vis the wild type N22 was carried out for 40 DUS traits to confirm the phenotypic similarity of the mutant to the wild type.

Genetic Analysis of Herbicide Tolerance

To analyze the inheritance pattern, the herbicide tolerant mutant, HTM-N22 was crossed with Pusa 1656, a wild type parent for the herbicide tolerance trait. The F_2 population comprising of 254 plants was raised in the field and screened for tolerance to Imazethapyr @ 2.5 ml/lit spray on 30th DAS. Data on number of plants dried and survived was recorded after 15 days of herbicide spray. The observed phenotypic data of the F_2 generation was analyzed for the goodness of fit with the expected Mendelian segregation ratio using chi-square (χ^2) test (Fisher, 1936).

Confirmation of Genomic Similarity Between the Mutant and its Wild Type and Molecular Mapping of Herbicide Tolerance

Since EMS mutagenesis is known to induce multiple mutations in the genome (Abe et al. 2012), testing with a large number of microsatellite markers is recommended before proceeding with mapping and or other applications (Wu et al. 2005; Lima et al. 2015). Hence, the herbicide tolerant mutant identified was subjected to comparative microsatellite analysis with the WT, N22 following the earlier protocol (Lima et al. 2015). A total of 127 genome wide SSR (Simple Sequence Repeat) loci including nine HvSSRs (hyper variable SSRs), which

Fig. 2 Mutant progenies 15 days after imazethapyr treatment in M_4 generation. **a**: HTM-N22 (25th Day after sowing and at the time of treatment); **b**: HTM-N22 emerging from the weeds 15 days after treatment; **c**: Efficacy of the weed control (20th day after treatment); **d**: Nagina 22 killed along with the weeds

covered all 12 rice chromosomes, were used to study their allelic constitution between the WT and the HTM. The SSR fragments were amplified and separated on 3.5% agarose gels, and stained with ethidium bromide. The gels were documented using VersaDoc Imaging System Model3000 (Bio-Rad Laboratories, Inc. USA) and allelic pattern was scored based on the amplicon size.

In order to map the gene(s) responsible for herbicide tolerance in the HTM, Bulked Segregant Analysis (BSA) proposed by Michelmore et al. (1991) was carried out using SSR markers in the vicinity of ALS gene which is well established as the target of the herbicide, Imazethapyr. BLAST search against rice Pseudomolecule ver. 7.0, revealed that rice has three candidate genes namely, LOC_Os06g51280, encoding ALS, LOC_Os02g30630 and LOC_Os04g32010, both encoding AHAS. Parental polymorphism survey was conducted between HTM and Pusa 1656, using SSR markers located within 0.5 Mb region of the above three loci.

DNA (isolated from leaves sampled before herbicide spray) from 10 tolerant and 10 sensitive plants from the F_2 generation, which was used for inheritance studies and phenotyped based on mortality upon herbicide spray, were pooled in equi-molar concentration to constitute the resistant and susceptible bulks, respectively. Based on inheritance pattern observed, the markers, which were polymorphic between the parents, and showing heterozygous (co-dominant) pattern in tolerant bulk and homozygous pattern in susceptible bulk, were considered to be putatively linked with the herbicide tolerance trait. The putatively linked markers were then used for genotyping of the individual F_2 plants to study linkage relationship.

The genotypic and the phenotypic data of the F_2 population were analyzed for segregation distortion, if any, using chi-square (χ^2) test for goodness of fit. Linkage between the putatively linked marker and the target gene was estimated using genotypic data and the phenotypic data with the help of MAPMAKER v3.0 software (Lander et al. 1987). Linkage was considered significant if the Logarithm of Odds (LOD) score was ≥3.0. The Kosambi mapping function (Kosambi, 1943) was used to convert recombination frequency into map distance between the marker and the target.

Identification of Causal Gene and Validation in the Mapping Population

The candidate gene which was linked to the microsatellite marker and co-segregated with herbicide resistance was amplified using the flanking markers in both the mutant and the wild type. The amplified fragments were purified, cloned in pGEM-T vector, propagated in *E.coli* and were sequenced (Eurofins Genomics India Pvt. Ltd., India). Obtained nucleotide sequences were compared

with reference *japonica* genome by multiple sequence alignment and all the nucleotide sequences were translated into amino acid sequences. Comparisons were made between the amino acid sequences of mutant plants and wild type, and then compared with the reference *japonica* genome sequence in the NCBI/TIGR database.

All the non-synonymous substitutions identified between the WT and HTM, and barring those with Nipponbare, were chosen for validation in the mapping population. Two sets of primers were designed to amplify the genomic regions that covered all the five single nucleotide polymorphisms of the candidate gene. The PCR product from each of the 96 genotypes (two parents as well as a subset of 94 individuals from F_2 population) was purified and sequenced directly using Sanger's dideoxy method on an automated capillary-based DNA sequencer (ABI 3730xl, Applied Biosystem, USA) in both forward and reverse directions as per standard ABI sequencing protocol. The trace files were base called and checked for quality using the internal software of the sequencer (ABI sequencer software v5.4). Trimming option was used to edit the poor quality sequences and Quality value (QV) of 20 was fixed for base calling for 99% accuracy. Contig formation using forward and reverse sequences, and sequence alignment so as to call SNPs were performed using BioEdit software.

Protein Modelling of the Causal Gene Between the WT and the HTM

Since neither X-ray crystallographic nor NMR (Nuclear Magnetic Resonance) spectroscopy) structure of the protein encoded by the causal gene were available, reference based structural modeling of the protein from WT and HTM was carried out using Modeler 9.17 programme (Webb and Sali, 2014). The developed 3-D model structure was verified with veriry3D. The motifs in the WT and HTM protein sequences were predicted using PROSITE programme (http://prosite.expasy.org/prosite.html; date of access 11th Nov. 2016).

Results
Mutant Recovery

A mutant population of about 100,000 single M_2 plants generated by bulking the progeny of M_1 plants raised from about 1 Kg of EMS treated N22 seeds were grown in the field and sprayed with herbicide, Imazethapyr. Three mutant plants putatively resistant to Imazethapyr were identified from the field in the M_2 generation. However, only a single mutant plant was found to be tolerant to subsequent herbicide spray in the re-potting experiments which was named as, 'HTM-N22'. Tolerance against the herbicide imazethapyr exhibited by the mutant HTM-N22 (HTM) was confirmed across subsequent filial generations up to M_5 and found to be true breeding (Figs. 1 and 2).

Genetic Analysis of Herbicide Tolerance

Phenotyping of F_2 progenies for their tolerance to the herbicide (Imazethapyr) could classify them into two distinct groups namely, herbicide tolerant and susceptible plants, based on survival after herbicide spray. Of the 254 individual F_2 plants, 197 were tolerant to Imazethapyr spray while 57 were sensitive. The frequency distribution of F_2 population with respect to herbicide tolerance suggested a dominant monogenic inheritance pattern. The observed data showed a goodness of fit to Mendelian segregation ratio of 3:1 ratio ($p = 0.3463$), when tested by χ^2 test, indicating that the herbicide tolerance trait in HTM is controlled by a single dominant gene (Table 1).

Genomic Integrity of the HTM in Comparison With the WT

Genotyping of the HTM and N22 with 127 SSR markers including 9 HvSSRs (Additional file 1: Table S1A and Additional file 2: Figure S1) revealed complete monomorphism, which confirmed the high degree of genetic similarity between the mutant and the WT. The mutant was also identical to the WT for all the 40 DUS characteristics as well as agro-morphological traits (Additional file 3: Table S1B), reiterating the results of marker analysis (Additional file 4: Table S1C and Additional file 5: Figure S2).

Molecular Mapping of Causal Mutation

Out of the 20 SSR markers selected from the genomic regions flanking the three putative candidate genes (Table 2), three markers namely RM20767 for ALS (LOC_Os06g51280), RM 6844 and RM5749 for AHAS in chromosomes 2 and 4 (LOC_Os02g30630 and LOC_Os04g32010) were found to be polymorphic between the parents, Pusa 1656 and HTM.

BSA was carried out using the three polymorphic markers between Pusa 1656 and HTM, with the herbicide tolerant and susceptible bulks, constituted from the plants of respective F_2 populations, based on their survival on Imazethapyr spray. In the BSA, only one SSR marker, RM6844, was able to differentiate the herbicide tolerant and susceptible bulks and also exhibit the expected pattern, similar to the parents, while the remaining two markers failed to differentiate the herbicide tolerant and susceptible bulks. In case of RM6844,

the herbicide tolerant bulk was heterozygous and the susceptible bulk was homozygous indicating that it is putatively linked with the herbicide tolerance trait (Fig. 3). The genotyping of 254 F_2 individuals from the cross, Pusa 1656/HTM was undertaken with the putatively linked marker, RM6844 for molecular mapping of the gene governing herbicide tolerance (Fig. 4).

The goodness of fit for the genotypic segregation of the F_2 population for the marker loci RM6844 was tested using χ^2 test against the expected ratio 1:2:1. The observed ratio fitted well with the expected ratio with respect to RM6844 marker with a p-value of 0.4392 (Table 3). Linkage analysis between phenotypic data on survival upon herbicide spray with genotypic data from RM6844 marker of 254 individuals in the F_2 population from the cross, Pusa 1656/HTM revealed a genetic distance of 1.2 cM (LOD score - 51.89) between the causal gene and the marker.

Sequence Analysis of the Causal Gene in the HTM and WT

The locus, AHAS (LOC_Os02g30630) was amplified from both the HTM and the WT by designing appropriate primers (Forward: TCGCCCAAACCCAGAAA; Reverse: ACATCATAGGCATACCACTCTT), cloned and sequenced. Since LOC_Os02g30630 encoding AHAS is an intronless gene, it was amplified and sequenced in a single reaction (both forward and reverse) using a 50 cm long capillary. Nucleotide sequences of HTM and WT were translated to amino acid sequence and compared with each other as well as Nipponbare (*japonica* genome) as reference (Fig. 5). The sequence analysis revealed a total of 16 point mutations which resulted in amino acids substitutions across the three genotypes (Table 4). Out of these, three amino acids substitutions (at positions 30, 50 and 627) were specific to the HTM when compared to WT and Nipponbare, while eight changes (at positions 11, 67, 71, 293, 318, 357, 400 and 643), could be ascribed to the differences in subspecies – *japonica* vs. *aus* genotypes (WT and HTM). Interestingly, at three positions (118, 146 and 569) the amino acids in HTM were identical to the Nipponbare but different from the WT. At positions 81 and 152, the amino acid substitutions in HTM were completely different from references, WT as well as *japonica*. Thus the non-synonymous mutations in positions 30, 50 and 627 which were exclusively different between HTM and WT as well as those different from both WT and Nipponbare (81 and 152) were considered as potential causal mutations.

Table 1 Segregation pattern of herbicide tolerance in F_2 population from cross Pusa 1656/HTM-N22

Cross	Total no. of plants	No. of F_2 plants		Expected Genetic ratio	χ^2 value	P-value
		Survived	Died			
Pusa 1656/HTM-N22	254	197	57	3:1	0.887	0.3463

Table 2 SSR markers identified in the vicinity of candidate genes for polymorphism survey between the HTM-N22 and Pusa 1656

Candidate gene	SSR	Position (bp)	Primers (Forward/Reverse)	Polymorphism
Aceto-lactate synthase (LOC_Os06g51280 at 31049803 in chromosome 6)	RM20765	30593302	CCAGCTCACCTCAGCTTCATCAGC CATCACCATCACCACCACCATGC	monomorphic
	RM20767	30593847	TCGATCGATCCTAGACTCCTTCC GACTCCACGAACAGCAGGTTAGC	polymorphic
	RM20768	30611463	CAGGGAATAAACAGGAGGAAGAGG CCAACCTCAACCTAGTTAACCTCACC	monomorphic
	RM20769	30653429	CAGATGCGGAGGATGAAGAGC CTTCCGGAATTTCAACTCAACG	monomorphic
	RM20771	30665825	CAAACCTGCGTCTCTGTCTCTCTCC GACGAGCACGACCCATCACC	monomorphic
	RM20773	30692412	TTGCCAATATTCCCTCCAGTGC GTTGTGTTGGGACCTTGATTCG	monomorphic
AcetoHydroxy Acid Synthase (LOC_Os02g30630 at 18236025 bp in chromosome 2)	RM 6844	18169413	AGTCCAAGAAAGGCACGAGAGG CTGCATCGAAGAAGAAGAAGAAGC	polymorphic
	RM13263	18187191	AAGATTGCACACTGGTGTTCTCC AGAAGAGCCGGTCTTTGTCTCC	monomorphic
	RM13264	18192480	ATCTCCATCGTCTCCTTCCTTGG CGTACAGCCATATCCAGCAAACG	monomorphic
	RM13267	18220791	TGAGGCGACGACCACCTTCG CCAAAGCCGCAGGTTAAGCATCC	monomorphic
	RM13268	18223511	CCCAAACATCCAATACGACACC GACGAGCCACCACGTTAGTACG	monomorphic
	RM13269	18237768	GCTTCGGTAATTTGGTTTCGTGATCC GACCACGTGACGTTCCAAGACG	monomorphic
AcetoHydroxy Acid Synthase (LOC_Os04g32010 at 19169130 bp in chromosome 4)	RM16844	19095807	CACCGACTGGTTCGTCTACAGG GAGAAGATATGCAGGTGGAACTGG	monomorphic
	RM16850	19172772	TGCTTGTAAGAGAGGTCAAGAGAGG CCCACCATCTCGTAGAGCTAACC	monomorphic
	RM16867	19452211	GGAACGTAGCTGAAGTCATGAACC ATGAGCCTTGCTTGCTGTAGTCC	monomorphic
	RM16873	19485808	GCATATGCATGCAGGAATTGACC TGCACTCCAGCATTAAAGAACAGG	monomorphic
	RM1205	19643755	CAATCACAGAGCAACACGTACCC GCAGAGGCAGCTGAGAAGTATAGC	monomorphic
	RM1359	20041155	CGACTTGCCAAAGGTCAACG GATTCTACGGGCCACAAGTCC	monomorphic
	RM3643	20128692	GCTAAGCTAATCTGACCGGATCTACG GATGGGCCGATTAACAAATTCC	monomorphic
	RM5749	20131193	GCTCGTTTCTCTCGATCACTCG GCAAGGTTGGATCAGTCATTTCG	polymorphic

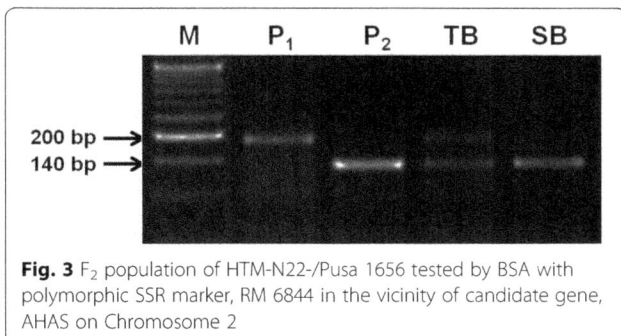

Fig. 3 F₂ population of HTM-N22-/Pusa 1656 tested by BSA with polymorphic SSR marker, RM 6844 in the vicinity of candidate gene, AHAS on Chromosome 2

Validation of Potential Causal Mutations in the Mapping Population

Since SSR marker, RM6844 found to be linked to the causal gene, LOC_Os02g30630, encoding AHAS was 1.2 cM away, sequencing of the causal gene was done in a subset of mapping population (94 F_2 individuals) for further validation. Two pairs of primers, one for amplifying region comprising amino acids 30, 50, 81 and 152 (Forward: CATCACCCACCATGGCTAC; Reverse: CTA TGGGCGTCTCCTGGAA) and the other for the amino acid 627 (Forward: CAGTCCGTGTAACAAAGAAGAG;

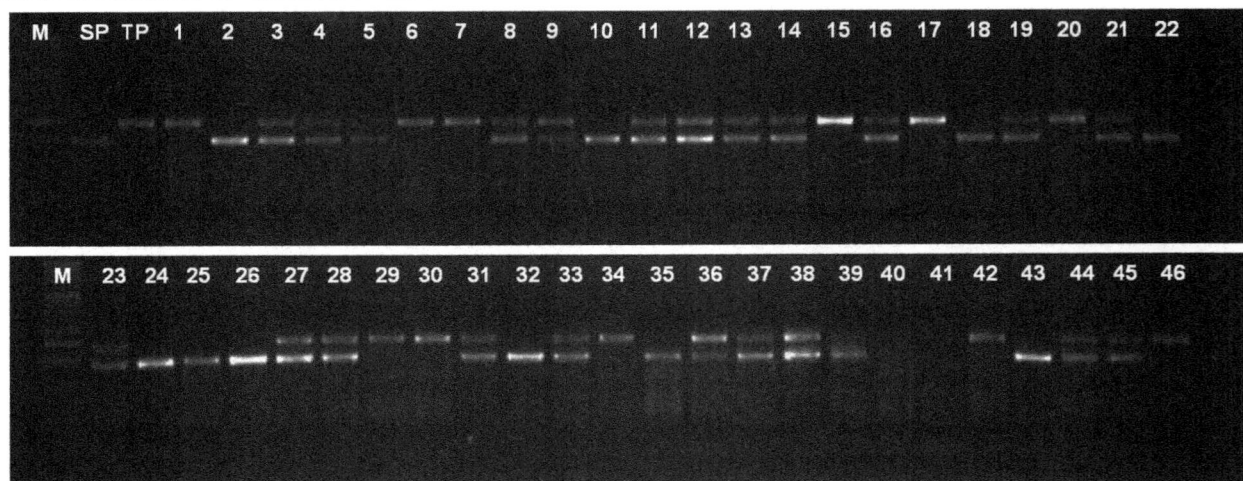

Fig. 4 A representative gel picture showing the segregation pattern of RM 6844 in the F$_2$ mapping population from the cross, Pusa 1656/HTM-N22. M -50 bp Ladder, SP – Herbicide susceptible parent, Pusa 1656; TP- Herbicide tolerant mutant, HTM-N22; 1– 46 progenies in F$_2$ population

Reverse: GGGTCATTCAGGTCAAACATAG) were designed and PCR products were purified and sequenced. The sequencing results of all the five putative mutations were in complete agreement with the expected phenotyping data. Though, we could confirm the causal gene, the causal mutation could not be ascertained as we could not identify any recombinants in the material tested among the five potential causal mutations, owing to small population size.

Reference Based AHAS Protein Modeling of WT and HTM
Three dimensional structure of AHAS protein in the WT was modeled based on four different templates, 3E9Y A (*Arabidopsis thaliana* ALS in complex with Monosulfuron), 1YBH A (*A. thaliana* ALS in complex with Sulfonylurea), 1JSC A (Yeast ALS as a target for herbicidal inhibitors), 1NOH A (Yeast ALS in complex with Sulfonylurea). BLASTP showed the corresponding identity with the *A. thaliana* templates to be 76%, while that of yeast to be 41%. The 3-D models of the AHAS from WT and HTM developed using Modeler had the 3D-1D score > = 0.2 score with the variation in WT and HTM in the range of 76.55 – 79.66% suggesting the models developed were appropriate. The DOPE (Discrete Optimized Protein Energy) score of the protein structure of WT and HTM was similar (-71451.57031 and -71297.08594 respectively) suggesting the high quality of the models. The superimposition of both the models did

not reveal any major structural differences other than a coil generated in the HTM around position 627 which was not there in the WT (Fig. 6). Despite the major similarities observed between the WT and HTM, hydrophobicity chart showed a difference in hydrophobicity around 152nd amino acid (Additional file 6: Figure S3).

Motif Prediction in WT and HTM AHAS Proteins
Functional motif prediction of AHAS protein in WT and HTM identified four distinct patterns viz., protein kinase C phosphorylation, Amidation, N-myristoylation and Casein kinase II phosphorylation sites (Table 5). Comparison of WT and HTM for individual motifs predicted (within each of the four patterns) revealed abolition of a N-myristoylation site motif in HTM involving amino acids 152–157 which was present in the WT. Comparison of the amino acid sequences between WT and HTM (Fig. 5) revealed glutamic acid at 152nd position in the HTM while it was glycine in the WT. Thus in the HTM, a small non-polar aliphatic amino acid (glycine) was replaced by a polar (acidic) amino acid, glutamic acid abolishing an N-myristoylation site. In Nipponbare also the 152nd position was occupied by another small aliphatic and non-polar amino acid, alanine. Thus the motif prediction results were corroborated by the hydrophobicity chart (Additional file 6: Figure S3). Hence, the SNP/mutation in the 152nd amino acid could be the causal mutation conferring tolerance against imazethapyr in the HTM.

Table 3 Genotypic segregation of F$_2$ population from the cross, Pusa1656/HTM-N22 for RM6844 marker

Cross	Total no. of plants	Allele (observed)			Expected Genetic ratio	χ^2 value	P-value
		Mutant type	Heterozygote	WT			
Pusa 1656/HTM-N22	254	55	135	64	1:2:1 (63.5:127:63.5)	1.646	0.4392

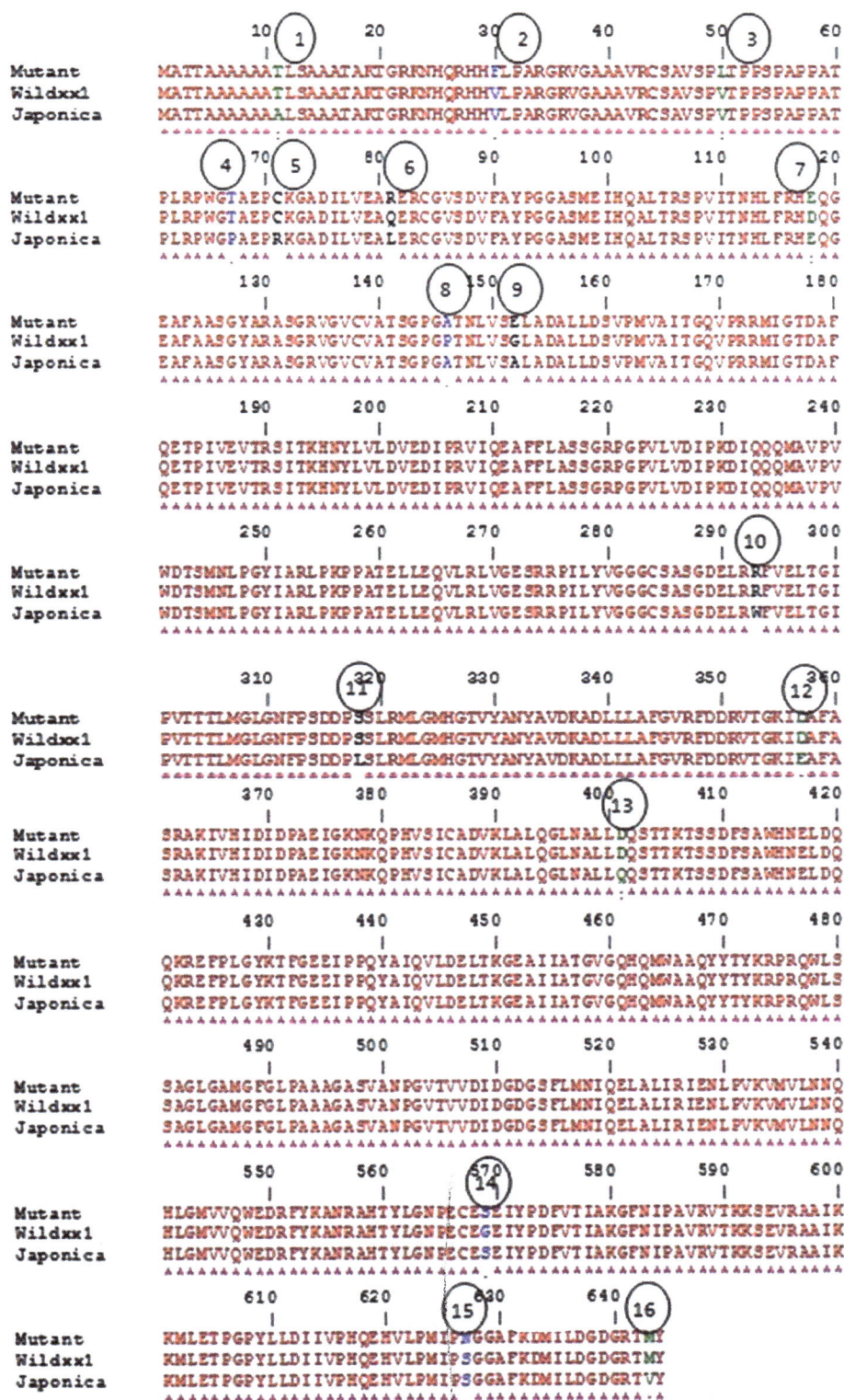

Fig. 5 Amino acid sequence alignment of AcetoHydroxy Acid Synthase (LOC_Os02g30630) between wild type Nagina 22 and herbicide tolerant mutant along with Nipponbare

Table 4 Summary of amino acid substitutions in HTM-N22 as compared to WT with Nipponbare as reference

Amino acid position	Amino acid substitutions in			Remarks on amino acid substitutions in the HTM-N22 with respect WT and Nipponbare
	HTM-N22	WT	Nipponbare	
11	T	T	A	N22 type
30	F	V	V	Mutation
50	L	V	V	Mutation
67	T	T	P	N22 type
71	C	C	R	N22 type
81	R	Q	L	Different from both references
118	E	D	E	Nipponbare type
146	A	P	A	Nipponbare type
152	E	G	A	Different from both references
293	R	R	W	N22 type
318	S	S	L	N22 type
357	D	D	E	N22 type
400	D	D	Q	N22 type
569	S	G	S	Nipponbare type
627	N	S	S	Mutation
643	M	M	V	N22 type

Discussion

Among various methods of weed management in rice cultivation, chemical control using broad spectrum herbicides is an effective and economic alternative management strategy. A few herbicide tolerant genotypes have been developed using non-transgenic approach, primarily through induced mutagenesis, in crops viz., maize, canola, wheat, rice and sunflower (Anderson and Georgeson, 1989; Swanson et al. 1989; Newhouse et al. 1992; Croughan 2002; Shaner et al. 1996; Al-Khatib et al. 1998) though transgenic approach is more prevalent (Green, 2009; Duke and Powles 2009). Imidazolinone herbicides, which include imazapyr, imazapic, imazethapyr, imazamox, imazamethabenz and imazaquin and belong to major class of Group B herbicides, are used commonly in soybean and pulse cultivation but not in rice owing to sensitivity of rice crop to this group of herbicides. In the present study, we have successfully identified an induced mutant in rice tolerant to a broad spectrum non-selective herbicide, imazethapyr, considering the ease in registering and commercializing the material developed through non-transgenic approaches.

The true breeding and monogenic nature of the herbicide tolerant mutant identified in the present study was evident from the results of progeny testing (Fig. 2), inheritance studies (Table 3), and SSR and DUS characterization (Additional file 2: Figure S1 and Additional file 1: Tables S1A and B). While genetic characterization of HTM using 127 markers, at an average distance of one marker per 3 Mb, revealed a very high degree of genetic similarity with the WT, Nagina 22, the causal gene sequencing results indicated multiple point mutations in HTM resulting in eight amino acid changes in a distance of just 1.88 kb (Fig. 5). EMS induced mutagenesis is known to induce multiple mutations and this is one of the reasons as to why DNA fingerprinting is recommended apriori to their genetic characterization (Wu et al. 2005; Lima et al. 2015). Since the causal gene for Imazethapyr tolerance has been identified in our study, it is possible to develop a CAPS marker using appropriate primers, (F: CATCACCCAC CATGGCTAC; R: CTATGGGCGTCTCCTGGAA) and restriction enzyme, BstEII-HF. However, use of CAPS marker is tedious in marker assisted breeding (MAB). Hence, RM6844, a codominant SSR, located 0.16 Mb upstream of AHAS (LOC_Os02g30630) can serve as a robust marker for MAB of this trait.

AHAS in *Arabidopsis* comprises of 670 amino acids while in rice it is 644 amino acids long (Sathasivan et al. 1990; http://rice.plantbiology.msu.edu/). At least seventeen alleles of AHAS gene, harbouring mutations in

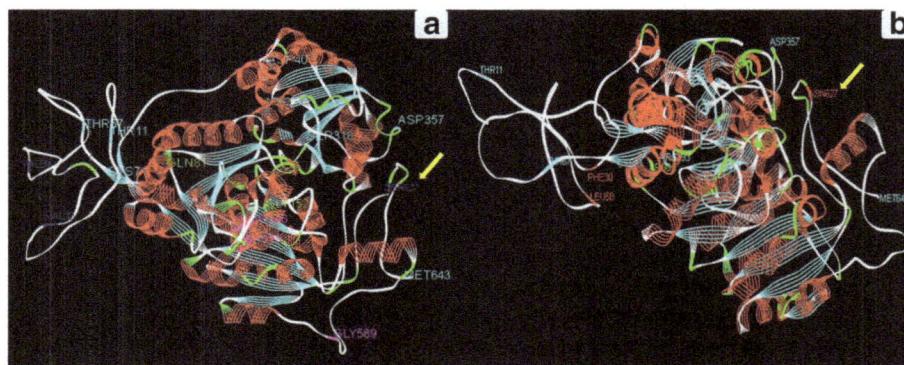

Fig. 6 Line ribbon representation of the homology based protein model of WT (**a**) and HTM-N22 (**b**)

Table 5 Functional motifs predicted in ALS proteins of WT and HTM-N22

S. No.	Motif sites	WT		HTM-N22	
		Position of the amino acid	Features	Position of the amino acid	Features
1	PS00005 PKC_PHOSPHO_SITE Protein kinase C phosphorylation site	17 – 19	TaK	17 – 19	TaK
		20 – 22	TgR	20 – 22	TgR
		132 – 134	SgR	132 – 134	SgR
		218 – 220	SgR	218 – 220	SgR
		273 – 275	SrR	273 – 275	SrR
		319 – 321	SlR	319 – 321	SlR
		353 – 355	TgK	353 – 355	TgK
		404 – 406	TtK	404 – 406	TtK
		471 – 473	TyK	471 – 473	TyK
		590 – 592	TkK	590 – 592	TkK
2	PS00009 AMIDATION Amidation site	20 – 23	tGRK	20 – 23	tGRK
3	PS00008 MYRISTYL N-myristoylation site	136 – 141	GVcvAT	136 – 141	GVcvAT
		152 – 157	GLadAL		
		281 – 286	GGgcSA	281 – 286	GGgcSA
		282 – 287	GGcsAS	282 – 287	GGcsAS
		283 – 288	GCsaSG	283 – 288	GCsaSG
		299 – 304	GlpvTT	299 – 304	GlpvTT
		310 – 315	GNfpSD	310 – 315	GNfpSD
		324 – 329	GMhgTV	324 – 329	GMhgTV
		327 – 332	GTvyAN	327 – 332	GTvyAN
		490 – 495	GLpaAA	490 – 495	GLpaAA
		496 – 501	GAsvAN	496 – 501	GAsvAN
		563 – 568	GNpeCE	563 – 568	GNpeCE
4	PS00006 CK2_PHOSPHO_SITE Casein kinase II phosphorylation site	287 – 290	SgdE	287 – 290	SgdE
		407 – 410	TssD	407 – 410	TssD
		431 – 434	TfgE	431 – 434	TfgE
		449 – 452	TkgE	449 – 452	TkgE
		505 – 508	TvvD	505 – 508	TvvD

amino acids at single or multiple positions such as, 121, 122, 124, 197, 199, 205, 206, 376, 574, 653 and 654 (with respect to *Arabisopsis*) and conferring tolerance to imidazolinone, have been reported across crops (rice, maize, canola, sunflower and wheat) and 87 weed species including prickly lettuce, pigweed, livid amaranth and weedy rice (Shaner, 1999; Duggleby and Pang, 2000; Heap, 2016 from http://www.weedscience.org/). The occurrence of multiple alleles and all of them conferring tolerance to Imidazolinone and/or sulfonylurea treatment suggests that mutations across AHAS could render them unresponsive to these groups of herbicides. In weeds, Ala122, Pro197, Ala205, Asp$_{376}$, Arg$_{377}$, Trp574, Ser653 and Gly$_{654}$ are the eight confirmed target sites reported across genera and countries so far (http://www.weedscience.com/Mutations/MutationDisplayAll.aspx; accessed on 21 Nov 2016). In weedy rice alone, three

mutations, G654E, S653D and A122T are known (Roso et al. 2010). Among them, the change S653D (S627D in rice) is common across crops and weeds (Croughan 2002) whereas the rice Clearfield events released in USA had mutation G654E (G628E in rice) as known from the patent search (US 20070028318 A1). Thus, the HTM rice, identified in the current study with multiple amino acid substitutions (Table 4), all different from the rice Clearfield event, and is free from IPR issues.

As compared to the previous results, the causal mutations in the HTM identified in our study are novel with the presence of five possible candidates in 30, 50, 81, 152 and 627 amino acid positions. The non-synonymous substitution, V30F is part of the chloroplast transit peptide (US patent 57311810) and hence could not be the causal mutation for Imazethapyr tolerance. Using I-TASSER based protein modeling, an additional coil was found to be present in V50L in the HTM (data not shown) which however could not be validated by the rigourous reference based modeling. Similarly, our *insilico* analysis could not find any difference in protein structure due to Q81R. Thus, there are two potential causal mutataions in HTM namely, S627D which has been established beyond doubt (Croughan 2002; Heap, 2016) and G152E as indicated by protein modeling and motif prediction analyses (Table 5 and Additional file 5: Figure S2). The N-myristoylation site abolished in the HTM owing to G152E substitution is known to be involved in signal transduction under the environmental stress condition in plants (Podell and Gribskov, 2004). This latter mutation is so far not reported in any of the crop plants and weeds (Heap 2016), thus making HTM-N22, a novel resource for herbicide tolerance. However, in order to conclusively demonstrate the role of this point mutation leading to substitution G152E, further detailed experimentation is needed either through transformation of the ALS gene with this point mutation/targeted mutation through CRISPR/CAS9 or through isolation of intra-genic recombinants with only one mutation namely S627D and G152E, by generating a sufficiently large F_2 population from the cross N22/HTM.

N-myristoylation is known to cause lipid modification of proteins resulting in their proper intracellular trafficking and correct sub cellular targeting (Bhatnagar et al. 2001). N-myristoylation facilitates protein folding and improves their thermostability (de-Jonge HR et al. 2000). N-myristoylation was also found to be important for proper functioning of *Arabidopsis* gene, *SOS3* (*Salt overly sensitive 3*) having important role in plant salt tolerance (Ishitani et al. 2000). About 29 predicted N-myristoylation sites were reported in SbASR1 gene of *Salicornia brachiata* (halophyte), which is known for enhanced salinity and drought endurance, as against two in rice (Tiwari et al. 2015).

The novel imazethapyr tolerant mutant which has been developed indigenously has opened up the possibilities of its extensive usage, without fear of infringement of any IPR, in the public rice breeding programmes in India and elsewhere. A stable herbicide tolerant source generated and characterized for the causal mutation and the gene linked marker developed for marker assisted selection in this study is expected to benefit the rice breeding programmes widely to protect rice cultivation from weeds and weedy rice. Already the breeding populations have been synthesized between HTM and CO51, Pusa Basmati 1121, Pusa Basmati 1509 and Pusa 44 and the progenies developed through the linked marker RM6844 are in the second or third generation of backcross. Polymorphism studies conducted between the HTM-N22 and several *indica* upland cultivars under cultivation viz., Sahbhagi dhan, Naveen, Pooja and SwarnaSub1

using RM6844 (Additional file 7: Figure S4) indicate the robust nature of the marker across diverse genetic backgrounds. The scope of this trait would be enormous if the upland rice system which is highly vulnerable to weed and weedy rice infestation, derives the benefit out of this gene and the identified marker.

Conclusions

In the present study, a novel herbicide (Imazethapyr) tolerant mutant resource was developed by EMS mutagenesis approach from a drought tolerant variety Nagina22. The genetic analysis established herbicide tolerance in the HTM as a monogenic trait. The tightly linked marker identified in our study has proven to be helpful in introgression of this trait in popular rice cultivars. Since this is a novel HTM in rice without any IPR infringements, it can be used in rice breeding by the public funded organizations in the country and elsewhere.

Additional Files

Additional file 1: Table S1A. SSR primers used for testing the genomic similarity between the WT and HTM-N22.

Additional file 2: Figure S1. Genotyping results of WT and HTM-N22 with selected SSR markers for testing genomic similarity.

Additional file 3: Table S1B. DUS characterization of WT and HTM-N22 for testing genetic similarity.

Additional file 4: Table S1C_DUS characterization for genetic integrity testing.

Additional file 5: Figure S2. Comparison of WT and HTM – N22 (A) Single Plant; (B) Panicle.

Additional file 6: Figure S3. Hydrophobicity chart of WT (upper panel) and HTM-N22 (lower panel).

Additional file 7: Figure S4. Polymorphism between HTM-N22 and upland cultivars for the herbicide tolerance trait linked marker, RM 6844.

Acknowledgements

The work funded by the Department of Biotechnology, Ministry of Science and Technology, Government of India as a network project on "Generation, Characterization, and Use of EMS Induced Mutants of Upland Variety Nagina22 for Functional Genomics of Rice (BT/PR 9264/AGR/02/406(06)/2007)" is greatly acknowledged.

Competing Interests

The authors declare that they have no competing interest.

Author's Contributions

SRP, RoS, RM conceived and planned the experiment; SD, US conducted the mutagenesis; SD, RoS identified the mutant, confirmed, stabilized it; SAK, GKS, NG, DD, SG, RaS, MS made genetic studies in segregating generations in diverse parents; RM, VR carried the gene sequencing, analyzed the sequence information; PJ, PK, RMK did the protein modeling; TM, KS, SNK, SN, SMS periodically guided the experiments; KMK carried the breeding part in upland rice; RoS, MR, ASV, GKS and SRP compiled the results and wrote the manuscript. All authors read and approved the final manuscript.

Author details

[1]Tamil Nadu Agricultural University, Coimbatore 641003, India. [2]Division of Genetics, ICAR-Indian Agricultural Research Institute, New Delhi 110012, India.

[3]ICAR-National Research Centre on Plant Biotechnology, Pusa, New Delhi 110012, India. [4]Indian Council of Agriculture Research, New Delhi 110 001, India. [5]Punjab Agricultural University, Ludhiana 141004, India. [6]Indian Institute of Rice Research, Rajendranagar, Hyderabad 500030, India. [7]University of Agricultural Sciences, Bengaluru 560065, India. [8]National Rice Research Institute, Cuttack, Odisha 753006, India. [9]INSA Honorary Scientist, NRCPB, IARI, Pusa, New Delhi 110012, India. [10]Present address: ICAR-National Bureau of Plant Genetic Resources, Pusa, New Delhi 110012, India.

References

Abe A, Kosugi S, Yoshida K, Natsume S, Takagi H, Kanzaki H, Innan H (2012) Genome Sequencing Reveals Agronomically Important Loci in Rice Using MutMap. Nature Biotechnol 30(2):174–178. doi:10.1038/nbt.2095

Al-Khatib K, Baumgartner JR, Peterson DE, Currie RS (1998) Imazethapyr Resistance in Common Sunflower (*Helianthus annuus*). Weed Sci 46:403–407

Amitha Mithra SV, Kar MK, Mohapatra T, Robin S, Sarla N, Seshashayee M, Singh K, Singh AK, Singh NK, Sharma RP (2016) DBT Propelled National Effort in Creating Mutant Resource for Functional Genomics in Rice. Curr Sci 110:543–48

Anderson PC, Georgeson M (1989) Herbicide Tolerant Mutants of Corn. Genome 31:994–999

Bhatnagar RS, Ashrafi K, Futterer K, Waksman G, Gordon JI (2001) Biology and Enzymology of Protein N-Myristoylation. In: Tamanoi F, Sigman DS (eds) The Enzymes. Academic Press San Diego, CA, pp 241–286

Chaleff RS, Mauvais CJ (1984) Acetolactate Synthase is the Site of Action of two Sulfonylurea Herbicides in Higher Plants. Science 224:1443–1445

Croughan TP (1998) Herbicide Resistant Rice. US Patent 5:773,704

Croughan TP (2002) Herbicide Resistant Rice., US Patent Application 20020019313

de-Jonge HR, Hogema B, Tilly BC (2000) Protein N-myristoylation: critical role in apoptosis and salt tolerance. Sci Signal 2000 (63), pe1–pe1. doi:10.1126/stke.2000.63.pe1.

Duggleby RG, Pang SS (2000) Acetohydroxyacid Synthase. J Biochem Mol Biol 33: 1–36

Duke SO, Powles SB (2009) Glyphosate-Resistant Crops and Weeds: now and in the Future. AgBio Forum 12:346–357

Endo M, Toki S (2013) Creation of Herbicide-Tolerant Crops by Gene Targeting. J Pestic Sci 38(2):49–59. doi:10.1584/jpestics.D12-073

Fisher RA (1936) Has Mendel's Work Been Rediscovered? Ann Sci 1:115–137

Gerwick BC, Subramanian MV, Loney-Gallant VI (1990) Mechanism of Action of the 1,2,4-Triazolo[1,5-*a*] Pyrimidines. Pestic Sci 29:357–364

Green JM (2009) Evolution of Glyphosate-Resistant Crop Technology. Weed Sci 57:108–117. doi:10.1614/WS-08-030.1

Green JM, Owen DK (2010) Herbicide-Resistant Crops: Utilities and Limitations for Herbicide-Resistant Weed Management. J Agric Food Chem 59:5819–5829. doi:10.1021/jf101286h

Hassan S, Rao AN (1996) Weed Management in Rice in the Near Eas. In: Auld BA, Kim KU (eds) Weed Management in Rice. FAO Plant Production and Protection Paper 139, Rome, pp 143–156

Heap I (2016) The International Survey of Herbicide Resistant Weeds. http://www.weedscience.org/. Accessed: 21 Nov 2016.

Ishitani M, Liu J, Halfter U, Kim CS, Shi W, Zhu JK (2000) SOS3 Function in Plant Salt Tolerance Requires N-Myristoylation and Calcium Binding. Plant Cell 12(9):1667–1677, http://dx.doi.org/10.1105/tpc.12.9.1667

Kosambi DD (1943) The Estimation of map Distances from Recombination Values. Ann Eugen 12:172–175

Lander ES, Green P, Abrahamson J, Barlow A, Daly MJ, Lincoln SE, Newburg L (1987) MAPMAKER: An Interactive Computer Package for Constructing Primary Genetic Linkage Maps of Experimental and Natural Populations. Genomics 1(2):174–181, http://dx.doi.org/10.1016/0888-7543(87)90010-3

Lima JM, Nath M, Dokku P, Raman KV, Kulkarni KP, Vishwakarma C, Sahoo SP, Mohapatra UB, Amitha Mithra SV, Chinnusamy V, Robin S, Sarla N, Seshashayee M, Singh K, Singh AK, Singh NK, Sharma RP, Mohapatra T (2015) Physiological, Anatomical and Transcriptional Alterations in a Rice Mutant Leading to Enhanced Water Stress Tolerance. AoB Plants 7:plv023. doi:10.1093/aobpla/plv023

Livore AB (2003) Rice Plants Having Increased Tolerance to Imidazolinone Herbicides: International Application pPublished Under the Patent Cooperation Treaty (PCT)., n.WO2005/020673A1

Livore AB, Prina AB, Birk I, Singh BK (2011) Rice Plants Having Increased Tolerance to Imidazolinone Herbicides., European Patent Applic 101840406.0 Filed 2004

Michelmore RW, Paran I, Kesseli RV (1991) Identification of Markers Linked to Disease Resistance Genes by Bulked Segregant Analysis: A Rapid Method to Detect Markers in Specific Genomic Regions by Using Segregating Populations. Proc Natl Acad Sci U S A 88:9828–9832. doi:10.1073/pnas.88.21.9828

Mohapatra T, Robin S, Sarla N, Sheshahsayee M, Singh AK, Singh K, Singh NK, Amitha Mithra SV, Sharma RP (2014) EMS Induced Mutants of Upland Rice Variety Nagina22: Generation and Characterization. Proc Indian Natl Sci Acad 80(1):163–172

Newhouse K, Smith WA, Starrett MA, Schaefer TJ, Singh BK (1992) Tolerance to Imidazolinone Herbicides in Wheat. Plant Physiol 100:882–886, http://dx.doi.org/10.1104/pp.100.2.882

Podell S, Gribskov M (2004) Predicting N-Terminal Myristoylation Sites in Plant Proteins. BMC Genomics 5(1):37. doi:10.1186/1471-2164-5-37

Rao AN, Johnson DE, Sivaprasad B, Ladha JK, Mortimer AM (2007) Weed Management in Direct-Seeded Rice. Adv Agron 93:153–255. doi:10.1016/S0065-2113(06)93004-1

Rao AN, Wani SP, Ramesha M, Ladha JK (2015) Weeds and Weed Management of Rice in Karnataka State, India. Weed Technol 29(1):1–17. doi:10.1614/WT-D-14-00057.1

Rathore M, Singh R, Kumar B (2013) Weedy Rice: An Emerging Threat to Rice Cultivation and Options for its Management. Curr Sci 105(8):1067–1072

Roso AC, Merotto A Jr, Delatorre CA, Menezes VG (2010) Regional Scale Distribution of Imidazolinone Herbicide-Resistant Alleles in red Rice Determined Through SNP Markers. Field Crop Res 119(1):175–182, http://dx.doi.org/10.1016/j.fcr.2010.07.006

Sathasivan K, Haughn GW, Murai N (1990) Nucleotide Sequence of a Mutant Acetolactate Synthase Gene from an Imidazolinone-Resistant Arabidopsis Thaliana *var*. Columbia. Nucleic Acids Res 18(8):2188

Sathasivan K, Haughn GW, Murai N (1991) Molecular Basis of Imidazolinone Herbicide Resistance in Arabidopsis Thaliana *var*. Columbia Plant Physiol 97(3):1044–1050

Shaner DL (1999) Resistance to Acetolactate Synthase (ALS) Inhibitors in the United States: History, Occurrence, Detection and Management. Weed Sci 44(4):405–411

Shaner DL, Anderson PC, Stldham MA (1984) Imidazolinones: Potent Inhibitors of Acetohydroxyacid Synthase. Plant Physiol 76:545–546

Shaner DL, Bascomb NF, Smith W (1996) Imidazolinone Resistant Crops: Selection, Characterization and Management. In: Duke SO (ed) Herbicide Resistant Crops. CRC Press, Boca Raton, pp 143–157

Singh BK, Shaner DL (1995) Biosynthesis of Branched Chain Amino Acids: From Test Tube to Field. Plant Cell 7(7):935–944. doi:10.1105/tpc.7.7.935

Stidham MA (1991) Herbicides That Inhibit Acetohydroxyacid Synthase. Weed Sci 39:428–434

Swanson EB, Herrgesell MJ, Arnoldo M, Sippell DW, Wong RSC (1989) Microspore Mutagenesis and Selection: Canola Plants With Field Tolerance to Imidazolinones. Theor Appl Genet 78:525–530

Tan SY, Evans RR, Dahmer ML, Singh BK, Shaner DL (2005) Imidazolinone-Tolerant Crops: History, Current Status and Future. Pest Manag Sci 61(3):246–257. doi:10.1002/ps.993

Tiwari V, Chaturvedi AK, Mishra A, Jha B (2015) Introgression of the SbASR-1 gene cloned from a halophyte *Salicornia brachiata* enhances salinity and drought endurance in transgenic groundnut (*Arachis hypogaea*) and acts as a transcription Factor. PLoS One 10(7):e0131567, http://dx.doi.org/10.1371/journal.pone.0131567

Webb B, Sali A (2014) Comparative protein structure modeling using Modeller. In: Curr Protoc Bioinformatics 47, p 5.6.1-5.6.32, John Wiley and Sons. doi:10.1002/0471250953.bi0506s47.

Wu JL, Wu C, Lei C, Baraoidan M, Bordeos A, Madamba MRS, Pamplona MR, Mauleon R, Portugal A, Ulat VJ, Bruskiewich R, Wang GL, Leach J, Khush GS, Leung H (2005) Chemical- and Irradiation- Induced Mutants of Indica Rice IR64 for Forward and Reverse Genetics. Plant Mol Biol 59:85–97. doi:10.1007/s11103-004-5112-0

8

Phenology, sterility and inheritance of two environment genic male sterile (EGMS) lines for hybrid rice

R. El-Namaky[1,2] and P.A.J. van Oort[3,4*]

Abstract

Background: There is still limited quantitative understanding of how environmental factors affect sterility of Environment-conditioned genic male sterility (EGMS) lines. A model was developed for this purpose and tested based on experimental data from Ndiaye (Senegal) in 2013-2015. For the two EGMS lines tested here, it was not clear if one or more recessive gene(s) were causing male sterility. This was tested by studying sterility segregation of the F2 populations.

Results: Daylength (photoperiod) and minimum temperatures during the period from panicle initiation to flowering had significant effects on male sterility. Results clearly showed that only one recessive gene was involved in causing male sterility. The model was applied to determine the set of sowing dates of two different EGMS lines such that both would flower at the same time the pollen would be completely sterile. In the same time the local popular variety (Sahel 108, the male pollen donor) being sufficiently fertile to produce the hybrid seeds. The model was applied to investigate the viability of the two line breeding system in the same location with climate change (+2oC) and in two other potential locations: in M'Be in Ivory Coast and in the Nile delta in Egypt.

Conclusions: Apart from giving new insights in the relation between environment and EGMS, this study shows that these insights can be used to assess safe sowing windows and assess the suitability of sterility and fertility period of different environments for a two line hybrid rice production system.

Keywords: Environment-conditioned genic male sterility (EGMS), Simulation model, Inheritance

Background

Rice hybrids often have higher yields than high-yielding inbred varieties, often between 15% and 20% higher (Virmani et al. 2003). Where local seed markets are well functioning, hybrids can play an important contribution to farmers' livelihoods, local and regional food security. In 2010, AfricaRice initiated breeding for hybrid rice (El-Namaky and Demont 2013). To produce hybrids, a line with male sterility is crossed with a local popular variety (the male pollen donor). Resulting F_1 seed (hybrids) benefits from the positive effects of heterosis and benefits from genes from the local popular variety, which ideally makes the F_1 seed higher yielding yet still well adapted to the local environment. Environment-sensitive genic male sterility (EGMS), also called Photoperiod-thermo-sensitive genic male sterile (PTGMS), has been extensively used for preventing self-pollination in the production of hybrid seeds in various crops (Virmani et al. 2003, Xu et al. 2011). Compared with three-line sterile lines in a hybrid rice system, EGMS can maintain sterile line production without using restorer lines. Furthermore, a two-line hybrid rice system by application of EGMS has many advantages, including a wider range of germplasm resources used as breeding parents, higher yields, and simpler procedures for breeding and hybrid seed production (Virmani et al. 2003; Zhou et al. 2012). With the discovery of the photoperiod sensitive genic male sterility (PGMS) line Nongken 58S in rice (Shi 1985), there has been great progress in two-line hybrid rice breeding in China.

* Correspondence: Pepijn.vanoort@wur.nl; p.vanoort@cgiar.org
[3]Africa Rice Center (AfricaRice), 01 B.P. 2551 Bouaké, Côte d'Ivoire
[4]Crop & Weed Ecology Group, Centre for Crop Systems Analysis, Wageningen University, P.O. Box 4306700, AK, Wageningen, The Netherlands
Full list of author information is available at the end of the article

Much research has been conducted into the molecular and genetic mechanisms causing sterility (Horner and Palmer 1995; Li et al. 2007; Chen et al. 2010; Xu et al. 2011, Huang et al. 2014). Much less is known about how exactly the environment affects male sterility (Lopez and Virmani 2000; Latha et al. 2004). In a qualitative sense it is known that long daylength and high temperature can cause male sterility. But very limited research has been conducted quantifying the relationship between daylength, temperature and sterility. Such quantitative understanding is needed when advising on sowing dates. For producing hybrids, it is important to be absolutely sure that the male sterile parent is 100% sterile and to have flowering of the male sterile parent and the local popular variety at the same date. With temperature varying between years the same sowing date could in one year give 100% sterility and less than 100% in another year. By simulating with for example 20 years of weather data, we can identify those sowing dates for which regardless of weather variability, a 100% sterility can be guaranteed. Models for predicting flowering date and sterility for normal rice varieties have been developed before, for example see van Oort et al. (2011); Julia and Dingkuhn (2013); Dingkuhn et al. (2015a,b) and van Oort et al. (2015a). A key difference between these normal varieties and the EGMS lines is that in the EGMS lines a long daylength at flowering can cause sterility, while in normal varieties no such effect of daylength on sterility has been reported. To date no simulation models have been developed for hybrid rice phenology and sterility. Once a model has been calibrated and shown to be sufficiently accurate, we can use the model to answer practical questions such as:

1. Between which start and end date is the safe sowing window for producing hybrids? (i.e. 100% sterility is simulated) Taking into account weather variability within and between years
2. Will the safe sowing window change with climate change?
3. What is the safe sowing window for producing hybrids in another location?
4. Which sowing dates are best suited for multiplying (selfing) the Male Sterile Parent (sterility <50%)?
5. What combination of sowing dates will guarantee that a local popular parent line and an EGMS parent line will flower at the same date, with the EGMS line being 100% sterile and the local popular parent sufficiently fertile (sterility <50%)?

Inheritance of male sterility is important in breeding programs for locally well adapted hybrids. In the first round, an EGMS parent is crossed with a local popular variety to produce the F_1 seeds. The F_2 generations were obtained from self-pollination of F_1 hybrids. From this segregated population, breeders will be able to select a subset of plants which performed best, i.e. combining the most desirable genes from the original EGMS line and the local popular variety. This subset is used as the second generation EGMS. In comparison with the first generation EGMS and F_1 hybrids, the second generation will have more desirable genes inherited from the local popular variety as well as the genes responsible for male sterility and needed to produce hybrids. This process can be repeated, leading to hybrids ever better performing in the test environment. Clearly, for this type of breeding it is desirable that only one recessive gene causes male sterility, in which case 25% of the F_2 population will express the EGMS trait. If two recessive genes are involved, then a much smaller fraction of the F_2 population will have the EGMS trait (1/16 = 6.25%), thus making breeding much more cumbersome. For this reason, it is important to understand inheritance of the EGMS trait.

The first objective of this paper is to develop a model for predicting flowering date and sterility of two EGMS lines as a function of sowing date, location data and weather data. The second objective is to study the F_2 population, to investigate how many gene(s) are involved in causing male sterility.

Methods

We first describe our data. Next the method of studying inheritance of the EGMS trait. The third main section of the Methods describes a new phenology and sterility model and the methods used for testing and comparing models.

Data
Experiments for model development
Two Environmental Genetic Male Sterile lines, IR75589-31-27-8-33 (EGMS1) and IR77271-42-5-4-36 (EGMS2) were provided by the International Rice Research Institute (IRRI). The two lines were evaluated at the Experimental Farm of the AfricaRice Sahel Regional Center, Ndiaye, Senegal (16.22 N, 16.29 W). Both lines were sown in 2013 and 2014 at a 15 days interval (two dates per month), with 3 replicates per date, to study phenology and sterility. Sowing started on 1 January 2013 and ended on 16 December 2014. The lines were well watered, well fertilised and kept free from pests and diseases. Three observations were made: flowering date, pollen sterility and spikelet sterility. The pollen sterility percentage was measured as follows. Spikelets were collected from each primary panicle and fixed in 70% ethyl alcohol. From these spikelets, 5 to 6 anthers were collected at random and smeared in iodine potassium iodide solution (1%) and examined under light microscope (40 × 10). About 200 pollen grains were examined

in three different slides. All the unstained pollens were considered as sterile and the stained ones as fertile. The pollen sterility per plant was computed and expressed as percentage for each single plant and parental lines as follows:

$$\text{Pollen sterility (\%)} = 100\% \times \frac{\text{No.of sterile pollen grains}}{\text{Total No.of pollen grains}} \tag{1}$$

This dataset contained $n = 288$ data points (48 sowing dates × 3 replicates per date × 2 EGMS lines). This was the dataset used for model development. The dataset was split in two: The first half of the data (sowing dates in 2013) were used for model calibration, the second half (sowing in 2014) was used for model validation.

In 2014, an additional alternative method of measuring sterility was tried, in which sterility was calculated from the number of filled spikelets. The spikelet fertility percentage was measured as follows. Two main panicles of each plant were bagged by glassine bag just before panicle emergence to avoid out-crossing. The total of sterile and fertile spikelet was counted from the bagged panicles of all the plants in each testcross. Spikelet fertility was calculated as:

$$\text{Spikelet stertility (\%)} = 100\% \times \frac{\text{No.of unfilled spikelets}}{\text{Total No.of spikelets}} \tag{2}$$

Sterility calculated with the two methods Eq. (1) and Eq. (2) was statistically compared using the chi-square test.

Experiments for studying inheritance of EGMS sterility

In 2014, Sahel108 (popular rice variety in West Africa) was used to develop two breeding populations with both EGMS lines (Sahel108/IR75589-31-27-8-33) and (Sahel108/IR77271-42-5-4-36). The F_1 were sown in 20 August 2014 (Wet season) and selfed to produce the F_2 populations. In the dry season of 2015 (sowing 15 April) both parents and F_2 populations were sown for investigating the number of genes causing EGMS sterility. For 520 F_2 plants (Sahel108/IR75589-31-27-8-33) and 460 F_2 plants (Sahel108/IR77271-42-5-4-36) we measured pollen fertility and spikelet fertility as described above. Next each plant was reclassified to 0 or 1, 0 (sterile) if sterility was larger than 99% and 1 (fertile) if sterility was less than 99%. A chi-square test was used to analyse the segregation pattern and determine the genes that control both EGMS. The chi-square test for fixed ratios was applied (Gomez and Gomez 1984, p. 464, Yang 1997), with the chi-square value calculated as:

$$\chi^2 = \frac{(|O_1 - E_1| - 0.5)^2}{E_1} + \frac{(|O_0 - E_0| - 0.5)^2}{E_0} \tag{3}$$

Where O_0 and E_0 are the observed and expected number of sterile plants and O_1 and E_1 are the observed and expected number of fertile plants. If one gene is controlling EGMS, we would expect a 1:3 segregation, i.e. $E_0 = 1/4 \times (O_0 + O_1)$ and $E_1 = 3/4 \times (O_0 + O_1)$. If 2 recessive genes are causing male sterility then we expect a 1:15 segregation, i.e. $E_0 = 1/16 \times (O_0 + O_1)$ and $E_1 = 15/16 \times (O_0 + O_1)$.

Weather data/study sites

Daily weather data were taken from the AfricaRice weather stations which have over time been operational at the Ndiaye site. Weather data were used to analyse the experimental data as well as for simulating effect of weather variability on sterility over a longer period of 25 years. The weather station time series showed many gaps. Missing data were filled in with data from the POWER database (http://power.larc.nasa.gov). The POWER database is known to have systematic errors in its minimum and maximum temperature while radiation values compare well with station data (White et al. 2008). Therefore daily T_{min} and T_{max} were corrected as follows. First, bias correction parameters b_0 and b_1 were estimated from linear regression between available station and POWER data (Eq. 4)

$$T_{min}(\text{station}) = b_0 + b_1 \times T_{min}(\text{POWER}) \tag{4}$$

For dates where $T_{min}(station)$ was available we used these values. For dates with missing station data we estimated T_{min} from POWER (Eq. 5), using b_0 and b_1 determined in Eq. (4).

$$T_{min}(\text{POWERcorrected}) = b_0 + b_1 \times T_{min}(\text{POWER}) \tag{5}$$

The same bias correction procedure was applied for the daily maximum temperature T_{max}.

After model development, simulations were conducted with weather data for four environments:

- Ndiaye, Senegal, 1990–2015
- Ndiaye, Senegal, 1990–2015, with 2 °C added to daily temperatures (climate change scenario)
- Cairo, Egypt, 1983–2012
- M'Be, Ivory Coast, 1998–2015

These locations were chosen to compare simulations for sites with contrasting environments. Rice is an important crop in the Nile delta, where input levels are high and yield gaps small (van Oort et al. 2015b), where an interesting opportunity could be to raise the yield ceiling. It is of interest to know if the same EGMS lines successfully grown in Senegal can also be grown in the Nile delta, so that they could be used to develop hybrids with in Egypt popular varieties in a breeding program located in the Nile delta. Ideally we would have used weather data from a station inside the delta, instead of Cairo which is just south of it, but no weather dataset was available from inside the delta with sufficiently long time series of weather data. Therefore Cairo weather data were used. We expect this station is fairly well representative for weather in the delta.

The hybrid rice breeding program of AfricaRice is located in the Sahel station in Ndiaye in Senegal. AfricaRice has another station with facilities and expertise for breeding located in M'Be in Ivory Coast. This would be a convenient location for making crosses between the EGMS lines and locally popular varieties in Ivory Coast (rather than importing hybrids from the AfricaRice station at Senegal, because these might be less well adapted to environment and consumer preferences in Ivory Coast). We were interested in finding out if the Cairo and M'Be sites would also be suitable for hybrid production and if so, what would be the best sowing windows for hybrid production and multiplication of the EGMS lines.

Phenology and sterility model
A model was developed for investigating which climatic variables contribute to male sterility and for predicting sterility as a function of sowing date, weather variables and genetic parameters. The model consists of three sub-models which are described in the following sections.

Phenology sub-model
We developed a simplified phenology model in which we considered only two phases: "SPI", from sowing to panicle initiation and "PIFL" as the phase from panicle initiation to flowering. The panicle initiation date was not observed. We estimated the panicle initiation (PI) date as:

$$\text{Date}_{PI} = \text{Date}_S + 0.65 \times (\text{Date}_{FL} - \text{Date}_S) \qquad (6)$$

We simulated on a numerical scale the Development Stage (DS) with $DS = 0$ for sowing, $DS = 0.65$ for panicle initiation and $DS = 1$ for flowering. The number 0.65 is taken from the ORYZA2000 model (Bouman et al. 2001); previous phenological research has shown that panicle initiation occurs at around this stage. Development starts on sowing day d_S. DS_d on day d is simulated as:

$$DS_d = \begin{cases} 0 & d < d_S \\ DS_{d-1} + DVR_{SPI} \times TI_d & 0 \leq DS < 0.65 \\ DS_{d-1} + DVR_{PIFL} \times TI_d & 0.65 \leq DS < 1.0 \end{cases} \qquad (7)$$

Were DS_{d-1} is the development stage the previous day, DVR_{SPI} and DVR_{PIFL} are model parameters estimated through model calibration and daily thermal time increment TI_d is a variable calculated from daily weather data. Daily TI_d is calculated as the average of hourly thermal time increment $TI_{h,d}$:

$$TI_d = \frac{1}{24} \sum_{h=1}^{24} TI_{h,d} \qquad (8)$$

Hourly thermal time increment $TI_{h,d}$ is calculated from hourly temperature $T_{h,d}$, using the so-called cardinal temperatures: the base temperature TBD and the optimum temperature TOD (above which development is fastest):

$$TI_{h,d} = \min\left(1, \max\left(0, \frac{T_{h,d} - TBD}{TOD - TBD}\right)\right) \qquad (9)$$

Note that according to these equation development is fastest at $T_{h,d} \geq TOD$ (leading to $TI_{h,d} = 1$). The assumption of no delay in development above TOD is based on van Oort et al. (2011) and Zhang et al. (2016) who showed models with no delay in development above TOD are consistently more accurate than models with slower development above TOD. Hourly temperature on day d, $T_{h,d}$, was calculated from daily minimum and daily (d) maximum temperature $T_{max}(d)$ and $T_{min}(d)$ as:

$$T_{h,d} = \frac{(T_{max}(d) + T_{min}(d))}{2} + (T_{max}(d) + T_{min}(d)) \times \cos(0.2618 \times (h-14)) \qquad (10)$$

Parameters TBD, TOD, DVR_{SPI} and DVR_{PIFL} were simultaneously estimated with the pheno_opt_rice2 phenology calibration program (van Oort et al. 2011) from the experimental data. Parameter sets were calibrated separately for the two EGMS lines. Starting from the sowing day d_S the phenology sub-model simulates for each day d the development stage DS_d using Eqs. 7–10. The simulated panicle initiation day d_{PI} is the day d for which $DS_d = 0.65$ and the simulated flowering day d_{FL} is the day d for which $DS_d = 1$. In summary the phenology sub-model predicts the panicle initiation day d_{PI} and flowering day d_{FL} as a function of the sowing day d_S, daily minimum and maximum temperatures T_{min} and T_{max} and parameters TBD, TOD, DVR_{SPI} and DVR_{PIFL}.

EGMS sterility sub-model
Three types of genic male sterility exist (Virmani et al. 2003, Xu et al. 2011): PGMS lines are completely or highly sterile under long day lengths, TGMS lines are completely or highly sterile under high or low

temperatures and in photothermosensitive genetic male sterile (PTGMS) lines sterility is determined by both day-length (=photoperiod) and temperature. The more general term EGMS lines (E stands for environment) is used to represent all types. Sterility was measured on a numerical scale from 0 to 1 (=0-100%). From the experimental data we fitted logistic regression models to predict sterility S_{EGMS} from different logit functions $g(X)$, where X refers to different sets of explanatory variables:

$$S_{EGMS} = \frac{e^{g(X)}}{1 + e^{g(X)}} \qquad (11)$$

Available research suggests that S_{EGMS} is most strongly affected by environmental conditions in the phase from panicle initiation to the 50% flowering date (Lopez and Virmani 2000; Latha et al. 2004), also in other crops with EGMS (Yuan et al. 2008). We therefore calculated average values over the period from panicle initiation to flowering (PIFL) for the following explanatory variables: $T_{min}(PIFL)$, $T_{avg}(PIFL)$, $T_{max}(PIFL)$ and $DAYL_{-6}(PIFL)$, where T_{min}, T_{max} and T_{avg} are daily minimum, daily maximum and daily average temperature and $DAYL_{-6}$ is the daily daylength including civil twilight (sun angle 6° below horizon). We also tested for differences between the two EGMS lines (recoded to 0 for EGMS1 and 1 for EGMS2). The logit function with minimum temperature $T_{min}(PIFL)$, daylength and EGMS line as explanatory variables is defined as:

$$g = b_{E0} + b_{E1} \times DAYL_{-6}(PIFL) + b_{E2} \\ \times T_{min}(PIFL) + b_{E3} \times EGMS \qquad (12)$$

For all models we tested using the standard t-test if parameters b_{E1}, b_{E2} and b_{E3} differed significantly from zero. Daylength $DAYL_{sa}(d)$ on day d with sun angle sa was calculated using equations presented in Goudriaan and van Laar (1994). The sun angle sa is used for twilight: 0° means no twilight, −6° is civil twilight, i.e. including the time before sunrise and after sunset when the sun is less than 6 degrees below the horizon. Daylength at latitude LAT on day d was calculated as:

$$a = \sin\left(LAT \times \frac{\pi}{180}\right) \times \sin(\delta) \qquad (13)$$

$$b = \cos\left(LAT \times \frac{\pi}{180}\right) \times \cos(\delta) \qquad (14)$$

$$\delta(d) = -\text{asin}\left(\sin\left(23.45\,\frac{\pi}{180}\right) \times \cos\left(2\,\pi\frac{(d+10)}{365}\right)\right) \qquad (15)$$

$$DAYL_{sa}(d) = 12\left(1 + \frac{2}{\pi} \times \text{asin}\left(\left(-\sin\left(sa \times \frac{\pi}{180}\right) + a\right)/b\right)\right) \qquad (16)$$

We simulated sterility for three different locations at different latitudes. Figure 1 shows daylength for Ndiaye Senegal, M'Be in Ivory Coast and Cairo in Egypt.

Cold sterility sub-model

When the varieties were flowering in December, January and February the EGMS sterility sub-models predicted zero sterility while observed sterility was 10-50%. We hypothesised this would be due to cold induced sterility, a phenomenon that has been documented before for the same study site for other varieties flowering in these months (Dingkuhn et al. 2015a, van Oort et al. 2015a). From these previous studies we know cold induced sterility is most strongly correlated with minimum temperature in the period from panicle initiation to flowering. Water temperature (which was not measured) can during part of this phase be a better predictor than air temperature because initially the panicle meristem is a ground height. With lack of data on water temperature we used air temperature. For parameter estimation we used the subset of data with flowering in the period of December to February, to avoid confounding effects of the EGMS type of sterility discussed in the previous section. Also here, we considered three candidate explanatory variables:

- $T_{min}(PIFL)$ = Average of daily minimum temperatures, averaged over the period from panicle initiation to flowering
- $T_{avg}(PIFL)$ = Average of daily average temperatures, averaged over the period from panicle initiation to flowering

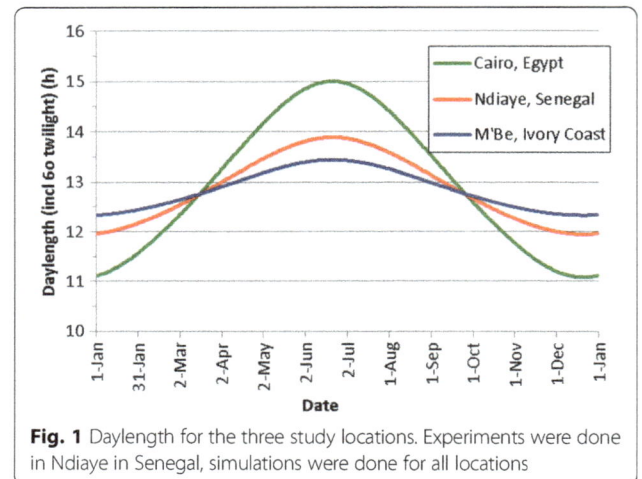

Fig. 1 Daylength for the three study locations. Experiments were done in Ndiaye in Senegal, simulations were done for all locations

- $T_{max}(PIFL)$ = Average of daily maximum temperatures, averaged over the period from panicle initiation to flowering

For each, a logistic regression model was fitted. For example for $T_{min}(PIFL)$:

$$S_{Cold} = \frac{e^{b_{C0}} + b_{C1} \times T_{min}(PIFL)}{1 + e^{b_{C0}} + b_{C1} \times T_{min}(PIFL)} \qquad (17)$$

Full model
The full model is a combination of the above three sub-models:

1. Phenology sub-model: predict panicle initiation date d_{PI} and flowering date d_{FL}
2. EGMS sterility sub-model: predict photoperiod and/or temperature induced sterility S_{EGMS}
3. Cold sterility sub-model: predict cold induced sterility S_{Cold}
4. Total sterility S_{Total} as the maximum of the two sterilities:

$$S_{Total} = \max(S_{EGMS}, S_{Cold}) \qquad (18)$$

Table 1 lists the input and output variables and the model parameters. The final model selected consisted of 9 parameters for each EGMS line and was used to predict PI date, Flowering date, S_{Cold}, S_{EGMS} and S_{Total} for n = 144 observations per EGMS line (48 sowing dates × 3 replicates). Of these 9 parameters 8 parameters turned out to be identical for the two EGMS lines. In total therefore, 10 parameters were used to predict phenology and sterility of n = 288 observations. Half of the observations was used for calibration, half for validation, the procedure for calibration/validation and accuracy assessment is discussed in the following section.

Model accuracy
The experimental data set used for model calibration and validations consisted of 24 sowing dates in 2013 and 24 sowing dates in 2014, with 2 EGMS lines and 3 replicates per sowing date. The data for the sowing dates in 2013 were used for calibration and the data for sowing dates in 2014 for validation. Accuracy in simulated days from sowing to flowering and simulated sterility was measured as root mean square error (RMSE) and as modelling efficiency (EF) (Nash and Sutcliffe, 1970):

$$RMSE = \sqrt{\frac{\sum(S_i - O_i)^2}{n}} \qquad (19)$$

$$EF = 1 - \frac{\sum(S_i - O_i)^2}{\sum(O - O_i)^2} \qquad (20)$$

Where S_i is the simulated variable (sterility or days from sowing to flowering) in treatment i, O_i is the observed variable in treatment i, $\overline{O}$ is the average of observed variables and n the number of observations. A value of 1 for EF indicates perfect prediction. For comparison of the logistic regression models we also used the Akaike Information Criterion (AIC), a standard measure for the goodness of fit of logistic regression models. A lower AIC value indicates a more accurate model.

Model applications
We used the full model to answer the following questions:

1. In Ndiaye, Senegal, which sowing give a simulated male sterility S_{EGMS} = 1 = 100%? (period for producing hybrids by crossing with local popular varieties)
2. In Ndiaye, Senegal, which sowing dates give a simulated sterility S_{Total} < 50%? (period for multiplication of the EGMS lines through self-pollination)
3. In Ndiaye, Senegal, will simulated sterility change with 2 °C temperature increase?
4. Cairo, Egypt: can the same EGMS lines and hybrids also be produced there?
5. M'Be, Ivory Coast: can the same EGMS lines and hybrids also be produced there?

Results
Fertility behaviour of EGMS lines with daylength and temperature
In Fig. 2, we show the observed sterility at 48 sowing dates covering two years. In the top pane (Fig. 2a) we show in lines the simulated sterility for the full model. In the middle pane (Fig. 2b) we show average daylength in the simulated period from panicle initiation to flowering (PIFL). In the bottom pane (Fig. 2c) we show temperatures in the simulated period from panicle initiation to flowering. In the (Appendix 1: Fig. 9), we present the same figure but with flowering date instead of sowing date on the x-axis. In terms of sterility, four distinct periods can be identified from the Figure:

1. Sowing dates from 1 January to mid-July: Constant high sterility S_{EGMS} = 100%. This is a suitable period for sowing to produce hybrids.
2. Sowing dates from mid-July to 1 September: Decreasing sterility. S_{EGMS} is declining because of shortening daylengths and temperatures in the

Table 1 Input variables, output variables and parameters

Name	Type[a]	Description	Sub-model	Table/Figure
T_{min}	input	Daily minimum temperature (°C)	Phenology, EGMS Sterility, Cold Sterility	Fig. 3
T_{max}	input	Daily maximum temperature (°C)	Phenology, EGMS Sterility, Cold Sterility	Fig. 3
$T_{min}(PIFL)$	input	T_{min}, averaged over period from panicle initiation (PI) to flowering (FL)	Phenology, EGMS Sterility, Cold Sterility	
$T_{max}(PIFL)$	input	T_{max}, averaged over period from panicle initiation (PI) to flowering (FL)	Phenology, EGMS Sterility, Cold Sterility	
$DAYL_{sa}(d)$	input	Daylength (h) on day d, calculated with sun angle sa (°) to account for twilight	EGMS Sterility	Fig. 1
$DAYL_{-6}(PIFL)$	input	Daylength (h) including civil twilight ($sa = -6°$), averaged over the period from panicle initiation (PI) to flowering (FL)	EGMS Sterility	
LAT	input	Latitude (decimal degrees)	Phenology, EGMS Sterility	
d_S	input	Sowing day of year (Julian day, 1 to 365)	Phenology	Figs. 2 and 4
d_{PI}	output	Panicle initiation day	Phenology, EGMS Sterility, Cold Sterility	
d_{FL}	output	Flowering day	Phenology, EGMS Sterility, Cold Sterility	Fig. 4
S_{EGMS}	output	Sterility due to photoperiod and high temperature (Eqs. 11 and 12)	EGMS sterility	Figs. 2 and 5
S_{cold}	output	Sterility due to cold (Eq. 17)	Cold Sterility	Figs. 2 and 6
S_{Total}	output	Total sterility (Eq. 18)	Full model	Fig. 2
TBD	param.	Base temperature for development (°C)	Phenology	Table 4
TOD	param.	Optimum temperature for development (°C)	Phenology	Table 4
DVR_{SPI}	param.	Development rate for the phase from sowing (DS = 0) to panicle initiation DS = 0.65 (d^{-1})	Phenology	Table 4
DVR_{PIFL}	param.	Development rate for the phase panicle initiation DS = 0.65 (d^{-1}) to flowering (DS = 1)	Phenology	Table 4
b_{E0}	param.	EGMS Sterility parameter (Eqs. 11 and 12)	EGMS Sterility	Table 5
b_{E1}	param.	EGMS Sterility parameter (Eqs. 11 and 12)	EGMS Sterility	Table 5
b_{E2}	param.	EGMS Sterility parameter (Eqs. 11 and 12)	EGMS Sterility	Table 5
b_{C0}	param.	Cold Sterility parameter (Eq. 17)	Cold Sterility	Table 6
b_{C1}	param.	Cold Sterility parameter (Eq. 17)	Cold Sterility	Table 6

[a]input variable, output variable, parameter

September-November period (the panicle initiation to flowering period associated with these sowing dates).

3. Sowing dates from 1 September to mid-November: Moderate to low sterility. S_{Cold} is 10-50% due to cold, S_{EGMS} (yellow lines) is around 0%.

4. Sowing dates from mid-November to 1 January: Increasing sterility. S_{EGMS} is increasing because of increasing daylengths and temperatures in the February-April period (the panicle initiation to flowering period associated with these sowing dates).

The full model (Fig. 2a) predicted sterility with RMSE values of 8 to 12% and model efficiencies of 0.88 to 0.94 (Table 2). Model accuracies were consistently a bit higher for EGMS1 than for EGMS2 and consistently a bit higher for the calibration than for

validation. For cold sterility our model underestimated cold sterility for the October sowing dates and overestimated cold sterility in the November sowing dates. Those crops sown in November are right from the start of the growing period exposed to colder temperatures. This suggests that possibly some process of acclimation is occurring (as has also been hypothesised by Dingkuhn et al. 2015a). The sparsity of our data does not allow for further investigating this hypothesis. Given this period of cold we can identify two periods best suited for multiplication of the EGMS lines, i.e. where sterility is lowest. These are (1) sowing in mid august to mid September and (2) sowing in November. Associated flowering dates are November and March, just outside the coldest part of the year, the December-February period.

From Fig. 2 we can already roughly determine daylength and temperature thresholds. The period of complete sterility starts at early January. The

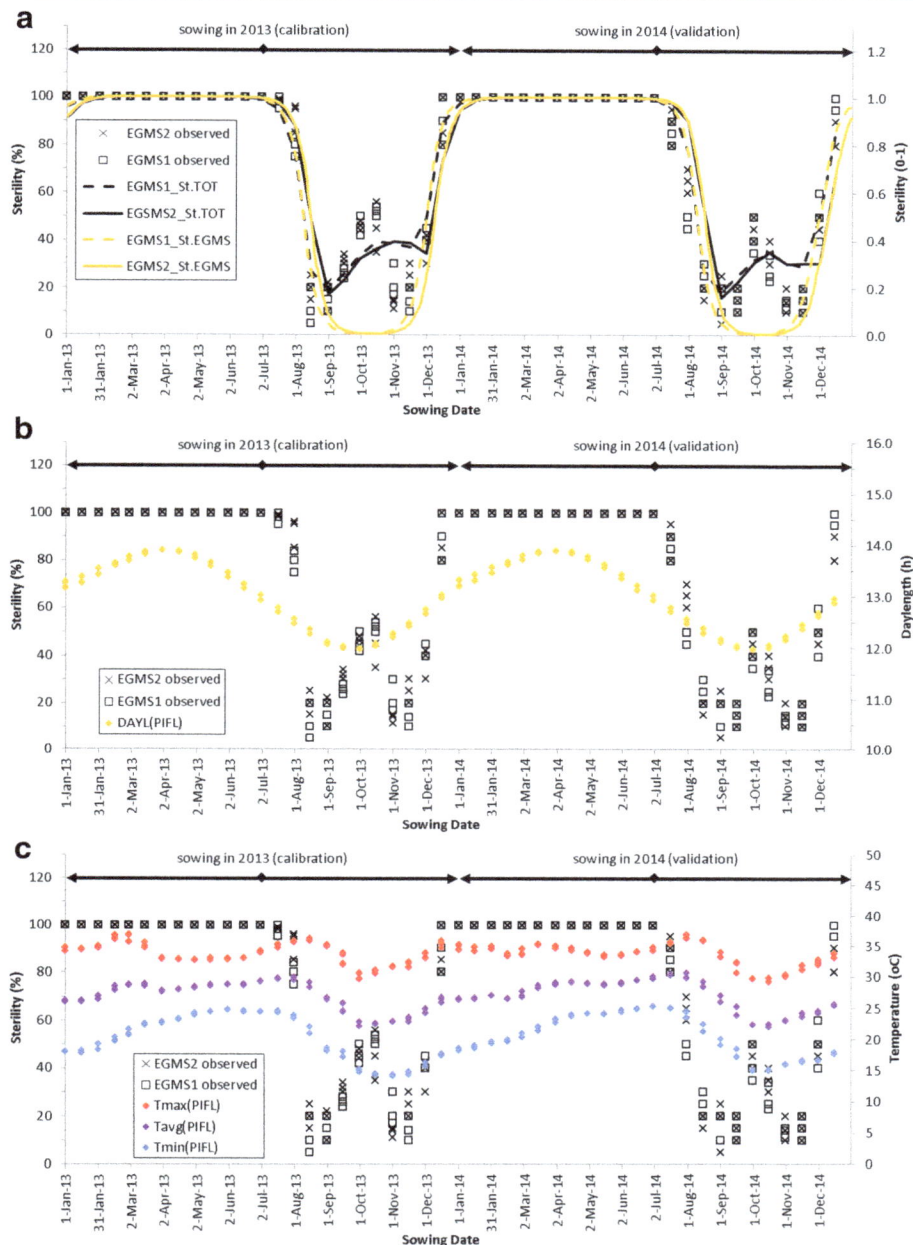

Fig. 2 Fertility behaviour of EGMS lines with daylength and temperature. *Black symbols* show observed sterilities for the two EGMS lines. In **a** *black lines* show the simulations with the full model, *yellow lines* show simulated sterility with the EGMS model (without cold sterility). In **b**, the right axis and *yellow symbols* show the average daylength in the period from panicle initiation to flowering (PIFL) for each sowing date. In **c** the right axis shows air temperature and *coloured symbols* show the average of daily maximum, daily average and daily minimum temperatures, averaged over the PIFL period associated with each sowing date

associated flowering date is early May. The associated daylength (incl twilight) in the PIFL period is 13.3 h (Fig. 2b) and average temperature (T_{avg}) in the PIFL period for this sowing date is 26 °C. The period with complete sterility ends with sowing early July (flowering early October), with daylength $DAYL_{-6}(PIFL) = 12.9$ h and $T_{avg}(PIFL)$

is 30 °C. Here temperature is higher and daylength is shorter than for the January sowing date. This suggests that both temperature and daylength play a role, that at higher temperatures the critical daylength will be shorter. We investigate this hypothesis in more detail in the following sections.

Table 2 Accuracy of total sterility simulations

		EGMS1	EGMS2
RMSE[a]	Calibration	8.2%	10.4%
	Validation	9.3%	12.0%
EF	Calibration	0.94	0.90
	Validation	0.93	0.88

[a]RMSE = Root mean square error (Eq. 19)
EF = Modelling efficiency (Eq. 20)

Inheritance of sterility

The F_1 hybrids of both populations of IR75589-31-27-8-33/Sahel108 (EGMS1/Sahel108) and IR77271-42-5-4-36/Sahel108 (EGMS2/Sahel108) sown in 20 August 2014 were completely fertile. Sterilities in the F_2 population sown in 15 April 2015 clearly indicated that only one recessive gene is involved in causing male sterility (Table 3). For IR77271-42-5-4-36/Sahel108 we found 23% pollen sterility and 27% spikelet sterility (104/460 and 125/460). For IR75589-31-27-8-33/Sahel108 we found 27% pollen sterility and 23% spikelet sterility. These fractions did not differ significantly from the expected sterility of 25% (*p*-value ranging from 0.21 to 0.31). Observed sterilities did differ significantly from the sterility of 6.25% which we would expect if two recessive genes were involved. For both F_2 populations, the difference between pollen sterility and spikelet sterility was statistically significant, but in absolute terms the difference was small (+4% and –4%).

Phenology and sterility model
Phenology

Figure 3 shows temperatures over the three years used for calibration and validation. Lower temperatures in December to February lead to longer duration from sowing to flowering when sown in November to January. Days from sowing to flowering differed significantly between the two EGMS lines, with EMGS1 having 7–14 days longer duration. A phenology model with a base temperature of 11 °C, an optimum temperature of 26 °C

and no delay in development above the optimum temperature accurately predicted the days to flowering (Table 4, Fig. 4). RMSE was 4 days and model efficiency EF was 0.89, both for the calibration and the validation and for both of the two EGMS lines (Table 3). Such accuracies are normal for phenology simulations (van Oort et al. 2011, Dingkuhn et al. 2015b, Zhang et al. 2016). A visual analysis revealed that for sowing in mid December to mid January, the model systematically predicted a too short duration, with maximum errors of 15 days.

EGMS sterility sub-model

Figure 5 shows the relation between daylength (incl civil twilight) from panicle initiation to flowering. Figure 5a shows the full dataset, including the 10-50% sterilities for flowering in December to February (when days are short) which are caused by cold sterility. To avoid confounding effects, these were not used for model calibration. For calibration we used the open symbols (sowing in 2013) from data shown in Fig. 5b. In Fig. 5b we can clearly see a sigmoid shape of sterility increasing with daylength, but we can also see from the colours that in the transition zone from 12.5 to 13.0 h daylength, sterility is higher when $T_{min}(PIFL)$ is higher.

Table 5 shows the fitted models with asterices (*) for significance for the parameters. Both $T_{min}(PIFL)$ and $DAYL_{-6}(PIFL)$ had statistically significant effects. Model accuracy of the $DAYL_{-6}(PIFL)$ model was much higher than that of the $T_{min}(PIFL)$ based model (compare models 1 and 2: AIC = 43 and 73). $T_{min}(PIFL)$ and $DAYL_{-6}(PIFL)$ were only weakly correlated (R^2 = 0.11) which means their effects can be separately investigated. In the model that incorporated both these variables (model 6, AIC = 25), both showed a significant effect, with a lower *p*-value for $DAYL_{-6}(PIFL)$ than for $T_{min}(PIFL)$. Next we investigated if models with interaction terms (models 10 and 15) would predict more accurately than the additive model. This was not the case. The model with only the interaction term $T_{min}(PIFL)$ x $DAYL_{-6}(PIFL)$ had

Table 3 Pollen and spikelet fertility of F_2 populations

	Total number of plants	Observed		Expected		χ^2 (*p*-value)
		Fertile	Sterile	Fertile (75%)	Sterile (25%)	
Pollen fertility						
IR75589-31-27-8-33/Sahel 108	520	378	142	390	130	1.36 (0.24)
IR77271-42-5-4-36/Sahel 108	460	356	104	345	115	1.28 (0.26)
Spikelet fertility						
IR75589-31-27-8-33/Sahel 108	520	403	117	390	130	1.60 (0.21)
IR77271-42-5-4-36/Sahel 108	460	335	125	345	115	1.05 (0.31)

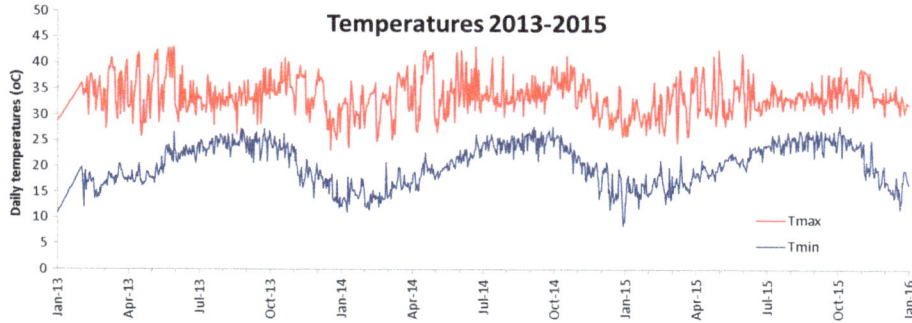

Fig. 3 Daily minimum and maximum air temperatures during the experimentation period

a higher AIC value than the best model without interaction (compare models 1 and 15, AIC = 43 and 61), thus model 1 is more accurate. In the model with both additive and interaction effects (model 10), none of the parameters differed significantly from zero and the AIC value (AIC = 24) is almost the same as that of model 6 (AIC = 25). We therefore consider the additive model 6 the most accurate, with $T_{min}(PIFL)$ and $DAYL_{-6}(PIFL)$ as the explanatory variables.

No statistically significant effects of $T_{max}(PIFL)$ and EGMS on sterility were found. Significant effects of $T_{avg}(PIFL)$ were found. But considering that T_{avg} is calculated as the average of T_{min} and T_{max} and given that the effect of T_{max} was not significant we think the significant effects of T_{avg} found here are the result of strong correlation between T_{min} and T_{avg} ($R^2 = 0.84$) and not an effect of T_{avg} per-se.

Cold sterility sub-model
Figure 6 shows two calibrated cold sterility sub-models, with parameters listed in Table 6. As with the EGMS sterility model, no significant differences were found between the EGMS lines in terms of cold sterility (result not shown). Although the regression

line in Fig. 6 showed increasing sterility at lower temperatures, statistical testing showed the regression parameter b_{C1} did not differ significantly from zero ($p > 0.1$). Possibly these high p-values are caused by lack of power (only $n = 56$ data points had flowering in December-February, of which $n = 29$ were sown in 2013 and used for calibration). Outside the December-February period temperatures were higher, but sterility was not lower (as we would expect from Fig. 6) because for flowering dates outside the December-February period sterilities increase again due to EGMS sterility. As a result of this, we could not find lower p-values by expanding the dataset with data from more sowing dates. Aside from this lack of power, low accuracies may be caused by uncertainties in the model. In particular, three uncertainties can be identified. Firstly, we know theoretically that water temperature is a better predictor than air temperature, but water temperature data were not available and therefore we used air temperature as a proxy. A second uncertainty is that we used estimated and not observed panicle initiation dates and this introduces some uncertainty in the estimation of temperature in the period from panicle initiation to flowering. A third uncertainty here is that the

Table 4 Phenology parameters and model accuracy

Parameter	EGMS1	EGMS2	Description
TBD (°C)	11	11	Base temperature for development
TOD (°C)	26	26	Optimum temperature for development[a]
DVR$_{SPI}$ (d^{-1})	0.011818	0.013265	Development rate for the phase from sowing (DS = 0) to panicle initiation DS = 0.65
DVR$_{PIFL}$ (d^{-1})	0.010000	0.010294	Development rate for the phase panicle initiation (DS = 0.65) to flowering (DS = 1)
Accuracy	EGMS1	EGMS2	
RMSE$_{SFL}$ (d)	4.0	4.2	Calibration Accuracy: Root mean square error for the duration from sowing to flowering
RMSE$_{SFL}$ (d)	4.0	4.2	Validation accuracy
EF	0.89	0.89	Calibration Accuracy: Model Efficiency for the duration from sowing to flowering
EF	0.89	0.88	Validation accuracy

[a]We assumed that above TOD, development rate remains optimal (Eq. 9)

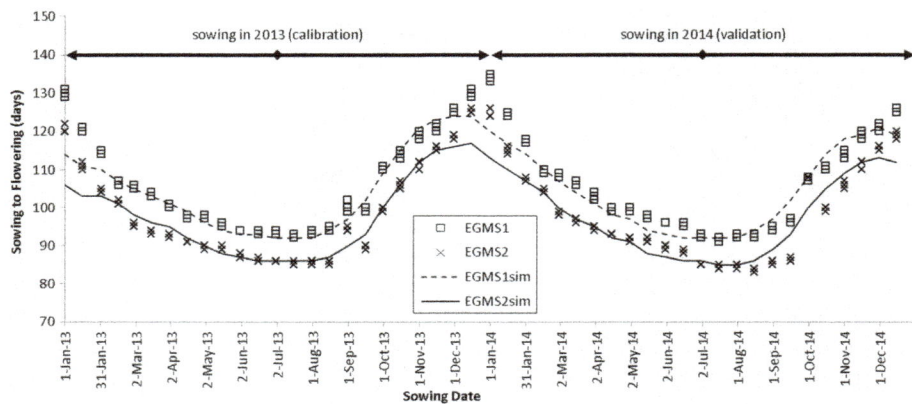

Fig. 4 Days from sowing to flowering. Lines show simulated sterility, points show observed data. Observed data for sowing dates in 2013 were used for model calibration, cata for sowing dates in 2014 were used for validation. The model predicts very accurately for the February to November period and less accurately for sowing dates in December and January

temporal window of the cold sensitive period was potentially wrong. We discuss this in more detail in the discussion section of this paper.

The results however confirm previous research on cold sterility effects and are our only plausible explanation for the spike in sterility for flowering dates in December – February, which cannot be explained from EGMS sterility (Figs. 2a, 5a). We therefore included the cold sterility sub-model in the full model despite the non-significant b_{C1} parameter. We included the model with $T_{min}(PIFL)$ because previous research for non-

EGMS varieties (Dingkuhn et al. 2015a, van Oort et al. 2015a) showed that T_{min} is a better predictor for cold sterility than T_{avg} or T_{max}.

Full model

For the full model, we used the three sub-models discussed in the above sub-sections. For EGMS sterility, we used EGMS model 6 (Table 5). For cold sterility we used Cold sterility model 1 (Table 6). Overall accuracy was already presented above.

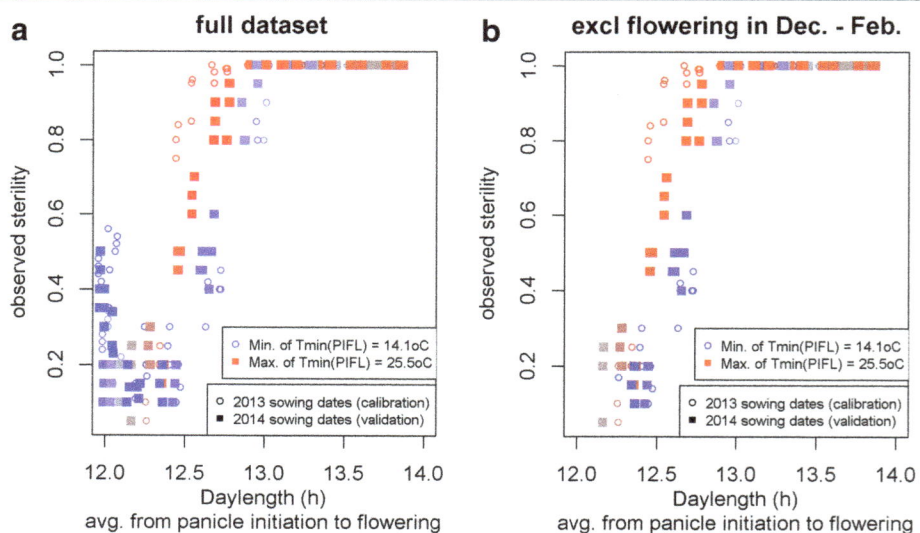

Fig. 5 Relation between observed sterility and daylength. The colours from *blue* to *red* show the average of daily minimum temperatures (T_{min}) averaged over the period from panicle initiation to flowering (PIFL). In **a** the full dataset is shown, with a spike in observed sterilities at short daylengths (12.0 hours) which is probably caused by cold sterility. In **b** the dataset is shown excluding data points with flowering dates in December to February. The dataset in **b** was used for calibration (*open circles*) and validation (*solid squares*) of the EGMS model

Table 5 EGMS models calibrated from the 2013 sowing dates

Model	Equation[1]	AIC[2]
1	$-84.272^{***} + 6.737^{***} \times DAYL_{-6}(PIFL)$	43
2	$-6.083^{**} + 0.400^{***} \times T_{min}(PIFL)$	73
3	$-14.571^{***} + 0.606^{***} \times T_{avg}(PIFL)$	76
4	$-3.545 + 0.158 \times T_{max}(PIFL)$	106
5	$1.940^{***} + 0.215xEGMS$	108
6	$-95.586^{***} + 7.120^{***} \times DAYL_{-6}(PIFL) + 0.330^{*} \times T_{min}(PIFL)$	25
7	$-100.627^{***} + 7.075^{***} \times DAYL_{-6}(PIFL) + 0.446^{*} \times T_{avg}(PIFL)$	27
8	$-108.282^{***} + 7.291^{***} \times DAYL_{-6}(PIFL) + 0.495 \times T_{max}(PIFL)$	36
9	$-84.328^{***} + 6.730^{***} \times DAYL_{-6}(PIFL) + 0.092 \times EGMS$	46
10	$316.130 - 25.809 \times DAYL_{-6}(PIFL) - 22.142 \times T_{min}(PIFL) + 1.799 \times DAYL_{-6}(PIFL) \times T_{min}(PIFL)$	24
11	$760.139 - 61.840 \times DAYL_{-6}(PIFL) - 31.919 \times T_{avg}(PIFL) + 2.592 \times DAYL_{-6}(PIFL) \times T_{avg}(PIFL)$	24
12	$430.004 - 35.931 \times DAYL_{-6}(PIFL) - 15.359 \times T_{max}(PIFL) + 1.273 \times DAYL_{-6}(PIFL) \times T_{max}(PIFL)$	36
13	$-64.587 + 5.160 \times DAYL_{-6}(PIFL) - 13.867 \times EGMS + 1.111 \times DAYL_{-6}(PIFL) \times EGMS$	47
14	$-95.704^{***} + 7.112^{***} \times DAYL_{-6}(PIFL) + 0.330^{*} \times T_{min}(PIFL) + 0.142 \times EGMS$	27
15	$-7.163^{***} + 0.035^{***} \times DAYL_{-6}(PIFL) \times T_{min}(PIFL)$	61

$DAYL_{-6}(PIFL)$ is the average of daylengths from panicle initiation to flowering, calculated with civil twilight (sun up to 6° below the horizon). $T_{min}(PIFL)$ is the average of daily minimum temperatures T_{min} from panicle initiation to flowering. EGMS is a binary: 0 for EGMS1, 1 for EGMS2
[1]Significance codes: * $p < 0.05$, ** $p < 0.01$, *** $p < 0.001$
[2]Akaike Information Criterion (lower is better)

Model applications

We present two applications of the model: site selection and synchronisation of flowering dates of two the male and female parents of F_1 seed.

Suitability of different sites

We used the full model to identify two distinct periods for production of hybrids, namely a period with 100% sterility and a period with <50% sterility. Fig. 7 shows simulated sterility for monthly sowing dates for different environments. Points above the red line are suitable sowing dates for producing hybrids (S_{EGMS} = 100%). Points in the shaded green area (S_{Total} < 50%) are suitable sowing dates for multiplication of the EGMS line. We show here only simulation results for EGMS1 as previous analyses (Fig. 2) revealed only small differences between the two EGMS lines. We can see from Fig. 7a, b that climate change changes sterilities a bit, but changes are too small to be consequential. With +2 °C climate change suitable sowing dates for hybrid production and EGMS multiplication (cross-pollination for producing F_1 seed and self-pollination to maintain the EGMS line) remain the same as in the current climate. Ndiaye in Senegal has a period of 6–7 sowing months suitable for production of hybrids and a 3–4 months period suitable for multiplication of the EGMS line. The AfricaRice station in M'be in Ivory coast has a period of 5–6 sowing months suitable for production of hybrids and a 3–4 months period suitable for multiplication of the

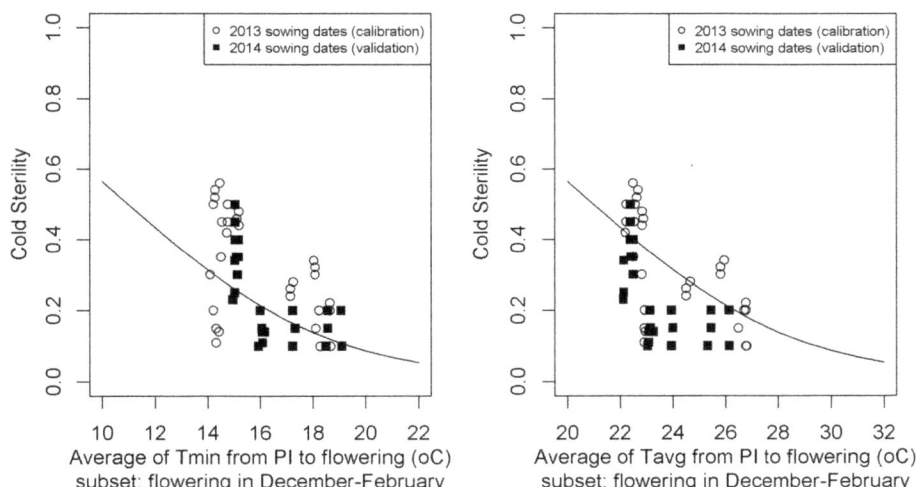

Fig. 6 Relation between cold sterility and temperature. Data points are observed values for 2 EGMS lines, sown at 24 dates (15 day interval) from 2012 to 2013, with 3 replicates per EGMS line and sowing date, for the subset of data points with flowering dates in December to February. Lines are regression lines based on parameters in Table 5

Table 6 Cold Sterility models calibrated from the 2013 sowing dates

Model	Equation[1]	AIC[2]
1	$2.5095 - 0.2056 \times T_{min}(PIFL)$	29
2	$5.4650 - 0.2604 \times T_{avg}(PIFL)$	29
3	$8.1892 - 0.2804 \times T_{max}(PIFL)$	30

$T_{min}(PIFL)$ is the average of daily minimum temperatures T_{min} from panicle initiation to flowering. $T_{avg}(PIFL)$ is the average of daily average temperatures and $T_{max}(PIFL)$ is the average of daily maximum temperatures over this period
[1]Significance codes: * $p < 0.05$, ** $p < 0.01$, *** $p < 0.001$
[2]Akaike Information Criterion (lower is better)

EGMS lines, thus also well suited. Application of the two line breeding system could be more difficult (but not impossible) in the Nile delta in Egypt, where only the August sowing date is predicted to give sterilities less than 50%. In this environment, multiplication of the EGMS line can be issue.

Synchronisation of flowering dates of two parents

To produce hybrid seeds, the EGMS line and the male parent must flower in exactly the same week, at a time when the EGMS line is 100% sterile and the local variety is sufficiently fertile. If phenology and sterility parameters of the local popular variety are known we can simulate for both varieties at any flowering date what their sterility will be. Once a

desirable flowering date is found, we can trace back from the simulated flowering date the associated sowing dates of the EGMS and the local variety. These can be different due to differences in phenological parameters.

In Fig. 8 we simulated sterility of the EGMS1 line (this paper) and a local popular variety in the Senegal river valley, Sahel108 (using Sahel108 phenology parameters reported in van Oort et al. 2011 and the model ORYZA2000v2n13s14 reported in van Oort et al. 2015a, with transplanted rice with a seedbed duration of 21 days). We can see in Fig. 8a that when flowering between 1 May and 1 September the EGMS1 line is 100% sterile and the local variety Sahel108 is fertile (<20% sterility). We can also see in this environment it is possible to have multiplication of the EGMS line in one part of the year and production of hybrid seeds in another part of the year. Say that for production of hybrids we want to have both parents flowering on 15 June. Then from phenology simulations (Fig. 8b) we can trace back that Sahel108 must be sown on 20 February and EGMS1 must be sown on 4 March, a difference of 12 days. This example shows how two independent phenology and sterility models can be used to fine tune sowing dates of two parent varieties for production of hybrids. In this case to arrive at the same (synchronised) flowering date, really different sowing dates are required. This shows that understanding of the phenology of the two parents

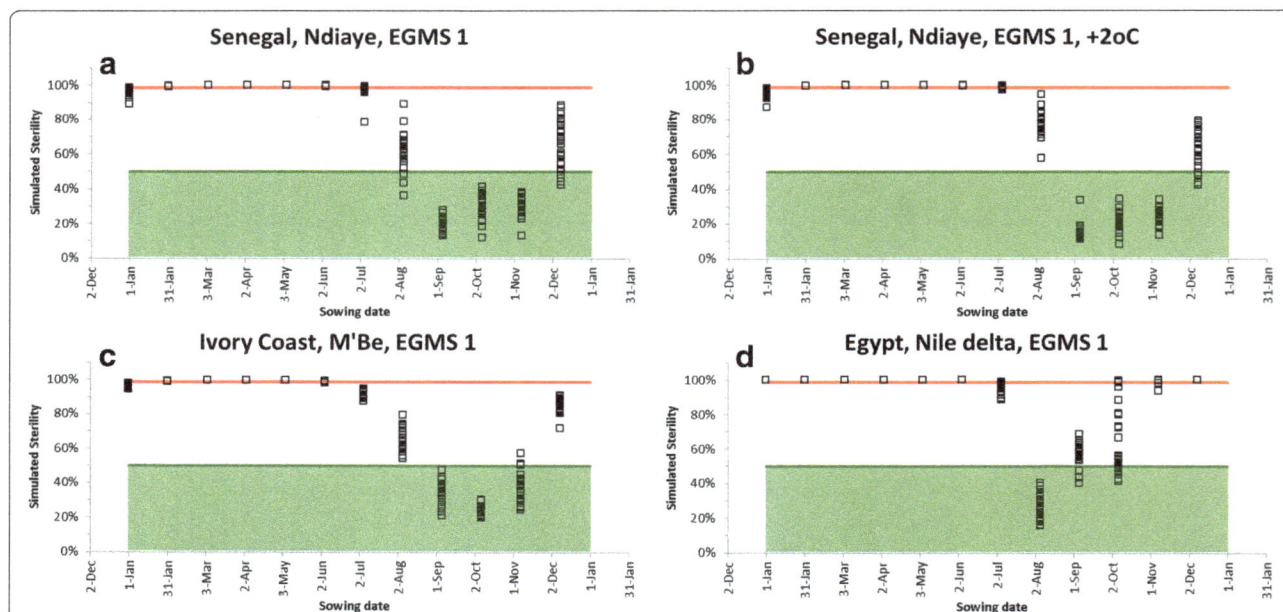

Fig. 7 Simulated total sterility at monthly sowing dates for 4 environments. **a** Ndiaye Senegal, **b** Ndiaye Senegal with 2oC temperature increase, **c** M'Be in AfricaRice and **d** Nile Delta in Egypt. Points at or above the red line are suitable for production of F₁ (hybrid) seed because the EGMS parent line is completely sterile. Points in the green area show suitable dates for multiplication (through self-pollination) of the EGMS line

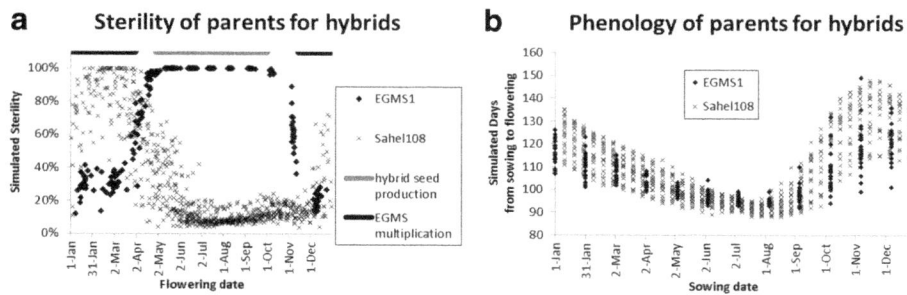

Fig. 8 Simulated sterility **a** and days from sowing to flowering **b** of EGMS1 (IR75589-31-27-8-33) and Sahel108. EGMS1 was simulated with sowing on a 31 days timestep Sahel 108 was simulated with sowing on a 10 days timestep. The data points show the variation caused by interannual weather variability, simulations were conducted for the years 1990 to 2015

is important because not per definition the two parents must be sown on the same date to have flowering on the same date.

Discussion

Practical implications

This paper shows that the two-line hybrid rice breeding system is feasible in the study site (Ndiaye, Senegal) because (1) we found a high (~25%) inheritance of the EGMS trait in the F_2 population and (2) observations and simulations showed that in this site a long period suitable for producing hybrids (F_1 seed) through the two line system exists and (3) also a long period suitable for EGMS multiplication exists.

We found a high (~25%) inheritance of the EGMS trait in the F_2 population. Such is desirable because from this population breeders can select the ~25% sterile plants, or a subset of these 25%, which can then serve as new parents for hybrids. The new parents and the new F_1 population of hybrids will have, in comparison with the first generation, more "locally desirable" genes. By repeating this process, breeders develop locally adapted hybrids. This whole process would be much less efficient if inheritance were less than 25%, for example 6.25% if two recessive genes were involved. The high (25%) inheritance allows for running a two line breeding system for hybrid rice, which is more efficient than the three line system. In the same time these two EGMS lines can be used directly as male lines with different local variety and breeding lines to produce high yielding F_1 hybrids.

We presented two possible applications of our model: (1) assessing the suitability of other sites for hybrid rice breeding programs using the same EGMS varieties as considered here and (2) synchronisation of flowering dates and from that selection of suitable sowing dates for the two parent lines for production of hybrids. The suitability analyses identified suitable sowing windows for producing hybrids and for multiplication of the

EGMS lines. This can provide useful guidance when establishing new breeding programs in these sites and also provide an opportunity for further validation of the model presented here. The synchronisation exercise showed that the model presented here can be used in combination with other independent existing models for 'normal' (=non-EGMS) varieties such as have already been developed and parameterised for many varieties (van Oort et al. 2011, van Oort et al. 2015a, Dingkuhn et al. 2015a,b, Zhang et al. 2016). It should be noted that these two applications serve as a first estimate, which involved applying the model outside the bounds of temperature and daylength for which it was calibrated. We are therefore less certain about the accuracy of these predictions. Validation in these sites is desirable and would be a logical next step if breeding programs are started in these sites. Apart from these two applications particular for rice we observe that the model presented here is very simple requiring estimation of only few parameters. We therefore expect that the same model can be quite easily applied to other EGMS lines in rice and to other crops.

Scientific implications

The number of genes involved in causing male sterility was not before investigated for the two EGMS lines considered in this paper. A single recessive gene has previously been reported to control male sterility in 5460S (Sun et al. 1989), R59TS (Yang and Wang, 1990), Norin PL 12 (Maruyama et al. 1991), Nongken 58 (Zhuping 1997) and Guangzhan63S (Xu et al. 2011). These lines are more attractive for breeding than those with two or more genes involved, possibly the latter receive less publicity in scientific literature. For the F_2 of EGMS lines 7001S, 5047S, 5088S previously Zhuping (1997) reported a 1:15 segregation, indicating two recessive genes can also be involved. Those relatively rare cases with two genes involved imply that we cannot take a one-gene controlled EGMS for granted and that it is relevant to check the number of genes involved, as we did in this paper.

Previously, similar models as presented here were developed for simulating phenology (van Oort et al. 2011, Dingkuhn et al. 2015b, Zhang et al. 2016). For photoperiod sensitive rice varieties development can already be delayed at daylengths longer than 10 h (e.g. Yin et al. 1997, Awan et al. 2014). Photoperiod in our study site varied from 12 to 14 h, thus long enough for detecting photoperiod sensitivity if present. Comparisons of model accuracies for phenology simulated with and without photoperiod sensitivity with the pheno_opt_rice2 phenology calibration program (van Oort et al. 2011) showed that for the EGMS lines considered here, accuracy could not be increased by adding photoperiodism. For other EGMS lines, it is relevant to test for photoperiodism and include this in the phenology model if necessary.

Previously, similar models as presented here were developed for heat and cold sterility and applied in the same study area for 'normal' varieties, i.e. without male sterility (van Oort et al. 2015a; Julia and Dingkuhn 2013; Dingkuhn et al. 2015a,b). There are two notable differences with the model presented here. Firstly, our EGMS lines have sterility at long days whereas normal varieties do not have this trait. And secondly, cold sterility for our EGMS lines was much smaller than for the varieties studied by van Oort et al. 2015a, Julia and Dingkuhn 2013, Dingkuhn et al. 2015a. For this reason we developed a new sterility model and tested for interaction between daylength and temperature effects. Especially for cold sterility our model was less accurate. Two hypotheses might possibly explain this lower accuracy, but were impossible to test here due to the sparsity of measurements in the cold period and due to lack of more detailed measured phenological data. The first hypothesis is that some process of acclimation is occurring, i.e. plants exposed to cold earlier on in their development might be more tolerant to cold later; this would explain the underestimation of cold sterility of October sown crops (less exposed to early cold, Fig. 2a) and the overestimation of cold sterility of November sown crops (more exposed to early cold, Fig. 2a). The second hypothesis is that we got the period in which rice is sensitive to cold not precisely right and were thus calculating temperature effect over the wrong period. It has been suggested that rice is especially sensitive to cold in a short period around microspore stage. Sensitivity to cold in such a short period is difficult to model, because predictive accuracy becomes strongly dependent on correct estimation of the start and end development stage in which the plant is sensitive.

Also for EGMS sterility it would for further research be interesting to investigate if a certain period within the phase from panicle initiation to flowering is more/less sensitive to daylength. This met with the same problems as noted above: lack of measurements on phenological stages. We used a logistic regression model for modelling the relation between sterility and daylength. The (sigmoid shaped) logisitic model is statistically more appropriate model when analysing binary response variables (sterility) as compared with previous studies that used correlation analysis and linear models (e.g. Latha et al. 2004) or threshold values (e.g. Lopez and Virmani 2000; Virmani et al. 2003). The appropriateness of the logistic model was also clearly shown in the sigmoidal shape of observed sterilities in Fig. 5b.

Cold sterility of the EGMS lines reached a maximum of 50% whereas previously van Oort et al. (2015a), Julia and Dingkuhn (2013) and Dingkuhn et al. (2015a) reported for tropical varieties sterilities of more than 80% when flowering in the same cold December-February period (see also Fig. 7a). The much lower sterility reported there is consistent in that in these previous studies tropical varieties were evaluated, whereas our EGMS lines originated from a cooler environment, thus better adapted to the cold temperatures. Consistently with this, we also found a lower base and optimum temperature for the EGMS lines (TBD = 11 °C, TOD = 26 °C) than for the tropical varieties normally evaluated for the study site, with TBD in the range of 14 to 18 °C and TOD in the range of 26 to 34 °C (van Oort et al. 2011). The low sterilities for flowering in the cold December-February period are convenient because they allow for multiplication of EGMS lines during the cold dry season, something which would not be possible with normal varieties (Fig. 8a).

Conclusions

A newly developed model could accurately simulate phenology and sterility of two EGMS lines grown in Senegal as a function sowing date, weather variables and a limited number of genetic parameters. Daylength from panicle initiation to flowering was the main explanatory variable, additionally also a significant effect of minimum temperatures during the same period was found. The model can be useful for identifying safe sowing windows for production of hybrids and multiplication of the male sterile parent. A statistical analysis of inheritance revealed that only one recessive gene is causing the male sterility. In Senegal, Dry season (February – July) is the suitable period for F_1 hybrid seed productions and Wet season (August – December) is the suitable period for multiplying the EGMS lines.

Appendix 1

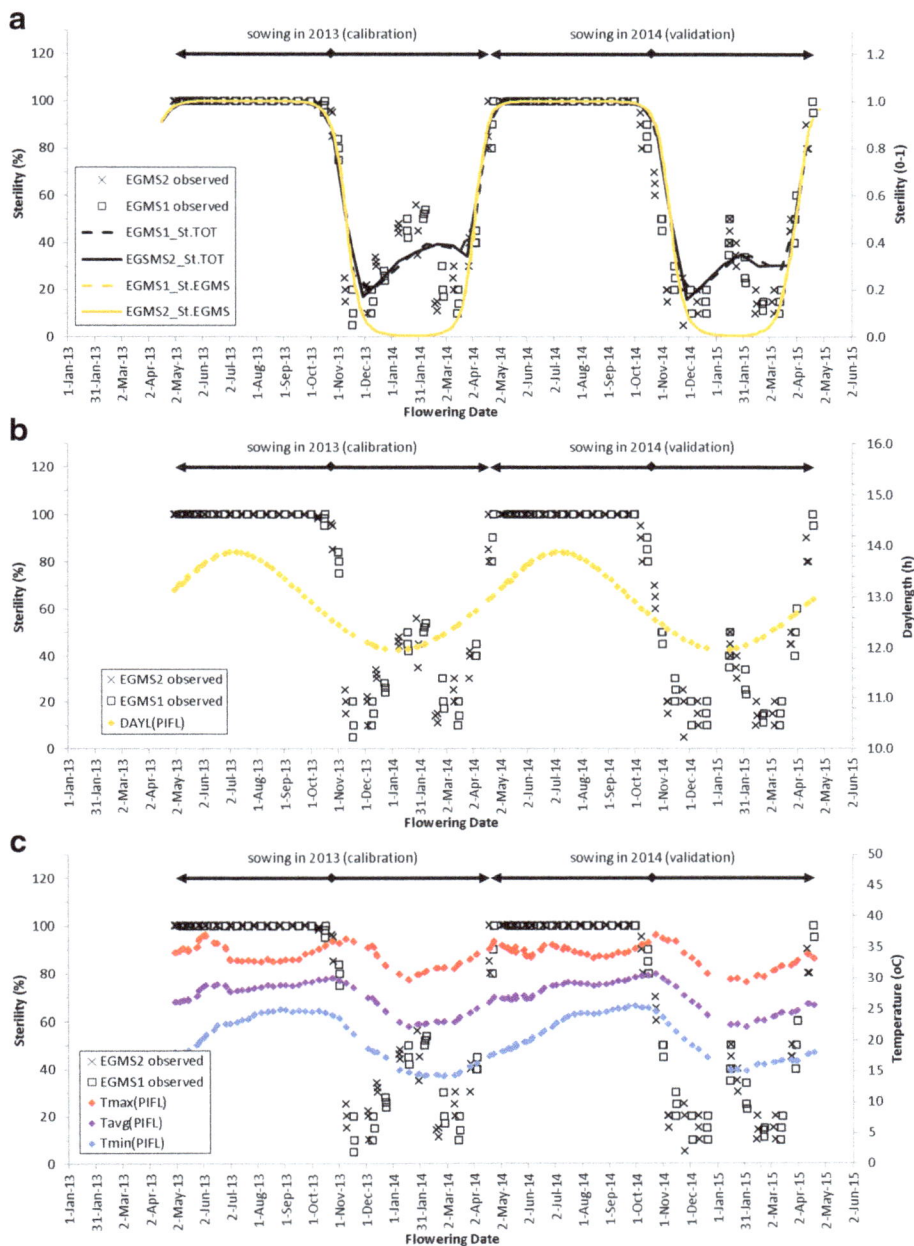

Fig. 9 Fertility behaviour of EGMS lines with daylength and temperature. *Black symbols* show observed sterilities for the two EGMS lines. In **a** *black lines* show the simulations with the full model, *yellow lines* show simulated sterility with the EGMS model (without cold sterility). In **b**, the *right* axis and *yellow* symbols show the average daylength in the period from panicle initiation to flowering (PIFL). In **c** the right axis shows air temperature and *coloured symbols* show the average of daily maximum, daily average and daily minimum temperatures, averaged over the PIFL period associated with each flowering date

Acknowledgements
We acknowledge funding from the Global Rice Science Partnership (GRiSP) and from the Green Super Rice (GSR) project.

Competing interests
We have no competing interests.

Authors' contributions
RE-N designed the research and conducted experiment. PvO developed the model and conducted the statistical analysis. We jointly wrote the paper. All authors read and approved the final manuscript.

Downloads
The model can be downloaded from http://models.pps.wur.nl/content/sterility-model-hybrid-rice-egms-lines

Author details
[1]Africa Rice Center (AfricaRice), P.3. 96, St. Louis, Senegal. [2]Rice Research & Training Center, 33717, Sakha Kafr Sheikh, Egypt. [3]Africa Rice Center (AfricaRice), 01 B.P. 2551 Bouaké, Côte d'Ivoire. [4]Crop & Weed Ecology Group, Centre for Crop Systems Analysis, Wageningen University, P.O. Box 4306700, AK, Wageningen, The Netherlands.

References
Awan MI, van Oort PAJ, Bastiaans L, van der Putten PEL, Yin X, Meinke H (2014) A two-step approach to quantify photothermal effects on pre-flowering rice phenology. Field Crop Res 155:14–22

Bouman, B.A.M., Kropff, M.J., Tuong, T.P., Wopereis, M.C.S., ten Berge, H.F.M., van Laar, H.H., 2001. ORYZA2000: Modeling Lowland Rice. International Rice Research Institute, Los Baños, Philippines and Wageningen University and Research Centre, Wageningen, The Netherlands.

Chen L-Y, Xiao Y-H, Lei D-Y (2010) Mechanism of sterility and breeding strategies for photoperiod/thermo-sensitive Genic male sterile Rice. Rice Sci 17(3):161–167

Dingkuhn M, Radanielina T, Raboin LM, Dusserre J, Ramantsoanirina A, Sow A, Manneh B, Balde AB, Soulié JC, Shrestha S, Ahmadi N, Courtois B (2015a) Field phenomics for response of a rice diversity panel to ten environments in Senegal and Madagascar. 2. Chilling-induced spikelet sterility. Field Crop Res 183:282–293

Dingkuhn M, Sow A, Manneh B, Radanielina T, Raboin LM, Dusserre J, Ramantsoanirina A, Shrestha S, Ahmadi N, Courtois B (2015b) Field phenomics for response of a rice diversity panel to ten environments in Senegal and Madagascar. 1. Plant phenological traits. Field Crop Res 183:342–355

El-Namaky, R.A. and Demont, M., 2013. Hybrid Rice in Africa: Challenges and Prospects. In: Wopereis, M.C.S., Johnson, D.E., Ahmadi, N., Tollens, E. and Jalloh, A., Realizing Africa's Rice Promise, CAB International. 173–178 http://www.africarice.org/publications/rice_promise/Chap13%209781845938123.pdf

Gomez KA, Gomez AA (1984) Statistical procedures for agricultural research. Wiley, New York, p 608

Goudriaan J, van Laar HH (1994) Modelling potential crop growth processes. Current issues in production ecology. Kluwer Academic Publishers, Dordrecht, p 239

Horner HT, Palmer RG (1995) Mechanisms of genic male sterility. Crop Sci 35(6):1527–1535

Huang J-Z, ZG E, Zhang H-L, Shu Q-Y (2014) Workable male sterility systems for hybrid rice: genetics, biochemistry, molecular biology, and utilization. Rice 7(1):1–14

Julia C, Dingkuhn M (2013) Predicting temperature induced sterility of rice spikelets requires simulation of crop-generated microclimate. Eur J Agron 49:50–60

Latha R, Thiyagarajan K, Senthilvel S (2004) Genetics, fertility behaviour and molecular marker analysis of a new TGMS line, TS6, in rice. Plant Breed 123(3):235–240

Li S, Yang D, Zhu Y (2007) Characterization and use of male sterility in hybrid rice breeding. J Integr Plant Biol 49(6):791–804

Lopez MT, Virmani SS (2000) Development of TGMS lines for developing two-line rice hybrids for the tropics. Euphytica 114(3):211–215

Maruyama K, Araki H, Kato H (1991) Thermosensitive genetic male sterility induced by irradiation. In: Rice genetics II. International Rice Research Institute, Manila, pp 227–235

Nash JE, Sutcliffe JV (1970) River flow forecasting through conceptual models part I - a discussion of principles. J Hydrol 10(3):282–290

van Oort PAJ, De Vries ME, Yoshida H, Saito K (2015a) Improved climate risk simulations for rice in arid environments. PLoS One 10(3):e0118114

van Oort PAJ, Saito K, Tanaka A, Amovin-Assagba E, Van Bussel LGJ, van Wart J, de Groot H, van Ittersum MK, Cassman KG, Wopereis MCS (2015b) Assessment of rice self-sufficiency in 2025 in eight African countries. Global Food Security 5:39–49

van Oort PAJ, Zhang TY, de Vries ME, Heinemann AB, Meinke H (2011) Correlation between temperature and phenology prediction error in rice (Oryza sativa L.) Agric For Meteorol 15:1545–1555

Shi MS (1985) The discovery and the study of the photosensitive recessive male sterile rice (Oryza sativa L. subsp. japonica). Sci Agric Sin 18(2):44–48 (in Chinese with English abstract)

Sun ZX, Min SK, Xiong ZM (1989) A temperature-sensitive male sterile line found in rice. Rice Genetics Newsletter 6:116–117 http://archive.gramene.org/newsletters/rice_genetics/rgn6/v6p116.html

Virmani SS, Sun ZX, Mou TM, Jauhar AA, Mao CX (2003) Two-line hybrid rice breeding manual. International Rice Research Institute, Los Baños, p 88

White JW, Hoogenboom G, Stackhouse PW, Hoell JM (2008) Evaluation of NASA satellite- and assimilation model-derived long-term daily temperature data over the continental US. Agric For Meteorol 148:1574–1584

Xu J, Wang B, Wu Y, Du P, Wang J, Wang M, Yi C, Gu M, Liang G (2011) Fine mapping and candidate gene analysis of ptgms2-1, the photoperiod-thermo-sensitive genic male sterile gene in rice (Oryza sativa L.) Theor Appl Genet 122(2):365–372

Yang ZP (1997) Inheritance of photoperiod genic male sterility and breeding of photoperiod sensitive genic male sterile lines in rice (Oryza sativa L.) through anther culture. Euphytica 94:93–99

Yang YQ, Wang L (1990) The breeding of thermosensitive male sterile rice R59TS. Sci Agric Sin 23(2):90

Yin X, Kropff MJ, Horie T, Nakagawa H, Centeno HGS, Zhu D, Goudriaan J (1997) A model for photothermal responses of flowering in rice I. Model description and parameterization. Field Crop Res 51:189–200

Yuan A-P, Hou A-B, Zhang F-Y, Guo Y-D (2008) Inheritance and effects of the photoperiod sensitivity in foxtail millet (Setaria italica P. Beauv). Hereditas 145(4):147–153

Zhang T, Li T, Yang X, Simelton E (2016) Model biases in rice phenology under warmer climates. Sci Rep 6:27355

Zhou G, Chen Y, Yao W, Zhang C, Xie W, Hua J, Xing Y, Xiao J, Zhang Q (2012) Genetic composition of yield heterosis in an elite rice hybrid. Proc Natl Acad Sci U S A 109(39):15847–15852

Zhuping Y (1997) Inheritance of photoperiod sensitive genic male sterility and breeding of photoperiod sensitive genic male-sterile lines in rice (Oryza sativa L.) through anther culture. Euphytica 94:93–99

9

Large-scale deployment of a rice 6 K SNP array for genetics and breeding applications

Michael J. Thomson[1*†], Namrata Singh[2†], Maria S. Dwiyanti[3,4†], Diane R. Wang[2], Mark H. Wright[2,5], Francisco Agosto Perez[2], Genevieve DeClerck[2,6], Joong Hyoun Chin[3,7], Geraldine A. Malitic-Layaoen[3], Venice Margarette Juanillas[3], Christine J. Dilla-Ermita[3,8], Ramil Mauleon[3], Tobias Kretzschmar[3] and Susan R. McCouch[2*] [iD]

Abstract

Background: Fixed arrays of single nucleotide polymorphism (SNP) markers have advantages over reduced representation sequencing in their ease of data analysis, consistently higher call rates, and rapid turnaround times. A 6 K SNP array represents a cost-benefit "sweet spot" for routine genetics and breeding applications in rice. Selection of informative SNPs across species and subpopulations during chip design is essential to obtain useful polymorphism rates for target germplasm groups. This paper summarizes results from large-scale deployment of an Illumina 6 K SNP array for rice.

Results: Design of the Illumina Infinium 6 K SNP chip for rice, referred to as the Cornell_6K_Array_Infinium_Rice (C6AIR), includes 4429 SNPs from re-sequencing data and 1571 SNP markers from previous BeadXpress 384-SNP sets, selected based on polymorphism rate and allele frequency within and between target germplasm groups. Of the 6000 attempted bead types, 5274 passed Illumina's production quality control. The C6AIR was widely deployed at the International Rice Research Institute (IRRI) for genetic diversity analysis, QTL mapping, and tracking introgressions and was intensively used at Cornell University for QTL analysis and developing libraries of interspecific chromosome segment substitution lines (CSSLs) between *O. sativa* and diverse accessions of *O. rufipogon* or *O. meridionalis*. Collectively, the array was used to genotype over 40,000 rice samples. A set of 4606 SNP markers was used to provide high quality data for *O. sativa* germplasm, while a slightly expanded set of 4940 SNPs was used for *O. sativa* X *O. rufipogon* populations. Biparental polymorphism rates were generally between 1900 and 2500 well-distributed SNP markers for *indica x japonica* or interspecific populations and between 1300 and 1500 markers for crosses within *indica*, while polymorphism rates were lower for pairwise crosses within U.S. *tropical japonica* germplasm. Recently, a second-generation array containing ~7000 SNP markers, referred to as the C7AIR, was designed by removing poor-performing SNPs from the C6AIR and adding markers selected to increase the utility of the array for elite *tropical japonica* material.

(Continued on next page)

* Correspondence: m.thomson@tamu.edu; srm4@cornell.edu
Michael J. Thomson, Namrata Singh and Maria S. Dwiyanti contributed equally to this work.
†Equal contributors
[1]Department of Soil and Crop Sciences, Texas A&M University, College Station, Houston, TX 77843, USA
[2]School of Integrative Plant Sciences, Plant Breeding and Genetics Section, Cornell University, Ithaca, New York 14853, USA
Full list of author information is available at the end of the article

(Continued from previous page)

Conclusions: The C6AIR has been successfully used to generate rapid and high-quality genotype data for diverse genetics and breeding applications in rice, and provides the basis for an optimized design in the C7AIR.

Keywords: *Oryza sativa, O. rufipogon*, Single nucleotide polymorphism (SNP), High-throughput genotyping, Rice diversity, SNP fingerprinting, CSSL development

Background

Future challenges to sustainably produce food for 9.5 billion people by 2050 using less land and fewer inputs will require higher yields in intensive systems under increasingly variable environments. Modern plant breeding and genetic research programs aim to utilize the latest technologies to accelerate the annual rate of genetic gain to keep up with rice demand. High-throughput molecular marker techniques have enabled routine, low-cost genotyping for both targeted and genome-wide approaches. Targeted methods, where a few markers (<100) are used to genotype a large number of samples, provide an efficient strategy for forward selection of major genes in breeding programs. Genome-wide methods, including fixed arrays and next generation sequencing, provide marker densities appropriate for genome-wide association studies, QTL mapping, diversity analysis, DNA fingerprinting, impact assessment studies, and breeding applications such as genomic selection (Thomson 2014; Varshney et al. 2014).

Single nucleotide polymorphisms (SNPs) are the markers of choice for most high throughput genotyping applications because they are ubiquitous in eukaryotic genomes, cost-effective to assay using automated platforms, and because allele calling, data analysis and data-basing are straightforward due to their biallelic nature. A number of medium- or high-resolution SNP arrays in rice have been deployed, primarily for genome-wide association studies, including a 44 K SNP chip (Zhao et al. 2011), 50 K SNP chips (Chen et al. 2013b; Singh et al. 2015), and the 700 K high-density rice array (HDRA, McCouch et al. 2016). These arrays provide automated platforms to dissect phenotype-genotype associations, while at the same time offering valuable datasets that can be used to validate high-quality SNP markers that are informative within and between key germplasm groups. The subsequent development of lower resolution detection platforms, including KASP, TaqMan, and Fluidigm that target individual SNPs, and the low-density SNP arrays, have made use of the wealth of information published from the higher-density arrays to extract informative SNPs and invariant SNP flanking sequences that convert well to other assays (McCouch et al. 2010; Tung et al. 2010; Chen et al. 2013a).

Historically, sets of 384, 768, or 1536 SNP markers were used for diversity analysis, QTL mapping, marker-assisted backcrossing, specialized genetic stock development, and pedigree confirmation among breeding lines in rice (Nagasaki et al. 2010; Zhao et al. 2010; Chen et al. 2011; Thomson et al. 2012; Ye et al. 2012; Rahman et al. 2016; Shah et al. 2016). Despite their utility across a variety of applications, the limited numbers of SNP markers in each assay required the development of multiple SNP sets in order to provide a high enough resolution of polymorphic markers for use with specific germplasm groups.

Combining SNP sets into larger arrays helps increase the number of potential users per array, which lowers cost while providing increased resolution across a diversity of germplasm. Previously, an Illumina Infinium 6 K array in rice (RICE6K) was developed in Wuhan, China to provide polymorphic SNPs within and between the *indica* and *japonica* subgroups for applications in background selection, mapping population genotyping, variety identification and purity tests, and bulk segregant analysis (Yu et al. 2013). With rapid genotyping turn-around times, ease of allele calling and data analysis provided by this and other 6 K SNP chips, breeders and geneticists can interact more directly and rapidly with the data and incorporate genotyping results in their programs without dependence upon bioinformatics specialists.

The primary alternative to fixed SNP arrays is reduced representation next-generation sequencing, such as restriction-site associated DNA (RAD) sequencing or genotyping by sequencing (GBS). These methods provide large numbers of genome-wide SNP markers at a low cost (Baird et al. 2008; Elshire et al. 2011; Peterson et al. 2012). While RAD-Seq and GBS have been very successful for certain applications, several limitations have become apparent as adoption has widened. In addition to reliance on complex protocols for library preparation, the requirement to multiplex hundreds of samples to minimize cost, and long delays in obtaining sequencing output, a challenge for many groups has been the costly bioinformatics infrastructure needed to support downstream analytical pipelines for accurate allele-calling. Although imputation techniques enable researchers to fill in gaps in data sets, GBS approaches typically suffer from large amounts of missing data, making it challenging to accurately call heterozygotes. More recently, core facilities are faced with the challenge of dealing with licensing of these technologies due to the KeyGene patent for Sequence-Based Genotyping (Truong et al. 2012; US Patent 8,815,512).

At the other end of the spectrum, targeted simplex SNP approaches, such as TaqMan and KASP-based genotyping, offer an alternative to fixed arrays for applications requiring a few, high-value markers across very large populations (Eathington et al. 2007; He et al. 2014; Semagn et al. 2014). These assays can be cost-effective for large sample sizes (1000 s-10,000 s) and are ideal when trait-predictive SNP markers are available for selection of large effect genes in breeding programs; however, their cost advantage is lost for applications involving small numbers of lines or requiring more than 100–200 genome-wide SNP markers. Thus, while no single genotyping system is ideal for all applications, the wide range of available genotyping platforms now offer solutions that can provide an optimal balance to meet the needs of different users, taking into account cost per sample, marker resolution, turnaround time, allele call rates, and data analysis requirements (Thomson 2014).

To replace multiple rice 384-SNP sets and provide a high-quality set of informative SNP markers for genetics and breeding applications, an Illumina Infinium 6 K SNP chip for rice, referred to as the Cornell_6K_Array_Infinium_Rice (C6AIR), was designed for use at both Cornell University and the International Rice Research Institute (IRRI). The design of the C6AIR includes 1571 SNP markers from legacy BeadXpress SNP sets and 4429 SNPs selected from whole genome re-sequencing data to be polymorphic within and between the target germplasm groups and mapping parents. This paper describes the efficacy of the C6AIR for QTL mapping, genetic diversity analysis, SNP fingerprinting of breeding lines, tracking of introgressions, and checking for recovery of recurrent parent background during marker-assisted backcrossing. Subsequently, an improved second-generation array was developed by removing SNPs that performed poorly on the 6 K array and increasing the number of bead types to just over 7000. This new 7 K rice array, referred to as the C7AIR, provides continuity with data sets from the C6AIR while increasing the number of high quality SNP loci for future use in genetics and breeding applications.

Results and discussion
Design of the Cornell 6 K SNP array
The C6AIR was designed and developed to be informative for *Oryza sativa* and *O. sativa*/*O. rufipogon* populations, incorporating markers from previous GoldenGate 384-SNP sets and selective SNPs from whole genome re-sequencing data. The custom-designed Infinium iSelect array consisted of 6000 attempted bead types, including 1571 SNP markers from legacy BeadXpress 384-SNP sets (Thomson et al. 2012) and 4429 SNPs selected from whole genome sequence data to be polymorphic within and between diverse germplasm groups and mapping

parents. The re-sequenced genomes used as a SNP discovery dataset were described by McCouch et al. (2016) and included 88 *O. sativa* accessions (21 *indicas*, 16 *aus*, 18 *tropical japonica*, 19 *temperate japonica*, 11 *aromatic*, 3 admixed), 9 *O. nivara*, 28 *O. rufipogon* accessions, one *O. meridionalis*, one *O. officianialis*, and one *O. punctata* (see Materials and Methods for details on SNP selection criteria).

Of the 6000 SNPs included in the initial design, 5274 genome-wide SNPs passed Illumina's production quality control, out of which 1695 SNP markers localized within MSUv7 gene models (Additional file 1: Fig. S1). The average gap between two adjacent SNPs was 79 kb, and more than 50% of SNPs were located within 60 kb of their closest neighbor (Fig. 1). For routine genotyping work at the Genotyping Services Lab at IRRI, a set of 4606 high quality SNPs was used, after filtering SNPs with (a) more than two alleles, (b) duplication of flanking sequences, and (c) high rates of missing data or "no calls" in the targeted *O. sativa* populations. A subset of 4940 SNPs providing high quality data were routinely called at Cornell University, with the higher number most likely due to the inclusion of several *O. rufipogon* and *O. meridionalis* accessions in the analysis.

Polymorphism rates across pairwise combinations
Pairwise comparisons of polymorphic SNPs between pairs of rice accessions from the same subgroup showed an average of 1347 well-distributed SNPs in *indica*, 1394 SNPs in *japonica*, and 1413 SNPs in wild populations. In contrast, an average of 2541 SNPs were detected between *indica* vs. *japonica* varieties, ~2500 SNPs between *indica* x *aromatic* varieties, ~1500 SNPs between *indica* and genetically divergent *aus* varieties, and an average of 1987 SNPs between either *indica* or *japonica* and the wild accessions evaluated in this study (Fig. 2).

While the average number of polymorphic markers across diverse germplasm was quite high, polymorphism rates decrease with more closely related germplasm (Fig. 3). For example, *indica* breeding lines at IRRI were often distinguished by only 500 SNPs, and polymorphism rates between two US breeding lines averaged 668 for long grain varieties and 450 SNPs for medium grain varieties (Additional file 2: Fig. S2). Nonetheless, the long and medium grain market classes could still be distinguished using the C6AIR. As can be seen by the number of polymorphic SNPs detected by C6AIR in 18 bi-parental populations commonly used for mapping in rice (Table 1), the pairwise polymorphism rate is more than sufficient for QTL mapping and diversity analyses for all but the most closely related accessions (such as *temperate japonica* X *temperate japonica*).

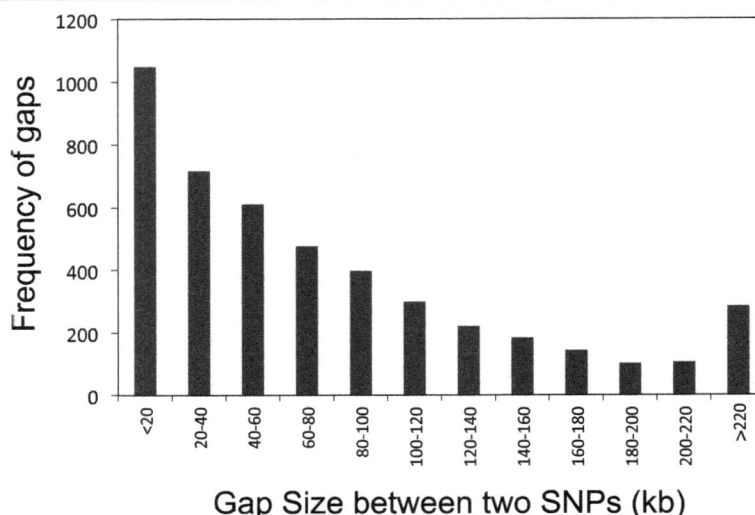

Fig. 1 Distribution of SNP distance to its neighboring SNP. More than 50% of SNPs are within 60 kb from a neighboring SNP. The average spacing between SNPs is 79 kb. About 10% of SNPs are >220 kb to the neighboring SNP

Applications of the Cornell 6 K rice array in genetics and breeding programs

The C6AIR has been used extensively in two institutions, Cornell University and IRRI. Cornell uses the C6AIR chip for pre-breeding, to develop introgression lines using the rice wild relatives *O. rufipogon* and *O. meridionalis,* for QTL mapping, and to genotype elite material used in US breeding programs, mainly *tropical japonica* varieties. On the other hand, IRRI's focus is mainly on *indica* varieties, though researchers in both institutions utilize all five *O. sativa* subgroups for genetics and breeding applications. Recent publications also highlight the utility of the Cornell 6 K rice chip for QTL analysis of heat, salinity and flooding tolerance (Ye et al. 2015; Gimhani et al. 2016; Gonzaga et al. 2017; Singh et al. 2017). Three additional applications are presented below: diversity analysis, tracking introgressions, and developing pre-breeding materials.

Diversity and genetic analysis

Using the C6AIR, a set of diverse germplasm consisting of 232 *O. sativa*, 23 *O. rufipogon* (AA genome), 2 *O. meriodionalis* (AA genome) and 1 *Oryza officinalis* (CC genome) accessions were genotyped (Additional file 3: Table S1). Diversity analysis using the neighbor-joining method defined the five *O. sativa* subgroups: *indica, aus, aromatic, temperate japonica* and *tropical japonica*, though the two *japonica* subgroups were poorly discriminated (Fig. 4). Consistent with previous analyses, *aromatic* varieties are closely related to *japonica* varieties (Garris et al. 2005; Zhao et al. 2010). We also genotyped *O. rufipogon* accessions which were distributed throughout the tree, many with long branch lengths (Fig. 4). The result confirmed the ability of the C6AIR to detect diversity within *O. rufipogon.*

Further analysis using Principal Component Analysis (PCA) of 232 *O. sativa* accessions also showed that genotype data from the C6AIR were able to distinguish the five major subgroups of *O. sativa* (Additional file 4: Fig. S3). The first PC (PC1) separated *indica* and *aus* varieties from *japonica* and *aromatic* varieties. The second PC (PC2) was able to distinguish *tropical japonica, temperate japonica* and *aromatic* varieties. Finally, the third PC (PC3) distinguished *indica* from *aus* varieties.

Tracking introgression of target QTLs during marker-assisted backcrossing

In marker assisted backcrossing (MABC), it is desirable to be able to rapidly introgress chromosomal regions containing a gene or QTL of interest, while at the same time selecting for small size of a target introgression to avoid linkage drag. We highlight two examples of tracking introgressions using C6AIR: one for the cloned and well-characterized *SUB1* gene and another for the *Xa7* gene. *SUB1* confers submergence tolerance for up to 14 days at the vegetative stage, and the gene responsible for the phenotype has been cloned and characterized (Xu et al. 2006; Septiningsih et al. 2009). *Xa7* is a bacterial leaf blight resistance gene from the *aus* variety DV85. It is effective at high temperature, which makes it a promising gene for bacterial leaf blight resistance in areas affected by high temperature (Webb et al. 2009). *Xa7* is located on chromosome 6 between 27.9–28.0 Mb, but the candidate gene has not yet been cloned.

The *SUB1* region contains three *SUB1* genes (*SUB1A, SUB1B, SUB1C*), located between 6.3–6.7 Mb on chromosome 9 (Fig. 5a). *SUB1C* (LOC_Os09g11460) and *SUB1B* (LOC_Os09g11480) are present in Nipponbare, but

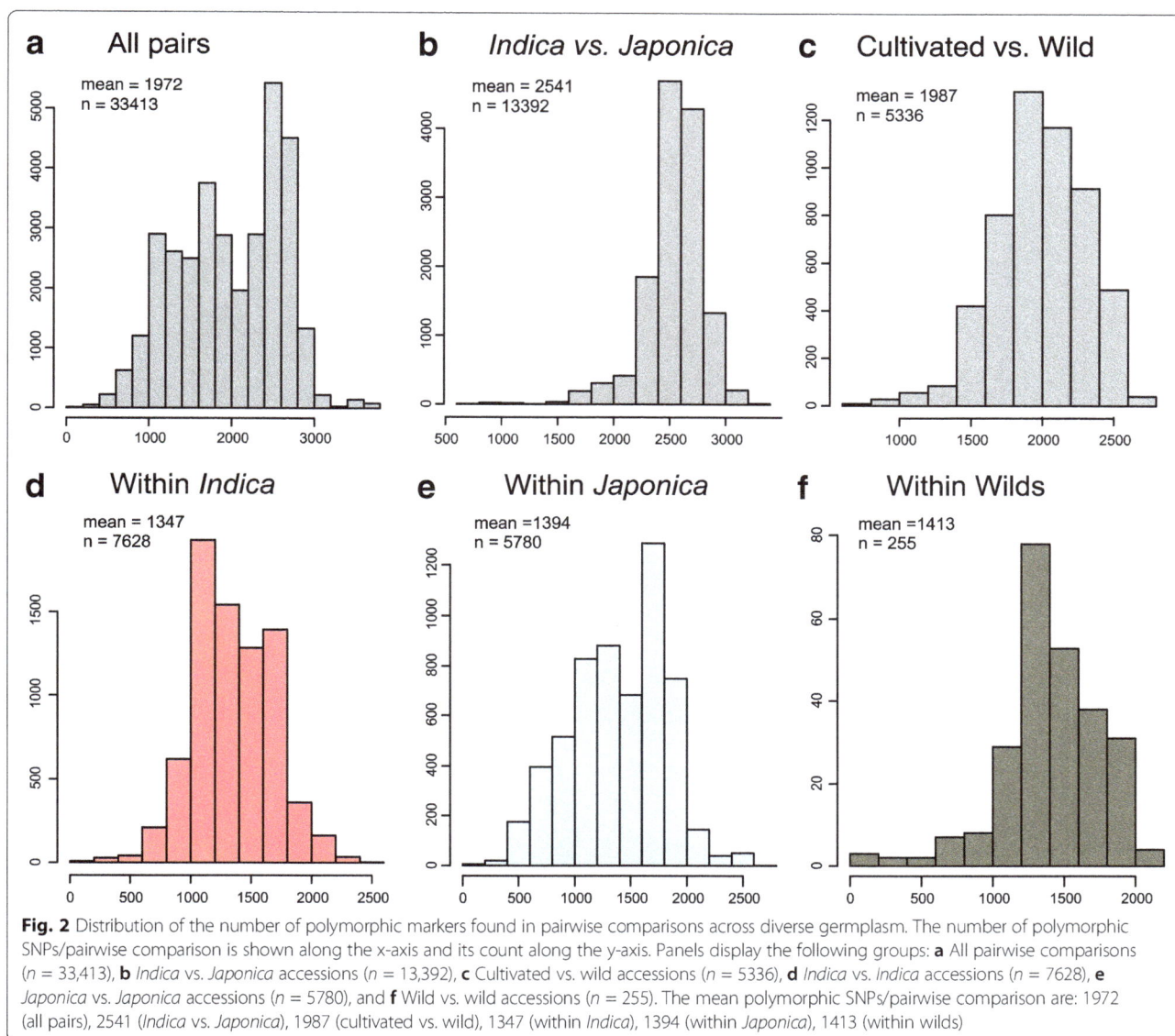

Fig. 2 Distribution of the number of polymorphic markers found in pairwise comparisons across diverse germplasm. The number of polymorphic SNPs/pairwise comparison is shown along the x-axis and its count along the y-axis. Panels display the following groups: **a** All pairwise comparisons (n = 33,413), **b** *Indica* vs. *Japonica* accessions (n = 13,392), **c** Cultivated vs. wild accessions (n = 5336), **d** *Indica* vs. *indica* accessions (n = 7628), **e** *Japonica* vs. *Japonica* accessions (n = 5780), and **f** Wild vs. wild accessions (n = 255). The mean polymorphic SNPs/pairwise comparison are: 1972 (all pairs), 2541 (*Indica* vs. *Japonica*), 1987 (cultivated vs. wild), 1347 (within *Indica*), 1394 (within *Japonica*), 1413 (within wilds)

SUB1A, which is responsible for submergence tolerance, is missing from the reference genome (Xu et al. 2006). Based on a comparison of *SUB1* and non-*SUB1* varieties, two SNPs on the C6AIR located at 6,360,984 bp and 6,774,928 bp are able to clearly distinguish varieties that do or do not carry *SUB1A* (Fig. 5b). In addition, varieties with different size introgressions can be distinguished; BR11-*Sub*1 and Sambha-Mahsuri-*Sub*1 have smaller introgressed regions around the *SUB1* locus than other varieties examined (Fig. 5c).

The *Xa7* introgression from DV85 could be readily tracked using the C6AIR in backcross lines with the recurrent parent, IR24 (i.e., the IRBB isolines). The introgression from DV85 into IR24 spanned between 26.3 Mb to 28.3 Mb (Fig. 5d). Lines carrying the resistant *Xa7* allele could be distinguished from susceptible

lines by 16 SNPs that mapped within the introgressed region, including one SNP that is located in the previously reported *Xa7* QTL region (Chen et al. 2008). These SNPs, although not functional, are in perfect linkage disequilibrium (LD) with the causal polymorphism in this material, and therefore can be utilized to identify varieties carrying the favorable allele. Trait-specific haplotypes defined by SNP markers discovered using the C6AIR can also be readily converted to other genotyping platforms, such as KASP, Taqman, or Fluidigm, which may be more cost effective for forward selection in breeding programs that target a few, large effect loci.

In addition to foreground selection, the C6AIR is ideally suited for comprehensive background selection in a marker assisted backcrossing (MABC) scheme, where the objective is to rapidly return the genetic background

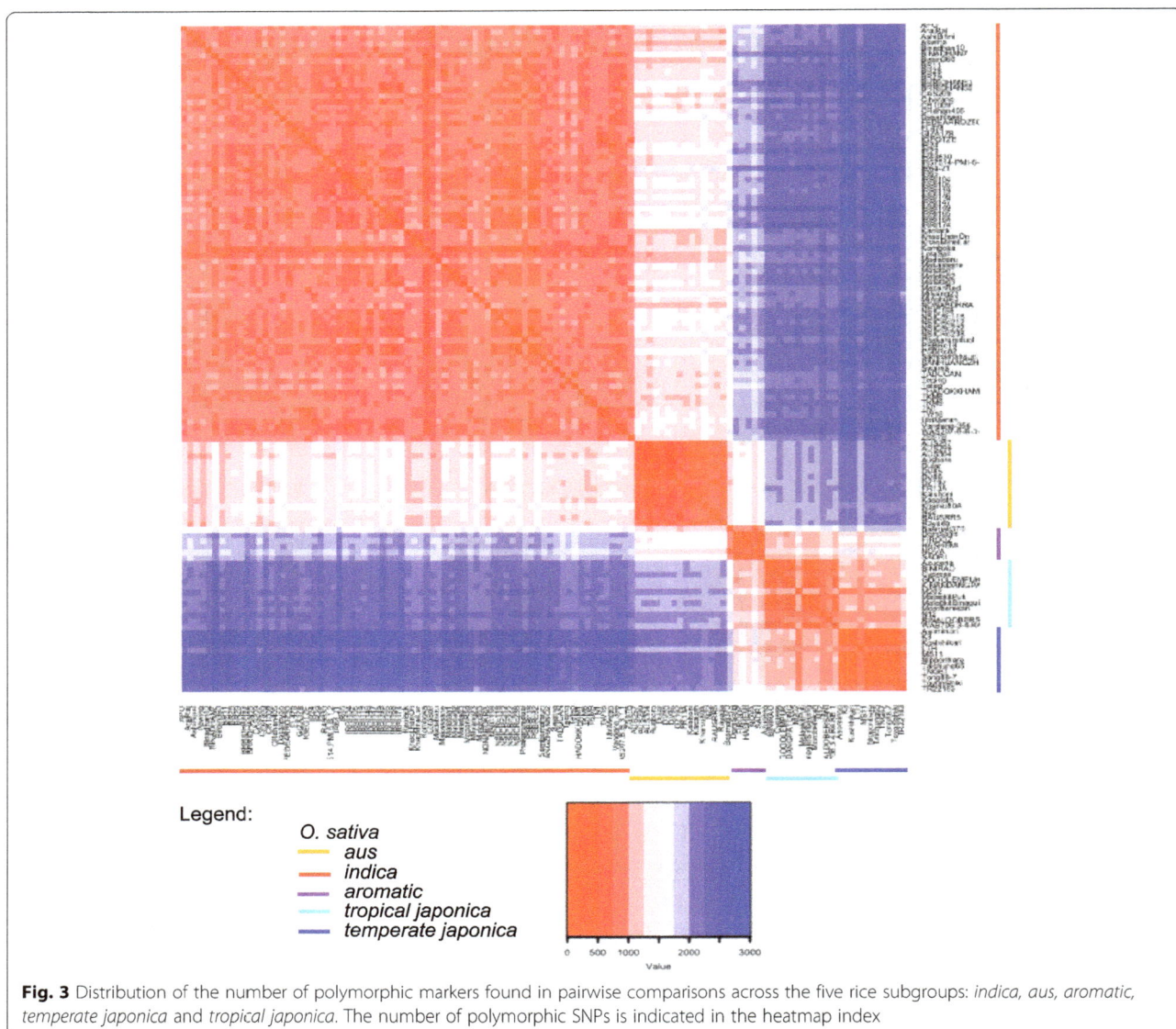

Fig. 3 Distribution of the number of polymorphic markers found in pairwise comparisons across the five rice subgroups: *indica, aus, aromatic, temperate japonica* and *tropical japonica*. The number of polymorphic SNPs is indicated in the heatmap index

to the recurrent parent type. For example, in an MABC program aiming to transfer the *Pup1* QTL for phosphorous uptake from IR74-*Pup1* to TR22183, BC$_2$F$_1$ individuals having the target introgression were screened with the C6AIR to identify the lines with the fewest donor introgressions in the background (Additional file 5: Fig. S4). Combining low-cost foreground and recombinant selection (using KASP markers) with high-resolution background selection (using the C6AIR) provides a powerful strategy for rapid and precise MABC transfer of introgressions into elite genetic backgrounds.

Developing sets of CSSLs between *O. sativa* and *O. rufipogon*

Six rice CSSL libraries were developed from crosses between three *O. rufipogon* accessions ('Khao Pa', W1944,

IRGC105567) and two elite recurrent parents, the *indica* IR64 and *tropical japonica* Cybonnet. Genotyping with the C6AIR provided between 1311 to 1952 polymorphic genome-wide SNPs per library, and was used in each backcross generation to identify lines that carried the *O. rufipogon* introgression of interest, to precisely delimit the size of the target introgression, and to select against unwanted donor introgressions in the background. Although CSSL development was initiated using 384-SNP GoldenGate assays, the C6AIR proved to be much more efficient and informative for CSSL development than the lower resolution assays that preceded it.

A comparison between 384-SNP OPA 6.1 and C6AIR genotyping platform was done by using the informative SNPs for foreground and background selection during the development of CSSLs between Cybonnet and

Table 1 Number of polymorphic SNPs detected by C6AIR in 18 bi-parental populations commonly used for mapping in rice. On average a 1 Mb region contains around 3–6 SNPs

Parent1	Subgroup - Parent1	Parent2	Subgroup - Parent2	Number of polymorphic markers between the two parents using C6AIR
DV85	*aus*	IR24	*indica*	1760
DJ123	*aus*	IR64	*indica*	1051
N22	*aus*	IR64	*indica*	1173
N22	*aus*	Swarna	*indica*	1017
Kasalath	*aus*	Nipponbare	*temperate japonica*	1555
DJ123	*aus*	Nipponbare	*temperate japonica*	1506
FR13A	*aus*	M202	*tropical japonica*	1636
Khao Hlan On	*indica*	IR64	*indica*	810
93–11	*indica*	Nipponbare	*temperate japonica*	1742
IR64	*indica*	Azucena	*tropical japonica*	1760
Teqing	*indica*	Lemont	*tropical japonica*	1713
Minghui 63	*indica*	Azucena	*tropical japonica*	1578
Jasmine85	*indica*	Lemont	*tropical japonica*	1629
IR64	*indica*	Jefferson	*tropical japonica*	1718
Nipponbare	*temperate japonica*	FR13A	*aus*	1627
Nipponbare	*temperate japonica*	IR64	*indica*	1982
Geumobyeo	*temperate japonica*	Moroberekan	*tropical japonica*	866
Kinandang Patong	*tropical japonica*	IR64	*indica*	1889

O.rufipogon parent IRGC105567 (NSF490) (Fig. 6). The low resolution 384-OPA identified 260 polymorphic SNPs sparsely distributed across the 12 rice chromosomes whereas the C6AIR detected 1868 well-distributed polymorphic SNPs (Fig. 6a). Each line was selected to contain a target 5–7 Mb introgression; using the 384-OPA, a 5 Mb region had an average of 1–3 informative SNPs while the C6AIR detected 5–8 informative SNPs across the same size region, thus providing a better approximation of introgression size.

Eight CSSLs provided coverage of chromosome 1 in the Cybonnet X NSF490 population (C4–1 to C4–8), each line carrying a homozygous donor segment in the background of the recurrent parent (Fig. 6b). During selection of lines representing C4–7 and C4–8, several segregating plants were found to have unfilled panicles. Genotyping analysis of the lines that produced seeds showed them to be either heterozygous towards the end of Chr 1 (38–42 Mb) or homozygous for the recurrent parent. To narrow down this putative incompatibility or sterility region, polymorphic SNPs between the two platforms were compared alongside the Chr1-CSSLs. The 384-OPA 6.1 detected 38 polymorphic SNPs on chromosome 1 and identified the putative incompatibility region at a resolution of 4.38 Mb (~17 cM) (between positions 38.58–42.96 Mb). Using the C6AIR, there were 221

informative SNPs on chr1, and the region of interest was narrowed down to 0.875 Mb (~3 cM) (between positions 41.04–41.91 Mb) (Fig. 6b).

Design and characteristics of the improved C7AIR

To further improve the C6AIR, a second-generation Infinium rice array was recently developed by eliminating SNPs that performed poorly on the 6 K array and increasing the number of attempted bead types. This new 7 K–SNP array, referred to commercially as the Cornell-IR LD Rice Array, and by the user community as the Cornell_7K_Array_Infinium_Rice (C7AIR), provides continuity with the C6AIR, while increasing the number of high quality SNP loci that can be interrogated. The C7AIR includes 4007 SNPs from the 6 K array, 2056 SNPs from the 700,000-SNP HDRA selected to be informative for US rice germplasm, 910 SNPs from the 384-SNP GoldenGate sets (OPA2.1, 3.1, 4.0, 5.0, 6.2 and 7.0), 189 SNPs from the 44 K array selected to be informative for US rice germplasm, and 21 gene-based SNPs from IRRI. SNPs on the C7AIR are expected to be informative for detecting genome-wide polymorphism between individuals within the *indica*, *aus*, and *tropical japonica* subpopulations, between any pairwise combination of accessions from the *indica*, *aus*, *tropical japonica*, *temperate japonica*, or *aromatic* (Group V) subpopulations, and between *O. sativa* and *O. rufipogon*. The array is expected to be

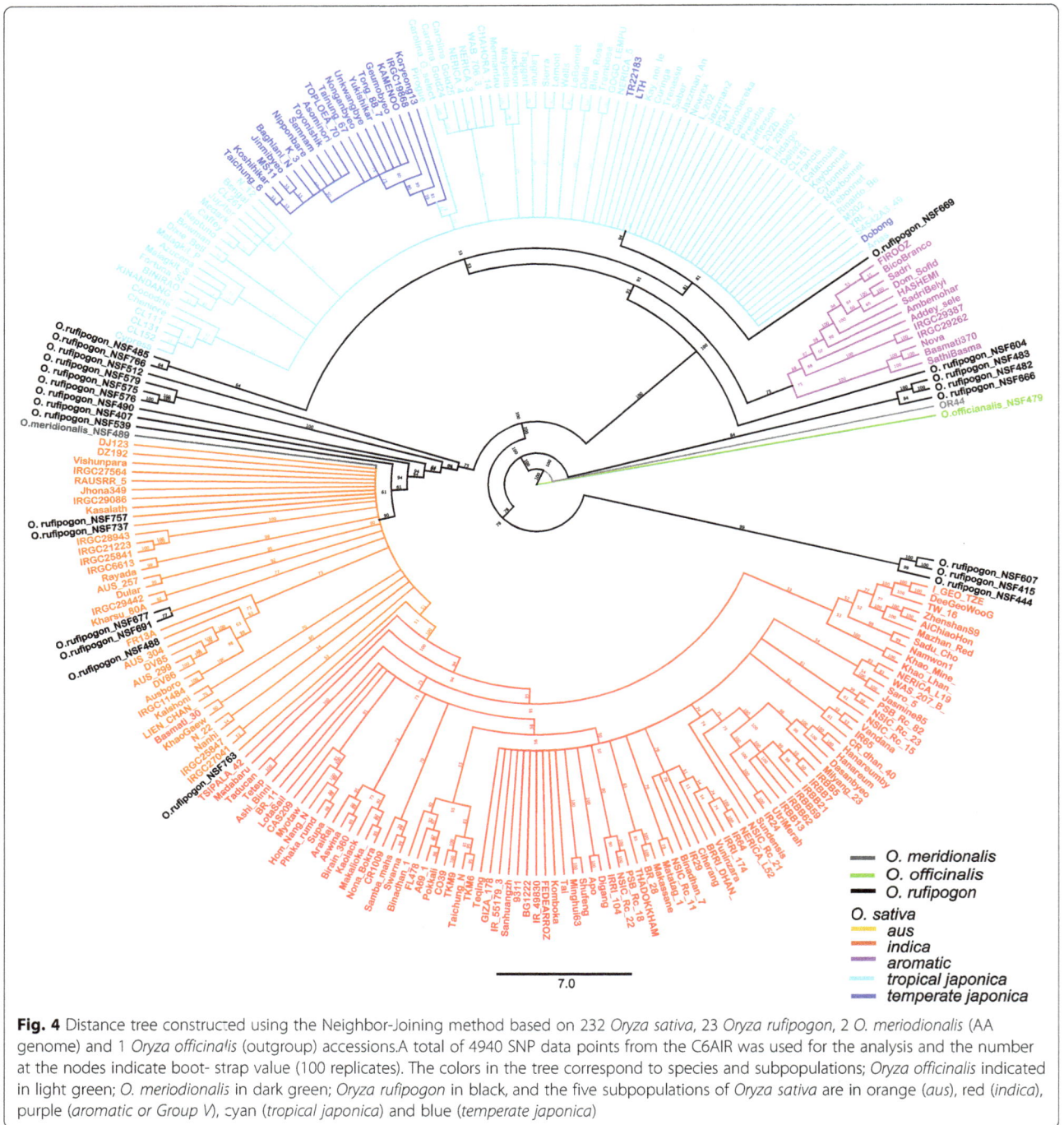

Fig. 4 Distance tree constructed using the Neighbor-Joining method based on 232 *Oryza sativa*, 23 *Oryza rufipogon*, 2 *O. meriodionalis* (AA genome) and 1 *Oryza officinalis* (outgroup) accessions. A total of 4940 SNP data points from the C6AIR was used for the analysis and the number at the nodes indicate boot-strap value (100 replicates). The colors in the tree correspond to species and subpopulations; *Oryza officinalis* indicated in light green; *O. meriodionalis* in dark green; *Oryza rufipogon* in black, and the five subpopulations of *Oryza sativa* are in orange (*aus*), red (*indica*), purple (*aromatic or Group V*), cyan (*tropical japonica*) and blue (*temperate japonica*)

moderately informative for detecting polymorphism within the *aromatic* (Group V) subpopulation, and least informative for detecting polymorphism within the *temperate japonica* subpopulation. Trait-specific markers include diagnostic SNPs for the SUB1A gene, grain quality characteristics, and loci for resistance to bacterial leaf blight, blast, brown planthopper, and tungro to enhance C7AIR utility for foreground

selection and QTL profiling. The C7AIR is being beta-tested at this time and will be manufactured by Illumina as a consortium array for future use in rice genetics and breeding worldwide.

Conclusions

The C6AIR has proven to be an effective genotyping system for rice diversity analysis, QTL mapping, tracking

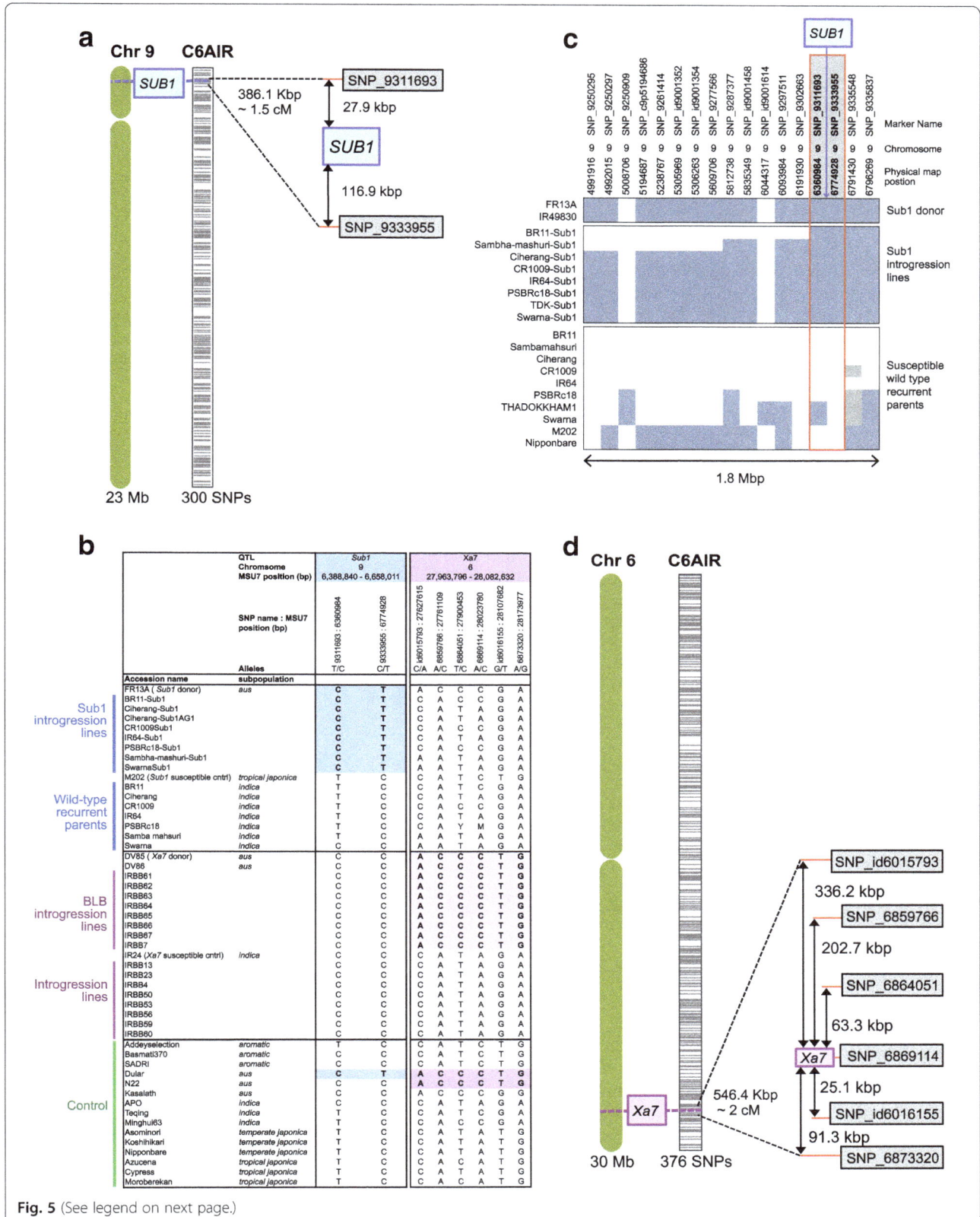

Fig. 5 (See legend on next page.)

(See figure on previous page.)

Fig. 5 Tracking QTL introgressions on chromosomes 6 and 9 of rice using the C6AIR. **a** QTL for *Sub1* was previously mapped to chromosome 9 at 6,388,840–6,658,011 bp (MSUv7), (Xu et al. 2006; Septiningsih et al. 2009). Two C6AIR markers that localize close to *Sub1* (~1.5 cM region) can be used to track the presence or absence of the QTL for development of *Sub1* introgression lines (blue box = *Sub1*, gray boxes = nearby markers). **b** Table shows genotypes of *Xa7* and *Sub1* predictive markers in popular rice varieties (control) and their derived introgression lines that carry *Xa7* and *Sub1*. **c** C6AIR genotype calls around the *Sub1* QTL region (~1.8 Mb) *Sub1* donors, Sub1-introgression lines and wild-type recurrent parents. Introgression size of varieties carrying a functional *SUB1A* gene vary, i.e. BR11-*Sub1* and Sambha-Mahsuri-*Sub1* have the smallest introgressed regions among the *SUB1* varieties (blue box = Sub1 genotype, white box = wild type genotype, gray boxes = missing data). **d** QTL for *Xa7* is located on chromosome 6 at MSU7 position 27,963,796–28,082,632 bp (Chen et al. 2008). Five C6AIR markers localize close to *Xa7* and one SNP is within the *Xa7* region. These markers can be used to track the presence or absence of the QTL for development of *Xa7* introgression lines (purple box = *Xa7*, gray boxes = nearby markers)

introgressions, genetic stock development, and finger-printing studies at both Cornell University and at the Genotyping Services Lab at IRRI over the last 4 years. Over 40,000 samples were run successfully, providing genotyping data for a large number of genetics, breeding and impact assessment projects of importance to people throughout the world. Arrays such as the C6AIR provide a relatively small but sufficient number of SNPs so that data can be handled without massive bioinformatics pipelines, while at the same time providing high enough resolution for most genetics and breeding applications. The SNPs on the C6AIR consistently have high call rates, including for heterozygotes, facilitating data management and integration of genotyping data across runs and populations. Because the C6AIR can readily distinguish the five major subgroups of *O. sativa* as well as *O. rufipogon*, it is especially useful for fingerprinting studies and, as summarized in this paper, finds a broad

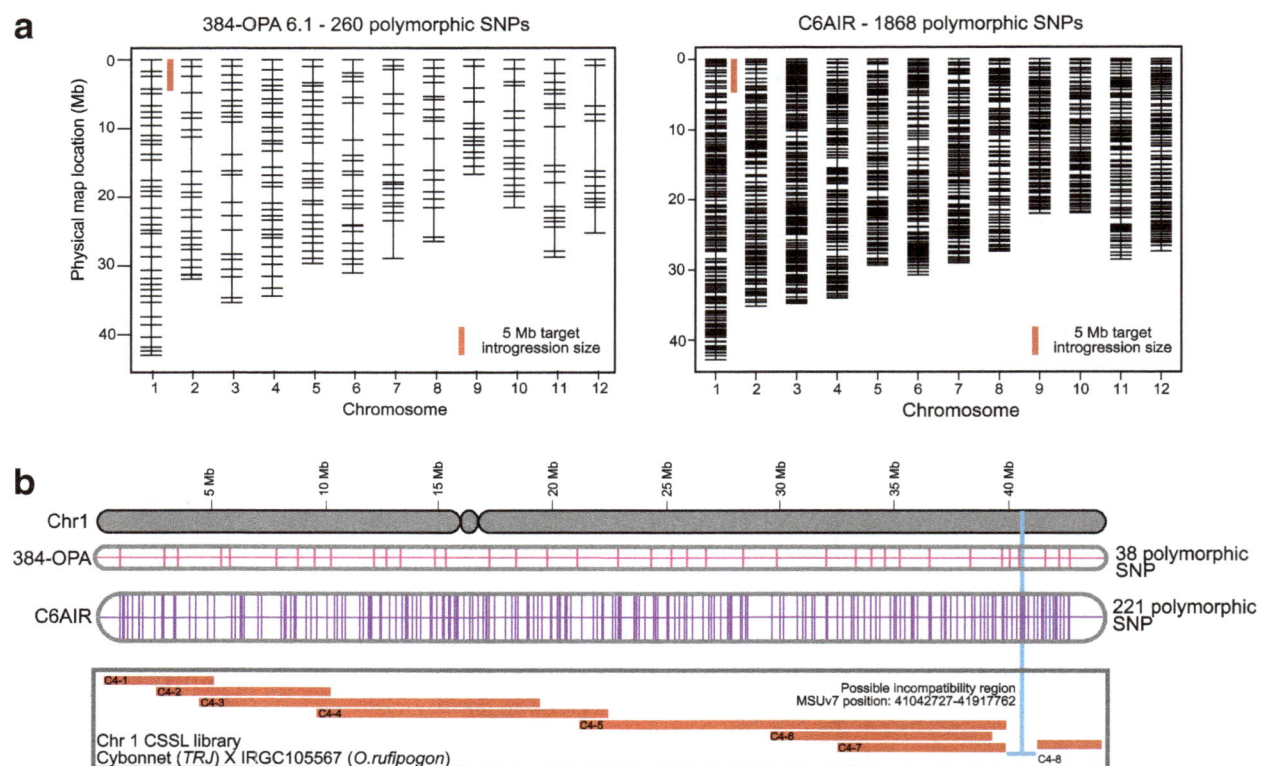

Fig. 6 Comparison of the use of 384-SNP Golden Gate assay (OPA 6.1) and C6AIR genotyping platforms for foreground and background selection. CSSLs were developed between the elite *tropical japonica* variety Cybonnet as the recurrent parent, and *O. rufipogon*, IRGC105567, as the donor parent. **a** Distribution of 260 informative SNPs across 12 rice chromosomes detected using the 384-SNP assay (OPA 6.1) (left) and 1868 polymorphic SNPs using the C6AIR (right). The red bar indicates 5 Mb target introgression from the donor parent in the background of Cybonnet. Using the 384-OPA, an average 5 Mb region harbors ~1–3 informative SNPs; using the C6AIR, a 5 Mb region harbors ~5–8 informative SNPs. **b** Graphical representation of CSSL selection targeting overlapping introgressions on Chromosome 1 in the Cybonnet X IRGC105567 library using the 384-OPA 6.1 platform (38 polymorphic SNPs detected) (pink), and the C6AIR (221 polymorphic SNPs) (purple). Blue line indicates a region of putative incompatibility or sterility identified in this population

range of applications in rice breeding programs. The improved second generation C7AIR offers all of the benefits of the 6 K array with additional high quality SNP markers. These resources provide the rice community with rapid, cost-effective tools for low-density, genome-wide SNP genotyping across a wide range of rice germplasm, and with the potential to dramatically increase resolution via imputation when integrated with other publicly available high-density rice genome datasets.

Methods

Plant materials

Rice accessions used in this study are listed in Additional file 3: Table S1. Genomic DNA (gDNA) was extracted from leaf tissue of single plants using the CTAB, Qiagen or SBEadex methods (Fulton et al. 1995). The quality of DNA was checked visually on 1% agarose gels, and the quantity was checked using a Nano-Drop spectrophotometer (read at 260/280 nm) and/or a Qubit 2.0 Fluorometer. For target probe preparation, 5 µl of gDNA was used. The concentration of each DNA sample was adjusted to 50–100 ng/µl.

Design of the C6AIR and the C7AIR

The custom-designed Infinium iSelect C6AIR consisted of 6000 attempted bead types, including 1571 SNP markers from legacy BeadXpress 384-SNP sets (Thomson et al. 2012) and 4429 SNPs selected from re-sequencing data available in the McCouch lab, as described in McCouch et al. (2016). Of these, 2000 SNPs were selected that segregated at mid frequencies within each of the 5 subpopulations of O. sativa (400 non-overlapping SNPs/subpopulation), and 2429 SNPs were selected to be polymorphic for specific bi-parental cross combinations (including the O. sativa and O. rufipogon parents of the targeted CSSL populations). A larger candidate set of 800,468 SNPs were initially scored by the Illumina Assay Design Tool (ADT). After scoring the candidate set of SNPs, the subsequent selection of markers was done through several levels of filtering, including basic filtering: i) there were no SNPs within +/− 10 bp of target on either side; ii) no SNPs with minor homozygote count of more than 4 within +/− 35 bp occurring on both sides; iii) INDELS were removed; iv) SNPs with repetitive sequence and low minor allele frequencies in the set of 128 re-sequenced genomes were omitted; and specific filtering: the remaining SNPs were then screened for segregation properties of specific cross combinations, and for segregation within each of the 5 major subpopulations at minor allele frequency > 20% and observation rate within subpopulation >80%. The union set of SNPs, which satisfy either the subpopulation allele frequency criteria or cross segregation criteria were those that became candidates for the array. The final filtering was done to optimize genome spacing and polymorphism detection within and between species and subpopulations. The minimum resolution was at least 1 informative SNP per Mb for the intra- and interspecific bi-parental cross combinations used during the design phase. The SNP information is provided in Additional file 6: Table S2.

The C7AIR, referred to by Illumina as the Cornell-IR LD Rice Array, represents an improved version of the C6AIR. It was designed by eliminating SNPs that performed poorly on the C6AIR (Additional file 7: Fig. S5) and selecting high quality SNPs from arrays previously designed in the McCouch lab (Zhao et al. 2010; Zhao et al. 2011; McCouch et al. 2016) to increase the number of SNPs that would be informative for interrogating elite *tropical japonica* breeding material.

Genotyping and SNP allele calling

DNA amplification was performed following the manufacturer's protocol. PCR products were hybridized to the Infinium II BeadChip and fluorescently stained following the manufacturer's protocol. The fluorescence intensity of the beadchip was scanned using an Illumina BeadArray Reader. The raw data of the scanned Infinium 6 K BeadChips from the BeadArray Reader were decoded to generate SNP data using the GenomeStudio Software. At Cornell a subset of 4940 SNPs were filtered based on call rate (80%) from 5274 SNPs that passed Illumina's QC. Whereas, at IRRI a customized cluster file was used to generate the forward strand SNP data. The cluster file used at IRRI consists of 4606 high quality SNP markers, which had undergone stringent filtering. Representative samples coming from the 5 subgroups of cultivated rice (ie. *indica, tropical japonica, temperate japonica, aus* and *aromatic*) were used to determine the efficiency of the filtered markers in segregating the aforementioned groups. Final SNP data were merged with SNP map (Nipponbare MSU7) information and encoded with the physical position and chromosome number of the 4606 SNP markers (Additional file 6: Table S2).

Tree construction

To resolve the genetic relationships among the different rice subgroups, 258 rice accessions were analyzed. A total of 5274 SNPs were filtered in TASSEL GUI 5.2.37 using minor allele count (MAC) = 5, call rate = 0.8, resulting in 4940 filtered SNPs that were aligned and used for tree construction using the Geneious Tree Builder. The tree was constructed using the Tamura-Nei genetic distance model, the Neighbor-Joining tree building method, No Outgroup, the Bootstrap Resampling method; and Number of replicates = 100. The tree was visualized using Geneious 8.1.6 to generate a circular cladogram.

Additional files

> **Additional file 1: Fig. S1.** Genome-wide marker coverage. a) Distribution of 5,274 markers successfully converted from C6AIR, and b) distribution of 1,695 markers that localize within MSUv7 gene models
>
> **Additional file 2: Fig. S2.** Distribution of the number of polymorphic markers found in pairwise comparisons across US rice germplasm
>
> **Additional file 3: Table S1.** Germplasm information
>
> **Additional file 4: Fig. S3.** Principal Components Analysis of diverse cultivated Asian rice accessions: PCA of 232 *O. sativa* accessions
>
> **Additional file 5: Fig. S4.** Use of C6AIR for background selection during TR22183-*Pup*1 development
>
> **Additional file 6: Table S2.** Information on the 6,000 SNPs used for the C6AIR design
>
> **Additional file 7: Fig. S5.** Distribution of the frequency of "no call" (NC) alleles in (a) cultivated and (b) wild accessions using the Cornell 6 k Array Infinium Rice

Acknowledgements

The authors are grateful to Endang Septiningsih and Bertrand Collard for providing DNA for testing of the *Sub1* varieties, to the group of Casiana Vera Cruz for providing DNA for testing of the IRBB lines, and to Cindy Lawley, Mike Thompson and Rebecca Enigk at Illumina for their support during the development of the C6AIR as a consortium array under the Global Agriculture Consortia program at Illumina, Inc.

Funding

This work was supported by the National Science Foundation (NSF) with a grant from the Plant Genome Research Program, Award #1026555 to SRM, by the Global Rice Science Partnership (GRiSP) and the Bill and Melinda Gates Foundation (BMGF) with funding for IRRI's Genotyping Services Lab to MJT and TK, and by the USDA National Institute of Food and Agriculture Hatch project 1009299 to MJT.

Authors' contributions

MJT, MSD and NS wrote the manuscript. NS, MSD, DRW, JHC, GAML, VMJ, GD, RM and CJDE provided DNA samples for testing, helped analyze the 6 K data, and prepared figures for case studies in using the C6AIR. MHW, SRM and NS performed SNP selections and designed the C6AIR; FAP and DRW performed SNP selections and designed the C7AIR. SRM, MJT and TK supervised running of the arrays and provided intellectual guidance for the project. All authors read and approved the final manuscript.

Competing interests

The authors declare that they have no competing interests.

Author details

[1]Department of Soil and Crop Sciences, Texas A&M University, College Station, Houston, TX 77843, USA. [2]School of Integrative Plant Sciences, Plant Breeding and Genetics Section, Cornell University, Ithaca, New York 14853, USA. [3]International Rice Research Institute, Los Baños, Philippines. [4]Present address: Research Faculty of Agriculture, Hokkaido University, Sapporo, Hokkaido 060-8589, Japan. [5]Department of Genetics, Stanford School of Medicine, Stanford, California 94305, USA. [6]Present address: DeClerck Design, LLC, Freeville, NY, USA. [7]Present address: Graduate School of Integrated Bioindustry, Sejong University, 209 Neungdong-ro, Gwangjin-gu, Seoul 05006, South Korea. [8]Present address: Department of Plant Pathology, Washington State University, Pullman, Washington 99164, USA.

References

Baird NA, Etter PD, Atwood TS, Currey MC, Shiver AL, Lewis ZA, Selker EU, Cresko WA, Johnson EA (2008) Rapid SNP discovery and genetic mapping using sequenced RAD markers. PLoS One 3:e3376

Chen S, Huang Z, Zeng L, Yang J, Liu Q, Zhu X (2008) High-resolution mapping and gene prediction of *Xanthomonas Oryzae* pv. *Oryzae* resistance gene *Xa7*. Mol Breeding 22:433–441

Chen H, He H, Zou Y, Chen W, Yu R, Liu X, Yang Y, Gao YM, Xu JL, Fan LM, Li Y (2011) Development and application of a set of breeder-friendly SNP markers for genetic analyses and molecular breeding of rice (*Oryza sativa* L.) Theor Appl Genet 123:869

Chen H, He H, Zhou F, Yu H, Deng XW (2013a) Development of genomics-based genotyping platforms and their applications in rice breeding. Curr Opin Plant Biol 16:247–254

Chen H, Xie W, He H, Yu H, Chen W, Li J, Yu R, Yao Y, Zhang W, He Y, Tang X, Zhou F, Deng XW, Zhang Q (2013b) A high-density SNP genotyping array for rice biology and molecular breeding. Mol Plant 7:541–553

Eathington SR, Crosbie TM, Edwards MD, Reiter RS, Bull JK (2007) Molecular markers in a commercial breeding program. Crop Sci 47:S154

Elshire RJ, Glaubitz JC, Sun Q, Poland JA, Kawamoto K, Buckler ES, Mitchell S (2011) A robust, simple genotyping-by-sequencing (GBS) approach for high diversity species. PLoS One 6(5):e19379

Fulton TM, Chunwongse J, Tanksley SD (1995) Microprep protocol for extraction of DNA from tomato and other herbaceous plants. Plant Mol Biol Report 13:207–209

Garris AJ, Tai TH, Coburn J, Kresovich S, McCouch S (2005) Genetic structure and diversity in *Oryza sativa* L. Genetics 169:1631–1638

Gimhani DR, Gregorio GB, Kottearachchi NS, Samarasinghe WLG (2016) SNP-based discovery of salinity-tolerant QTLs in a bi-parental population of rice (*Oryza sativa*). Mol Gen Genomics 291:2081–2099

Gonzaga ZJC, Carandang J, Singh A, Collard BCY, Thomson MJ, Septiningsih EM (2017) Mapping QTLs for submergence tolerance in rice using a population fixed for *SUB1A* tolerant allele. Mol Breeding 37:47

He C, Holme J, Anthony J (2014) SNP genotyping: the KASP assay. In: crop breeding: methods and protocols, 75-86

McCouch SR, Zhao K, Wright M, Tung CW, Ebana K, Thomson M, Reynolds A, Wang D, DeClerck G, Ali ML, McClung A, Eizenga G, Bustamante C (2010) Development of genome-wide SNP assays in rice. Breeding Sci 60:524–535

McCouch S, Wright M, Tung C-W, Maron L, McNally K, Fitzgerald M, Singh N, DeClerck G, Agosto Perez F, Korniliev P, Greenberg A, Nareda ME, Mercado SM, Harrington S, Shi Y, Branchini D, Kuser-Falção LH, Ebana K, Yano M, Eizenga G, McClung A, Mezey J (2016) Open access resources for genome wide association mapping in rice. Nat Commun 7:10532

Nagasaki H, Ebana K, Shibaya T, Yonemaru J, Yano M (2010) Core single-nucleotide polymorphisms—a tool for genetic analysis of the Japanese rice population. Breeding Sci 60:648–655

Peterson BK, Weber JN, Kay EH, Fisher HS, Hoekstra HE (2012) Double digest RADseq: an inexpensive method for *de novo* SNP discovery and genotyping in model and non-model species. PLoS One 7:e37135

Rahman MA, Thomson MJ, Shah-E-Alam M, de Ocampo M, Egdane J, Ismail AM (2016) Exploring novel genetic sources of salinity tolerance in rice through molecular and physiological characterization. Ann Bot 117:1083–1097

Semagn K, Babu R, Hearne S, Olsen M (2014) Single nucleotide polymorphism genotyping using Kompetitive allele specific PCR (KASP): overview of the technology and its application in crop improvement. Mol Breeding 33:1–14

Septiningsih EM, Pamplona AM, Sanchez DL, Neeraja CN, Vergara GV, Heuer S, Ismail AM, Mackill DJ (2009) Development of submergence-tolerant rice cultivars: the *Sub1* locus and beyond. Ann Bot 103:151–160

Shah SM, Arif M, Aslam K, Shabir G, Thomson MJ (2016) Genetic diversity analysis of Pakistan rice (*Oryza Sativa*) germplasm using multiplexed single nucleotide polymorphism markers. Genet Resour Crop Evol 63:1113–1126

Singh N, Jayaswal PK, Panda K, Mandal P, Kumar V, Singh B, Mishra S, Singh Y, Singh R, Rai V, Gupta A, Sharma TR, Singh NK (2015) Single-copy gene based 50 K SNP chip for genetic studies and molecular breeding in rice. Sci Rep 5:11600

Singh A, Carandang J, Gonzaga ZJC, Collard BYC, Ismail AM, Septiningsih EM (2017) QTL mapping of stagnant flooding tolerance in rice: yield and important agronomic traits. Rice 10:15

Thomson MJ (2014) High-throughput SNP genotyping to accelerate crop improvement. Plant Breeding Biotech 2:195–212

Thomson MJ, Zhao K, Wright M, McNally KL, Rey J, Tung CW, Reynolds A, Scheffler B, Eizenga G, McClung A, Kim H, Ismail AM, de Ocampo M,

Monica C, Reveche MY, Ermita CJ, Mauleon R, Leung H, Bustamante C, McCouch SR (2012) High-throughput single nucleotide polymorphism genotyping for breeding applications in rice using the BeadXpress platform. Mol Breeding 29:875–886

Truong HT, Ramos AM, Yalcin F, de Ruiter M, van der Poel HJA, Huvenaars KHJ, Hogers RCJ, van Enckevort LJG, Janssen A, van Orsouw NJ, van Eijk MJT (2012) Sequence-based genotyping for marker discovery and co-dominant scoring in germplasm and populations. PLoS One 7:e37565

Tung CW, Zhao K, Wright MH, Ali ML, Jung J, Kimball J, Tyagi W, Thomson M, McNally K, Leung H, Kim H, Ahn SN, Reynolds A, Scheffler B, Eizenga G, McClung A, Bustamante C, McCouch SR (2010) Development of a research platform for dissecting phenotype-genotype associations in rice (*Oryza* spp.) Rice 3:205–217

Varshney RK, Terauchi R, McCouch SR (2014) Harvesting the promising fruits of genomics: applying genome sequencing technologies to crop breeding. PLoS Biol 12:e1001883

Webb KM, Ona I, Bai J, Garrett KA, Mew T, Vera Cruz CM, Leach JE (2009) A benefit of high temperature: increased effectiveness of a rice bacterial blight disease resistance gene. New Phytol 185:568–576

Xu K, Xu X, Fukao T, Canlas P, Maghirang-Rodriguez R, Heuer S, Ismail AM, Bailey-Serres J, Ronald PC, Mackill DJ (2006) *Sub1A* is an ethylene-response-factor-like gene that confers submergence tolerance to rice. Nature 442:705–708

Ye C, Argayoso MA, Redona ED, Sierra SN, Laza MA, Dilla CJ, Mo Y, Thomsom MJ, Chin J, Delavina CB, Diaz GQ, Hernandez JE (2012) Mapping QTL for heat tolerance at flowering stage in rice using SNP markers. Plant Breed 131:33–41

Ye C, Tenorio FA, Argayoso MA, Laza MA, Koh HJ, Redoña ED, Jagadish KSV, Gregorio GB (2015) Identifying and confirming quantitative trait loci associated with heat tolerance at flowering stage in different rice populations. BMC Genet 16:41

Yu H, Xie W, Li J, Zhou F, Zhang Q (2013) A whole-genome SNP array (RICE6K) for genomic breeding in RICE. Plant Biotech J 12:28–37

Zhao K, Wright M, Kimball J, Eizenga G, McClung A, Kovach M, Tyagi W, Ali ML, Tung CW, Reynolds A, Bustamante CD (2010) Genomic diversity and introgression in *O. sativa* reveal the impact of domestication and breeding on the rice genome. PLoS One 5:e10780

Zhao K, Tung CW, Eizenga GC, Wright MH, Ali ML, Price AH, Norton GJ, Islam MR, Reynolds A, Mezey J, McClung AM, Bustamante CD, McCouch SR (2011) Genome-wide association mapping reveals a rich genetic architecture of complex traits in *Oryza sativa*. Nat Commun 2:467

10

Effect of *qGN4.1* QTL for Grain Number per Panicle in Genetic Backgrounds of Twelve Different Mega Varieties of Rice

Vijay Kumar Singh[1,4], Ranjith Kumar Ellur[2], Ashok Kumar Singh[2], M. Nagarajan[3], Brahma Deo Singh[4] and Nagendra Kumar Singh[1*] (iD)

Abstract

Background: Rice is a major source of food, particularly for the growing Asian population; hence, the utilization of genes for enhancing its yield potential is important for ensuring food security. Earlier, we have mapped a major quantitative trait loci (QTL) for the grain number per panicle, *qGN4.1*, using biparental recombinant inbred line (RIL) populations involving a new plant type Indica rice genotype Pusa 1266. Later, three independent studies have confirmed the presence of a major QTL for spikelet number by three different names (*SPIKE*, *GPS* and *LSCHL4*) in the same chromosomal region, and have implicated the overexpression of *Nal1* gene as the causal factor for high spikelet number. However, the effect of *qGN4.1* in different rice genetic backgrounds and expression levels of the underlying candidate genes is not known.

Results: Here, we report the effect of *qGN4.1* QTL in the genetic backgrounds of 12 different high-yielding mega varieties of rice, introgressed by marker assisted-backcross breeding (MABB) using two QTL positive markers for foreground selection and two QTL negative flanking markers for recombinant selection together with phenotypic selection for the recovery of recipient parent genetic background. Analysis of the performance of BC_2F_3 plants showed a significant increase in the average number of well-filled grains per panicle in all the backgrounds, ranging from 21.6 in CSR 30-*GN4.1* to 147.6 in Samba Mahsuri-*GN4.1*. Furthermore, *qGN4.1* caused a significant increase in flag leaf width and panicle branching in most backgrounds. We identified BC_3F_3 *qGN4.1* near-isogenic lines (NILs) with 92.0–98.0% similarity to the respective recipient parent by background analysis using a 50 K rice SNP genotyping chip. Three of the NILs, namely Pusa Basmati 1121-*GN4.1*, Samba Mahsuri-*GN4.1* and Swarna-*GN4.1*, showed a significant yield superiority to their recipient parents. Analysis of differential gene expression revealed that high grain number in these QTL-NILs was unlikely due to the overexpression of *Nal1* gene (LOC_Os04g52479). Instead, another tightly linked gene (LOC_Os04g52590) coding for a protein kinase domain-containing protein was consistently overexpressed in the high grain number NILs.

Conclusion: We have successfully introgressed the *qGN4.1* QTL for high grain number per panicle into 12 different mega varieties of rice using marker-assisted backcross breeding. The advanced near-isogenic lines are promising for the development of even higher yielding versions of these high-yielding mega varieties of rice.

Keywords: Marker-assisted backcross breeding (MABB), Grain number, *qGN4.1*, Rice, Near-isogenic lines (NILs), Yield

* Correspondence: nksingh@nrcpb.org; nksingh4@gmail.com
[1]ICAR-National Research Centre on Plant Biotechnology, Pusa Campus, New Delhi 110012, India
Full list of author information is available at the end of the article

Background

Food shortage is becoming a serious global problem as the rate of increase in world population exceeds the rate of increase in food production. In this regard, rice, which is one of the most important staple food crops feeding more than half of the human population worldwide, grown on the most productive irrigated land has achieved nearly maximum production with the current varieties. There is immense dependence on rice to satiate the food need of a large population and decrease the hunger index. With the completion of rice genome project, efforts are on to characterize the genes responsible for yield component traits such as the number of panicle-bearing tillers per plants, number of well-filled grains (spikelets) per panicle, and thousand grain weight with profound implications on yield improvement. Although the mechanisms that regulate each component trait are not yet fully understood, the genes and the available knowledge on the associated molecular markers offer a set of tools that can be combined to achieve higher grain-yielding varieties.

Grain yield is a complex trait controlled by several quantitative trait loci (QTLs), most of which have minor effects but some also have a major effect. Due to its complexity, different allelic combinations can give rise to similar phenotypes. It is still unclear whether a single QTL has the potential to enhance grain yield in diverse genetic backgrounds of elite rice varieties, though there is an example of *sd1* gene showing a consistent effect on grain yield enhancement of Indica rice varieties, leading to green revolution (Sasaki et al. 2002). Although many QTLs have been identified for the yield component traits (www.gramene.org/archive/QTL data), only a few of them have been evaluated for their impact on grain yield. Most of the QTLs for grain yield have been detected in crosses between Japonica and Indica cultivar groups because inter-cultivar group diversity is significantly higher than intra-cultivar group diversity. QTL-based marker-assisted breeding has been advocated for the transfer of beneficial genes into elite cultivars for food security (Takeda and Matsuoka 2008).

Many QTLs and genes affecting grain yield have been identified in rice during the last decade. A few examples are as follows: *Gn1a* and *APO1* for number of grains per panicle (Ashikari et al. 2005; Ikeda-Kawakatsu et al. 2009; Terao et al. 2010); *GS3*, *GW2*, and *qSW5* for grain size (Fan et al. 2006; Song et al. 2007; Shomura et al. 2008); *DEP1* and *WFP* for panicle architecture (Huang et al. 2009; Miura et al. 2010); *SCM2* for strong culm (Ookawa et al. 2010); *Ghd7* for late heading and number of grains (Xue et al. 2008) and *NAL1* (*SPIKE*, *GPS* and *LSCHL4*) for plant architecture and photosynthesis rate (Fujita et al. 2013; Takai et al. 2013; Zhang et al. 2014). *DEP1*, *SCM2* and *APO1* genes have been found to enhance grain yield

in Japonica rice backgrounds in field experiments (Terao et al. 2010; Huang et al. 2009; Ookawa et al. 2010). A QTL *qGN4.1* with major effect on grain number per panicle was stable across 3 years with high LOD scores of 13, 6.8 and 5.3, explaining 27%, 16% and 12% of phenotypic variation, respectively (Deshmukh et al. 2010). The QTL *qGN4.1* was first mapped on the long arm of rice chromosome 4 in two different recombinant inbred line (RIL) populations (Pusa 1266/Pusa Basmati 1 and Pusa 1266/Jaya) derived from a new plant type (NPT) Indica rice genotype Pusa 1266 (Deshmukh et al. 2010; Marathi et al. 2012). This QTL is co-located with other important QTLs for the number of primary and secondary branches per panicle, number of tillers per plant, and flag leaf length and width, which may be due to either a tight genetic linkage or pleiotropic effects of the same gene on multiple traits (Deshmukh et al. 2010). No major gain in yield potential has been reported with the new plant type (NPT) breeding lines developed by IRRI due to poor grain filling and low biomass production. These drawbacks may be due to low crop growth rate during vegetative stage in the NPT lines as compared to Indica cultivars (Yamagishi et al. 1996), dense arrangement of spikelets on the panicle (Khush and Peng 1996), a limited number of large vascular bundles for assimilate transport, and source constraint due to early leaf senescence (Ladha et al. 1998). The introduction of high grain number trait from new plant type cultivars into recipient lines has also been initiated to broaden the genetic background of the NPT germplasm and refine the original ideotype design for increasing grain filling percentage and biomass production (Peng et al. 1999).

In the present study, we used marker-assisted backcross breeding (MABB) approach to introgress *qGN4.1* QTL for high grain number per panicle into genetic backgrounds of nine Indica and three Basmati rice varieties. A 50 K SNP genotyping chip was used for background selection in all the backcross-derived lines to ensure maximum recipient parent genome (RPG) recovery. The effect of *qGN4.1* QTL on grain yield in diverse genetic backgrounds was evaluated in field trials.

Methods

Plant Material

The donor parents (DP) used in this study, namely HG28 and HG67, were *qGN4.1* QTL-positive RILs derived from the cross between an NPT Indica rice genotype Pusa 1266 and Pusa Basmati 1 (Deshmukh et al. 2010). Recipient parents (RP) were all *qGN4.1* QTL-negative mega varieties of rice that are being cultivated in large acreage in India, namely Pusa Basmati 1121, Samba Mahsuri, Swarna, IR 64, MTU 1010, HUR 105, Sarjoo 52, Pusa 44, CSR 30, Ranjit, CR 1009, and Pusa

Basmati 1. Pusa Basmati 1121, CSR 30 and Pusa Basmati 1 are Basmati cultivars, whereas the other RP varieties belong to Indica cultivar group. The seeds were obtained from the collection maintained at ICAR-NRCPB and Division of Genetics, ICAR-IARI, New Delhi.

Plant DNA Extraction

The genomic DNA was isolated from the leaves of field-grown plants three to 4 weeks after transplanting. The leaf tissues from individual plants were frozen in liquid nitrogen, and DNA was extracted using the method of Murray and Thompson (1980). The quality and quantity of DNA were checked by electrophoresis in 0.8% agarose gel in 1× TBE buffer and comparison with known amounts of lambda DNA.

Foreground and Recombinant Selection Using qGN4.1 Flanking SSR Markers

The qGN4.1 QTL was transferred into 12 different mega varieties of rice using the previously described MABB strategy (Fig.1, Singh et al. 2015). In the first step,

qGN4.1 QTL-NILs

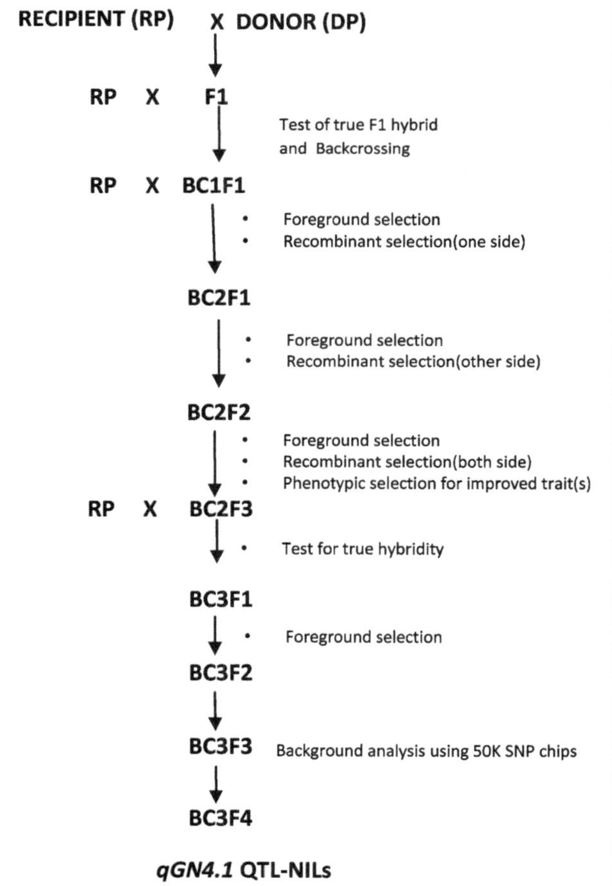

Fig. 1 Crossing and selection scheme for the transfer of high grain number QTL qGN4.1 in to popular high yielding mega varieties of rice

foreground selection for the presence of qGN4.1 was performed using two QTL-positive markers (nkssr 04-11 and nkssr 04-19), which reduced the number of BC_nF_1 plants by 50%. In the second step, two QTL-negative recombinant selection markers (RM2441 and HVSSR 4-49) flanking the QTL-positive markers on both sides were used to identify lines with minimum linkage drag of the DP (Additional file 1: Table S6). The qGN4.1 QTL has been narrowed down to a 360 kb interval between QTL-positive SSR markers nkssr 04-11 and nkssr 04-19 (Deshmukh et al. 2010, Sharma 2013). The PCR reaction consisted of 1.0 μl of 10× reaction buffer, 0.15 μl of 10 mM dNTPs (133 μM), 1.0 μl each of forward and reverse primers (10 pmol), and 2.0 μl of template genomic DNA (20-30 ng), 0.10 μl of Taq DNA polymerase in a final reaction volume of 10 μl. The PCR was carried out using Bio-Rad thermal cycler with an initial denaturation for 5 min at 94 °C, followed by 35 cycles for denaturation at 94 °C for 30 s, annealing at 55 °C for 30 s, extension at 72 °C for 1 min, and a final extension at 72 °C for 10 min. The PCR amplicons of nkssr 04-11, nkssr 04-19 and RM2441 were resolved by electrophoresis in 3.5–4% Metaphor™ Agarose gel and visualized on UV transilluminator (Gel Doc™ XR + Imager, Bio-Rad Laboratories Inc., U.S.A). However, because of the small size difference between donor and recipient alleles of HvSSR 04-49, it was analyzed by capillary electrophoresis of fam dye-labeled PCR products using ABI 3730 XL Genetic Analyzer.

Analysis of Recipient Parent Genome Recovery Using 50 K SNP Chip

In early generations of backcrossing, the background selection for RPG recovery was performed on the basis of morphological similarity of the backcross-derived lines with the RP (Singh et al. 2015). However, at BC_3F_3 stage, marker-assisted background analysis was performed on multiple QTL-near-isogenic lines (NILs) selected for maximum phenotypic similarity with the RP. For genotypic background selection, the short-listed QTL-NILs were compared with respective RPs using the Affymetrix 50 K Axiom® 2.0 SNP chip 'OsSNPnks' on GeneTitan® instrument (Singh et al. 2015). High-quality DNA with $OD_{260/280}$ and $OD_{260/230}$ in the range of 1.8–2.0 was used for SNP genotyping. Genomic DNA amplification, fragmentation, chip hybridization, washing, single-base extension through DNA ligation, and signal amplification was carried out according to Affymetrix Axiom® 2.0 Assay Manual Target Prep Protocol. Approximately 0.75–1.00 μg genomic DNA was labeled overnight at 25 °C using three volumes of the BioPrime DNA labeling reaction. The labeled DNA was ethanol precipitated, resuspended in 40 μl H_2O and then added to Affymetrix SNP

6.0 hybridization cocktail. Staining, washing, and scanning were performed using GeneTitan integrated platform (http://www.affymetrix.com). SNP genotypes were called using the Affymetrix Genotyping Console™ v4.1 (AGC) software package. SNPs with low call rates across all samples were removed from the dataset and high-performing SNPs with a development quality check (DQC) score of >0.85, and call rates of >95.0% were used for further analyses (Singh et al. 2015). Graphical representation of 50 K SNP genotyping based RPG recovery was done using Phenogram software from Ritchie lab, Penn. State University, Pennsylvania, USA (http://visualization.ritchielab.psu.edu/phenograms/plot). The RPG similarity based on SNP markers was calculated using the formula, RPG (%) = (R + 1/2H) × 100/P, where R = number of markers homozygous for RP allele, H = number of heterozygous markers, P = total number of SNP markers used for background selection (Ellur et al. 2015).

Measurement and Analysis of Field Phenotyping Data

Phenotypic traits were evaluated on BC_2F_3 plants using augmented field design. All the 12 RPs and six selected NILs for each recipient were planted across five blocks without replication with four checks repeated in each block. Measurements were taken on five plants for each entry for the following ten traits. Plant height was measured in cm from the base of the main stem to the tip of the primary panicle of the plant. The total number of tillers was recorded as both productive and unproductive tillers in a single plant. Productive tillers per plant referred to the number of tillers-bearing panicle at harvest. Flag leaf length was measured as the length of leaf-bearing primary panicle from leaf base to the tip. Flag leaf width was the maximum width of the leaf-bearing primary panicle. Panicle length was the length of primary panicles measured from the base of peduncle to the tip excluding awns. Primary branches per panicle referred to the total number of branches coming out directly from the peduncle. Secondary branches per panicle referred to the total number of branches coming out from the primary braches of the panicle. Spikelet fertility was measured as the number of well-filled grains/total number of grains per panicle × 100. The total number of grains per panicles was measured as the total number of florets (spikelets) present in the primary panicle. All the NILs and their respective RP varieties were planted in two-meter rows with row-to-row spacing of 30 cm and plant-to-plant spacing of 20 cm. The best five BC_2F_3 plants showing the maximum phenotypic similarity with the RP were used for measuring plant height, flag leaf length, flag leaf width, tiller number and number of grain per panicle. Pollen from one of the plants most similar to RP was used for backcrossing to produce BC_3 seeds.

Field evaluation of yield and morphological traits was carried out in the experimental fields of Rice Breeding and Genetics Centre, Aduthurai (Tamil Nadu) and Indian Agricultural Research Institute, New Delhi. Evaluation of yield-associated traits was done on NILs, their RPs and four checks using an augmented randomized complete block design (RCBD) with five blocks, each having a maximum of 21 entries including four checks. The data for each character were analyzed as per the procedure of augmented RCBD using the Statistical Package for Augmented Designs. The treatment sum of squares was partitioned into the sum of squares among NILs, among checks and between NILs and checks. The adjusted means of NILs and checks were obtained. For making all possible pair-wise treatment comparisons, critical difference was calculated at 5% level of significance ($P = 0.05$) for each of the four comparisons, namely between checks, between NILs and checks, between NILs of the same block, and between NILs of different blocks (Federer 1956, 1961; Parsad and Gupta 2000). The backcross-derived lines along with the respective RPs were evaluated for yield (kg/ha) at BC_3F_4 generation in an RCBD with two replications.

RNA Extraction and qRT-PCR Analysis of Candidate Gene Expression

Total RNA was extracted from panicle primordia measuring 2–3 cm in length using Promega's SV Total RNA Isolation System. The concentration of each RNA sample was measured using NanoDrop ND-1000 spectrophotometer (NanoDrop Technologies). Only the RNA samples with 260/280 ratio between 1.9 and 2.1 and 260/230 ratio greater than 2.0 were used for the analysis. First-strand cDNA was synthesized from 6 μg of total RNA using oligo dT as a primer and reverse transcriptase mix (AffinityScript QPCR cDNA Synthesis Kit; Agilent Technologies) in a 20 μl reaction volume. The PCR mixture contained 2 μl of template cDNA, 5 μl of 2× SYBR Green qPCR Master Mix (Agilent Technologies), 0.4 μl of gene-specific forward and reverse primers, 0.15 μl of ROX reference dye1, and 2.05 μl of nuclease-free H_2O. Gene expression was normalized against eEF1a as an internal control. Control PCR reaction with no template was also performed for each primer pair. The RT-PCR was performed using Agilent Technologies Stratagene, Mx3005P Detection System and software. The PCR amplification condition included one cycle of 95 °C (3 min), followed by 40 cycles of denaturation at 95 °C (30 s), 60 °C (20 s), and 72 °C (20 s), interrupted by the dissociation curve with denaturation at 95 °C (1 min), cooling at 55 °C (30 s) and gradually heating up to 95 °C (30 s) in 96-well optical reaction plate. The amplicon purity was determined when a single melting peak was reached. Two biological replicates for each sample and three technical replicates for each biological replicate were used for RT-PCR analysis. The rice eEF1a gene was used as internal control (primer pair 5′TTTCA CTCTTGGAGTGAAGCAGAT3′ and 5′GACTTCCTT CACGATTTCATCGTAA3′). Gene-specific primers were designed using the online PrimerQuest tool of IDT-Integrated DNA Technologies (https://eu.idtdna.com/ PrimerQuest/Home/Index) using full-length cDNA sequences of the three candidate genes in the QTL interval: (i) *Nal1* (LOC_Os04g52479); (ii) Retrotransposon gene (LOC_Os04g52540), and (iii) Protein kinase domain containing protein gene (LOC_Os04g52590), (http://rice.plantbiology.msu.edu/cgi-bin/gbrowse/rice/ #search). Primers for qRT-PCR expression analysis were: 5′GGTGTTTGGTGTGATGTTGATG3′ and 5′ CCTGAGAGCCTGAACCAATAC3′ for LOC_Os04g 52479 (product size 136 bp); 5′TCAGGTCTTCGC ATTGGCACAC3′ and 5′GTACCGTTGAGCGCAT TGCTTC3′ for LOC_Os04g52540 (product size 105 bp); and 5′GCAGCTCTCTCAGTCATCTAATC3′ and 5′ TCTCAAGGCAGCAACCATAC3′ for LOC_Os04g52590 (product size 113 bp). Relative expression levels of the three genes were assayed on a real-time PCR system (Agilent Technologies Stratagene, Mx3005P). The relative gene expression of the target gene was calculated using the equation: $Exp = 2^{-\Delta Ct}$, where $\Delta Ct = Ct_{targetgene} - Ct_{eEF1a}$ (Livak and Schmittgen 2001).

Results and Discussion

qGN4.1 QTL-NILs Developed by Marker-Assisted Backcross Breeding

We transferred *qGN4.1*, a major QTL for grain number per panicle, into 12 different mega varieties of rice by marker-assisted backcross breeding (Fig. 1). In addition to marker-assisted selection using two tightly linked QTL-positive foreground selection markers and two flanking QTL-negative recombinant selection markers, the backcross progenies were also selected for phenotypic similarities with the respective RP at each filial generation. The background selection was initially restricted to this visual screening of phenotypic traits for similarity with the recipient parent. The backcross progeny was selected for the high grain number donor parent (DP) alleles of the foreground selection markers

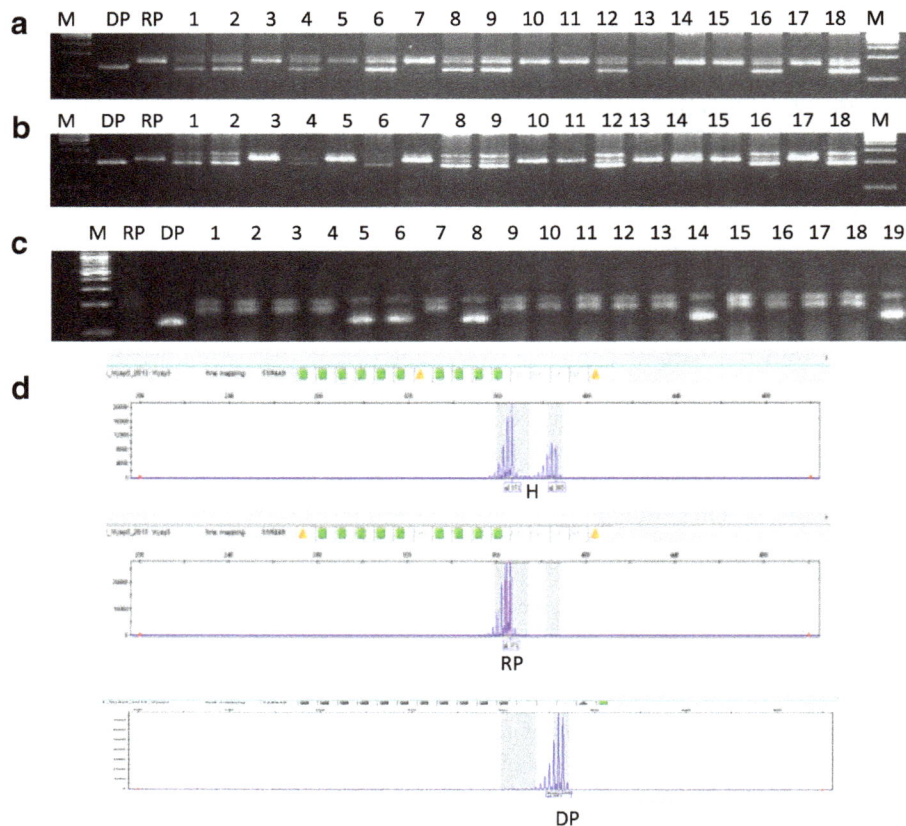

Fig. 2 Foreground (**a**, **b**) and recombinant (**c**, **d**) selection in backcross F₁ lines of variety Ranjit with linked SSR markers **a**. nkssr 04-11 **b**. nkssr 04-19. **c**. RM2441 using 3.5-4% metaphor agarose gel electrophoresis, and **d** HVSSR 04-49 by capillary electrophoresis. M = DNA size markers, RP = recipient parent, DP = donor parent, H = F1 heterozygote, 1-18 BC1F1 plants. The bands marked RP and DP on the gel picutre are the main bands

nkssr 04-11 and nkssr 04-19 tightly linked to *qGN4.1*, and the recombinant selection was for the presence of RP alleles of QTL flanking markers RM 2441 and HvSSR 04-49 (Fig. 2, Additional file 1: Table S6). Details of the number of backcross progenies planted in each generation for all the 12 RP varieties and the number of plants selected positive for both foreground and recombinant selection markers for each cross are shown in Additional file 6: Table S1. Out of the 12 crosses, both side recombinants were recovered in BC_2F_1 generation for ten crosses, except for HUR 105 and MTU 1010 for which both side recombinants were selected in the BC_2F_3 and BC_3F_1 generations, respectively (Additional file 2: Table S1). The selected lines recombinant on both sides of the QTL was taken further to the next generation for selection of lines with maximum visual similarity to the respective RP.

Finally, six lines were selected for each RP variety for the presence of DP alleles of QTL-positive foreground selection markers, both side RP parent alleles of the QTL-negative recombinant selection marker and phenotypic similarity with RP for background RPG recovery and designated RP-*qGN4.1-1* to RP-*qGN4.1-6* (Additional file 7: Table S2).

Phenotypic Characteristics and Yield Performance of *qGN4.1* QTL-NILs at BC_2F_3

Six selected *qGN4.1* QTL-NILs each for the 12 different RP genetic backgrounds at BC_2F_3 generation were planted along with their respective RPs for evaluation in an augmented block design (Additional file 3: Table S2). Phenotypic data were recorded for ten traits, including plant height (PH), total number of tillers (TT), number of productive tillers (PT), flag leaf length (FLL), flag

Table 1 Adjusted Means for Different Morphological Traits for the *qGN4.1* QTL-NILs Compared With Respective Recipient Parents in BC_2F_3 Generation

Genotype	PH	TT	PT	FLL	FLW	PL	PB	SB	TG	FG	SF
PB1121	101.7	20.8	16.6	27.2	1.0	21.8	10.1	37.2	99.5	87.5	87.9
PB1121 + *qGN4.1*	117.1*	11.1*	9.3*	29.1	1.4*	25.3	12.2*	55.2*	128.7*	117.5*	91.3
Samba Mahsuri	84.5	12.6	9.3	25.9	1.3	17.6	11.5	55.2	284.1	235.9	83.0
SM + *qGN4.1*	87.5	8.5	7.1	30.3	2.3*	20.1	14.3*	68.1*	409.7*	383.5*	93.6*
Swarna	89.7	12.9	9.5	23.8	1.0	21.2	12.9	56.7	197.9	174.5	88.2
SW + *qGN4.1*	92.9	11.4	9.4	31.6	1.6*	24.7	14.9*	59.3	221.1*	208.1*	94.1*
IR 64	85.1	16	10.6	32.5	1.2	21.7	10.1	37.8	158.4	142.1	89.7
IR64 + *qGN4.1*	106.3*	9.9*	8.4	33.4	1.3	23.2	11.9*	54.5*	211.8*	197.7*	93.3
MTU1010	80.9	11.6	8.8	31.5	1.4	25.1	12.1	46.8	180.2	153.5	85.2
MTU + *qGN4.1*	98.9*	11.1	9.8	24.9	1.5	20.1	12.9	66.8*	210.9*	200.3*	95.0*
HUR 105	90.8	21	15.9	23.4	1.5	20.8	12.8	54.9	185.3	165.2	89.2
HUR + *qGN4.1*	93.2	15.6*	14.5	31.6*	1.2	23.7	13.4	60.3	208.5*	190.0*	91.1
Sarjoo 52	79	18.4	15.3	23.8	1.3	22.1	12.8	59.6	199.9	178.2	89.1
Sarjoo52 + *qGN4.1*	84.5	10.4*	9.6*	33.4*	1.5	24.3	13.9	69.1	239.7*	218.2*	91.0
PUSA 44	84.4	9.6	6.9	31.3	1.2	24.9	12.6	57.9	183.5	162.6	88.6
PUSA44 + *qGN4.1*	91.8	7.0	5.8	33.8	1.5*	25.8	13.0	61.6	226.7*	208.8*	92.1
CSR 30	105.2	14.7	10.4	26.0	1.0	25.1	10.4	31.5	121.4	105.1	86.6
CSR30 + *qGN4.1*	115.6	12.5	11.8	27.7	1.1	26.8	12.6*	36.5	134.4*	126.7*	94.3*
Ranjit	121.6	13.4	9.0	28.1	1.8	23.2	12	51.9	215.8	204.9	94.9
Ranjit + *qGN4.1*	125.8	11.1	9.7	28.5	2.2*	25.7	12.7	67.5*	317.6*	303.7*	95.6
CR 1009	93.1	15.5	10.9	21.2	1.3	19.4	12.1	56.7	175.1	152.0	86.8
CR1009 + *qGN4.1*	99.5	12.9	11.1	24.9	1.4	23.7	13.3*	60.5	215.1*	199.6*	92.8*
Pusa Basmati 1	96.5	12.2	10.4	22.8	0.6	23.7	11.7	52.6	154.5	137.5	89.0
PB1 + *qGN4.1*	101.6	11.6	10.3	29.4*	1.0*	20.3	14.5*	57.9	198.1*	180.3*	91.0
CD(0.05)	14.5	5.2	5.0	6.0	0.3	4.5	1.5	10.0	11.7	7.4	5.0

PH plant height in cm, *TT* total tillers, *PT* productive tillers, *FLL* flag leaf length in cm, *FLW* flag leaf width in cm, *PL* panicle length in cm, *PB* primary branches per anicle, *SB* secondary branches per panicle, *SF* percent spikelet fertility, *FG* filled grain per panicle, *TG* total grain number per panicle, *SM* Samba Mahsuri, *SW* Swarna. *Significantly from respective recipient parent (*P* > 0.05)

leaf width (FLW), panicle length (PL), number of primary branches per panicle (PB), number of secondary branches per panicle (SB), total number of grains per panicle (TG), number of well-filled grains per panicle (FG) and spikelet fertility (SF) (Table 1). The mean values of data from five plants for each QTL-NIL were adjusted based on the performance of the check varieties in that block, resulting in adjusted means. For each of the 12 RP varietal backgrounds plant morphology of the NILs was similar to the respective RP because of good RPG background recovery in the NILs (Fig. 3; Additional file 4: Figure S1; Additional file 5: Figure S2; Additional file 6: Figure S3). For the ten phenotypic traits compared there was no statistically significant difference between NILs and their respective parents for panicle length, but plant height increased significantly in case of PB 1121-*qGN4.1*, IR 64-*qGN4.1* and MTU 1010-*qGN4.1*. The total number of tillers decreased significantly in PB 1121-*qGN4.1*, IR 64- *qGN4.1*, HUR 105-*qGN4.1* and Sarjoo 52-*qGN4.1* but the number of productive tillers decreased significantly only in PB 1121-*qGN4.1* and Sarjoo 52-*qGN4.1*. Flag leaf length increased significantly in HUR 105-*qGN4.1*, Sarjoo 52-*qGN4.1* and PB 1-*qGN4.1* while flag leaf width increased significantly in PB 1121-*qGN4.1*, Samba Mahsuri-*qGN4.1*, Swarna-*qGN4.1*, Pusa 44-*qGN4.1*, Ranjit-*qGN4.1*, and Pusa Basmati 1-*qGN4.1* (Table 1). The number of primary branches per panicle increased significantly in PB1121-*qGN4.1*, Samba Mahsuri-*qGN4*.1, Swarna-*qGN4.1*,

IR64-*qGN4.1*, CSR30-*qGN4.1*, Ranjit-*qGN4.1* and PB1-*qGN4.1* while number of secondary branches per panicle increased significantly in PB1121-*qGN4.1*, Samba Mahsuri-*qGN4.1*, IR 64-*qGN4.1* MTU1010-*qGN4.1*, and Ranjit-*qGN4.1*. The total number of grains (spikelets) and the number of well-filled grains per panicle both increased significantly in the QTL-NILs of all 12 RP genetic backgrounds. Panicle length, number of primary and secondary branches per panicle, total number of grains per panicle and spikelet fertility directly influence the number of well-filled grains per panicle, which has a direct relevance to enhancing of yield potential (Fig. 4). The increase in number of well-filled grains per panicle ranged from 21.6 in CSR 30 to 147.6 in Samba Mahsuri (Table 1).

Plant height increased significantly in PB 1121, IR 64 and MTU 1010 genetic backgrounds leading to 15.4-21.2 cm taller plants with higher biomass accumulation without causing any lodging problem (Fig.3; Additional file 4: Figure S1; Additional file 5: Figure S2; Additional file 6: Figure S3). The total number of tillers decreased in the *qGN4.1* QTL-NILs at the cost of unproductive tillers, a typical characteristic of the new plant type donor variety. The number of productive tillers decreased significantly only in the Pusa 1121 and IR 64 background. Thus, there were similar numbers of healthy and sturdy productive tillers in most of the *qGN4.1* QTL-NILs as in the respective RPs. Comparison of the original new plant type (NPT) lines with the highest yielding Indica rice

Fig. 3 Plant architecture of *qGN4.1* QTL-NILs (left side) of rice as compared to their recipient parents (right side): **a** Pusa Basmati 1121(PB1121) **b** Samba Mahsuri **c** Swarna (scale bars: 10 cm)

Fig. 4 Panicle structure of *qGN4.1* QTL-NILs (right side) along with respective recipient parents (left side): **a** Pusa Basmati 1121 (PB 1121) **b** Samba Mahsuri, **c** Swarna **d** IR 64 **e** MTU 1010 **f** HUR 105 **g** Sarjoo 52 **h** Pusa 44 **i** CSR 30 **j** Ranjit **k** CR 1009 **l** Pusa Basmati 1(PB 1), Scale bars- 5 cm

cultivars has shown that the original NPT lines had significantly lower tillering capacity (Peng et al. 1994). Sufficient tillering capacity is needed for biomass production to improve compensation when tillers are lost due to insect damage or other stresses during vegetative stage (Peng et al. 1999). Primary branches, secondary branches and the total number of filled grains per panicle are important parameters which would directly affect the grain yield of rice. All of these characteristics were directly correlated to each other leading to a larger panicle size (Fig. 4). Larger panicle size of *qGN4.1* QTL-NILs without significant change in panicle length was because of increase in panicle branching without more compact arrangement of spikelets which may result in poor grain filling. It is very useful that *qGN4.1* QTL-NILs in Sambha Mahsuri, Swarna, MTU 1010, Ranjit and CR 1009 backgrounds have significant improvement in percent spikelet fertility. The spikelet fertility was enhanced by as much as 10.3% in Samba Mahsuiri-*qGN4.1*, which might help develop an even higher yielding version of this popular mega variety of rice. Thus, this study clearly shows that introgression *qGN4.1* QTL significantly increased grain numbers per panicle in all the 12 QTL-NILs as compared to

the respective RP without any reduction in spikelet fertility.

The QTL *qGN4.1* for grain number per panicle was first identified in our laboratory in an Indica/Indica RIL population of Pusa 1266/Pusa Basmati 1, including eight differentially expressed candidate genes in the QTL interval (Deshmukh et al. 2010). Three subsequent studies have also identified a QTL for high grain (spikelet) number in the same region of rice chromosome 4 but with different QTL names, viz. *SPIKE*, *GPS*, *LSCHL4* using Indica/Japanica crosses (Fujita et al. 2013; Takai et al. 2013; Zhang et al. 2014). The latter three studies have implicated the overexpression of previously described *NAL1* gene for narrow leaves mutation for the QTL effect (Qi et al. 2008). Here, we introgressed the *qGN4.1* QTL in the genetic backgrounds of 12 different Basmati and Indica rice cultivars to see its effect on different yield parameters. Thus, we can presume that *NAL1* allele could also be involved in enhancing flag leaf area in the *qGN4.1* QTL-NILs for enhanced photosynthetic efficiency, and grain number per panicle traits, but as shown below, the expression of *NAL1* gene was down-regulated in the high grain number lines in our material. The present study employed a MABB approach to demonstrate a consistent effect of $qGN4.1$ QTL in diverse genetic backgrounds. We believe that $qGN4.1$ QTL-NILs in the genetic backgrounds of different Indica and Basmati cultivars have improved photosynthetic efficiency through coordination of leaf morphological and physiological traits, which has great potential for use in breeding for higher-yielding rice varieties.

Recipient Parent Genome Recovery and Yield Performance of the $qGN4.1$ QTL-NILs

RPG recovery of the $qGN4.1$ QTL-NILs was analyzed by comparing the percentage of marker allele similarity with the respective RP at BC_3F_3 generation using a 50 K SNP genotyping chip (Singh et al. 2015). We analyzed the six best QTL-NILs each for six RP varieties, namely Samba Mahsuri, Swarna, MTU 1010, Sarjoo 52, PUSA 44 and Pusa Basmati 1 (PB 1), and two best NILs each for the remaining six varieties, namely Pusa Basmati 1121, IR 64, HUR 105, CSR 30, Ranjit and CR 1009 due to resource limitations (Additional file 7: Table S3). During early generations of backcrossing until BC_3F_2 plants were selected for RPG recovery phenotypically, more attention was paid to foreground selection for the presence of donor QTL allele using QTL-positive markers and recombinant selection for the presence RP alleles using QTL-negative markers to minimize the linkage drag on the carrier chromosome. The QTL-NILs with the highest overall RPG recoveries for the 12 recipient rice varieties were Pusa Basmati 1121 (97.6%), Samba Mahsuri (96.5%), Swarna (93.8%), IR 64 (98.0%), MTU 1010 (93.8%), HUR 105 (96.5%), Sarjoo 52 (92.0%), Pusa 44 (94.3%), CSR 30 (97.4%), Ranjit (97.7%),

CR 1009 (96.0%), and Pusa Basmati 1 (93.4%). RPG recoveries for QTL $qGN4.1$ career chromosome 4 in these lines were: 94.6%, 93.1%, 89.6%, 95.4%, 90.6%, 93.3%, 78.6%, 91.3%, 85.8%, 91.3%, 90.3% and 91.5%, respectively (Table 2), which was significantly lower than the overall RPG recovery due to linkage drag of the donor genome in the QTL region. The recipient genome was almost fully recovered in the selected NILs at BC_3F_4 generation, and these lines were used for evaluating the effect of $qGN4.1$ QTL on yield performance in different genetic backgrounds. Although the selected NILs still have 2-8% genomic content from the DP, it can be reduced further by additional backcrossing if needed.

The RPG recovery was variable, but DP alleles were commonly present in the $qGN4.1$ region of chromosome 4 in all the QTL-NILs as visualized by graphical genotyping (Fig. 5; Additional file 8: Figure S4; Additional file 9: Figure S5; Additional file 10: Table S5). A similar pattern of increased grain number per panicle and related traits across all the 12 RP backgrounds validate the effect of $qGN4.1$ QTL on the grain number per panicle. Graphical representation of the high-resolution SNP genotyping data of QTL-NILs revealed the substantial size of DP chromosome segments introgressed at the $qGN4.1$ QTL region on chromosome 4, viz. Pusa Basmati 1121 (2.2 Mb), Samba Mahsuri (2.7 Mb), Swarna (4.1 Mb), IR 64 (2.4 Mb), MTU 1010 (3.4 Mb), HUR 105 (4.2 Mb), Sarjoo 52 (1.2 Mb), Pusa 44 (2.5 Mb), CSR 30 (2.8 Mb), Ranjit (4.3 Mb), CR 1009 (2.3 Mb), and Pusa Basmati 1 (2.8 Mb) despite conscious efforts to reduce the linkage drag by recombinant selection using QTL-negative flanking markers (Table 2). In addition to the consistent presence of 1.2-

Table 2 Percentage Genome Similarity of The Best $qGN4.1$ QTL-NILs With Recipient Parent at BC3F2 Generation in the Genetic Background of Twelve Different Mega Varieties of Rice Analysed Using OsSNPnks SNP Genotyping Chip

Sr. no.	Recipient variety	Rice chromosome												Overall similarity
		1	2	3	4	5	6	7	8	9	10	11	12	
1.	PB 1121	99.1	95.0	98.5	94.6	97.1	98.8	95.8	98.8	95.3	98.2	98.6	97.2	97.6
2.	Samba Mahsuri	96.6	97.0	96.1	93.1	96.7	98.0	99.6	96.0	96.9	96.7	95.0	95.7	96.5
3.	Swarna	95.5	93.4	93.3	89.6	94.4	97.1	99.4	93.9	92.2	91.5	86.1	92.1	93.8
4.	IR 64	98.8	96.2	98.9	95.4	98.9	99.4	98.1	94.5	99.3	96.1	99.3	98.0	98.0
5.	MTU 1010	95.6	94.7	93.0	90.6	94.7	93.5	96.8	92.9	94.5	91.3	91.7	92.2	93.8
6.	HUR105	91.4	96.3	97.9	93.3	97.9	92.7	98.5	98.2	97.9	97.6	97.6	97.5	96.5
7.	Sarjoo 52	95.6	90.4	90.3	78.6	93.8	93.9	98.1	93.2	89.5	90.9	93.5	93.9	92.0
8.	Pusa 44	93.0	95.1	93.7	91.3	95.0	96.8	99.5	95.3	95.1	92.5	91.4	93.3	94.3
9.	CSR 30	99.6	95.9	97.8	85.8	98.5	99.6	96.9	92.6	99.6	99.4	99.3	99.2	97.4
10	Ranjit	97.9	99.2	94.9	91.3	99.5	99.5	99.4	99.5	98.7	93.9	99.2	99.6	97.7
11	CR1009	94.5	99.5	97.9	90.3	99.0	94.1	99.4	89.6	92.7	98.8	99.3	94.9	96.0
12	PB 1	97.2	94.2	92.1	91.5	95.3	96.3	97.1	95.2	89.2	89.2	90.1	93.3	93.4
	Average	96.2	95.6	95.4	90.5	96.7	96.6	98.2	95.0	95.1	94.7	95.1	95.6	95.5

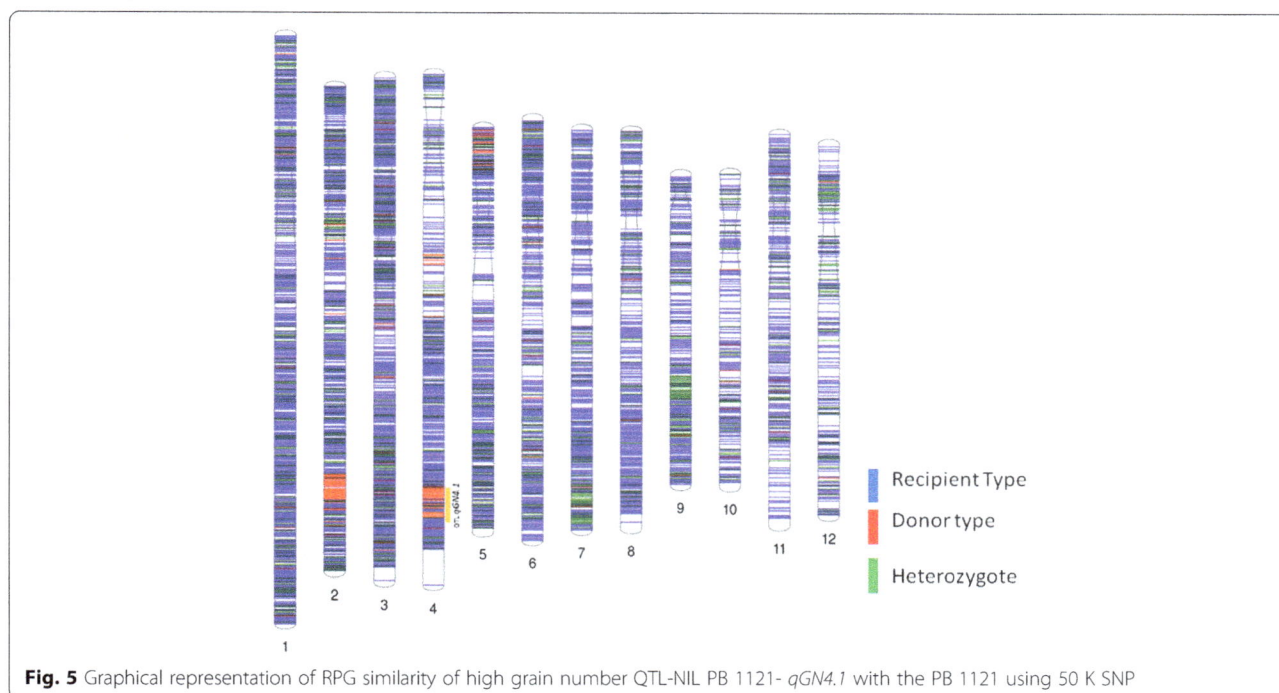

Fig. 5 Graphical representation of RPG similarity of high grain number QTL-NIL PB 1121- *qGN4.1* with the PB 1121 using 50 K SNP

4.3 Mb of DP chromosome segment in the QTL region, the QTL NILs also showed other random large donor chromosome segments (Additional file 11: Table S4). For example, PB 1121-*qGN4.1* carries 2.6 Mb, 0.7 Mb and 2.5 Mb segments of DP on chromosomes 2, 4 and 5, respectively (Fig. 5, Additional file 12: Table S7). Samba Mahsuri-*qGN4.1* has 48 DP segments of >0.2 Mb on chromosomes 1, 2, 8, 11 and 12; Swarna-*qGN4.1* has 75 DP segments of >0.2 Mb on chromosomes 1, 3, 5, 11 and 12; IR 64-*qGN4.1* has five large DP segments on chromosomes 1 (0.4 Mb), 2 (1.8 Mb), 7 (0.6 Mb), 8 (3.7 Mb) and 12 (2.2 Mb); MTU 1010-*qGN4.1* has 163 DP segments of >0.2 Mb through out the genome except chromosomes 7; HUR 105-*qGN4.1* has 43 DP segments of >0.2 Mb on chromosome 1, 2, 11 and 12; Sarjoo 52-*qGN4.1* has a large 8.7 Mb DP segment on chromosome 4 and 50 smaller segments of >0.2 Mb on chromosomes 2, 9, 11 and 12; Pusa 44-*qGN4.1* has 104 DP segments on chromosomes 1, 2, 3, 5, 11 and 12; CSR 30-*qGN4.1* has three large DP segments on chromosomes 3 (1.4 Mb), 5 (0.3 Mb) and 8 (2.4 Mb); Ranjit-*qGN4.1* has a large DP segment on chromosome 1(1.1 Mb), four separate large segments of >0.2 Mb on chromosome 3, and 0.2 Mb on chromosome 8; CR 1009-*qGN4.1* has two large DP segments 1.1 Mb and 1.9 Mb on chromosome 1, 0.9 Mb on chromosome 3; 1.0 Mb and 0.9 Mb on chromosome 6, 1.4 Mb on chromosome 9 and 1.1 Mb on chromosome 12; and Pusa Basmati 1-*qGN4.1* has 85 DP segments of 0.2 Mb on chromosomes 2, 3, 6, 8, 9,10, 11 and 12. Thus,

the high-resolution background analysis using 50 K SNP chip showed high RPG recovery in Pusa Basmati 1121, IR 64, CSR 30, Ranjit and CR 1009, and need for additional back crossing in Samba Mahsuri, Swarna, MTU 1010, HUR 105, Sarjoo 51, Pusa 44 and Pusa Basmati 1 for the elimination of the remaining large DP chromosome segments to develop QTL-NILs with uniform high RPG recovery.

The QTL-NILs with the highest RPG recovery in 11 different varietal backgrounds were evaluated for yield performance in large replicated field plots. Yield data could not be obtained for CSR 30 because of poor seed setting due to environmental factors at late maturity. There was a significant increase in yield with standard agronomic practices with the introgression of *qGN4.1* QTL in PB 1121, Samba Mahsuri and Swarna backgrounds as compared to their respective RPs, but the maximum gain was observed in Samba Mahsuri genetic background where the yield was enhanced from 6690 kg/ha in the RP Samba Mahsuri to 8667 kg/ha in Samba Mahsuri-*qGN4.1*. Significant yield increases were also observed with Pusa Basmati 1121-*qGN4.1* from 5808 kg/ha for RP to 6833 kg/ha for QTL-NIL, and with Swarna-*qGN4.1* from 3667 kg/ha for RP to 5250 kg/ha for QTL-NIL. Other eight varieties also showed the numerical superiority of yield for *qGN4.1* QTL-NILs as compared to the RP control, but the gain was not statistically significant (Fig. 6; Additional file 7: Table S4). After further multilocation evaluation of yield performance, some of these QTL-NILs have the potential to be

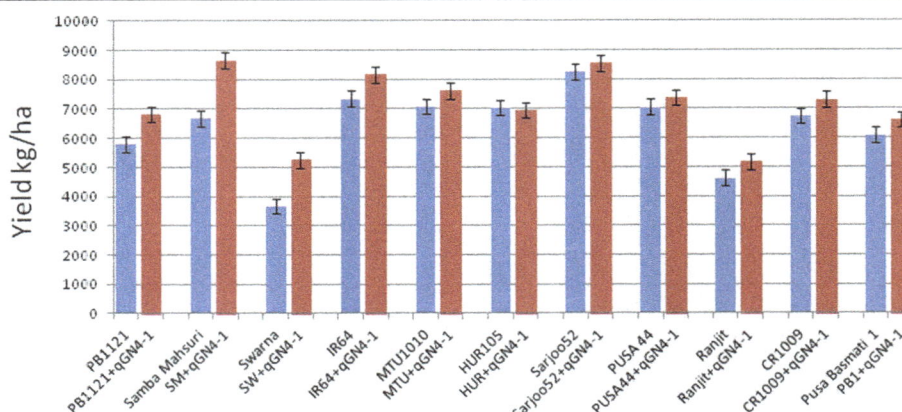

Fig. 6 Bar diagram showing yield performance of *qGN4.1* QTL-NILs in 11 different backgrounds of rice in comparison to their recipient parents. Values are average of two replications, with the whiskers showing S.E. of means. SM = Samba Mahsuri; SW = Swarna

released for commercial cultivation because of high acceptability of the RP varieties.

Differential Expression of *Nal1* and Other Candidate Genes in the *qGN4.1* QTL Interval

The *qGN4.1* QTL in the marker interval nkssr 04-02 and nkssr 04-19 (11.1 cM, 0.78 Mb) was identified as the most important region controlling the grain number per panicle in the NPT genotype Pusa 1266 (Deshmukh et al. 2010, Marathi et al. 2012). Other studies have also shown the involvement of this chromosomal region on photosynthetic efficiency and plant architecture including the number of spikelets per panicle (Takai et al. 2013; Fujita et al. 2013; Zhang et al. 2014). This region has the evidence for the presence of 148 expressed genes in the EST database (TIGR Version 7.0, http://rice.plantbiology.msu.edu/). This list was too large for identification of causal gene(s) underlying *qGN4.1* QTL. Therefore, QTL mapping was complemented with genome-wide transcriptome profiling to identify differentially expressed genes at early panicle development stage (Deshmukh et al. 2010). Microarray-based transcriptome profiling revealed eight genes in the *qGN4.1* region that were differentially expressed between the two parents, including *Nal1* gene (LOC_Os04g52479) that codes for a serine and cysteine protease. Of the eight differentially genes, only one coding for protein kinase domain-containing protein (LOC_Os04g52590) showed overexpression in the high grain number genotype; *Nal1* gene was actually downregulated by −4.0 and −1.1 folds in microarray and qRT-PCR studies, respectively (Deshmukh et al. 2010). Hence, we analyzed differential expression of these two potential candidate genes along with a third neutral gene coding for retrotransposon protein (LOC_Os04g52540) in all the 12 QTL-NILs developed in the present study. Consistent with our earlier results (Deshmukh et al. 2010), the LOC_Os04g52590 gene

coding for protein kinase domain-containing protein was consistently overexpressed in QTL-NILs of all the 12 RP backgrounds (Table 3). Also, consistent with our earlier results (Deshmukh et al. 2010), *Nal1* (LOC_Os04g52479) and retrotransposon (LOC_Os04g52540) genes actually showed down regulation in the *qGN4.1* QTL-NILs (Table 3). This is different from the situation with *SPIKE*, *GPS* and *LSCHL4* QTLs where overexpression of *Nal1* gene is shown to be the causal factor for high spikelet number (Fujita et al. 2013; Takai et al. 2013; Zhang et al. 2014). The role of overexpressed gene (LOC_Os04g52590) coding for protein kinase domain-containing protein in *qGN4.1* QTL-NILs is being further investigated by genetic recombination and transformation studies.

Table 3 Fold Change in the Expression Level of Three Genes Located in the *qGN4.1* QTL Interval in the Panicle Primordia of Twelve QTL Introgression Lines in Comparison to the Respective Recipient Parents as Detrmined Using qRT-PCR

Genotype	Gene Id.		
	LOC Os04g52479	LOC_ Os04g52540	LOC Os04g52590
PB1121 + *qGN4.1*	−0.1	−0.9*	+4.2**
Samba Mahsuri + *qGN4.1*	−0.7*	−0.1	+2.2**
Swarna + *qGN4.1*	−0.7 *	−0.8*	+1.9**
IR64 + *qGN4.1*	0.6 *	−0.3	+2.1**
MTU1010 + *qGN4.1*	0.5*	−0.6*	+1.7**
HUR105 + *qGN4.1*	0.1	−0.4	+0.9**
Sarjoo52 + *qGN4.1*	−0.3	−0.2	+3.2**
PUSA44 + *qGN4.1*	−0.4	−0.2	+1.6**
CSR30 + *qGN4.1*	−0.3	−0.3	+0.9**
Ranjit + *qGN4.1*	−0.6 *	−0.5*	+2.3**
CR1009 + *qGN4.1*	−0.5*	−0.8*	+2.6 **
PB1 + *qGN4.1*	−0.5*	−0.5*	+2.2**

Significant at *$P < 0.01$, **$P < 0.001$ by t test

Conclusions

This study describes the development and evaluation of NILs with *qGN4.1* QTL for high grain number per panicle in the backgrounds of 12 different mega varieties of rice. Introgression of *qGN4.1* resulted in a significant increase in the number of grains per panicle in all the 12 backgrounds. The associated morphological traits such as flag leaf width and the number of primary and secondary branches in the panicle also increased significantly, but the total number of tillers was reduced. However, the number of productive tillers remained unchanged in all backgrounds except Pusa Basmati 1121 and Sarjoo 52, which showed a significant reduction in the number of productive tillers per plant. Even with the normal planting density, the QTL-NILs showed significantly higher yield as compared to the respective RPs in the field experiments for three varieties, namely Pusa Basmati 1121, Samba Mahsuri and Swarna. Our results showed that the *qGN4.1* QTL can be effectively deployed to further enhance the yield potential of popular mega varieties of rice.

Additional file

Additional file 1: Table S6. The sequence of DNA marker and its physical position using pseudomolecule release 7 of TIGR on chromosome 4 used for introgression of QTL *qGN4.1* into 12 different mega rice varieties.

Additional file 2: Table S1. Number of plants after foreground and recombinant selection for *qGN4.1* QTL for high grain number into 12 different mega varieties of rice at different backcross generations.

Additional file 3: Table S2. Adjusted means for different morphological agronomic traits of the six different *qGN4.1* QTL-NILs for each variety along with respective recipient parents in BC2F3 generation.

Additional file 4: Figure S1. Plant architecture of *qGN4.1* QTL-NILs (left side) of rice as compared to their recipient parents (right side): (D) IR 64 (E) MTU 1010 (F) HUR 105 (scale bars: 10 cm).

Additional file 5: Figure S2. Plant architecture of *qGN4.1* QTL-NILs (left side) of rice as compared to their recipient parents (right side): (G) Sarjoo 52 (H) Pusa 44 (I) CSR 30 (scale bars: 10 cm).

Additional file 6: Figure S3. Plant architecture of *qGN4.1* QTL-NILs (left side) of rice as compared to their recipient parents (right side): (J) Ranjit (K) CR 1009 (L) Pusa Basmati 1(PB 1) (scale bars: 10 cm). (TIFF 4098 kb)

Additional file 7: Table S3. Percentage genome similarity of the best eight *qGN4.1* QTL-NILs of Samba Mahsuri, Swarna, MTU 1010, Sarjoo 52, Pusa 44, Pusa Basmati 1 and best two lines of Pusa Basmati 1121, IR 64, HUR 105, CSR 30, Ranjit, CR 1009 with their respective recipient parent at BC3F2 generation in the genetic background of rice analysed using OsSNPnks SNP genotyping chip.

Additional file 8: Figure S4. Graphical representation of RPG similarity with *qGN4.1* QTL-NILs of (A) Samba Mahsuri, (B) Swarna (C) IR 64 (D) MTU 1010 (E) HUR 105(F) Sarjoo 52 (Blue colors denotes recipient segment, Red denotes donor segment and Green denotes Heterozygote).

Additional file 9: Figure S5. Graphical representation of RPG similarity with *qGN4.1* QTL-NILs of (G) Pusa 44 (H) CSR 30(I) Ranjit (J) CR 1009 (K) Pusa Basmati 1(PB 1) (Blue colors denotes recipient segment, Red denotes donor segment and Green denotes Heterozygotes).

Additional file 10: Table S5. Analysis of all the affymetrix 50 K SNP calls of 12 RPG along with *qGN4.1* QTL-NILs on separate sheets.

Additional file 11: Table S4. Yield performance of *qGN4.1* QTL-NILs in 11 different backgrounds in comparison to their recipient parents (data not available for CSR 30). Values are average of two replications, with LSD at 5%. SM = Samba Mahsuri; SW = Swarna.

Additional file 12: Table S7. Number of donor parent genome segments of larger than 0.2 Mb in the *qGN4.1*-NILs in recipient genetic backgrounds of 12 different mega varieties of rice.

Abbreviations

BC: Backcross; CD: Critical difference; DP: Donor parent; MABB: Marker-assisted backcross breeding; NIL: Near-isogenic line; NPT: New pant type; qRT-PCR: Quantitative Reverse transcriptase-polymerase chain reaction; QTL: Quantitative trait locus; RIL: Recombinant inbred line; RP: Recipient parent; RPG: Recipient parent genome; SNP: Single nucleotide polymorphism; SSR: Simple sequence repeat

Acknowledgements

We are thankful to Council of Scientific & Industrial Research (CSIR), Human Resource Development Group, Indian Council of Agricultural Research (ICAR) and Department of Biotechnology, Government of India for financial support.

Author's Contributions

The study is part of the Ph.D. thesis research of the first author. VKS: Conducted all experiments, SSR and SNP genotyping, morphological trait measurements, data analysis and manuscript writing; RKE did the statistical analysis and SNP genotyping; MN, AKS: Guiding field experiments; BDS Manuscript editing; NKS: Conceptualization of the study, planning, supervision, editing and finalization of the manuscript. All authors read and approved the final manuscript.

Competing Interests

The authors declare that they have no competing interest.

Author details

[1]ICAR-National Research Centre on Plant Biotechnology, Pusa Campus, New Delhi 110012, India. [2]Division of Genetics, ICAR-Indian Agricultural Research Institute (ICAR-IARI), Pusa Campus, New Delhi 110012, India. [3]ICAR-IARI-Rice Breeding and Genetics Research Centre, Aduthurai, Tamil Nadu 612101, India. [4]School of Biotechnology, Banaras Hindu University, Varanasi 221005, India.

References

Ashikari M, Sakakibara H, Lin S, Yamamoto T, Takashi T, Nishimura A, Angeles ER, Qian Q, Kitano H, Matsuoka M (2005) Cytokinin oxidase regulates rice grain production. Science 309(5735):741–745

Deshmukh R, Singh A, Jain N, Anand S, Gacche R, Singh A, Gaikwad K, Mohapatra T, Singh N (2010) Identification of candidate genes for grain number in rice (Oryza sativa L.) Funct Integr Genomics 10:339–347

Ellur RK, Khanna A, Yadav A, Pathania S et al (2015) Improvement of basmati rice varieties for resistance to blast and bacterial blight diseases using marker assisted backcross breeding. Plant Sci 242(2016):330–341

Fan C, Xing Y, Mao H, Lu T, Han B, Xu C, Li X, Zhang Q (2006) GS3, a major QTL for grain length and weight and minor QTL for grain width and thickness in rice, encodes a putative transmembrane protein. Theor App Genet 112(6):1164–1171

Federer WT (1956) Augmented designs. Hawaiian Planters Record 55:191–208

Federer WT (1961) Augmented designs with one way elimination of heterogeneity. Biometrics 17:447–473

Fujita D, Trijatmiko KR, Tagle AG, Sapasap MV et al (2013) NAL1 allele from a rice landrace greatly increases yield in modern indica cultivars. Proc Natl Acad Sci U S A 110:20431–20436

Huang X, Qian Q, Liu Z, Sun H, He S, Luo D, Xia G, Chu C, Li J, Fu X (2009) Natural variation at the DEP1 locus enhances grain yield in rice. Nat Genet 41(4):494–497

Ikeda-Kawakatsu K et al (2009) Expression level of ABERRANT PANICLE ORGANIZATION1 determines rice inflorescence form through control of cell proliferation in the meristem. Plant Physiol 150(2):736–747

Khush GS, Peng S (1996) Breaking the yield frontier of rice, pp 36–51

Ladha JK, Kirk GJD, Bennett J, Peng S, Reddy CK, Reddy PM, Singh U (1998) Opportunities for increased nitrogen-use efficiency from improved lowland rice germplasm. Field Crops Res 56:41–71

Livak KJ, Schmittgen TD (2001) Analysis of relative gene expression data using real time quantitative PCR and the 2ΔΔC(T) method. Methods 25(4):402–408

Marathi B, Guleria S, Mohapatra T, Parsad R, Mariappan N, Kurungara VK, Atwal SS, Prabhu KV, Singh NK, Singh AK (2012) QTL analysis of novel genomic regions associated with yield and yield related traits in new plant type based recombinant inbred lines of rice (Oryza sativa L.) BMC Plant Biol 12:137

Miura K et al (2010) OsSPL14 promotes panicle branching and higher grain productivity in rice. Nat Genet 42(6):545–549

Murray HG, Thompson WF (1980) Rapid isolation of high molecular weight DNA. Nucleic Acids Res (8):4321–4325

Ookawa T et al (2010) New approach for rice improvement using a pleiotropic QTL gene for lodging resistance and yield. Nat Commun 1:132

Parsad R, Gupta VK (2000) A note on augmented designs. Indian J Plant Genet Resources 13(1):53–58

Peng S, Cassman KG, Virmani SS, Sheehy J, Khush GS (1999) Yield potential trends of tropical rice since the release of IR8 and the challenge of increasing rice yield potential. Crop Sci 39(6):1552–1559

Peng S, Khush GS, Cassman KG (1994) Evolution of the new plant ideotype for increased yield potential. In: Cassman KG (ed) Breaking the yield barrier. IRRI, Los Baños, pp 5–20

Qi J, Qian Q, Bu QY, Li SY, Chen Q, Sun JQ, Liang WX, Zhou YH, Chu CC, Li XG et al (2008) Mutation of the rice Narrow leaf1 gene, which encodes a novel protein, affects vein patterning and polar auxin transport. Plant Physiol 147: 1947–1959

Sasaki A, Ashikari M, Ueguchi-Tanaka M, Itoh H et al (2002) A mutant gibberellin-synthesis gene in rice- new insight into the rice variant that helped to avert famine over thirty years ago. Nature 416:701

Sharma A (2013) Identification of Candidate Gene (s) for High Grain Number from a Major QTL Region on Rice Chromosome 4. Ph.D Thesis, Bansthali University Rajasthan, India http://shodhganga.inflibnet.ac.in/handle/10603/141928

Shomura A et al (2008) Deletion in a gene associated with grain size increased yields during rice domestication. Nat Genet 40(8):1023–1028

Singh N, Jayaswal PK, Panda K, Mandal P, Kumar V et al (2015) Single-copy gene based 50 K SNP chip for genetic studies and molecular breeding in rice. Sci Rep 5(2015):11600

Singh R, Singh Y, Xalaxo S, Verulkar S et al (2015) From QTL to variety-harnessing the benefits of QTLs for drought, flood and salt tolerance in mega rice varieties of India through a multi-institutional network. Plant Sci 242:278–287

Song XJ, Huang W, Shi M, Zhu MZ, Lin HX (2007) A QTL for rice grain width and weight encodes a previously unknown RING-type E3 ubiquitin ligase. Nat Genet 39(5):623–630

Takai T, Adachi S et al (2013) A natural variant of NAL1, selected in high-yield rice breeding programs, pleiotropically increases photosynthesis rate. Sci Rep 3:2149

Takeda S, Matsuoka M (2008) Genetic approaches to crop improvement: responding to environmental and population changes. Nat Rev Genet 9:444–457

Terao T, Nagata K, Morino K, Hirose T (2010) A gene controlling the number of primary rachis branches also controls the vascular bundle formation and hence is responsible to increase the harvest index and grain yield in rice. Theor Appl Genet 120(5):875–893

Xue W et al (2008) Natural variation in Ghd7 is an important regulator of heading date and yield potential in rice. Nat Genet 40(6):761–767

Yamagishi T, Peng S, Cassman KG, Ishii R (1996) Studies on grain filling characteristics in "New Plant Type" rice lines developed in IRRI. Japanese J Crop Sci 65 (Extra issue No. 2):169–170

Zhang GH, Li SY, Wang L, Ye WJ et al (2014) Lschl4 from japonica cultivar, which is allelic to nal1, increases yield of indica super rice 93-11. Mol Plant 7:1350–1364

11

QTL mapping and candidate gene analysis of peduncle vascular bundle related traits in rice by genome-wide association study

Laiyuan Zhai[1], Tianqing Zheng[2], Xinyu Wang[1], Yun Wang[2], Kai Chen[2,3], Shu Wang[1], Yun Wang[1*], Jianlong Xu[2,3*] [iD] and Zhikang Li[2,4]

Abstract

Background: The vascular bundle especially in the peduncle is one of crucial limiting factors of rice yield, and it determines how plants efficiently transport photosynthetic products, mineral nutrients and water from leaf and root to panicle. However, the genetic base of rice vascular bundle related traits in the peduncle still remains unknown.

Results: The 423 panel showed substantial natural variations of peduncle vascular bundle. In total, 48 quantitative trait loci/locus (QTL) affecting the eight traits were identified throughout the genome by applying a significance threshold of $P < 1.0 \times 10^{-4}$. Combined determining linkage disequilibrium (LD) blocks associated with significant SNPs and haplotype analyses allowed us to shortlist six candidate genes for four important QTL regions affecting the peduncle vascular bundle traits, including one cloned gene (*NAL1*) and three newly identified QTL (*qLVN6*, *qSVN7*, and *qSVA8.1*). Further the most likely candidate genes for each important QTL were also discussed based on functional annotation.

Conclusions: Genetic base on peduncle vascular bundle related traits in rice was systematically dissected, and most likely candidate genes of the known gene *NAL1* and the three newly identified QTL (*qLVN6*, *qSVN7*, and *qSVA8.1*) were analyzed. The results provided valuable information for future functional characterization and rice breeding for high yield through optimizing transportation efficiency of photosynthetic products by marker-assisted selection.

Keywords: GWAS, Vascular bundle, SNPs, Candidate genes, Quantitative trait loci/locus (QTL)

Background

Rice (*Oryza sativa* L.) is one of the most important cereal crops and a staple food for more than one-half of the world's population. In retrospect of rice breeding history, good harmony of source, sink and translocation capacity (i.e. flow) of assimilates plays an important role in improvement of rice yield potential (Donald 1968; Lafitte and Travis 1984; Ashraf et al. 1994). Among them, three top leaves especially flag leaf are main source traits of assimilate synthesis while spikelets including grain number and size are main traits of accumulation of assimilate. The vascular bundle especially in the peduncle is the transport system that links source and sink, and it determines how plants efficiently transport photosynthetic products, mineral nutrients and water from source to sink (Housley and Peterson 1982). Rice breeding history indicated that good balance among source, sink and flow of assimilates, namely, large sink, sufficient source and flow fluency is prerequisite for high yield potential of rice (Ashraf et al. 1994).

The vascular bundle systems in plants play a significant role in delivering water and nutrients as well as other substances needed for growth and defense (Lucas et al. 2013). Vascular bundles, which interconnect all parts of the plant, extending from the stem into leaves, and down into the root system. It consists of xylem and phloem, two differentiated conductive tissues, as well as

* Correspondence: wangyunli55555@163.com; xujianlong@caas.cn
[1]Rice Research Institute, Shenyang Agricultural University/Key Laboratory of Northern Japonica Rice Genetics and Breeding, Ministry of Education, Shenyang 110866, China
[2]Institute of Crop Sciences/National Key Facility for Crop Gene Resources and Genetic Improvement, Chinese Academy of Agricultural Sciences, 12# South Zhong-Guan-Cun Street, Haidain District, Beijing 100081, China
Full list of author information is available at the end of the article

undifferentiated cambial or procambial stem cells. Xylem transports water and dissolved minerals, whereas phloem is required for the distribution of mainly photosynthetic products (sugars, RNA, proteins, and other organic compounds) from source to sink organs (Dettmer et al. 2009). A significant positive correlation between grain yield and the number of vascular bundles and the area of phloem in peduncle have been reported in rice (Ashraf et al. 1994), wheat (Evans et al. 1970) and oat (Housley and Peterson. 1982). The capacity of the vascular bundle system for efficiently transporting assimilates from the source to the sink has been shown as a limiting factor for yield potential realization (Housley and Peterson 1982). Additionally, the differentiation in vascular bundle system is an important parameter that defines differentiation between *indica* and *japonica* rice (Fukuyama et al. 1996). Generally, *indica* varieties tend to have more number of large vascular bundle (LVN) than *japonica* varieties, though there is no significant difference in the number of small vascular bundle (SVN) between them. The exploitation of the great heterosis that exists in the inter-subspecific crosses between *indica* and *japonica* rice has long been considered as a promising way to increase the yield potential. It is therefore of importance to study the genetic basis of peduncle vascular bundle to provide useful information for the genetic improvement of rice yield potential using heterosis between *indica* and *japonica*.

The use of QTL mapping has contributed to a better understanding of the genetic basis of many agronomic important traits. For past decades, many researchers have identified QTL for peduncle vascular bundle related traits in rice (Sasahara et al. 1999; Zhang et al. 2002; Cui et al. 2003; Bai et al. 2012). Notably, several QTL affecting the vascular bundle system in rice have been further cloned. For instance, genes such as *APO1* (Terao et al. 2010), *AVB* (Ma et al. 2017) and *NAL1* (Qi et al. 2008; Fujita et al. 2013) have been reported to involve in enhanced translocation capacity of vascular bundles.

Cloning QTL affecting complex traits has been a major challenge to plant geneticists and molecular biologists since the classical strategy using map-based cloning of QTL is extremely troublesome and time-consuming. In recent years, GWAS, based on the genome sequencing and SNP chip technology, makes large-scale high-precision identification to the alleles and the variation which are widely distributed in natural groups of rice become reality (Huang et al. 2010; Zhao et al. 2011; Han and Huang. 2013). In the present study, we reported the first GWAS for vascular bundle system in any agricultural crop using a panel of germplasm resources with a broad genetic diversity. A diverse panel consisting of 423 accessions selected from 3 K Rice Genome Project (3 K RGP) (3K RGP 2014; Sun et al. 2017) was evaluated in 2015 and 2016. GWAS was performed using 27 K SNPs generated from 3 K RGP through high-throughput sequencing technologies (Zheng et al. 2015). And then the candidate genes based on the SNPs significantly correlated to the vascular bundles through GWAS can be identified.

Results

Trait variance and correlations

In general, most of the vascular bundle related traits in peduncle appeared to be normally distributed, but a few traits, phloem area of large vascular bundle (LVPA) and phloem area of small vascular bundle (SVPA) showed skewed distributions in 2016 (Fig. 1a). The panel presented substantial variations for all the measured traits. The phenotype pairwise correlations between the measured traits were similar in both years. In 2015 and 2016, strong positive correlations were observed for all area related traits (total area, phloem area, and xylem area) between large and small vascular bundle, indicating the consistent effects of vascular bundle on these traits across all accessions. As expected, the vascular bundle total area had significant positive correlations with its corresponding component traits (phloem area and xylem area) whereas no significant correlations were detected between the vascular bundle total area and the number of vascular bundle (Fig. 1b).

Basic statistics of markers

For the 26,097 high quality SNPs, the number of markers per chromosome ranged from 1600 on chromosome 9 to 3101 on chromosome 1. The size of chromosome varied from 22.93 Mb for chromosome 9 to 43.25 Mb for chromosome 1. The whole genome size was 372.50 Mb. The average marker spacing was 14.27 kb with spacing ranging from 12.68 kb for chromosome 10 to 15.50 kb for chromosome 4 (Additional file 1: Figure S1).

Population structure

The screen plot showed that two principal component (PC) can be reserved as informative (Fig. 2a–b; Additional file 2: Table S1), where downhill changed gradually (Fig. 2a). The score plot of PC showed the distribution of each accession in the rice diversity panel (Fig. 2b). Additionally, the highest log likelihood scores of the population structure were observed when the number of populations was set at 2 (K = 2; Fig. 2c), indicating that there were two distinct subpopulations (Pop I and Pop II) in the current panel (Fig. 2d; Additional file 2: Table S1). And the heat

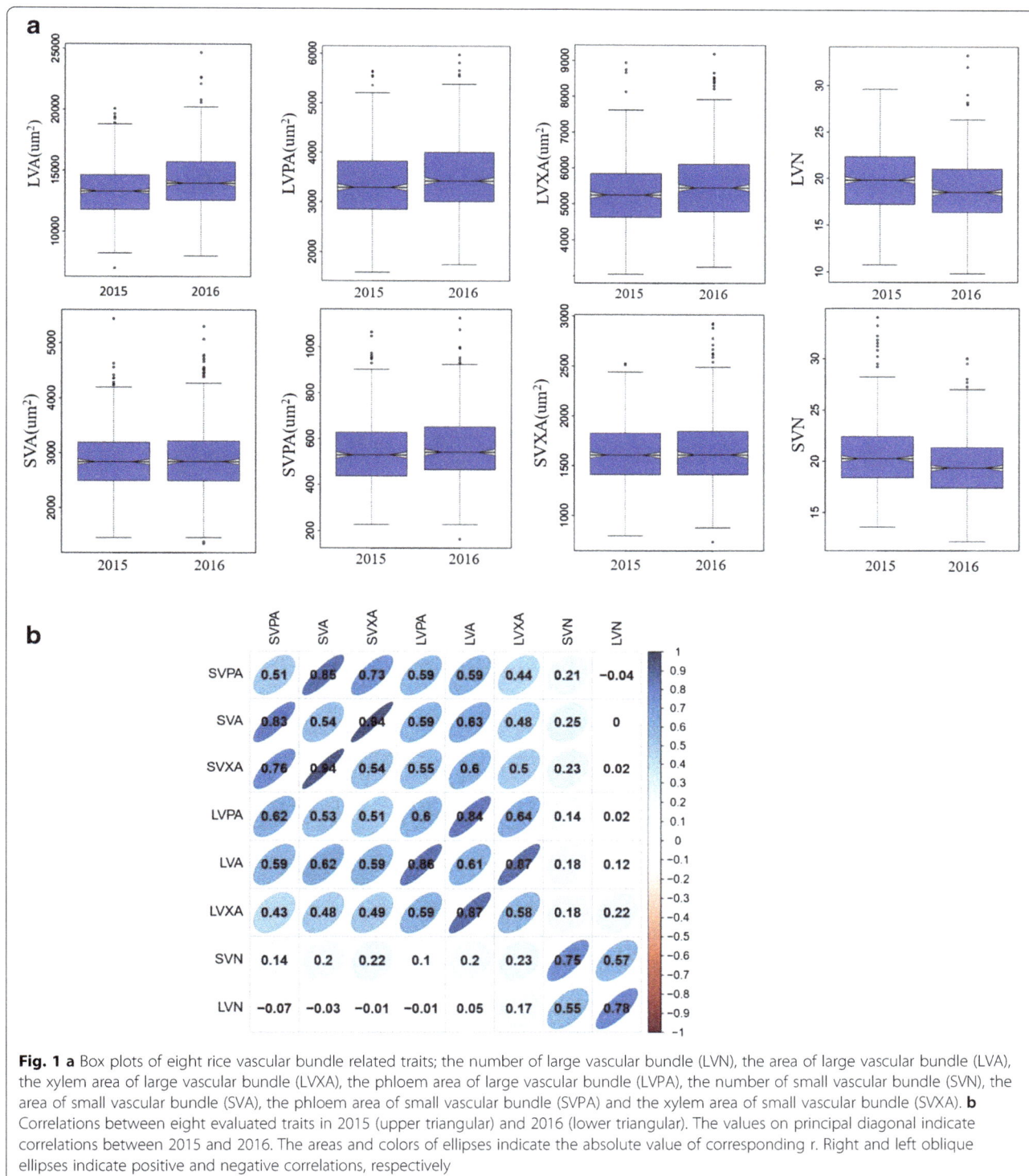

Fig. 1 a Box plots of eight rice vascular bundle related traits; the number of large vascular bundle (LVN), the area of large vascular bundle (LVA), the xylem area of large vascular bundle (LVXA), the phloem area of large vascular bundle (LVPA), the number of small vascular bundle (SVN), the area of small vascular bundle (SVA), the phloem area of small vascular bundle (SVPA) and the xylem area of small vascular bundle (SVXA). **b** Correlations between eight evaluated traits in 2015 (upper triangular) and 2016 (lower triangular). The values on principal diagonal indicate correlations between 2015 and 2016. The areas and colors of ellipses indicate the absolute value of corresponding r. Right and left oblique ellipses indicate positive and negative correlations, respectively

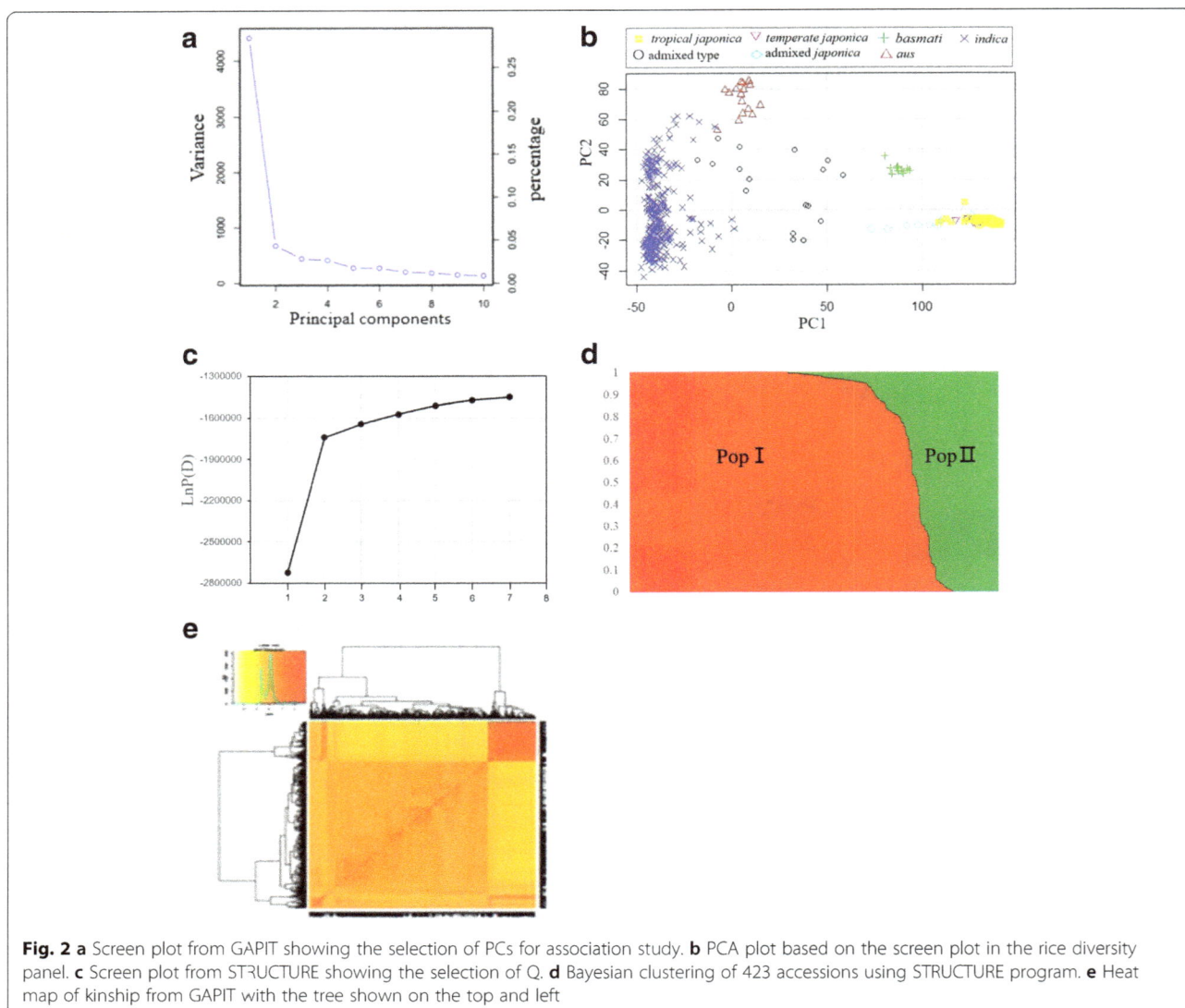

Fig. 2 a Screen plot from GAPIT showing the selection of PCs for association study. **b** PCA plot based on the screen plot in the rice diversity panel. **c** Screen plot from STRUCTURE showing the selection of Q. **d** Bayesian clustering of 423 accessions using STRUCTURE program. **e** Heat map of kinship from GAPIT with the tree shown on the top and left

map of kinship relatedness matrix (KI) showed the same result (Fig. 2e).

Detection of QTL by GWAS

A total of 48 different QTL for all traits were identified in 2015 and 2016, ranging from two QTL for SVPA to 10 QTL for LVPA. Among them, 23 (37) QTL were detected in 2015 (2016) and 12 QTL were commonly detected in two years (Table 1; Additional file 3: Figure S2).

For LVN, seven QTL were detected on chromosomes 1, 2, 4, 6 and 11. Among them, *qLVN2.1* and *qLVN4* were detected in 2015, and accounted for 2.4% and 2.2% of phenotypic variance, respectively. Four QTL, *qLVN1*, *qLVN2.2*, *qLVN2.3* and *qLVN11*, were detected in 2016, and explained 2.2% to 3.0% of phenotypic variance. Only *qLVN6* was detected in both 2015 and 2016, and

explained the average phenotypic variance of 2.2% (Table 1; Additional file 3: Figure S2).

For total area of large vascular bundle (LVA), seven QTL were detected on chromosomes 1, 3, 4, 6, 8 and 11. The *qLVA6* with 3.4% of phenotypic variance was only detected in 2015 and three QTL (*qLVA1*, *qLVA3*, and *qLVA4.2*) were only detected in 2016, which were accounted for phenotypic variance from 3.5% to 3.9%. Three QTL, *qLVA4.1*, *qLVA8* and *qLVA11*, were detected in both years and explained the average phenotypic variance of 2.6%, 3.1% and 3.2%, respectively (Table 1; Additional file 3: Figure S2).

Ten QTL for LVPA were detected on chromosomes 1, 2, 4, 8 and 11. Three QTL (*qLVPA8.2*, *qLVPA8.3* and *qLVPA11*) were detected in 2015, and accounted for 3.2% to 3.6% of phenotypic variance. Five QTL (*qLVPA1.1*, *qLVPA1.2*, *qLVPA2*, *qLVPA4.2* and *qLVPA8.1*)

Table 1 QTL identified for eight peduncle vascular bundle related traits in two years

Trait[a]	QTL	Year	Peak SNP	Allele[b]	P	Effect[c]	R^2 (%)[d]
LVN	qLVN1	2016	Chr1-13,801,593	T/A	7.8E-05	−0.82	2.4
	qLVN2.1	2015	Chr2–31,993	A/G	4.8E-05	−0.89	2.4
	qLVN2.2	2016	Chr2–25,871,069	A/G	5.5E-05	1.64	2.5
	qLVN2.3	2016	Chr2–29,521,254	T/C	8.8E-05	−1.56	2.2
	qLVN4	2015	Chr4–5,193,957	G/A	8.1E-05	0.77	2.2
	qLVN6	2015	Chr6–28,662,981	C/T	9.4E-05	−0.67	2.0
		2016	Chr6–28,639,502	G/C	9.5E-05	−0.51	2.4
	qLVN11	2016	Chr11–1,063,214	C/T	1.3E-05	1.37	3.0
LVA	qLVA1	2016	Chr1–2,179,066	A/G	4.4E-05	− 617.1	3.8
	qLVA3	2016	Chr3–33,078,718	T/C	8.2E-05	− 714.3	3.5
	qLVA4.1	2015	Chr4–31,081,174	C/A	9.6E-05	− 392.1	1.4
		2016	Chr4–31,068,550	G/C	4.1E-05	− 1073.3	3.8
	qLVA4.2	2016	Chr4–32,189,616	T/C	3.8E-05	− 960.8	3.9
	qLVA6	2015	Chr6–20,057,366	C/T	6.1E-05	1362.6	3.4
	qLVA8	2015	Chr8–3770	G/A	6.4E-05	− 524.4	2.5
		2016	Chr8–69,600	C/T	6.4E-05	− 620.0	3.6
	qLVA11	2015	Chr11–21,798,323	T/G	7.1E-05	463.6	3.0
		2016	Chr11–21,798,323	T/G	9.0E-05	582.6	3.4
LVPA	qLVPA1.1	2016	Chr1–1,806,093	C/A	6.4E-05	252.9	3.5
	qLVPA1.2	2016	Chr1–2,179,066	A/G	8.7E-05	− 175.1	3.4
	qLVPA2	2016	Chr2–30,012,958	C/G	8.0E-05	− 508.7	3.4
	qLVPA4.1	2015	Chr4–31,081,174	C/A	6.7E-05	−83.1	3.0
		2016	Chr4–31,068,550	G/C	3.0E-05	− 322.3	3.9
	qLVPA4.2	2016	Chr4–32,189,616	T/C	6.1E-05	−277.5	3.6
	qLVPA8.1	2016	Chr8–3770	G/A	4.5E-05	− 219.5	3.7
	qLVPA8.2	2015	Chr8–2,393,328	T/A	3.8E-05	−524.5	3.6
	qLVPA8.3	2015	Chr8–27,101,483	C/T	9.5E-05	220.4	3.2
	qLVPA8.4	2015	Chr8–27,737,011	C/T	2.9E-05	187.6	3.7
		2016	Chr8–27,707,779	C/T	9.2E-05	−190.4	2.8
	qLVPA11	2015	Chr11–25,611,048	C/A	3.9E-05	−360.2	3.5
LVXA	qLVXA1	2016	Chr1–2,758,104	A/G	5.8E-05	− 514.4	3.7
	qLVXA5	2015	Chr5–4,823,527	C/T	8.4E-05	− 224.9	3.2
		2016	Chr5–4,823,527	C/T	3.4E-05	− 287.8	3.9
	qLVXA6	2016	Chr6–21,044,730	C/A	5.9E-05	415.6	3.7
	qLVXA8.1	2015	Chr8–87,741	T/C	5.4E-05	220.6	2.8
		2016	Chr8–69,600	C/T	1.4E-06	− 325.9	5.3
	qLVXA8.2	2015	Chr8–220,938	C/T	3.0E-05	195.8	2.3
		2016	Chr8–220,938	C/T	2.7E-05	282.4	4.0
	qLVXA8.3	2015	Chr8–27,031,242	T/A	8.6E-05	439.9	3.2
	qLVXA11	2015	Chr11–21,798,323	T/G	1.6E-05	245.6	3.9
		2016	Chr11–21,798,323	T/G	2.7E-05	222.9	2.7
SVN	qSVN1	2016	Chr1–9,724,654	T/C	4.3E-05	−1.76	3.6
	qSVN2	2016	Chr2–25,915,457	C/A	5.5E-05	−1.98	3.5
	qSVN5.1	2015	Chr5–24,874,414	A/G	3.3E-05	−1.69	3.9

Table 1 QTL identified for eight peduncle vascular bundle related traits in two years *(Continued)*

Trait[a]	QTL	Year	Peak SNP	Allele[b]	P	Effect[c]	R² (%)[d]
		2016	Chr5–24,874,414	A/G	2.7E-06	−1.77	4.7
	qSVN5.2	2016	Chr5–25,915,442	G/A	5.7E-05	1.97	3.5
	qSVN7	2015	Chr7–29,604,076	C/T	4.7E-05	1.27	2.8
		2016	Chr7–29,604,076	C/T	1.9E-06	1.69	4.9
SVA	qSVA2.1	2015	Chr2–10,294,412	G/A	8.0E-05	139.9	3.7
	qSVA2.2	2016	Chr2–11,856,232	C/A	3.7E-05	−342.0	3.7
	qSVA8.1	2016	Chr8–2,031,279	C/T	3.0E-05	− 258.4	3.8
	qSVA8.2	2016	Chr8–20,982,368	G/T	7.9E-05	−251.3	3.4
	qSVA11	2015	Chr11–25,611,048	C/A	6.67E-05	−303.8	3.6
	qSVA12	2016	Chr12–3,630,490	C/T	5.17E-05	− 234.6	3.5
SVPA	qSVPA4.1	2016	Chr4–13,422,276	C/T	6.53E-05	110.0	3.7
	qSVPA4.2	2016	Chr4–31,878,154	A/G	3.56E-05	38.7	3.9
SVXA	qSVXA2	2016	Chr2–30,012,958	C/G	6.4E-05	−265.8	3.5
	qSVXA8	2016	Chr8–2,031,279	C/T	9.74E-05	− 145.4	3.4
	qSVXA9	2015	Chr9–16,464,178	C/T	1.28E-05	−149.5	4.4
	qSVXA11	2015	Chr11–25,611,048	C/A	1.13E-05	−88.2	4.5

[a]LVN the number of large vascular bundle, LVA the total area of large vascular bundle; LVPA the phloem area of large vascular bundle, LVXA the xylem area of large vascular bundle, SVN the number of small vascular bundle, SVA the total area of small vascular bundle, SVPA the phloem area of small vascular bundle, SVXA the xylem area of small vascular bundle

[b]Major/Minor allele

[c]Effect: Allele effect with respect to the minor allele

[d]R² (%): Phenotypic variance explained

were detected in 2016 and accounted for phenotypic variance from 3.4% to 3.7%. Another two QTL, *qLVPA4.1* and *qLVPA8.4* were commonly detected in the two years, and explained the average phenotypic variance 3.5% and 3.3%, respectively (Table 1; Additional file 3: Figure S2).

For xylem area of large vascular bundle (LVXA), seven QTL were detected on chromosomes 1, 5, 6, 8, and 11. Among them, only *qLVXA8.3* was detected in 2015 and explained 3.2% of phenotypic variance. Two QTL, *qLVXA1* and *qLVXA6*, were detected in 2016 and accounted for both 3.7% of phenotypic variance. Four QTL, namely, *qLVXA5*, *qLVXA8.1*, *qLVXA8.2* and *qLVXA11*, were simultaneously detected in the two years and explained the average phenotypic variance of 3.6%, 4.1%, 3.2% and 3.3%, respectively (Table 1; Additional file 3: Figure S2).

For SVN, five QTL were detected on chromosomes 1, 2, 5, and 7, including three (*qSVN1*, *qSVN2* and *qSVN5.2*) detected in 2016 and two (*qSVN5.1* and *qSVN7*) in the two years. The first three QTL accounted for phenotypic variance from 3.5% to 3.6% while the latter two QTL explained the average phenotypic variance of 4.3% and 3.9%, respectively (Table 1; Additional file 3: Figure S2).

For total area of small vascular bundle (SVA), a total of six QTL were detected on chromosomes 2, 8, 11, and 12. Two QTL (*qSVA2.1* and *qSVA11*) were detected in

2015, and accounted for 3.7% and 3.6% of phenotypic variance, respectively. The other four QTL (*qSVA2.2*, *qSVA8.1*, *qSVA8.2* and *qSVA12*) were detected in 2016, and accounted for 3.4% to 3.8% of phenotypic variance (Table 1; Additional file 3: Figure S2).

For SVPA, two QTL, *qSVPA4.1* and *qSVPA4.2* were detected on chromosome 4 in 2016, and accounted for 3.7% and 3.9% of phenotypic variance, respectively (Table 1; Additional file 3: Figure S2).

For xylem area of small vascular bundle (SVXA), four QTL were detected on chromosomes 2, 8, 9 and 11. Of them, two QTL (*qSVXA9* and *qSVXA11*) were detected in 2015 and accounted for 4.4% and 4.5% of phenotypic variance, respectively. Another two QTL (*qSVXA2* and *qSVXA8*) were detected in 2016 and explained 3.5% and 3.4% of phenotypic variance, respectively (Table 1; Additional file 3: Figure S2). There was no any QTL simultaneously detected in the two years.

Candidate gene analysis

For eight peduncle vascular bundle related traits, we conducted gene-based association and haplotype analysis to identify candidate genes for important QTL identified using the whole population. In addition, considering that among the eight peduncle vascular bundle related traits LVN, SVA, SVXA and SVPA were both significantly different between *indica* and *japonica* in 2015 and 2016 (Additional file 4: Figure S3), we conducted haplotype

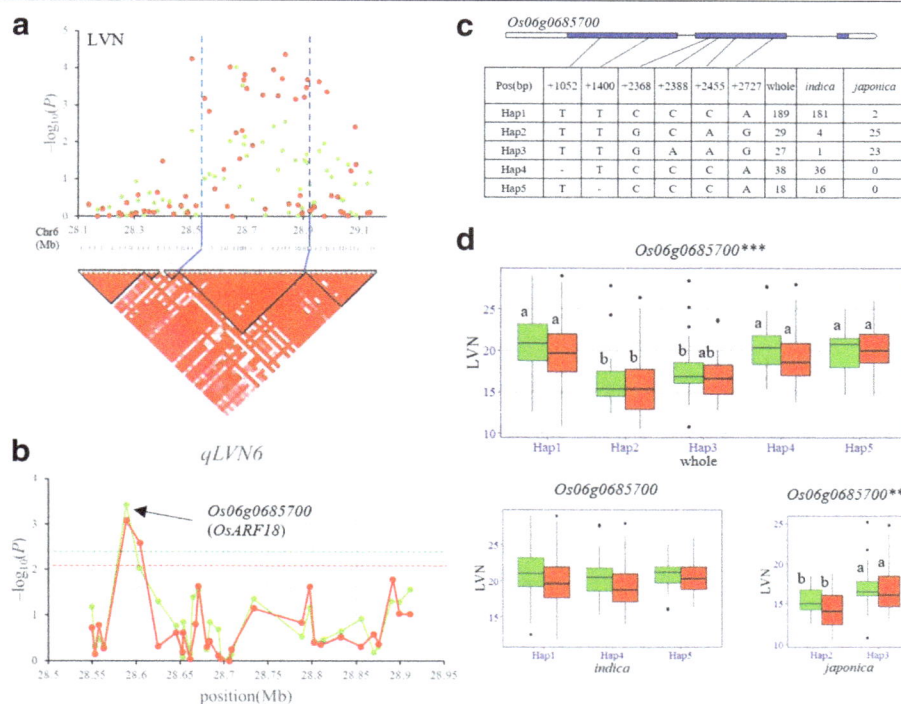

Fig. 3 Gene-based association analysis and haplotype analysis of targeted genes related to *qLVN6*. **a** Local Manhattan plot (top) and LD block (bottom) surrounding the peak on chromosome 6. Dashed lines indicate the candidate region for the peak. **b** Gene-based association analysis of targeted genes related to *qLVN6*. Each point is a gene indicated by one SNP having largest −log10 (*P*) value. Dash line show the threshold to determine significant SNP. **c** Exon-intron structure of *Os06g0685700* and DNA polymorphism in that gene. **d** Boxplots for LVN based on the haplotypes (Hap) for *Os06g0685700* in the whole, *indica* and *japonica* populations. The ** and *** suggest significance of ANOVA at *P* < 0.01 and *P* < 0.001, respectively. The letter on histogram (a, and b) indicate multiple comparisons result at the significant level 0.05. Green and red colors indicate in 2015 and 2016, respectively

analysis of targeted genes related to LVN, SVA, SVXA and SVPA in *indica* and *japonica* subpopulations, respectively.

Using gene-based association analysis and haplotype analysis of candidate gene, six important candidate genes were identified for the four QTL regions distributing on chromosomes 4, 6, 7 and 8, ranging from one to two candidate genes for each QTL. For *qLVN6* in the candidate region of 28.55–28.92 Mb (370 kb) on chromosome 6 (Fig. 3a), 616 non-synonymous SNPs of 52 genes were used for association analysis according to the Rice Genome Annotation Project Database (RAP-DB). The peak SNP was at 28588881 bp on chromosome 6 with −$\log_{10}$ (*P*) of 3.43 and 3.08 in 2015 and 2016, respectively. Again, SNPs with −$\log_{10}$ (*P*) above the threshold (2.43 in 2015 and 2.08 in 2016) were found to locate in one gene, *Os06g0685700* (Fig. 3b). Significant differences for LVN between different haplotypes were observed at *Os06g0685700* in whole population (Fig. 3c–d).In this study, haplotype was separately analyzed for *indica* and *japonica* varieties owing to the significant differentiation in LVN between the two subspecies.

Highly significant differences in LVN were detected between different haplotypes at the candidate gene (*Os06g0685700*) in *japonica*, though there was no significant difference for LVN between different haplotypes in *indica* accessions (Fig. 3c–d).

Similarly, the candidate region was predicted in the region from 31.07 to 31.26 Mb (190 kb) on chromosome 4 harboring *qLVA4.1* and *qLVPA4.1* (Fig. 4a), and 303 non-synonymous SNPs of 30 genes were used for association analysis. The peak SNP was at 31079854 bp with −$\log_{10}$ (*P*) of 3.73 (3.47) in 2015 (2016) for LVA, and 3.49 (3.21) in 2015 (2016) for LVPA. Two genes harboring SNPs with −$\log_{10}$ (*P*) larger than the thresholds were identified, including *Os04g0613000* and *Os04g0615000* (Fig. 4b). Highly significant differences for *qLVA4.1* and *qLVPA4.1* between different haplotypes were observed at the two genes (*Os04g0613000* and *Os04g0615000*) in whole population (Fig. 4c–d). Of the two genes, *Os04g0615000* is identical to *NAL1* (*NARROW LEAF1*), the gene regulating the development of vascular bundle in previously reported (Qi et al. 2008). Three haplotypes of *NAL1* were found and haplotype

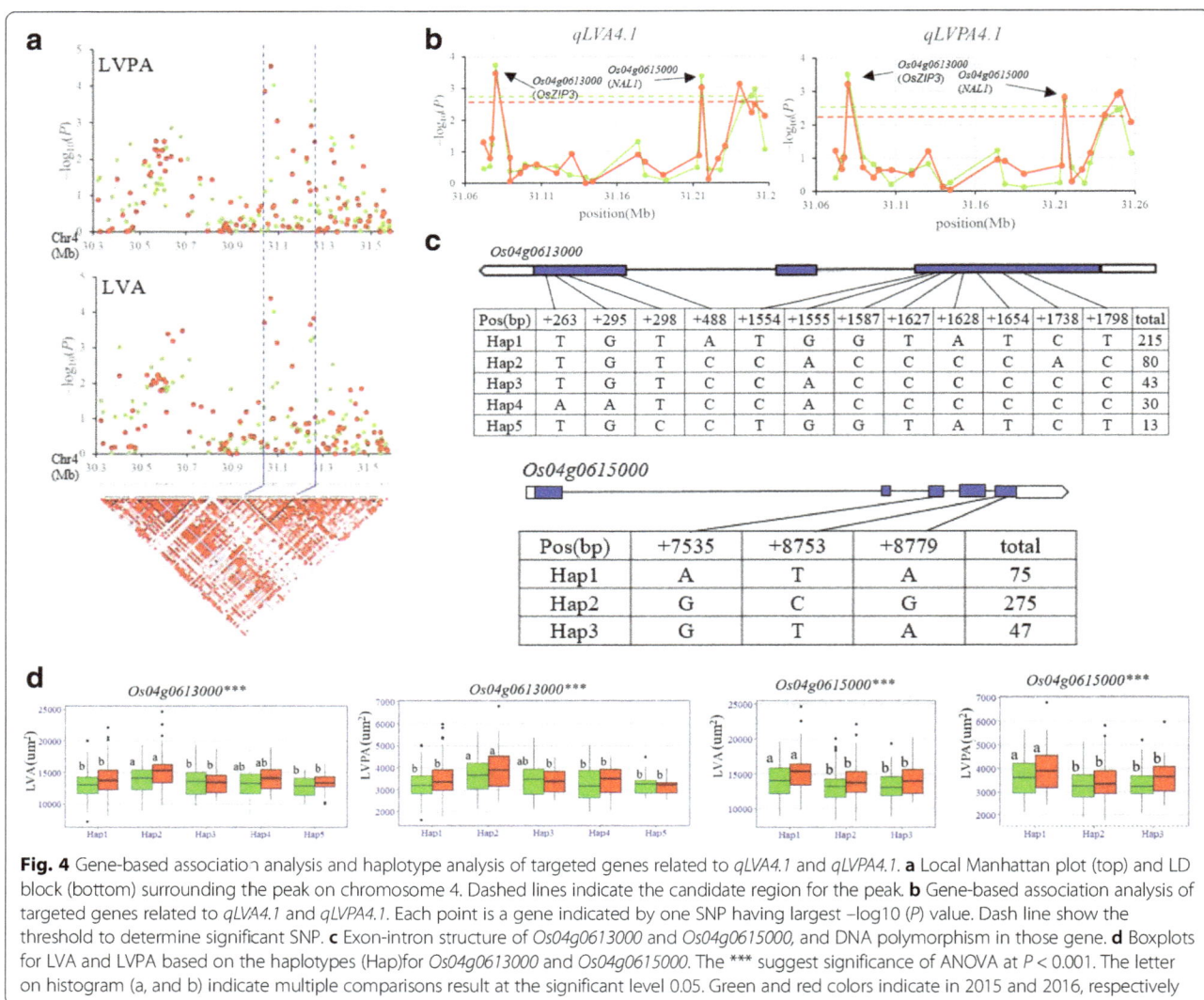

Fig. 4 Gene-based association analysis and haplotype analysis of targeted genes related to *qLVA4.1* and *qLVPA4.1*. **a** Local Manhattan plot (top) and LD block (bottom) surrounding the peak on chromosome 4. Dashed lines indicate the candidate region for the peak. **b** Gene-based association analysis of targeted genes related to *qLVA4.1* and *qLVPA4.1*. Each point is a gene indicated by one SNP having largest −log10 (*P*) value. Dash line show the threshold to determine significant SNP. **c** Exon-intron structure of *Os04g0613000* and *Os04g0615000*, and DNA polymorphism in those gene. **d** Boxplots for LVA and LVPA based on the haplotypes (Hap)for *Os04g0613000* and *Os04g0615000*. The *** suggest significance of ANOVA at *P* < 0.001. The letter on histogram (a, and b) indicate multiple comparisons result at the significant level 0.05. Green and red colors indicate in 2015 and 2016, respectively

ATA was associated with significantly larger LVA and LVPA than haplotypes GCG and GTA (Fig. 4c–d).

QSVN7 was fine-mapped in the region of 29.46–29.67 Mb (210 kb) on chromosome 7 (Fig. 5a), and 353 non-synonymous SNPs of 28 genes were used for association analysis. The peak SNP for *qSVN7* was at 29626817 bp on chromosome 7 with −log$_{10}$ (*P*) of 3.57 and 3.34 in 2015 and 2016, respectively. SNPs with −log$_{10}$ (*P*) above the thresholds were located in the two candidate genes, *Os07g0695100* and *Os07g0694700* (Fig. 5b). No significant haplotypes were found in *Os07g0694700*. Five haplotypes were found for *Os07g0695100*. Haplotypes analysis revealed that significant differences for SVN were observed between haplotypes GCGGTAGCGG and CTGACGGTGA in both years (Fig. 5c–d).

On chromosome 8, a high peak of *qSVA8.1* was found to be overlapped with the peak of *qSVXA8*. We predicted the candidate region from 1.99–2.17 Mb (180 kb)

(Fig. 6a), and 277 non-synonymous SNPs of 29 genes were used for association analysis. The peak SNP was at 2079621 bp on chromosome 8 with −log$_{10}$ (*P*) of 3.56 for *qSVA8.1* and 3.50 for *qSVXA8* in 2016. Two genes harboring the SNPs with −log$_{10}$ (*P*) larger than the threshold (2.56 and 2.50 for *qSVA8.1* and *qSVXA8*, respectively) were identified, including *Os08g0137100* and *Os08g0137250* (Fig. 6b). Significant differences were detected among different haplotypes at the two genes in whole population (Fig. 6c–d). However, no significant difference in SVA and SVXA was detected between different haplotypes at the two candidate genes in *indica* subpopulation, and only one prevalent haplotype of each gene was detected in *japonica* subpopulation (Additional file 5: Table S2). Among them, *Os08g0137100* is identical to *OsFIE2*, a gene affecting vascular bundle related traits in previously report (Liu et al. 2016).

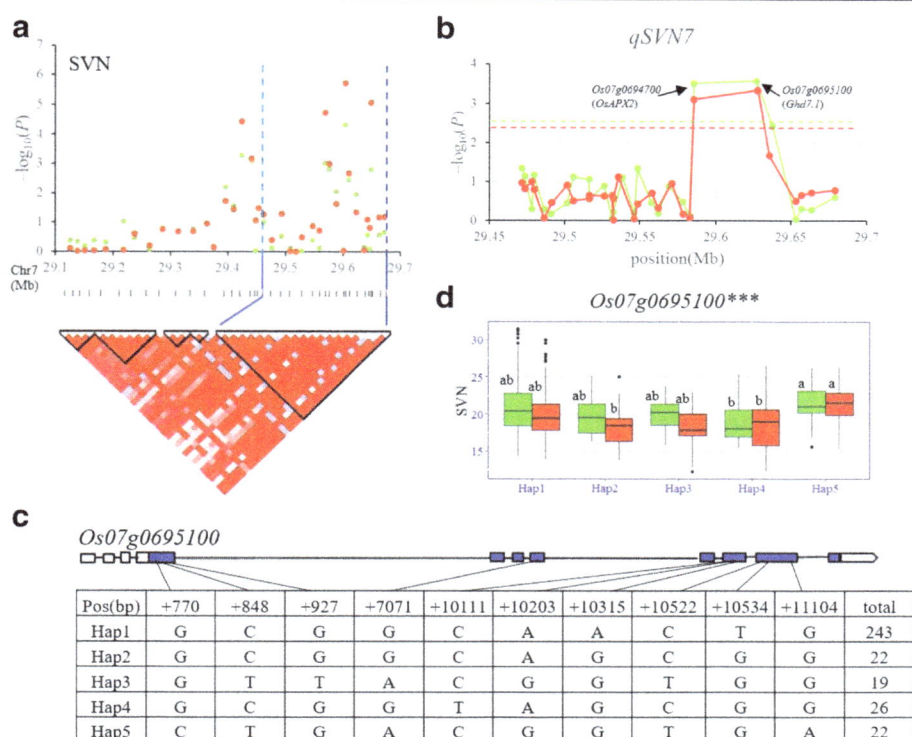

Fig. 5 Gene-based association analysis and haplotype analysis of targeted genes related to *qSVN7*. **a** Local Manhattan plot (top) and LD block (bottom) surrounding the peak on chromosome 7. Dashed lines indicate the candidate region for the peak. **b** Gene-based association analysis of targeted genes related to *qSVN7*. Each point is a gene indicated by one SNP having largest −log10 (*P*) value. Dash line show the threshold to determine significant SNP. **c** Exon-intron structure of *Os07g0695100* and DNA polymorphism in that gene. **d** Boxplots for LVN based on the haplotypes (Hap) for *Os07g0695100*. The *** suggest significance of ANOVA at *P* < 0.001. The letter on histogram (**a**, and **b**) indicate multiple comparisons result at the significant level 0.05. Green and red colors indicate in 2015 and 2016, respectively

Discussion

Characteristics of vascular bundle between *Indica* and *Japonica*

Indica and *japonica* are the two major types of Asian cultivated rice (*Oryza sativa* L.). There are significant differences between them in many morphological and physiological traits associated with the processes of *indica-japonica* differentiation (Morishima and Oka 1981). Among them, the differentiation in vascular bundle system is an important parameter that defines differentiation between the two subspecies. *Indica* varieties tend to have more large vascular bundles in peduncle than *japonica* varieties, and have significant higher ratio of the LVN to the number of primary rachis branches (ranged from 1.3 to 2.2) than *japonica* (nearly1.0) (Fukuyama and Takayama 1995). Our studies showed that LVN in peduncle of *indica* (averaged 21 in 2015 and 19.9 in 2016) was also significantly higher than those of *japonica* (averaged 16.1 in 2015 and 15.5 in 2016) (Additional file 4: Figure S3). In *japonica*-type cultivars, each of the large vascular bundles is directly connected to a primary rachis branch (PRB), resulting almost in same quantity of LVN in the peduncle and number of

PRBs. On the other hand, in *indica*-type some large vascular bundles are directly connected to the secondary rachis branches due to surplus large vascular bundles (Fukushima and Akita 1997; Terao et al. 2010). The different vascular bundle characteristics between *indica* and *japonica* probably explained inherently shorter grain development phase (Osada et al. 1983) or shorter grain filling period (Nagato and Chaudhry 1969) in *indica* than *japonica* resulting from more rapid multiplication of endosperm cells (Nagato and Chaudhry 1969) and dominant transportation efficiency of photosynthetic products (Weng and Chen 1987) in *indica*. Rice grain yield is highly positively significantly correlated with spikelet number on first rachis branches especially secondary rachis branches. So, genetic mechanisms and their genetic relationships between the number of rachis branches and vascular bundle related traits in rice need further explored.

Candidate gene identification of the important QTL

Cloning QTL affecting complex traits has been a major challenge to plant geneticists and molecular biologists since the classical strategy using map-based cloning is

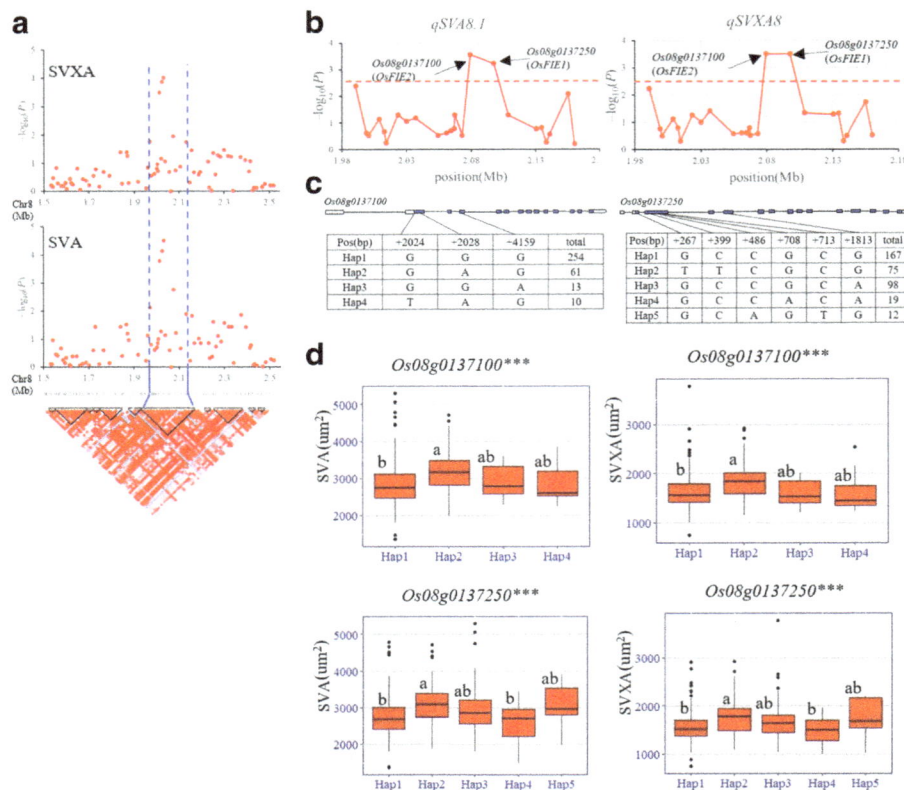

Fig. 6 Gene-based association analysis and haplotype analysis of targeted genes related to *qSVA8.1* and *qSVXA8*. **a** Local Manhattan plot (top) and LD block (bottom) surrounding the peak on chromosome 8. Dashed lines indicate the candidate region for the peak. **b** Gene-based association analysis of targeted genes related to *qSVA8.1* and *qSVXA8*. Each point is a gene indicated by one SNP having largest −log10 (*P*) value. Dash line show the threshold to determine significant SNP. **c** Exon-intron structure of *Os08g0137250* and *Os08g0137100*, and DNA polymorphism in those gene. **d** Boxplots for SVA and SVXA based on the haplotypes (Hap) for *Os08g0137250* and *Os08g0137100*. The *** suggest significance of ANOVA at *P* < 0.001. The letter on histogram (a, and b) indicate multiple comparisons result at the significant level 0.05. Green and red colors indicate in 2015 and 2016, respectively

extremely troublesome and time-consuming. Using GWAS and haplotype analysis of candidate genes, we were able to shortlist six candidate genes governing the four important QTL affecting vascular bundles related traits in the peduncle.

In the region of 31.07–31.26 Mb on chromosome 4, a QTL cluster (*qLVA4.1* and *qLVPA4.1*) was detected, containing two candidate genes. One of them was *NAL1* (*Os04g0615000*) regulating the development of vascular bundle which may be involved in polar auxin transport (Qi et al. 2008) and contributed to grain yield (Fujita et al. 2013; Xu et al. 2015). Culms of *nal1* showed a defective vascular system, in which the number and distribution pattern of vascular bundles was altered (Qi et al. 2008). Additionally, the Takanari allele of *NAL1*, which was found to be a partial loss-of-function allele of *NAL1*, increased mesophyll cell number between vascular bundles (Takai et al. 2013). Haplotypes analysis revealed haplotype ATA was associated with significantly larger LVA and LVPA than haplotypes GCG and GTA

(Fig. 4c–d). Further, we found one nucleotide variation located in the third exon which results in the substitution of a histidine (CAT) in the trypsin-like serine and cysteine protease domain of the large LVA and LVPA by an arginine (CGT) in the small LVA and LVPA (Additional file 6: Fig. S4). And an allele variation produced less panicles per plant, more spikelet numbers and more width leaf blades than varieties containing G allele (Yano et al. 2016). Taguchi-Shiobara et al. (2015) reported that the flag leaf width of transgenic plants of *NAL1* varied depending on the ratio of A allele to G allele. Thus, we presumed that the nucleotide variation (A/G) in the third exon exerted a critical function on determining panicles per plant, spikelet numbers per panicle, flag leaf width and area of large vascular bundle in the peduncle.

In the present study, two QTL (*qLVN6* and *qSVN7*) were detected in both two years. The *qLVN6* affecting LVN was identified in the region of 28.55–28.92 Mb on chromosome 6. The only candidate gene for *qLVN6* was *OsARF18* (*Os06g0685700*), a similar to auxin response

QTL mapping and candidate gene analysis of peduncle vascular bundle related traits in rice...

factor gene. The deregulation of the *OsmiR160* target gene *OsARF18* leads to rice abnormal growth and development, including dwarf stature, rolled leaves, and small seeds through affecting the auxin signaling (Huang et al. 2016). Compared with wild type, numbers of bulliform cell and total abaxial epidermal cells between two vascular bundles were significantly reduced in *mOsARF18* leave (Huang et al. 2016). A number of studies have shown that polar auxin transport is required for continuous vascular pattern formation and establishment of procambial strands (Mattsson et al. 2003; Scarpella et al. 2006; Berleth et al. 2007). The classic auxin signal flow canalization hypothesis predicts that the distribution of auxin is narrowed down from a wide field of cells to a subset of cells with high auxin transport, which then become the site of the procambium (Sachs 1981). Then gene function in *PINT*, *MP* and *ATHB8*, which is a positive feedback loop of auxin–*MP*–*ATHB8*–*PIN1*, confirms the hypothesis (Paciorek et al. 2005; Scarpella et al. 2006; Donner et al. 2009). In addition, *AVB* gene affecting vascular bundles involved in procambium establishment following auxin signaling in lateral primordial (Ma et al. 2017). Previously, researches considered that *NAL1* (*Os04g0615000*) regulating the development of vascular bundle played its role in leaf morphogenesis by regulating polar auxin transport, and plays a regulatory role in the development of plant type in rice (Qi et al. 2008; Fujita et al. 2013). Therefore, *OsARF18* (*Os06g0685700*) is considered as a possible candidate gene of *qLVN6*. Transgenic experiments are under way to verify the functionalities of the candidate genes. The only candidate gene for *qSVN7*, *Os07g0695100* (*Ghd7.1*) was a major QTL with pleiotropic effects on grain number per panicle, plant height, and heading date in rice (*Oryza sativa* L.) in previously report (Liu et al. 2013; Gao et al. 2014). In this study, compared with haplotype GCGGTAGCGG, haplotype CTGACGGTGA delayed heading and increased SVN, grain number per panicle and plant height in 2016 Sanya (Additional file 7: Figure S5), hinting that *Ghd7.1* probably affects vascular bundle in rice. However and so far there has been no reported *Ghd7.1* gene relevant to vascular bundle. Thus, as to whether *Ghd7.1* is candidate gene for *qSVN7*, it must be verified using reverse genetic approaches.

Of the two candidate genes for *qSVA8.1*, *OsFIE2* (*Os08g0137100*) belongs to the PcG protein family and is the most likely candidate gene. Compared to wild-type, the number of large and small vascular bundles decreased in second internode of *osfie2*–1 stem, as well as cell number and cell size in spikelet hulls (Liu et al. 2016). However, no significant difference was detected between different haplotypes at *Os08g0137100* in *indica* subpopulation, and only one haplotype was observed in *japonica* subpopulation (Additional file 5: Table S2),

though there were significant differences for SVA and SVXA between different haplotypes in whole population (Fig. 6d), suggesting that *indica-japonica* differentiation probably led to significant phenotypic differences in SVA and SVXA between different haplotypes. Thus, it must be further studies that *OsFIE2* is candidate gene for *qSVA8.1* or evolution gene of *indica* and *japonica*.

Application in Rice breeding for improved vascular bundle system

The grain yield of crop is highly dependent on the source-sink relationship in which grain or kernel is the photosynthetic sink while the leaves are the primary source. The vascular bundle is the transport system that links source and sink, and it determines how plants efficiently transport photosynthetic products, mineral nutrients and water from source to sink (Housley and Peterson 1982). Efficient transport of assimilates from leaves and stems to developing spikelets is required for better grain filling and high yield. *SPIKE* gene increases grain yield of *indica* rice through enlarging sink size, source size, and translocation capacity. Compared with that of IR64 isogenic control, the grain fertility of NIL-*SPIKE* was improved, presumably owing to a strengthening of assimilate supply to the larger number of spikelets by an increase in vascular bundle number of panicle neck (Fujita et al. 2013). *APO1* gene controlling the number of primary rachis branches also controls the vascular bundle formation and hence is responsible to increase the harvest index and grain yield in rice (Terao et al. 2010). The effect on vascular bundle system development regulated by the *HI1* allele of *APO1* limited reduction in ripening percentage (Terao et al. 2010). In a word, the capacity of transporting assimilates from source to sink could limit grain filling (Ashraf et al. 1994). What rice breeders concern is how to develop new variety with an advanced vascular bundle system and a strong ability to transport assimilates from source to sink.

Our results suggested that in the region of 28.55–28.92 Mb on chromosome 6, a QTL (*qLVN6*) was detected in both two years affecting LVN. According to gene function annotation, *Os06g0685700*, a similar to auxin response factor gene is considered as the most likely candidate gene of *qLVN6*. Haplotype analysis revealed that significant differences for LVN were between different haplotypes at the gene in *japonica* accessions. Two tropical *japonica* varieties, PLUS (averaged 20.4 in 2015 and 23.5 in 2016) and ITA 221 (averaged 21.7 in 2015 and 23.7 in 2016) with more LVNs were found in this panel. Haplotype TTGAAG of the *Os06g0685700* from the two *japonica* accessions exhibited additive effect for increased number of large vascular bundles in peduncle in *japonica* and this favorable QTL allele would be introgressed into *indica* and other *japonica* varieties with superior agronomic traits for improved large vascular bundle by marker-assisted selection (MAS). Specifically, for example, hybrid between *indica* and *japonica* holds large panicle with more primary rachis branches especially secondary rachis branches but its grain yield is seriously limited by insufficient grain filling due to disharmony of source, sink and flow of assimilates. So, it is possible to improve grain filling of hybrid between two subspecies by reasonably deploying favorable alleles of QTL for peduncle vascular bundle related traits in rice as indicated in this study.

Conclusion

We identified 48 QTL for eight peduncle vascular bundle related traits via GWAS. A total of six candidate genes of four important QTL regions, including the cloned gene (*NAL1*), were identified by determining linkage disequilibrium blocks associated with significant SNPs, and haplotype analyses. In addition, the most likely candidate genes for three new QTL (*qLVN6*, *qSVN7*, and *qSVA8.1*) were identified based on functional annotation. The results will enhance our knowledge of the genetic basis of vascular bundle traits in rice, facilitate exploring the molecular mechanisms underlying the traits, and provide valuable information for improving rice vascular bundle system by MAS breeding.

Methods

Plant materials

The 423 worldwide accessions were selected from 3 K RGP (Zheng et al. 2015). This panel consisted of seven types, including *indica* (301), *temperate japonica* (13), *tropical japonica* (57), admixed *japonica* (9), admixed type (17), *aus/boro* (16), and *basmati/sadri* (10) (Additional file 2: Table S1).

Phenotypic investigation

All of these accessions were grown in Sanya (18.3°N, 109.3°E) during Dec 2014–April 2015 and Dec 2015–April 2016 and each accession was planted in two rows with 10 individuals each row at a spacing of 25 cm between rows and 17 cm between plants with two replications for each accession. The field management followed the farmers' standard management practices. In order to minimize flowering time effects experiment-wise, batch sampling were performed based on heading date of each accession. At full heading stage, five uniform plants in the middle of each plot were selected. The transverse section of stem were made at uniformly 2 cm above the neck-panicle node and kept in formalin-acetic-alcohol (FAA) fixative solution. After safranin O staining, vascular bundle related traits in the peduncle, including the

number of large vascular bundle (LVN), the total area of large vascular bundle (LVA, um^2), the phloem area of large vascular bundle (LVPA, um^2), the xylem area of large vascular bundle (LVXA, um^2), the number of small vascular bundle (SVN), the total area of small vascular bundle (SVA, um^2), the phloem area of small vascular bundle (SVPA, um^2) and the xylem area of small vascular bundle (SVXA, um^2) (Additional file 8: Figure S6), were observed and measured using a microscopy (ZEISS AXIO, Germany). The average trait value of each accession was used in data analyses of GWAS.

Genotyping

The 27 K SNP genotype data of the 423 accessions was generated from the 3 K RGP (Zheng et al. 2015). For those SNPs, only the alleles of highest frequency in the 423 panels were retained and other alleles of low frequency were considered missing. The heterozygous alleles were also eliminated. SNPs with missing rate over 20% and minor allele frequency (MAF) less than 5% were removed. Finally, a total of 26,097 SNPs were used in the GWAS.

Population structure and kinship

Principal component (PC) and population structure (Q) were applied to infer population structure. For the 26,097 SNP, we further removed SNP loci with missing rate over 10% and MAF less than 0.10. Then, 6601 evenly distributed SNPs with average marker spacing around 50 kb were sampled to calculate population structure (Q). A model based Bayesian clustering analysis method implemented in STRUCTURE software version 2.3.4 (Pritchard et al. 2000) was used. The program was run with the following parameters: k, the number of groups in the panel varying from 1 to 7; 10 runs each k value; for each run, 10,000 burn in iterations followed by 10,000 MCMC (Markov Chain Monte Carlo) iterations. A principal component analysis (PCA) was performed the efficient mixed-model association (EMMA) method in the Genome Association and Prediction Integrated Tool (GAPIT) R package (Lipka et al. 2012) to examine the population structure. The K matrix (kinship matrix) was obtained from the results of the relatedness analysis using the EMMA method in GAPIT, a package of R software (Lipka et al. 2012). The PCA and K matrix were used in the following association analysis.

GWAS and identification of candidate genes

We performed a GWAS to excavate the SNPs significant associations with all measured traits using 26,097 SNPs and the mean trait values of the 423 accessions. In this study, the model of mixed linear (MLM), PCA + K, was used in the association analysis. And we performed the study by the software GAPIT. The MLM includes both fixed and random effects. Including individuals as random effects gave an MLM the ability to incorporate information about relationships among individuals. This information about relationships was conveyed through the KI matrix, which was used in an MLM as the variance-covariance matrix between the individuals. Then a genetic marker-based KI matrix was used jointly with PCA (Yu et al. 2006). After the GWAS, SNPs affecting the measured traits were claimed when the test statistics reached $P < 1.0 \times 10^{-4}$ in at least one of the two years.

Gene-based association analysis was carried out to detect candidate genes for important QTL. Here, QTL regions meeting at least one of the following criteria were considered as important: (1) consistently identified in both two years; (2) affecting more than one trait; and/or (3) close to previously reported cloned genes or fine mapped QTL. The following five steps were conducted to identify candidate genes for important QTL identified. Firstly, the LD block was analyzed using the software "Haploview v4.2" (Barrett et al. 2005). A LD block was created when the upper 95% confidence bounds of D' value exceeded 0.98 and the lower bounds exceeded 0.70 (Gabriel et al. 2002). The LD block where the significant trait-associated SNPs were situated was defined as the candidate gene regions. We can find all the genes located in the candidate regions for each important QTL from the Rice Annotation Project Database (RAP-DB).Secondly, all available non-synonymous SNPs located inside of these genes were searched from 32 M SNPs data generated from 3 K RGP in the Rice SNP-Seek Database (Alexandrov et al. 2015). Thirdly, the SNPs with minor allele frequency less than 0.05 and/or missing rate over 20% were removed and remaining the high quality SNPs inside of these candidate genes of each important QTL were used to perform gene-based association analyses through MLM using the PCA and K applied in GWAS. The threshold was defined as $-\log_{10}(P)$ of the peak SNP of the detected QTL minus 1 (Wang et al. 2017; Zhang et al. 2017). Fourthly, haplotype analysis were carried out for the candidate genes in each important QTL region using all non-synonymous SNPs located inside of the gene CDS region. Finally, candidate genes were determined by testing the significant differences among major haplotypes (containing more than 10 samples) for each important QTL through analysis of variance (ANOVA).

Additional files

Additional file 1: Figure S1. Distribution of SNP markers on chromosomes. The colors show the number of SNPs within 1 Mb window size.

Additional file 2: Table S1. List of the 423 accessions including country of origin, subpopulation (*indica*, *temperate japonica*, *tropical japonica*, admixed *japonica*, admixed type, *aus/boro*, and *basmati/sadri*), population structure (Q), PC score, vascular bundle related traits measured in 2015 and 2016, and heading date measured in 2016.

Additional file 3: Figure S2. Genome-wide association results for 8 vascular bundle related traits. Manhattan plots (left) and quantile-quantile plots (right) associated with LVN (a), LVA (b), LVPA (c), LVXA (d), SVN (e), SVA (f), SVPA (g), and SVXA (h) in 423 accessions in 2015 and 2016. For the Manhattan plots, $-\log_{10}$ P-values from a genome-wide scan were plotted against the position of the SNPs on each of 12 chromosomes and the horizontal grey dashed lines show the suggestive threshold ($P = 1.0 \times 10^{-4}$). For the quantile-quantile plots, the horizontal axes indicate the $-\log_{10}$ -transformed expected P values, and the vertical axes indicate the $-\log_{10}$-transformed observed P-values. Arrows indicate QTL overlapping with the published QTL.

Additional file 4: Figure S3. Comparisons of eight vascular bundle traits between *indica* and *japonica* subpopulations. Blue and orange bar indicate *indica* and *japonica*, respectively. All data are presented as the mean ± SD. *, $P < 0.05$ and ***, $P < 0.001$.

Additional file 5: Table S2. Haplotype analysis of the targeted genes related to qSVA8.1 and qSVXA8 in the whole population, *indica* and *japonica* subpopulations, respectively. The letters (a, and b) indicate multiple comparison results at the significant level 0.05.

Additional file 6: Figure S4. Comparison of LVA and LVPA between A and G alleles located in the third exon of *Os04g0615000* (*NAL1*) gene. The letters on histogram (a, and b) indicate multiple comparisons result at the significant level 0.05. The value on the histogram is the number of individuals of each allele.

Additional file 7: Figure S5. Comparisons of heading date, grain number per panicle and plant height among different haplotypes of *Ghd7.1* gene in 2016 Sanya. The letters on histogram (a, b and c) indicate multiple comparison results at the significant level 0.05. The value on the histogram is the number of individuals of each haplotype.

Additional file 8: Figure S6. Cross-sections of large and small vascular bundle in peduncle stained with safranin O staining. P, phloem; XP, xylem parenchyma; PX, protoxylem element; MX, metaxylem element; PXL, protoxylem lacuna.

Abbreviations

3 K RGP: 3 K Rice Genome Project; ANOVA: Analysis of variance; EMMA: Efficient mixed-model association; FAA: Formalin-acetic-alcohol; GAPIT: Genome association and prediction integrated tool; GWAS: Genome-wide association study; KI: Kinship; LD: Linkage disequilibrium; LVA: Total area of large vascular bundle; LVN: Number of large vascular bundle; LVPA: Phloem area of large vascular bundle; LVXA: Xylem area of large vascular bundle; MAF: Minor allele frequency; MAS: Marker-assisted selection; MLM: Model of mixed linear; PCA: Principal component analysis; PRB: Primary rachis branch; QTL: Quantitative trait loci/locus; RAP-DB: Rice annotation project database; SNP: Single-nucleotide polymorphism; SVA: Total area of small vascular bundle; SVN: Number of small vascular bundle; SVPA: Phloem area of small vascular bundle; SVXA: Xylem area of small vascular bundle

Acknowledgements

This work was funded by the National Natural Science Foundation of China (31671602), the Key Laboratory Project of Department of Education of Liaoning Province, China (LZ2015065), the national key research and development program of China (2016YFD0300104), the "863" Key Project (2014AA10A601) from the Chinese Ministry of Science and Technology and the Bill and Melinda Gates Foundation project (#OPP1130530).

Authors' contributions

YW, JLX and ZKL designed and supervised the research. LYZ, XYW, and YW performed phenotyping. LYZ, TQZ, and SW analyzed data. LYZ, YW and JLX wrote the paper. All authors read and approved the final manuscript.

Competing interests

The authors declare that they have no competing interests.

Author details
[1]Rice Research Institute, Shenyang Agricultural University/Key Laboratory of Northern Japonica Rice Genetics and Breeding, Ministry of Education, Shenyang 110866, China. [2]Institute of Crop Sciences/National Key Facility for Crop Gene Resources and Genetic Improvement, Chinese Academy of Agricultural Sciences, 12# South Zhong-Guan-Cun Street, Haidain District, Beijing 100081, China. [3]Agricultural Genomics Institute at Shenzhen, Chinese Academy of Agricultural Sceinces, Shenzhen 518120, China. [4]Shenzhen Institute of Breeding and Innovation, Chinese Academy of Agricultural Sciences, Shenzhen 518120, China.

References

3K RGP (2014) The 3,000 rice genomes project. Gigascience 3:7

Alexandrov N, Tai S, Wang W, Mansueto L, Palis K, Fuentes RR, Ulat VJ, Chebotarov D, Zhang G, Li Z, Mauleon R, Hamilton RS, McNally KL (2015) SNP-seek database of SNPs derived from 3000 rice genomes. Nucleic Acids Res 43:1023–1027

Ashraf M, Akbar M, Salim M (1994) Genetic improvement in physiological traits of rice yield. In: Slafer GA (ed) Genetic improvement of field crops. Marcel Dekker Incorporates, New York, pp 413–455

Bai X, Wu B, Xing Y (2012) Yield-related QTLs and their applications in rice genetic improvement. J Integr Plant Biol 54:300–311

Barrett JC, Fry B, Maller J, Daly MJ (2005) Haploview: analysis and visualization of LD and haplotype maps. Bioinformatics 21:263–265

Berleth T, Scarpella E, Prusinkiewicz P (2007) Towards the systems biology of auxin-transport-mediated patterning. Trends Plant Sci 12:151–159

Cui K, Peng S, Xing Y, Yu S, Xu C, Zhang Q (2003) Molecular dissection of the genetic relationships of source. sink and transport tissue with yield traits in rice. Theor Appl Genet 106:649–658

Dettmer J, Elo A, Helariutta Y (2009) Hormone interactions during vascular development. Plant Mol Biol 69:347–360

Donald CM (1968) The breeding of crop ideotypes. Euphytica 17:385–403

Donner TJ, Sherr I, Scarpella E (2009) Regulation of preprocambial cell state acquisition by auxin signaling in Arabidopsis leaves. Development 136:3235–3246

Evans LT, Dunstone RL, Rawson HM, Williams RF (1970) The phloem of the wheat stem in relation to requirements for assimilate by the ear. Aust J Biol Sci 23:743–752

Fujita D, Trijatmiko KR, Tagle AG, Sapasap MV, Koide Y, Sasaki K, Tsakirpaloglou N, Gannaban RB, Nishimura T, Yanagihara S, Fukuta Y, Koshiba T, Slamet-Loedin IH, Ishimaru T, Kobayashi N (2013) Nal1 allele from a rice landrace greatly increases yield in modern indica cultivars. Proc Natl Acad Sci U S A 110:20431–20436

Fukushima A, Akita S (1997) Varietal differences of the course and differentiation time of large vascular bundles in the rachis of rice. Jpn J Crop Sci 66:24–28

Fukuyama T, Sasahara H, Fukuta Y (1996) New characters for identification of indica and japonica rice. Rice Genet Newsl 13:48–50

Fukuyama T, Takayama T (1995) Variations of the vascular bundle system in Asian rice cultivars. Euphytica 86:227–231

Gabriel SB, Schaffner SF, Nguyen H, Moore JM, Roy J, Blumenstiel B, Higgins J, DeFelice M, Lochner A, Faggart M, Liu-Cordero SN, Rotimi C, Adeyemo A, Cooper R, Ward R, Lander ES, Daly MJ, Altshuler D (2002) The structure of haplotype blocks in the human genome. Science 296:2225–2229

Gao H, Jin M, Zheng XM, Chen J, Yuan D, Xin Y, Wang M, Huang D, Zhang Z, Zhou K, Sheng P, Ma J, Ma W, Deng H, Jiang L, Liu S, Wang H, Wu C, Yuan L, Wan J (2014) Days to heading 7, a major quantitative locus determining photoperiod sensitivity and regional adaptation in rice. Proc Natl Acad Sci U S A 111:16337–16342

Han B, Huang X (2013) Sequencing-based genome-wide association study in rice. Curr Opin Plant Biol 16:133–138

Housley TL, Peterson DM (1982) Oat stem vascular size in relation to kernel number and weight. II. Field Environment1. Crop Sci 22:274–278

Huang J, Li Z, Zhao D (2016) Deregulation of the OsmiR160 target gene OsARF18 causes growth and developmental defects with an alteration of auxin signaling in rice. Sci Rep 6:29938

Huang X, Wei X, Sang T, Zhao Q, Feng Q, Zhao Y, Li C, Zhu C, Lu T, Zhang Z, Li M, Fan D, Guo Y, Wang A, Wang L, Deng L, Li W, Lu Y, Weng Q, Liu K, Huang T, Zhou T, Jing Y, Li W, Lin Z, Buckler ES, Qian Q, Zhang QF, Li J, Han B (2010) Genome-wide association studies of 14 agronomic traits in rice landraces. Nat Genet 42:961–967

Lafitte HR, Travis RL (1984) Photosynthesis and assimilate partitioning in closely related lines of rice exhibiting different sink: source relationships. Crop Sci 24:447–452

Lipka AE, Tian F, Wang Q, Peiffer J, Li M, Bradbury PJ, Gore MA, Buckler ES, Zhang Z (2012) GAPIT: genome association and prediction integrated tool. Bioinformatics 28:2397–2399

Liu T, Liu H, Zhang H, Xing Y (2013) Validation and characterization of Ghd7.1, a major QTL with pleiotropic effects on spikelets per panicle, plant height and heading date in rice (Oryza sativa L.) J Integr Plant Biol 55:917–927

Liu X, Wei X, Sheng Z, Jiao G, Tang S, Luo J, Hu P (2016) Polycomb protein OsFIE2 affects plant height and grain yield in rice. PLoS One 11:e0164748

Lucas WJ, Groover A, Lichtenberger R, Furuta K, Yadav SR, Helariutta Y, He XQ, Fukuda H, Kang J, Brady SM, Patrick JW, Sperry J, Yoshida A, López-Millán AF, Grusak MA, Kachroo P (2013) The plant vascular system: evolution, development and functions. J Integr Plant Biol 55:294–388

Ma L, Sang X, Zhang T, Yu Z, Li Y, Zhao F, Wang Z, Wang Y, Yu P, Wang N, Zhang C, Ling Y, Yang Z, He G (2017) ABNORMAL VASCULAR BUNDLES regulates cell proliferation and procambium cell establishment during aerial organ development in rice. New Phytol 213:275–286

Mattsson J, Ckurshumova W, Berleth T (2003) Auxin signaling in arabidopsis leaf vascular development. Plant Physiol 131:1327–1339

Morishima H, Oka HI (1981) Phylogenetic differentiation of cultivated rice, XXII. Numerical evaluation of the Indica-Japonica differentiation. Jpn J Breed 31:402–413

Nagato K, Chaudhry FM (1969) A comparative study of ripening process and kernel development in japonica and indica rice. Proc Crop Sci Soc Japan 38:425–433

Osada A, Ishizaki Y, Suzuki S (1983) Difference in the number of days for ripening of grains between japonica and indica rice varieties. Japan J. Trop Agric 27:59–66

Paciorek T, Zazimalova E, Ruthardt N, Petrasek J, Stierhof YD, Kleine-Vehn J, Morris DA, Emans N, Jurgens G, Geldner N, Friml J (2005) Auxin inhibits endocytosis and promotes its own efflux from cells. Nature 435:1251–1256

Pritchard JK, Stephens M, Donnelly P (2000) Inference of population structure using multilocus genotype data. Genetics 155:945–959

Qi J, Qian Q, Bu Q, Li S, Chen Q, Sun J, Liang W, Zhou Y, Chu C, Li X, Ren F, Palme K, Zhao B, Chen J, Chen M, Li C (2008) Mutation of the rice narrow leaf1 gene, which encodes a novel protein, affects vein patterning and polar auxin transport. Plant Physiol 147:1947–1959

Sachs T (1981) The control of the patterned differentiation of vascular tissues. Adv Bot Res 9:151–262

Sasahara H, Fukuta Y, Fukuyama T (1999) Mapping of QTLs for vascular bundle system and spike morphology in rice, Oryza sativa L. Breed Sci 49:75–81

Scarpella E, Marcos D, Friml J, Berleth T (2006) Control of leaf vascular patterning by polar auxin transport. Dev Biol 201:1015–1027

Sun C, Hu Z, Zheng T, Lu K, Zhao Y, Wang W, Shi J, Wang C, Lu J, Zhang D, Li Z, Wei C (2017) RPAN: rice pan-genome browser for 3000 rice genomes. Nucleic Acids Res 45:597–605

Taguchi-Shiobara F, Ota T, Ebana K, Ookawa T, Yamasaki M, Tanabata T, Yamanouchi U, Wu J, Ono N, Nonoue Y, Nagata K, Fukuoka S, Hirabayashi H, Yamamoto T, Yano M (2015) Natural variation in the flag leaf morphology of rice due to a mutation of the narrow leaf1 gene in Oryza sativa L. Genetics 201:795–808

Takai T, Adachi S, Taguchi-Shiobara F, Sanoh-Arai Y, Iwasawa N, Yoshinaga S, Hirose S, Taniguchi Y, Yamanouchi U, Wu J, Matsumoto M, Sugimoto K, Kondo K, Ikka T, Ando T, Kono I, Ito S, Shomura A, Ookawa T, Hirasawa T, Yano M, Kondo M, Yamamoto T (2013) A natural variant of NAL1, selected in high-yield rice breeding programs, pleiotropically increases photosynthesis rate. Sci Rep 3:2149

Terao T, Nagata K, Morino K, Hirose T (2010) A gene controlling the number of primary rachis branches also controls the vascular bundle formation and hence is responsible to increase the harvest index and grain yield in rice. Theor Appl Genet 120:875–893

Wang XQ, Pang YL, Wang CC, Chen K, Zhu YJ, Shen C, Ali J, Xu JL, Li ZK (2017) New candidate genes affecting rice grain appearance and milling quality detected by genome-wide and gene-based association analyses. Front Plant Sci 7:1998

Weng JH, Chen CY (1987) Differences between indica and japonica rice varieties in CO2 exchange rates in response to leaf nitrogen and temperature. Photosynth Res 14:171–178

Xu JL, Wang Y, Zhang F, Wu Y, Zheng TQ, Wang YH, Zhao XQ, Cui YR, Chen K, Zhang Q, Lin HX, Li JY, Li ZK (2015) SS1 (NAL1)- and SS2-mediated genetic networks underlying source-sink and yield traits in rice (*Oryza sativa* L.) PLoS One 10:e0132060

Yano K, Yamamoto E, Aya K, Takeuchi H, Lo PC, Hu L, Yamasaki M, Yoshida S, Kitano H, Hirano K, Matsuoka M (2016) Genome-wide association study using whole-genome sequencing rapidly identifies new genes influencing agronomic traits in rice. Nat Genet 48:927–934

Yu J, Pressoir G, Briggs WH, Vroh Bi I, Yamasaki M, Doebley JF, McMullen MD, Gaut BS, Nielsen DM, Holland JB, Kresovich S, Buckler ES (2006) A unified mixed-model method for association mapping that accounts for multiple levels of relatedness. Nat Genet 38:203–208

Zhang J, Chen K, Pang YL, Naveed SA, Zhao XQ, Wang XQ, Wang Y, Dingkuhn M, Pasuquin J, Li ZK, Xu JL (2017) QTL mapping and candidate gene analysis of ferrous iron and zinc toxicity tolerance at seedling stage in rice by genome-wide association study. BMC Genomics 18:828

Zhang ZH, Li P, Wang LX, Tan CJ, Hu ZL, Zhu YG, Zhu LH (2002) Identification of quantitative trait loci (QTLs) for the characters of vascular bundles in peduncle related to *indica-japonica* differentiation in rice (*Oryza sativa* L.) Euphytica 128:279–284

Zhao K, Tung CW, Eizenga GC, Wright MH, Ali ML, Price AH, Norton GJ, Islam MR, Reynolds A, Mezey J, McClung AM, Bustamante CD, McCouch SR (2011) Genome-wide association mapping reveals a rich genetic architecture of complex traits in *Oryza sativa*. Nat Commun 2:467

Zheng TQ, Yu H, Zhang HL, Wu ZC, Wang WS, Tai SS, Chi L, Ruan J, Wei CC, Shi JX, Gao YM, Fu BY, Zhou YL, Zhao XQ, Zhang F, Mcnally KL, Li ZC, Zhang GY, Li JY, Zhang DB, Xu JL, Li ZK (2015) Rice functional genomics and breeding database (RFGB): 3K rice SNP and InDel sub-database (in Chinese). Chin Sci Bull 60:367–371

12

Rice nitrate transporter *OsNPF7.2* positively regulates tiller number and grain yield

Jie Wang[1], Kai Lu[1], Haipeng Nie[1,2], Qisen Zeng[1,2], Bowen Wu[1], Junjie Qian[1] and Zhongming Fang[1,2*] [ID]

Abstract

Background: Rice tiller number is one of the most important factors that determine grain yield, while nitrogen is essential for the crop growth and development, especially for tiller formation. Genes involved in nitrogen use efficiency processes have been identified in the previous studies, however, only a small number of these genes have been found to improve grain yield by promoting tillering.

Results: We constructed over-expression (OX) lines and RNA-interference (Ri) lines, and selected a mutant of *OsNPF7.2*, a low-affinity nitrate transporter. Our analyses showed that rice tiller number and grain yield were significantly increased in OX lines, whereas Ri lines and mutant *osnpf7.2* had fewer tiller number and lower grain yield. Under different nitrate concentrations, tiller buds grew faster in OX lines than in WT, but they grew slower in Ri lines and mutant *osnpf7.2*. These results indicated that altered expression of *OsNPF7.2* plays a significant role in the control of tiller bud growth and regulation of tillering. Elevated expression of *OsNPF7.2* also improved root length, root number, fresh weight, and dry weight. However, reduced expression of *OsNPF7.2* had the opposite result on these characters. *OsNPF7.2* OX lines showed more significantly enhanced influx of nitrate and had a higher nitrate concentration than WT. The levels of gene transcripts related to cytokinin pathway and cell cycle in tiller bud, and cytokinins concentration in tiller basal portion were higher in OX lines than that in WT, suggesting that altered expression of *OsNPF7.2* controlled tiller bud growth and root development by regulating cytokinins content and cell cycle in plant cells. Altered expression of *OsNPF7.2* also was responsible for the change in expression of the genes involved in strigolactone pathway, such as *D27, D17, D10, Os900, Os1400, D14, D3,* and *OsFC1*.

Conclusion: Our results suggested that *OsNPF7.2* is a positive regulator of nitrate influx and concentration, and that it also regulates cell division in tiller bud and alters expression of genes involved in cytokinin and strigolactone pathways, resulting in the control over rice tiller number. Since elevated expression of *OsNPF7.2* is capable of improving rice grain yield, this gene might be applied to high-yield rice breeding.

Keywords: Rice, *OsNPF7.2*, Tiller bud, Cytokinin, Tiller number, Grain yield

Background

Rice (*Oryza sativa* L.) is one of the three major grain crops grown worldwide and is consumed by more than half of the world's population (Khush 2005). The rapid increase of the human population puts high demand on rice production, meanwhile high rice yield is a target pursued by plant breeders. Rice yield is mainly controlled by three factors: panicle number per plant, grain number per panicle, and thousand-grain weight. Panicle number per plant is dependent on the ability of plant to produce tillers (Liang et al. 2014). Starting with shoot branching, rice tiller experience two distinct stages in its development: the formation of an tiller bud at each leaf axil and the outgrowth of the tiller bud (Li et al. 2003; Xing and Zhang 2010). Therefore, final tiller number is determined not only by the number of tiller bud but also by outgrowth rate of tiller bud (Wang and Li 2011). In the past few years, many quantitative trait loci (QTLs) and genes involved in tiller bud formation and outgrowth in rice have been identified, such as *MOC1* (Li et al. 2003), *MOC2* (Koumoto et al. 2013), *MOC3/SRT1* (Lu et al. 2015; Mjomba et al. 2016), *TAD1/TE* (Xu et al. 2012; Lin et al. 2012), *LAX1* (Oikawa and Kyozuka 2009), *LAX2*

* Correspondence: zmfang@mail.hzau.edu.cn
[1]Center of Applied Biotechnology, Wuhan Institute of Bioengineering, Wuhan 430415, China
[2]National Key Laboratory of Crop Genetic Improvement, Huazhong Agricultural University, Wuhan 430070, China

(Tabuchi et al. 2011), *OsTB1/OsFC1* (Takeda et al. 2003; Minakuchi et al. 2010), especially, the genes responsible for strigolactone pathways, such as *D27* (Lin et al. 2009), *D17/OsCCD7/HTD1* (Zou et al. 2005; Zou et al. 2006; Kulkarni et al. 2014; Yang et al. 2017), *D10* (Arite et al. 2007), *D14* (Arite et al. 2009), *D3* (Ishikawa et al. 2005; Yoshida et al. 2012), and *D53* (Zhou et al. 2013; Jiang et al. 2013).

Tiller bud outgrowth is regulated not only by endogenous factors, but also by environmental signals (Xing and Zhang 2010). Nitrogen (N), as an important environmental factor, affects rice growth and development including rice tillering. Nitrate is the major form of N available in aerobic environments and many members of nitrate transporter gene families are found in rice, such as 80 NPFs (NRT1/PTRs: NRT1, low-affinity nitrate transporter; PTR, di/tripeptide transporter), 5 NRT2s, and 2 NAR2s members. To date, only a few NPF members have been characterized in rice (Li et al. 2017). *OsNRT1* (*OsNPF8.9*) was first described and found to function as a low affinity nitrate transporter (Lin et al. 2000). Afterwards, other *NPFs* were explored, such as *SP1* (*OsNPF4.1*) and *OsPTR9* (*OsNPF8.20*), however, their substrates remain unclear (Lin et al. 2000; Fang et al. 2013), Recently, *OsNPF2.4*, *OsNPF2.2*, and *OsNPF7.2* have been reported to serve as low-affinity nitrate transporters functioning under high nitrate concentrations (Li et al. 2015; Xia et al. 2015; Hu et al. 2016). Allelic differences in the dual-affinity nitrate transporter *NRT1.1B* (*OsNPF6.5*) have been reported between *indica* and *japonica* cultivars with high nitrogen-use efficiency and grain yield in the *NRT1.1B*–indica allele (Hu et al. 2015). OsPTR6 (*OsNPF7.3*) transports di/tripeptides Gly-His and Gly-His-Gly and its high levels of expression enhance rice growth (Fan et al. 2014). A recent study reveals that *OsNPF7.3* is induced by organic nitrogen, and that elevated expression of *OsNPF7.3* increases the number of panicles per plant, filled grain numbers per panicle, grain nitrogen content, and enhances grain yield (Fang et al. 2017). OsPTR7 (*OsNPF8.1*) shows dimethylarsenate (DMA) transport activity and is involved in the long-distance translocation of DMA into rice grain (Tang et al. 2017).

Of all the characterized NPF transporters to date, only *OsNPF8.20*, *OsNPF6.5*, and *OsNPF7.3* can moderate rice tiller number and enhance grain yield (Fang et al. 2013; Hu et al. 2015; Fang et al. 2017). It is unclear whether other NPF genes play a role in rice tillering, especially by regulating N and phytohormones in plant cells. One previous study showed that knock-out of *OsNPF7.2* retarded rice root growth under high nitrate supply (Hu et al. 2016). However, the effect of increased expression of *OsNPF7.2* on rice growth and development is yet unknown, neither is the influence agronomic traits. This study analysed over-expression lines (OX), RNA-interference lines (Ri), and a mutation of *OsNPF7.2* and found that over-expression of *OsNPF7.2* significantly increased rice tiller

number by promoting tiller bud elongation and by regulating cytokinin (CK) and strigolactone (SL) pathway in cells.

Results

Over-expression of *OsNPF7.2* improves rice tiller number and grain yield

OsNPF7.2 is mainly expressed in the roots of seedlings, and its protein transports nitrate at vacuolar membrane (Hu et al. 2016). In order to investigate the effects of the altered expression of *OsNPF7.2* on rice growth and development, we constructed over-expression (OX) lines and RNA-interference (Ri) lines; we also analysed knock-out mutant *osnpf7.2*. We found that tiller number increased in three OX lines at reproductive stage (Fig. 1b-d, r) compared to that in wild-type (WT) ZH11 (Fig. 1a, r), but it dramatically decreased in three Ri lines (Fig. 1e-g, r) and mutant *osnpf7.2* (Fig. 1h, r). Three OX lines also had a higher total grain number per plant than WT (Fig. 1i-l, t), whereas Ri lines had a lower total grain number per plant than WT (Fig. 1m-o, t). The total grain number of mutant *osnpf7.2* was less than half of that of WT (Fig. 1p, t). It was confirmed that the formation of phenotype resulted from the altered expression of *OsNPF7.2* by using qRT-PCR in different transgenic lines (Fig. 1a-h and q). Overall, our results indicated that elevated *OsNPF7.2* expression level significantly enhanced the total grain number per plant.

The number of panicles derived from rice tillers is one of the three key factors determining rice grain yield (Xing and Zhang 2010). The analysis of panicle number in the transgenic lines presented similar change trend of tiller number as described above (Fig. 1s). Moreover, there was no significant difference in 1000-grain weights among different transgenic lines of *OsNPF7.2* (Fig. 1u). Grain yield per plant in OX lines was significantly greater than that in WT (Fig. 1v). Thus, over-expression of *OsNPF7.2* significantly increased rice tiller number and total grain number per plant, while down-regulation of *OsNPF7.2* produced the opposite effects.

Elevated expression of *OsNPF7.2* promotes rice tiller bud outgrowth especially under high nitrate concentrations

We further investigated the regulatory effort of *OsNPF7.2* on rice tillering by analysing of the development of tiller buds in WT, OX lines, Ri lines, and mutant *osnpf7.2* grown under different nitrate concentrations (0.5–8 mM). Tiller buds grew more rapidly in OX lines than in WT under all nitrate concentrations with this phenomenon observed continuously for 34 days after germination (DAG); tiller buds growth was slower in line Ri and mutant *osnpf7.2* than in WT (Fig. 2a-e). OX lines had significantly longer tiller buds than WT when plants were treated with 0.5 mM nitrate at 19–34 DAG (Fig. 2f). However, no significant difference in buds length between

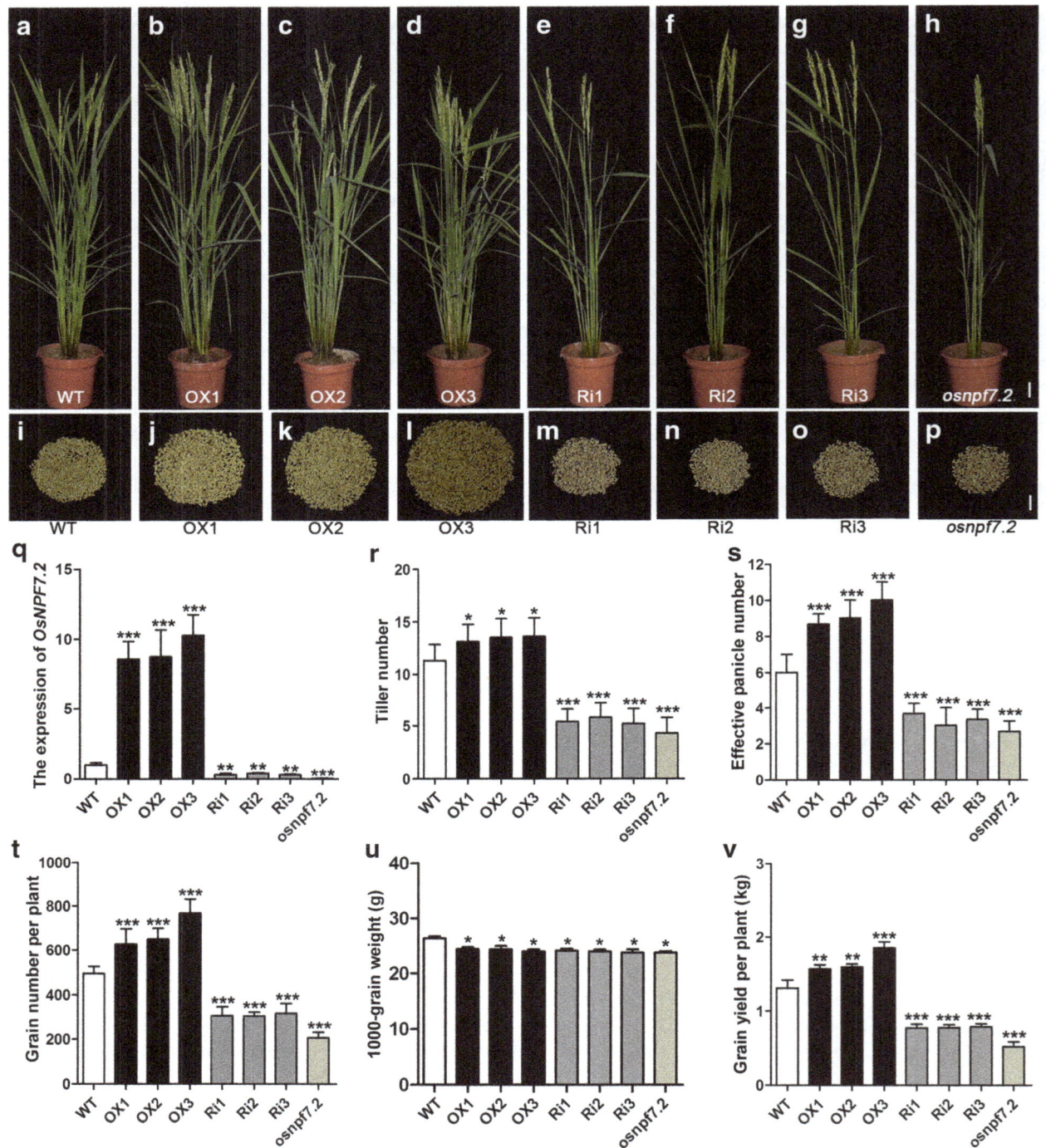

Fig. 1 Characterization of transgenic lines of *OsNPF7.2*. **a-h** Phenotypic analysis of WT, OX lines, Ri lines and mutant *osnpf7.2* at mature stage. Bar = 5 cm. **i-p** Comparison of total grain number among WT, OX lines, Ri lines and *osnpf7.2*. Bar = 5 cm. **q** Transcript abundance of *OsNPF7.2* in tiller buds among WT, OX lines, Ri lines and *osnpf7.2*. **r-v** Comparison of agronomic traits including tiller number (**r**), effective panicle number (**s**), grain number per plant (**t**), 1000-grain weight (**u**) and grain yield per plant (**v**) among WT, OX lines, Ri lines and *osnpf7.2*. WT, OX and Ri were abbreviations for wild type, over-expression and RNA-interference, respectively. Data in **q-v** are shown as means ± SD ($n = 10$) from three replicates. A student's t-test was used to generate P values: "*, ** and ***" indicate significance at $P < 0.05$, $P < 0.01$ and $P < 0.001$, respectively

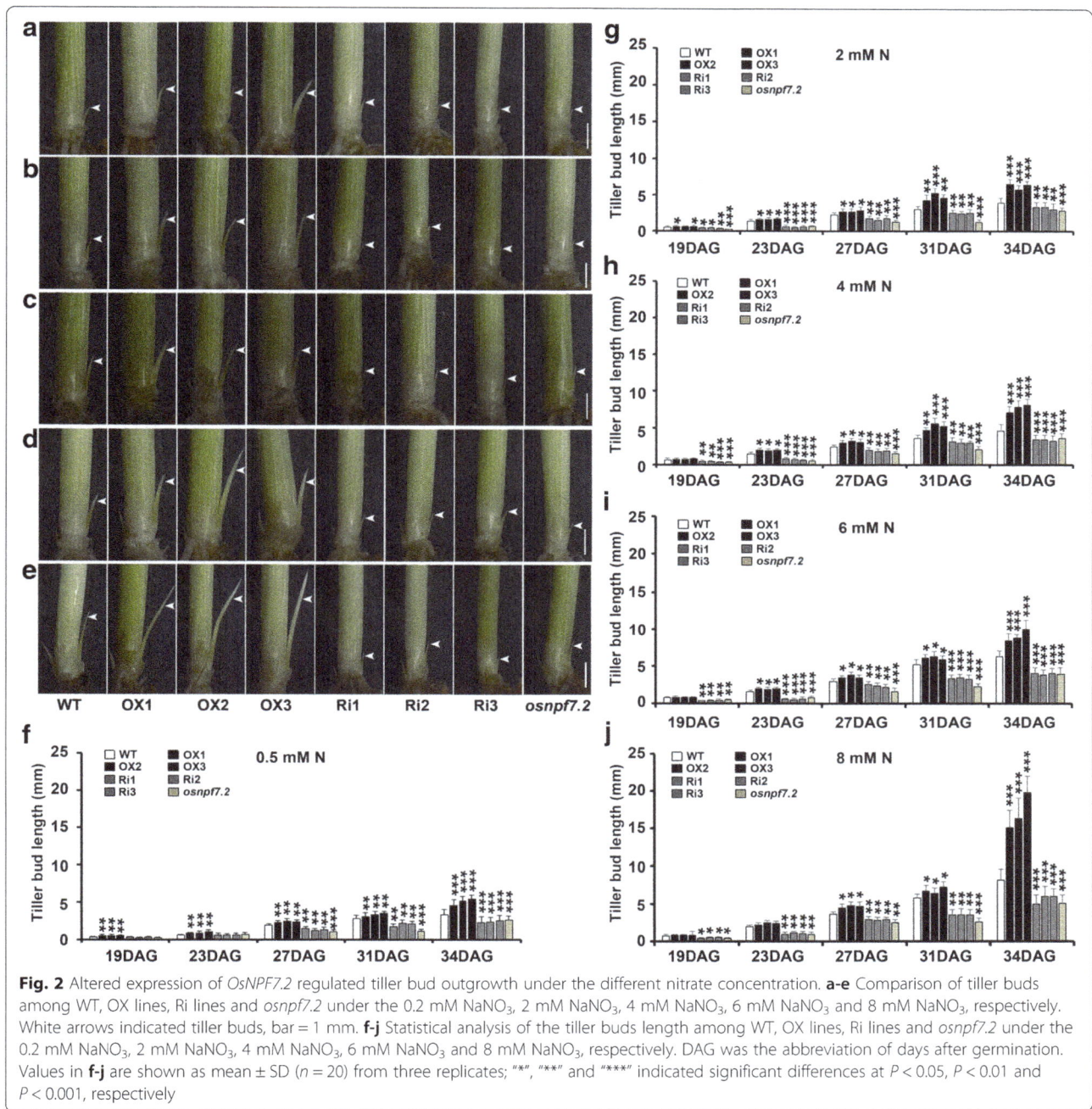

Fig. 2 Altered expression of *OsNPF7.2* regulated tiller bud outgrowth under the different nitrate concentration. **a-e** Comparison of tiller buds among WT, OX lines, Ri lines and *osnpf7.2* under the 0.2 mM NaNO₃, 2 mM NaNO₃, 4 mM NaNO₃, 6 mM NaNO₃ and 8 mM NaNO₃, respectively. White arrows indicated tiller buds, bar = 1 mm. **f-j** Statistical analysis of the tiller buds length among WT, OX lines, Ri lines and *osnpf7.2* under the 0.2 mM NaNO₃, 2 mM NaNO₃, 4 mM NaNO₃, 6 mM NaNO₃ and 8 mM NaNO₃, respectively. DAG was the abbreviation of days after germination. Values in **f-j** are shown as mean ± SD ($n = 20$) from three replicates; "*", "**" and "***" indicated significant differences at $P < 0.05$, $P < 0.01$ and $P < 0.001$, respectively

line OX and WT was observed under high nitrate concentrations (4–8 mM) at 19 DAG (Fig. 2h-j). OX lines had longer tiller buds than WT after 27 DAG at all nitrate concentrations, and the maximum length of tiller buds in OX lines were found at 34 DAG in plants treated with 8 mM nitrate (Fig. 2j). Significantly shorter tiller buds were observed in Ri lines than in WT (Fig. 2f-j). Based on these results, it could be concluded that elevated expression of *OsNFP7.2* promoted rice tiller bud outgrowth, especially under high nitrate concentrations, between 19 and 34 DAG.

Elevated expression of *OsNPF7.2* benefits rice seedling growth and root development

Next, we investigated the effect of up-regulation of *OsNPF7.2* on rice seedling growth and development in hydroponic cultures under different nitrate concentrations. Seedlings of OX lines under 8 mM NaNO₃ conditions produced stronger culms than seedlings of WT (Fig. 3a). However, Ri lines and mutant *osnpf7.2* showed the opposite result (Fig. 3a). Root morphology is important for plant to optimize N absorption from the soil

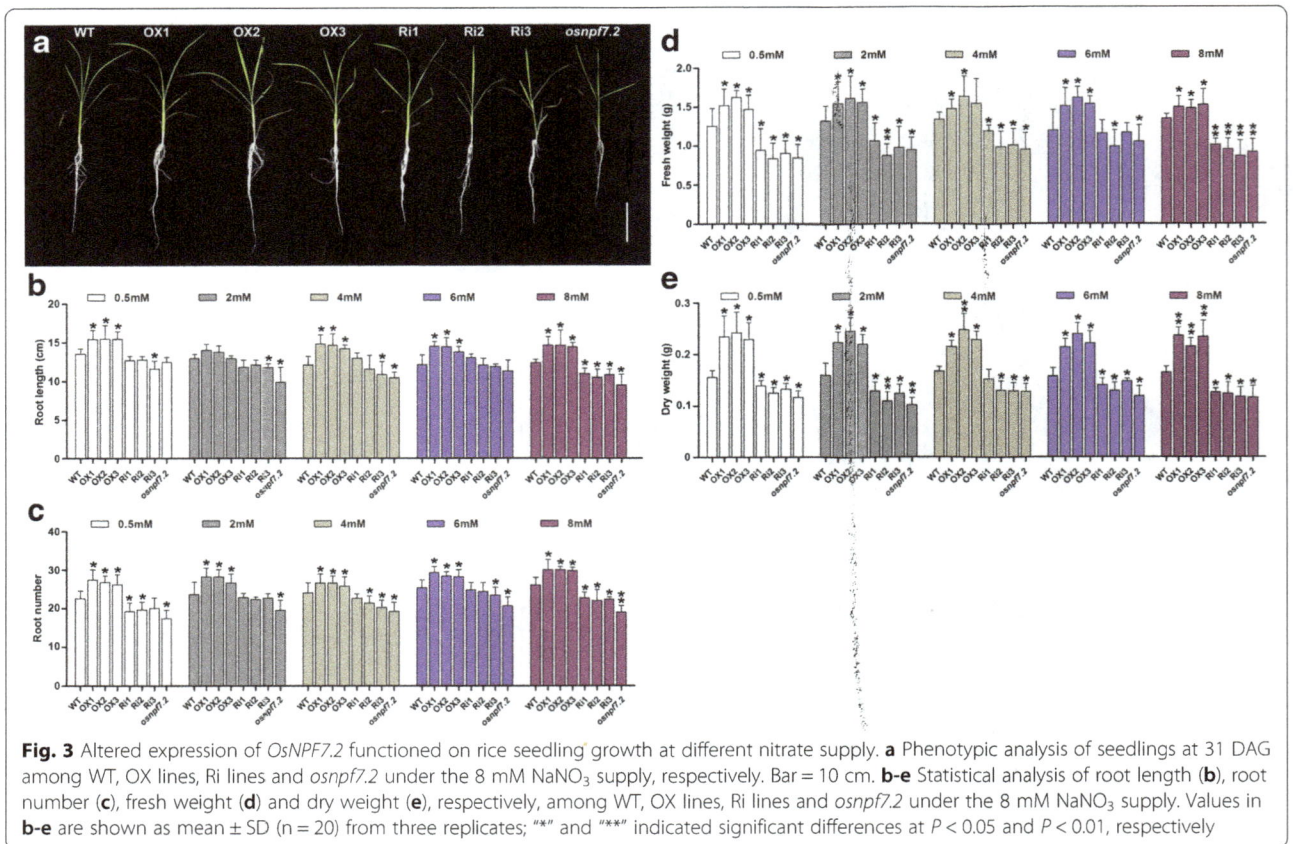

Fig. 3 Altered expression of *OsNPF7.2* functioned on rice seedling growth at different nitrate supply. **a** Phenotypic analysis of seedlings at 31 DAG among WT, OX lines, Ri lines and *osnpf7.2* under the 8 mM NaNO$_3$ supply, respectively. Bar = 10 cm. **b-e** Statistical analysis of root length (**b**), root number (**c**), fresh weight (**d**) and dry weight (**e**), respectively, among WT, OX lines, Ri lines and *osnpf7.2* under the 8 mM NaNO$_3$ supply. Values in **b-e** are shown as mean ± SD (n = 20) from three replicates; "*" and "**" indicated significant differences at $P < 0.05$ and $P < 0.01$, respectively

through its responses to nitrates (Hachiya and Sakakibara 2017), we examined root development in *OsNPF7.2* transgenic lines. In Ri lines and mutant *osnpf7.2*, root growth was inhibited resulting in their roots were shorter than those of WT seedlings (Fig. 3a-b). The comparison of root number revealed that elevated expression of *OsNPF7.2* caused a significant increase at all nitrate concentrations, while downregulated expression of *OsNPF7.2* result in a significant decrease of root number at 8 mM NaNO$_3$ (Fig. 3c). Compared with those of WT, the fresh weight and dry weight of OX lines were significantly increased (Fig. 3d-e). These results demonstrated that genetically modification of *OsNPF7.2* could significantly influence rice root development.

We also cultured the different lines under 4 mM (NH$_4$)$_2$SO$_4$, and found no significant difference in root length among WT, OX lines, Ri lines, and mutant *osnpf7.2* at the seedling stage (40 DAG, Additional file 1 Figure S1b and d). However, root length of OX lines exceeded that of WT when seedlings were treated with 8 mM NaNO$_3$, whereas root length of the seedlings with down-regulated *OsNPF7.2* expression decreased (Additional file 1 Figure S1a and c). These results indicated that *OsNPF7.2* transgenic seedlings responded to environmental nitrate by altering root growth and development, but they did not respond to ammonium.

Changes in expression of *OsNPF7.2* influence the rate of NO$_3^-$ influx and concentration

Seedlings with various expression lines were treated under nitrogen starvation conditions for a week, and then were cultured with a solution containing 8 mM nitrate for 24 h. The amount of NO$_3^-$ obsorbed by seedlings was then measured. The rate of NO$_3^-$ influx into roots of OX lines was higher than that of WT (Fig. 4a), indicating that elevated expression of *OsNPF7.2* enhanced nitrate uptake by roots. In OX lines, we also detected a higher rate of NO$_3^-$ influx into the leaf sheath and leaf blade, implying that elevated expression of *OsNPF7.2* promoted the translocation of NO$_3^-$ from roots to leaf sheath (Fig. 4a). Besides, we measured nitrate concentration of root, leaf sheath and leaf blade in the seedlings with different expression lines. The detected nitrate concentration was consistent with the rate of NO$_3^-$ influx in different lines (Fig. 4b). Total nitrogen concentrations in root, leaf sheath, and leaf blade did not differ significantly among WT, OX lines, Ri lines, and mutant *osnpf7.2* (data not shown). However, total nitrogen content in those lines with up-regulated expression became higher than that in WT, and repression lines exhibited lower total nitrogen content compared to WT (Fig. 4c). These results demonstrated that over-

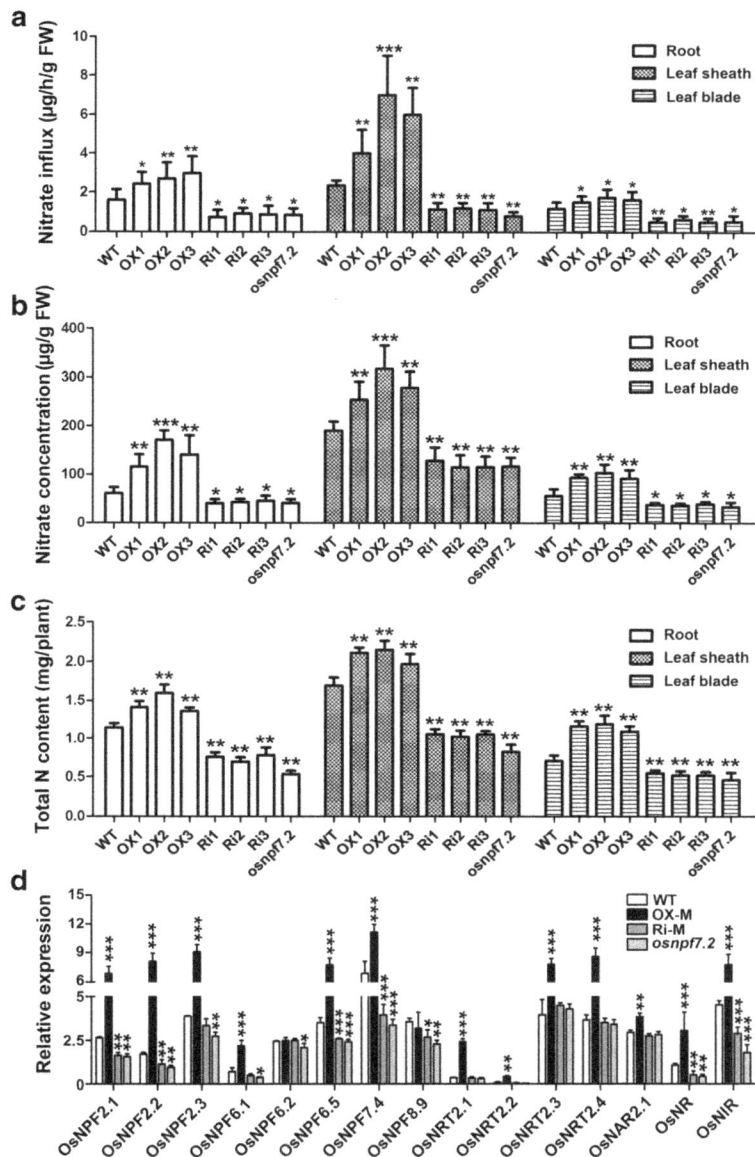

Fig. 4 *OsNPF7.2* influenced NO$_3^-$ influx, NO$_3^-$ concentration and total nitrogen content among transgenic lines. **a** Analysis of NO$_3^-$ influx rate among WT, OX lines, Ri lines and *osnpf7.2* cultured under 6 mM NaNO$_3$ supply. **b** NO$_3^-$ concentration of WT, OX lines, Ri lines and *osnpf7.2* cultured under 6 mM NaNO$_3$ supply. **c** Comparison of total nitrogen content among WT, OX lines, Ri lines and *osnpf7.2* cultured under 6 mM NaNO$_3$ supply. **d** Expression level of crucial genes involved in absorbing, transporting and assimilation of nitrate among WT, OX-M, Ri-M and *osnpf7.2* cultured under 6 mM NaNO$_3$ supply. OX-M and Ri-M indicated that mixed equal-amount RNA which extracted from 10 seedlings' tiller buds of each three OX lines and Ri lines, respectively. Date are shown as mean ± SD (n = 10) from three replicates; "*", "**" and "***" indicated significant differences at P < 0.05, P < 0.01 and P < 0.001, respectively

expression of *OsNPF7.2* promoted translocation of nitrate from roots to leaf sheath, and enhanced nitrate influx and concentration.

We detected the effect of *OsNPF7.2* various expression lines on expression levels of 18 genes, including low-affinity nitrate transporters, high-affinity nitrate transporters, nitrate and ammonium assimilation genes. The expression level of 7 NPF genes including *OsNPF6.5* in OX lines was increased compared to that in WT (Fig. 4d). Two of the 18 genes, *OsNPF6.5* and *OsGS1;2*, were reported to promote tiller bud outgrowth and the increase in rice tiller number (Hu et al. 2015; Ohashi et al. 2017). Expression of both *OsNPF6.5* and *OsGS1;2* was upregulated in OX lines and down-regulated in Ri lines and mutant *osnpf7.2* (Additional file 2 Figure S2). Based on these results, it could be concluded that elevated expression of *OsNPF7.2* promotes nitrate uptake and assimilation by regulating other nitrate-related transporters and enzymes.

OsNPF7.2 regulates cell proliferation in tiller bud by coordinating cytokinin and strigolactone pathways

Shoot branching (tiller) is regulated by plant hormones, particularly cytokinins (CKs) and strigolactones (SLs). It was reported that CKs promote tillering in rice, whereas SLs inhibit it (Leyser 2003; Ferguson and Beveridge 2009; Hayward et al. 2009; Shimizu-Sato et al. 2009; Xu et al. 2015). To investigate the possible interaction between nitrogen and plant hormones, we measured the expression levels of the important genes responding to CK and SL pathways in the rice tiller bud with various transgenic lines. Over-expression of *OsNPF7.2* resulted in the up-regulated expression of *OsIPTs*, *LONELY GUY* (*LOG*), and CK-response regulators (*OsARRs/OsRRs*) (Fig. 5a). Levels of cytokinins were regulated through the irreversible

oxidative cleavage of the N^6-side chain by CYTOKININ DEHYDROGENASE/OXIDASE (CKXs) (Zurcher and Muller 2016). Based on their report, we measured gene expression of 10 *CKXs*, and found half of *OsCKXs* exhibited lower expression level in OX lines compared to that in WT, whereas higher expression level in both Ri lines and mutant than that in WT (Fig. 5b). Additionally, we also detected the content of four CKs (iP, tZ, cZ and DZ) in tiller basal portion among different genetically modified lines. Compared to WT, OX lines exhibited a significant increase in the concentration of iP and tZ, while Ri lines and mutant showed a little but not significant decrease in the concentration of iP and a significant decrease in the concentration of tZ (Fig. 5c). These results suggested that altered expression of *OsNPF7.2* controlled tiller bud outgrowth possibly by regulating CKs content in the tiller bud. Cytokinins function mainly by stimulating cell division and growth and by promoting cell differentiation as

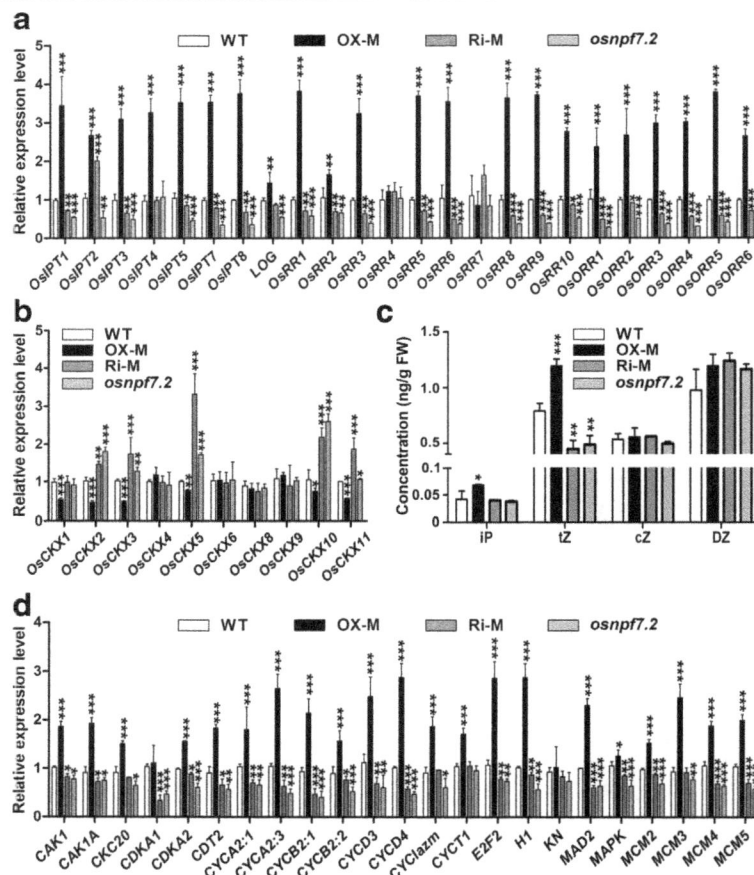

Fig. 5 Altered expression of OsNPF7.2 regulated CKs concentration and cell cycle. **a** Expression level of rice genes involved in cytokinin synthesis and response in tiller buds of seedlings at 34 DAG among WT, OX-M, Ri-M and *osnpf7.2*. **b** Expression of 10 *CKXs* at 34 DAG among WT, OX-M, Ri-M and *osnpf7.2*. **c** CK free base concentration in seedling tiller basal portion at 34 DAG among WT, OX-M, Ri-M and *osnpf7.2*. **c** Comparison of genes involved in cell cycle in tiller buds of seedlings at 34 DAG among different transgenic lines. OX-M and Ri-M in (**a**)-(**b**) and (**d**) was identical to that in fig. 4d. OX-M and Ri-M in (**c**) indicated tiller basal portion (about 0.5 cm) mixed from 30 seedlings of each OX lines and Ri lines, respectively. Date are shown as mean ± SD from three replicates; "*", "**" and "***" indicated significant differences at $P < 0.05$, $P < 0.01$ and $P < 0.001$, respectively

well (Zurcher and Muller 2016). However, cell proliferation is strictly controlled by the major regulators: cyclin-dependent kinases (CDKs) and their regulatory partner cyclins (Yamaguchi et al. 2003). We measured the expression levels of selected genes involved in cell cycle. Significant up-regulation of *OsNPF7.2* was associated with an increased expression level of *CDKs* and cyclin genes such as *CYCAs*, *CYCBs* and *CYCDs*. By contrast, Ri lines and mutant *osnpf7.2* showed decreased expression of these genes (Fig. 5d). We also found that the expression patterns of four minichromosome maintenance genes (*MCM2*, *MCM3*, *MCM4* and *MCM5*) were similar to those of cyclin genes in the transgenic lines (Fig. 5d). These results indicated that elevated expression of *OsNPF7.2* promoted tiller bud growth possibly by accelerating plant cell proliferation.

SL biosynthesis (*D27*, *D17*, *D10*, *Os900* and *Os1400*), perception (*D14* and *D3*), and signalling (*D53*) were reported to have participated in the regulation of tiller bud outgrowth (Jiang et al. 2013; Zhou et al. 2013; Liang et al. 2014; Zhang et al. 2014). These SL synthesis and signalling genes were also detected in tiller bud in the different transgenic lines. The genes (*D27*, *D17*, *D10*, *D14*, and *D3*) were down-regulated in OX lines but up-regulated in Ri lines (Fig. 6). Besides, two members of CYP711 enzymes (*Os900* and *Os1400*, Zhang et al. 2014) showed similar expression pattern to that of *D27*, *D17*, *D10*, *D14*, and *D3* (Fig. 6). However, expression level of *D53*, a repressor of SL signalling, exhibited no significant differences among WT, OX lines, Ri lines, and mutant (Fig. 6). It was reported that the interaction between OsMADS57 and OsFC1/OsTB1 targets *D14* to control the outgrowth of tiller bud in rice (Guo et al. 2013). We compared the expression level of *OsFC1* in tiller bud in the various transgenic lines and found that *OsFC1* was down-regulated in OX lines and was up-regulated in Ri lines and mutant (Fig. 6). SLs, as most of the germination stimulants identified so far, function on stimulating germination of root parasitic plants such as witchweeds (*Striga* spp.) and broomrapes (*Orobanche* and *Phelipanche*

spp., Yoneyama et al. 2010). We performed germination assay of *Orobanche Cumana* to estimate SLs level, and found that the germination rate of *Orobanche Cumana* seeds was higher when the seeds were treated with root exudates extracting from Ri lines and mutant *osnpf7.2* than that from WT, and the opposite results were found when the seeds were treated with root exudates from OX lines (Additional file 3 Figure S3). These results indicated that altered expression of *OsNPF7.2* might influence SL biosynthesis, which in turn influenced perception and signalling in rice tiller bud, therefore controlled rice tillering.

Based on these results, we propose a model in which altered expression of *OsNPF7.2* participates in CK and SL pathways to modify rice tillering (Fig. 7). In OX lines, elevated expression of *OsNPF7.2* is capable of enhancing CK accumulation and inhibiting SL accumulation. Hence, cell division in the tiller bud is promoted, which is favourable for rice tillering. However, down-regulation of *OsNPF7.2* induces the opposite effects. Additionally, we suggest that *OsNPF7.2* coordinates CK and SL pathways, and further regulates tiller bud, eventually controls rice tillering.

Discussion

OsNPF7.2 positively regulates rice tiller number and grain yield

Nitrogen is a crucial determinant of plant growth and grain yield (Hachiya and Sakakibara 2017; Li et al. 2017). Plants make use of transporters to take up nitrogen from the soil via the roots and transport it to other organs. Thus, the coordinated expression of transporter genes is essential to meet the plant growth's requirements for nitrogen. Up till now, only a few nitrate transporters of NPF family (*OsNPF8.9*, *OsNPF2.2*, *OsNPF2.4*, *OsNPF6.5*, and *OsNPF7.2*) have been characterized in rice (Lin et al. 2000; Li et al. 2015; Xia et al. 2015; Hu et al. 2016). Of these, only *OsNPF6.5* regulates rice tiller number and promotes grain yield (Hu et al. 2015). Our study revealed that elevated expression of *OsNPF7.2* significantly

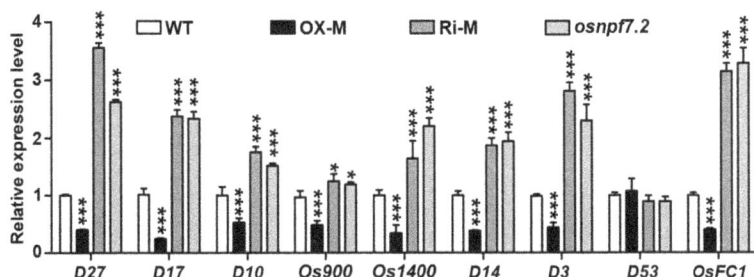

Fig. 6 Expression of genes involved in strigolactone biosynthesis, perception and signalling pathway. Genes' expression involved in SL biosynthesis (*D27*, *D17*, *D10*, *Os900* and *Os1400*), perception (*D14* and *D3*), and signalling (*D53*) were detected among different genetical modified lines. OX-M and Ri-M was same to that in fig. 4d. Date are shown as mean ± SD from three replicates; "*" and "***" indicated significant differences at $P < 0.05$ and $P < 0.001$, respectively

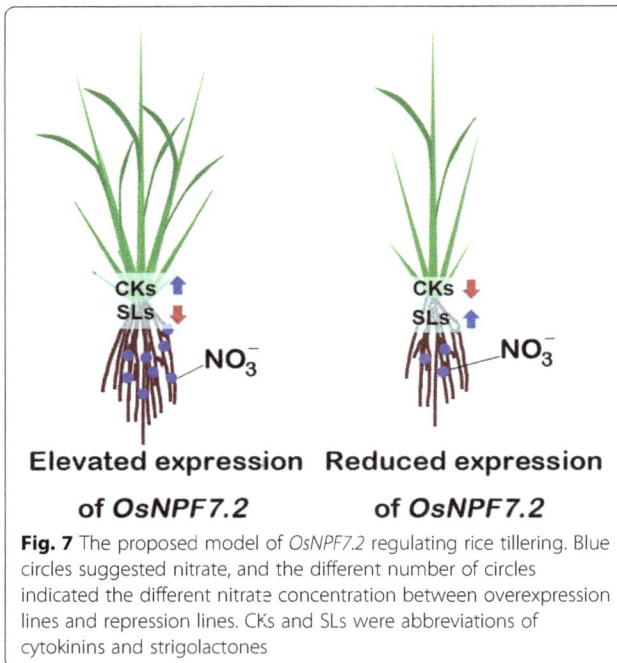

Fig. 7 The proposed model of *OsNPF7.2* regulating rice tillering. Blue circles suggested nitrate, and the different number of circles indicated the different nitrate concentration between overexpression lines and repression lines. CKs and SLs were abbreviations of cytokinins and strigolactones

enhances tiller number, whereas repressed expression causes a reduction in tiller number (Fig. 1). However, previous study only reported the retarded growth of rice root of plant mutant *osnpf7.2* (Hu et al. 2016). Rice tiller number was reported to be one of the most important agronomic traits determining panicle number and grain yield (Li et al. 2003; Xing and Zhang 2010). The increased number of panicles resulted in a larger number of filled grain per plant in OX lines than in WT, which gave rise to an improved rice grain yield per plant (Fig. 1r-t and v). Interesting, 1000-grain weight among all genetically modified lines exhibited slight decreased compared with that in WT (Fig. 1u), which indicated that appropriate expression of *OsNPF7.2* might be beneficial to increase grain weight. This study suggested that *OsNPF7.2* may be useful to culture high-yield rice varieties.

OsNPF7.2 influences NO$_3^-$ influx and concentration, and tiller bud growth

One previous study reported that vacuolar-membrane-localized OsNPF7.2 could transport nitrate at a low affinity (Hu et al. 2016). Our study showed that over-expression of *OsNPF7.2* enhanced the rate of NO$_3^-$ influx into roots, increased NO$_3^-$ concentration in root, and promoted the translocation of nitrate from roots to leaf sheath (Fig. 4a-c), indicating that elevated expression of *OsNPF7.2* contributes to nitrate allocation between roots and shoots. Recently, three NRT1/NPF family members (NPF5.11, NPF5.12 and NPF5.16) in *Arabidopsis* were reported to be localized at vacuolar membrane and to play a possible role in modulating nitrate allocation between roots and shoots (He et al. 2017).

NRT1.1B–indica allele was reported to increase tiller number per plant, and to enhance grain yield per plant (Hu et al. 2015). Our study found that up-regulated expression of *OsNPF7.2* significantly enhanced tiller bud growth, and that down-regulated expression of *OsNPF7.2* impaired tiller bud development (Fig. 2), suggesting that the enhanced translocation of nitrate into leaf sheath in OX lines might contribute to faster growth of tiller bud determining rice tiller number. In addition, the biomass of OX lines was higher than that of WT (Fig. 3d-e), which might mainly be attributed to the increased tiller number at vegetative stage. Therefore, it can be concluded that elevated expression of nitrate transporter *OsNPF7.2* not only promoted rice tillering at vegetative stage, but also played a potentially important role in increasing rice grain yield at reproductive stage.

Elevated of OsNPF7.2 promotes cell division and rice tiller formation through cytokinin and strigolactone pathways

It has been reported that elevated CK level might promote tiller bud outgrowth (Turnbull et al. 1997). Our study indicated that the expression of such CK crucial genes as *IPTs* and *LOG* was higher in OX lines than in WT, suggesting that CKs probably produced in larger amounts in OX lines than in WT; however, the opposite result was found in Ri lines and in mutant (Fig. 5a). Furthermore, the expression pattern of CK response genes (*OsARRs* and *OsRRs*) was similar to *IPTs* expression pattern in transgenic lines. Moreover, the content of iP and tZ was higher in OX lines than in WT (Fig. 5c), which indicated that over-expression of *OsNPF7.2* enhanced CK accumulation in tiller bud.

Nitrogen uptake, assimilation, and recycling in plant roots were reported to determine plant development and productivity (Yamaya and Kusano 2014). However, numerous plant developmental processes such as root meristem specification, vascular development, and shoot and root growth, are determined by CKs (Zurcher and Muller 2016). In addition, CKs are key phytohormones for cell division and growth (Riou-Khamlichi et al. 1999). This study illustrates that elevated expression of *OsNPF7.2* promotes up-regulation of crucial genes in cell cycle (Fig. 5d), which indicated that over-expression of *OsNPF7.2* might promote cell division. To explore the regulatory mechanism of *OsNPF7.2*, we analyse expression of the major regulators in cell cycle, namely, cyclin-dependent kinases (CDKs) and their regulatory partner cyclins. The analysis shows that up-regulation of *OsNPF7.2* significantly increases the expression level of *CDKs*, while down-regulation of *OsNPF7.2* reduces the level of expression of *CDKs* (Fig. 5c). Based on these results, we conclude that altered expression of *OsNPF7.2* controls rice tillering by regulating CK contents, and further regulating the cell cycle in tiller bud.

Recently, SLs have been reported to be important phytohormones inhibiting tiller bud outgrowth in various plant species (Gomez-Roldan et al. 2008; Umehara et al. 2008). Reduction of SL production, perception, and signalling results in faster outgrowth of tiller bud (Domagalska and Leyser 2011; Ruyter-Spira et al. 2013). This study found that *D27*, *D17*, *D10*, *Os900*, *Os1400*, *D14*, *D3*, and *OsFC1* were down-regulated in OX lines but up-regulated in Ri lines and mutant, compared to WT (Fig. 6). OX lines displayed reduced SL biosynthesis, perception, and signalling, whereas Ri lines and mutant showed increased SL signalling. Based on these results, we construct a model of altered expression of *OsNPF7.2* function in rice tillering, in which over-expression of *OsNPF7.2* enhances CK levels but might inhibit SL pathway, which results in increased tillering (Fig. 7). However, the reduced tiller number in Ri lines and mutant *osnpf7.2* might have resulted from weaker CK and stronger SL signalling in their tiller bud.

Conclusions

In this study, we constructed transgenic lines with different expression level of *OsNPF7.2* and found that elevated expression of *OsNPF7.2* contributed to the enhancement of NO_3^- influx rate and the increase of nitrate concentration in over-expression lines. Importantly, *OsNPF7.2* positively regulated tiller bud outgrowth, probably by coordinating CK and SL pathways in plant cells.

Methods

Generation of transgenic rice lines

To construct *OsNPF7.2*-overexpressing lines, a 1726-bp *OsNPF7.2* cDNA was inserted downstream of the *35S* promoter in pCAM1301 using BglII and AflII, to produce the *p35S-OsNPF7.2* plasmid. To generate the *OsNPF7.2*-RNAi lines, two 323-bp fragments of *OsNPF7.2* cDNA were amplified and inserted downstream of the *Ubi-1* promoter in vector pTCK303 (Wang et al. 2004). All of the constructed plasmids were transferred into *japonica* rice variety 'Zhonghua 11' (ZH11) by the *Agrobacterium*-mediated transformation method (Hiei et al. 1997). Homozygous T2 generation of each transgenic lines screened with hygromycin at final concentration of 50 mM for a week were chosen for further studies. Mutant *osnpf7.2* in ZH11 background was obtained from the Rice Mutant Database of Huazhong Agricultural University (http://rmd.ncpgr.cn/), which was the same to that used in the previous study (Hu et al. 2016). The corresponding primers are listed in Additional file 4 Table S1.

Plant cultivation and agronomic traits analysis

For basic agronomic traits analysis, rice plants were grown in the paddy field from June to October at the rice experimental station of the Wuhan Institute of Bioengineering. Ten plants at a spacing of 16.5 cm × 26.5 cm were planted

in a row and 5 rows of each line were planted. At reproductive stage, 10 plants of each lines were randomly chosen to detect agronomic traits. The grain number per panicle was measured as the total number of grains per plant divided by the number of panicles per plant. The 1000-grain weight was calculated as the weight of the total grains per plant and divided by the grain number, then converted to 1000-grain weight. Grain yield was measured as the weight of total grains per plant.

To analyse *OsNPF7.2* expression function in seedling growth and development under different nitrogen conditions, ZH11, OX, Ri, and *osnpf7.2* seedlings at 7 DAG were cultured in basic nutrient solution (pH = 5.8) for a week. The composition of the basic solution was as follows: 1 mM NH_4NO_3, 0.32 mM NaH_2PO_4, 0.51 mM K_2SO_4, 1 mM $CaCl_2$, 1.65 mM $MgSO_4$, 8.9 μM $MnSO_4$, 0.5 μM Na_2MoO_4, 18.4 μM H_3BO_3, 0.14 μM $ZnSO_4$, 0.16 μM $CuSO_4$ and 40 μM $FeSO_4$. Then seedlings at 14 DAG were transferred to basic nutrient solutions supplemented with the following sole nitrogen source: 0.5 mM $NaNO_3$, 2 mM $NaNO_3$, 4 mM $NaNO_3$, 6 mM $NaNO_3$, 8 mM $NaNO_3$ and 4 mM $(NH_4)_2SO_4$. Each nutrient solution was renewed every three days. Daytime conditions in the greenhouse were 32 °C, with light from a sodium lamp (400 W) for 14 h; night-time conditions were 25 °C, and dark for 10 h. At 34 DAG, root length, root number, fresh weight and dry weight of each lines were measured. Besides, tiller buds of different lines (34 DAG) were obtained to detect the expression level of *OsNPF7.2* and other phytohormone-related genes.

Measurement of nitrate influx, nitrate concentration, and total nitrogen content

To analyse the nitrate influx, nitrate concentration and total nitrogen content, ten seedlings at 7 DAG of ZH11, OX, Ri, and *osnpf7.2* were cultured in basic nutrient solution for a week. They were then placed in basic nutrient solution without nitrogen for a week for nitrogen-starvation treatment. The nitrogen-starved seedlings were transferred to culture solution containing 8 mM nitrate for 24 h. Free NO_3^- content analysis was carried out by homogenizing plant tissues in cold extraction buffer [50 mM Tris-HCl (pH 7.0), 10 mM imidazole, and 0.5% (w/v) β-mercaptoethanol]. The suspension was centrifuged at 12,000 rpm for 30 min and the supernatant was collected. Free NO_3^- content was determined from a standard curve of KNO_3 (Cai et al. 2009). NO_3^- influx was calculated as the difference in NO_3^- content between the 8 mM nitrate-treatment and nitrate-starved plants in an hour. Total nitrogen content was determined using the semi-micro Kjeldahl method using a nitrogen analyser (Smart Chem 200, Westco, Italy). Three replicates of each assay were performed.

RNA isolation and qRT-PCR

Total RNA was extracted from tiller buds using TRIzol reagent (Invitrogen, Beijing, China). First-strand cDNA was synthesized using random primers and MLV reverse transcriptase (TaKaRa Bio, Beijing, China). qRT-PCR reaction solution was prepared in a total volume of 20 μL, containing 2 μL of the cDNA, 0.2 mM of each primer, and 10 μL of 2 × SYBR green PCR master mix (Takara Co. Ltd., http://www.takarabiomed.com.cn/). Quantitative real-time PCR was performed using SYBR Green mix (TaKaRa Bio, Beijing, China) and the 7500 RT qPCR system (Applied Biosystems, Foster City, CA, United States). The rice *Actin* gene (LOC_Os03g50885) was used as the internal control, and three technical replicates were performed for each sample. Expression level was calculated using the relative quantification method (Carleton 2011). The primers used for qPCR are listed in Additional file 4.

Extraction of root exudates and germination assay of *Orobanche cumana* seeds

Rice seedlings at 7 DAG were cultured in basic nutrient solution with 6 mM $NaNO_3$ supply for a month, then root exudates of ZH11, OX lines, Ri lines and *osnpf7.2* seedlings were extracted using a modified method (Chen et al. 2017). The shoots (5 cm above the roots) were excised with a razor, and the xylem sap was collected for 12 h after decapitation of the shoots. Root exudates were then diluted with distilled water for 10 times to stimulate *Orobanche cumana* seeds germination.

Germination assay was performed according to Ma et al. (2005). *Orobanche cumana* seeds (20–40 seeds) were incubated on 8 mm moist glass-fiber filter paper at 30 °C for 7 days, and then 30 μl diluted root exudates were applied to glass-fiber filter paper to stimulate seeds germination. Germination of the treated seeds was recorded after incubated at 30 °C for another week. Three replicates of each assay were performed and germination data were statistically analyzed using SPSS software.

Determination of CKs concentration

The tiller basal portion (about 0.5 cm) from 30 seedlings of each OX line (OX1, OX2 and OX3) at 34 DAG were mixed, which were named as OX-M. Ri-M indicated the mixed tiller basal portion from 30 seedlings of each RNA interference lines (Ri1, Ri2 and Ri3) at 34 DAG. Then CKs content were measured by MetWare (http://www.metware.cn/) based at ABSciexQ-TRAP®4500LC-MS/MS platform among different genetically modified lines. Three replicates of each assay were performed.

Additional files

Additional file 1 Figure S1. *OsNPF7.2* responded to nitrate specially, not to (TIFF 4917 kb) ammonium. **a** Phenotypic analysis of seedlings (40 DAG) of transgenic lines.cultured under the 8 mM $NaNO_3$. Bar = 10 cm. **b** Seedlings (40 DAG) of WT, OX.lines, Ri lines and *osnpf7.2* cultured under the 4 mM $(NH_4)_2SO_4$. Bar = 10 cm. **c-d.**Statistical analysis of root length of transgenic lines cultured under the 8 mM. $NaNO_3$ and 4 mM $(NH_4)_2SO_4$, respectively. Date are shown as mean ± SD ($n = 10$). From three replicates; "*" and "**" indicated significant differences at $P < 0.05$ and $P. < 0.01$, respectively.

Additional file 2 Figure S2. Transcript abundance of two glutamine synthetase *GS1;2* and *GS2* in tiller buds among transgenic lines. Date are shown as mean ± SD from three replicates; "*", "**" and "***" indicated significant differences at $P < 0.05$, $P < 0.01$ and $P < 0.001$, respectively.

Additional file 3 Figure S3. Gemination rate of *Orobanche cumana* seeds. To estimated SLs levels among ZH11, OX lines, Ri lines and mutant *osnpf7.2*, root exudates of each line were applied to pre-incubated *Orobanche cumana* seeds. Date are shown as mean ± SD from three replicates; "*" indicated significant differences at $P < 0.05$.

Additional file 4 Table S1. Primers used in this study.

Abbreviations

CK: Cytokinin; DAG: Days after germination; OX: Over-expression; OX-M: Over-expression lines mixed; Ri: RNA-interference; Ri-M: RNA-interference lines mixed; SL: Strigolactone; WT: Wild type

Acknowledgments

The authors would like to thank Prof. Yongqing Ma and Lectuer Chao Zhang (North West Agriculture and Forestry University) for kindly providing *Orobanche cumana* seeds.

Statistical analysis

Two-tailed Student's t tests were performed using the SPSS 10 software (IBM, Inc.). "*, ** and ***" indicate significance at $P < 0.05$, $P < 0.01$ and $P < 0.001$, respectively.

Funding

This research was supported by grants from the National Natural Science Foundation (31301250, 31701990), the National Key Research and Development Program (2016YFD01007), the open project of State Key Laboratory of Rice Biology (160203), Hubei Natural Science Foundation (2017CFB696), Chinese Postdoctoral Science Foundation (2015 M582243), and the Scientific research projects of Hubei Education Department (B2017293).

Authors' contributions

ZF and JW designed this study. JW, KL and HN performed the experiments of NO_3^- influx and concentration, total N content, CKs concentration measurements, root exudates extraction, germination assay of *Orobanche cumana* seeds, and qRT-PCR analysis. QZ and JW investigated length of tiller buds in transgenic lines. ZF, QZ, BW and JQ constructed the transgenic lines and took care of rice lines. JW and ZF performed the statistical analysis and wrote the manuscript. All authors read and approved the final manuscript.

Competing interests

The authors declare that they have no competing interests.

References

Arite T, Iwata H, Ohshima K, Maekawa M, Nakajima M, Kojima M, Sakakibara H, Kyozuka J (2007) DWARF10, an RMS1/MAX4/DAD1 ortholog, controls lateral bud outgrowth in rice. Plant J 51:1019–1029

Arite T, Umehara M, Ishikawa S, Hanada A, Maekawa M, Yamaguchi S, Kyozuka J (2009) D14, a strigolactone-insensitive mutant of rice, shows an accelerated outgrowth of tillers. Plant Cell Physiol 50:1416–1424

Cai H, Zhou Y, Xiao J, Li X, Zhang Q, Lian X (2009) Overexpressed glutamine synthetase gene modifies nitrogen metabolism and abiotic stress responses in rice. Plant Cell Rep 28:527–537

Carleton KL (2011) Quantification of transcript levels with quantitative RT-PCR. Methods Mol Biol 772:279–295

Chen Z, Yamaji N, Horie T, Che J, Li J, An G, Ma J (2017) A magnesium transporter OsMGT1 plays a critical role in salt tolerance in Rice. Plant Physiol 174:1837–1849

Domagalska MA, Leyser O (2011) Signal integration in the control of shoot branching. Nat Rev Mol Cell Biol 12:211–221

Fan X, Xie D, Chen J, Lu H, Xu Y, Ma C, Xu G (2014) Over-expression of OsPTR6 in rice increased plant growth at different nitrogen supplies but decreased nitrogen use efficiency at high ammonium supply. Plant Sci 227:1–11

Fang Z, Bai G, Huang W, Wang Z, Wang X, Zhang M (2017) The rice peptide transporter OsNPF7.3 is induced by organic nitrogen, and contributes to nitrogen allocation and grain yield. Front plant Sci 8:1338

Fang Z, Xia K, Yang X, Grotemeyer MS, Meier S, Rentsch D, Xu X, Zhang M (2013) Altered expression of the PTR/NRT1 homologue OsPTR9 affects nitrogen utilization efficiency, growth and grain yield in rice. Plant Biotechnol J 11:446–458

Ferguson BJ, Beveridge CA (2009) Roles for auxin, cytokinin, and strigolactone in regulating shoot branching. Plant Physiol 149:1929–1944

Gomez-Roldan V, Fermas S, Brewer PB, Puech-Pages V, Dun EA, Pillot JP, Letisse F, Matusova R, Danoun S, Portais JC, Bouwmeester H, Becard G, Beveridge CA, Rameau C, Rochange SF (2008) Strigolactone inhibition of shoot branching. Nature 455:189–194

Guo S, Xu Y, Liu H, Mao Z, Zhang C, Ma Y, Zhang Q, Meng Z, Chong K (2013) The interaction between OsMADS57 and OsTB1 modulates rice tillering via DWARF14. Nat Commun 4:1566

Hachiya T, Sakakibara H (2017) Interactions between nitrate and ammonium in their uptake, allocation, assimilation, and signaling in plants. J Exp Bot 68: 2501–2512

Hayward A, Stirnberg P, Beveridge C, Leyser O (2009) Interactions between auxin and strigolactone in shoot branching control. Plant Physiol 151:400–412

He YN, Peng JS, Cai Y, Liu DF, Guan Y, Yi HY, Gong JM (2017) Tonoplast-localized nitrate uptake transporters involved in vacuolar nitrate efflux and reallocation in Arabidopsis. Sci Rep 7:6417

Hiei Y, Komari T, Kubo T (1997) Transformation of rice mediated by agrobacterium tumefaciens. Plant Mol Biol 35:205–218

Hu B, Wang W, Ou S, Tang J, Li H, Che R, Zhang Z, Chai X, Wang H, Wang Y, Liang C, Liu L, Piao Z, Deng Q, Deng K, Xu C, Liang Y, Zhang L, Li L, Chu C (2015) Variation in NRT1.1B contributes to nitrate-use divergence between rice subspecies. Nat Genet 47:834–838

Hu R, Qiu D, Chen Y, Miller AJ, Fan X, Pan X, Zhang M (2016) Knock-down of a tonoplast localized low-affinity nitrate transporter OsNPF7.2 affects Rice growth under high nitrate supply. Front plant Sci 7:1529

Ishikawa S, Maekawa M, Arite T, Onishi K, Takamure I, Kyozuka J (2005) Suppression of tiller bud activity in tillering dwarf mutants of rice. Plant Cell Physiol 46:79–86

Jiang L, Liu X, Xiong G, Liu H, Chen F, Wang L, Meng X, Liu G, Yu H, Yuan Y, Yi W, Zhao L, Ma H, He Y, Wu Z, Melcher K, Qian Q, Xu HE, Wang Y, Li J (2013) DWARF 53 acts as a repressor of strigolactone signalling in rice. Nature 504:401–405

Khush GS (2005) What it will take to feed 5.0 billion rice consumers in 2030. Plant Mol Biol 59:1–6

Koumoto T, Shimada H, Kusano H, She KC, Iwamoto M, Takano M (2013) Rice monoculm mutation moc2, which inhibits outgrowth of the second tillers, is ascribed to lack of a fructose-1, 6-bisphosphatase. Plant Biotech 30:47–56

Kulkarni KP, Vishwakarma C, Sahoo SP, Lima JM, Nath M, Dokku P, Gacche RN, Mohapatra T, Robin S, Sarla N, Seshashayee M, Singh AK, Singh K, Singh NK, Sharma RP (2014) A substitution mutation in OsCCD7 cosegregates with dwarf and increased tillering phenotype in rice. J Genet 93:389–401

Leyser O (2003) Regulation of shoot branching by auxin. Trends Plant Sci 8:541–545

Li H, Hu B, Chu C (2017) Nitrogen use efficiency in crops: lessons from Arabidopsis and rice. J Exp Bot 68:2477–2488

Li X, Qian Q, Fu Z, Wang Y, Xiong G, Zeng D, Wang X, Liu X, Teng S, Hiroshi F, Yuan M, Luo D, Han B, Li J (2003) Control of tillering in rice. Nature 422:618–621

Li Y, Ouyang J, Wang YY, Hu R, Xia K, Duan J, Wang Y, Tsay YF, Zhang M (2015) Disruption of the rice nitrate transporter OsNPF2.2 hinders root-to-shoot nitrate transport and vascular development. Scientific reports 5:9635

Liang WH, Shang F, Lin QT, Lou C, Zhang J (2014) Tillering and panicle branching genes in rice. Gene 537:1–5

Lin CM, Koh S, Stacey G, Yu SM, Lin TY, Tsay YF (2000) Cloning and functional characterization of a constitutively expressed nitrate transporter gene, OsNRT1, from rice. Plant Physiol 122:379–388

Lin H, Wang R, Qian Q, Yan M, Meng X, Fu Z, Yan C, Jiang B, Su Z, Li J, Wang Y (2009) DWARF27, an iron-containing protein required for the biosynthesis of strigolactones, regulates rice tiller bud outgrowth. Plant Cell 21:1512–1525

Lin Q, Wang D, Dong H, Gu S, Cheng Z, Gong J, Qin R, Jiang L, Li G, Wang JL, Wu F, Guo X, Zhang X, Lei C, Wang H, Wan J (2012) Rice APC/C (TE) controls tillering by mediating the degradation of MONOCULM 1. Nat Commun 3:752

Lu Z, Shao G, Xiong J, Jiao Y, Wang J, Liu G, Meng X, Liang Y, Xiong G, Wang Y, Li J (2015) MONOCULM 3, an ortholog of WUSCHEL in rice, is required for tiller bud formation. J Genet Genomics 42:71–78

Ma Y, Shui J, Inanaga S, Cheng J (2005) Stimulatory effects of Houttuynia cordata Thunb. On seed germination of Striga hermonthica (Del.) Benth. Allelopathy J 15:49–56

Minakuchi K, Kameoka H, Yasuno N, Umehara M, Luo S, Kobayashi K, Hanada A, Ueno K, Asami T, Yamaguchi S, Kyozuka J (2010) FINE CULM1 (FC1) works downstream of strigolactones to inhibit the outgrowth of axillary buds in rice. Plant Cell Physiol 51:1127–1135

Mjomba FM, Zheng Y, Liu H, Tang W, Hong Z, Wang F, Wu W (2016) Homeobox is pivotal for OsWUS controlling tiller development and female fertility in Rice. G3 (Bethesda) 6:2013–2021

Ohashi M, Ishiyama K, Kojima S, Kojima M, Sakakibara H, Yamaya T, Hayakawa T (2017) Lack of cytosolic glutamine Synthetase1;2 activity reduces nitrogen-dependent biosynthesis of Cytokinin required for axillary bud outgrowth in Rice seedlings. Plant Cell Physiol 58:679–690

Oikawa T, Kyozuka J (2009) Two-step regulation of LAX PANICLE1 protein accumulation in axillary meristem formation in Rice. Plant Cell 21:1095–1108

Riou-Khamlichi C, Huntley R, Jacqmard A, Murray JA (1999) Cytokinin activation of Arabidopsis cell division through a D-type cyclin. Science 283:1541–1544

Ruyter-Spira C, Al-Babili S, van der Krol S, Bouwmeester H (2013) The biology of strigolactones. Trends Plant Sci 18:72–83

Shimizu-Sato S, Tanaka M, Mori H (2009) Auxin-cytokinin interactions in the control of shoot branching. Plant Mol Biol 69:429–435

Tabuchi H, Zhang Y, Hattori S, Omae M, Shimizu-Sato S, Oikawa T, Qian Q, Nishimura M, Kitano H, Xie H, Fang X, Yoshida H, Kyozuka J, Chen F, Sato Y (2011) LAX PANICLE2 of rice encodes a novel nuclear protein and regulates the formation of axillary meristems. Plant Cell 23:3276–3287

Takeda T, Suwa Y, Suzuki M, Kitano H, Ueguchi-Tanaka M, Ashikari M, Matsuoka M, Ueguchi C (2003) The OsTB1 gene negatively regulates lateral branching in rice. Plant J 33:513–520

Tang Z, Chen Y, Chen F, Ji Y, Zhao FJ (2017) OsPTR7 (OsNPF8.1), a putative peptide transporter in Rice, is involved in Dimethylarsenate accumulation in Rice grain. Plant Cell Physiol 58:904–913

Turnbull CG, Raymond MA, Dodd IC, Morris SE (1997) Rapid increases in cytokinin concentration in lateral buds of chickpea (Cicer arietinum L.) during release of apical dominance. Planta 202:271–276

Umehara M, Hanada A, Yoshida S, Akiyama K, Arite T, Takeda-Kamiya N, Magome H, Kamiya Y, Shirasu K, Yoneyama K, Kyozuka J, Yamaguchi S (2008) Inhibition of shoot branching by new terpenoid plant hormones. Nature 455:195–200

Wang Y, Li J (2011) Branching in rice. Curr Opin Plant Biol 14:94–99

Wang Z, Chen C, Xu Y, Jiang R, Han Y, Xu Z, Chong K (2004) A practical vector for efficient knockdown of gene expression in rice (Oryza sativa L.) Plant Mol Bio Rep 22:409–417

Xia X, Fan X, Wei J, Feng H, Qu H, Xie D, Miller AJ, Xu G (2015) Rice nitrate transporter OsNPF2.4 functions in low-affinity acquisition and long-distance transport. J Exp Bot 66:317–331

Xing Y, Zhang Q (2010) Genetic and molecular bases of rice yield. Annu Rev Plant Biol 61:421–442

Xu C, Wang Y, Yu Y, Duan J, Liao Z, Xiong G, Meng X, Liu G, Qian Q, Li J (2012) Degradation of MONOCULM 1 by APC/C(TAD1) regulates rice tillering. Nat Commun 3:750

Xu J, Zha M, Li Y, Ding Y, Chen L, Ding C, Wang S (2015) The interaction between nitrogen availability and auxin, cytokinin, and strigolactone in the control of shoot branching in rice (Oryza sativa L.) Plant Cell Rep 34:1647–1662

Yamaguchi M, Kato H, Yoshida S, Yamamura S, Uchimiya H, Umeda M (2003) Control of in vitro organogenesis by cyclin-dependent kinase activities in plants. Proc Natl Acad Sci U S A 100:8019–8023

Yamaya T, Kusano M (2014) Evidence supporting distinct functions of three cytosolic glutamine synthetases and two NADH-glutamate synthases in rice. J Exp Bot 65:5519–5525

Yang X, Chen L, He J, Yu W (2017) Knocking out of carotenoid catabolic genes in rice fails to boost carotenoid accumulation, but reveals a mutation in strigolactone biosynthesis. Plant Cell Rep 36:1533–1545

Yoneyama K, Awad A, Xie X, Yoneyama K, Takeuchi Y (2010) Strigolactones as germination stimulants for root parasitic plants. Plant Cell Physiol 51:1095–1103

Yoshida S, Kameoka H, Tempo M, Akiyama K, Umehara M, Yamaguchi S, Hayashi H, Kyozuka J, Shirasu K (2012) The D3 F-box protein is a key component in host strigolactone responses essential for arbuscular mycorrhizal symbiosis. New Phytol 196:1208–1216

Zhang Y, Van Dijk AD, Scaffidi A, Flematti GR, Hofmann M, Charnikhova T, Verstappen F, Hepworth J, van der Krol S, Leyser O, Smith SM, Zwanenburg B, Al-Babili S, Ruyter-Spira C, Bouwmeester HJ (2014) Rice cytochrome P450 MAX1 homologs catalyze distinct steps in strigolactone biosynthesis. Nat Chem Biol 10:1028–1033

Zhou F, Lin Q, Zhu L, Ren Y, Zhou K, Shabek N, Wu F, Mao H, Dong W, Gan L, Ma W, Gao H, Chen J, Yang C, Wang D, Tan J, Zhang X, Guo X, Wang J, Jiang L, Liu X, Chen W, Chu J, Yan C, Ueno K, Ito S, Asami T, Cheng Z, Lei C, Zhai H, Wu C, Wang H, Zheng N, Wan J (2013) D14-SCFD3-dependent degradation of D53 regulates strigolactone signalling. Nature 504:406–410

Zou J, Chen Z, Zhang S, Zhang W, Jiang G, Zhao X, Zhai W, Pan X, Zhu L (2005) Characterizations and fine mapping of a mutant gene for high tillering and dwarf in rice (Oryza sativa L.) Planta 222:604–612

Zou J, Zhang S, Zhang W, Li G, Chen Z, Zhai W, Zhao X, Pan X, Xie Q, Zhu L (2006) The rice HIGH-TILLERING DWARF1 encoding an ortholog of Arabidopsis MAX3 is required for negative regulation of the outgrowth of axillary buds. Plant J 48:687–698

Zurcher E, Muller B (2016) Cytokinin synthesis, signaling, and function–advances and new insights. Int Rev Cell Mol Biol 324:1–38

13

Comparative whole genome re-sequencing analysis in upland New Rice for Africa: insights into the breeding history and respective genome compositions

Naoki Yamamoto[1]* [ORCID], Richard Garcia[1], Tomohiro Suzuki[2], Celymar Angela Solis[1], Yuichi Tada[3], Ramaiah Venuprasad[4] and Ajay Kohli[1]*

Abstract

Background: Increasing rice demand is one of the consequences of the steadily improving socio-economic status of the African countries. New Rice for Africa (NERICA), which are interspecific hybrids between Asian and African rice varieties, are one of successful breeding products utilizing biodiversity across the two different rice crop species. Upland NERICA varieties (NU) exhibit agronomic traits of value for the harsh eco-geography, including shorter duration, higher yield and stress tolerance, compared to local African varieties. However, the molecular basis of the traits in NU varieties is largely unknown.

Results: Whole genome re-sequencing was performed of four NU lines (3, 4, 5, and 7) and for the parental *Oryza sativa* WAB56–104 and *Oryza glaberrima* CG14. The *k*-mer analysis predicted large genomes for the four NU lines, most likely inherited from WAB56–104. Approximately 3.1, 0.10, and 0.40 million single nucleotide polymorphisms, multi nucleotide polymorphisms, and short insertions/deletions were mined between the parental lines, respectively. Integrated analysis with another four NU lines (1, 2, 8, and 9) showed that the ratios of the donor CG14 allelic sites in the NU lines ranged from 1.3 to 9.8%. High resolution graphical genotype indicated genome-level similarities and common genetic events during the breeding process: five *xyloglucan fucosyltransferase* from *O. glaberrima* were introgressed in common. Segregation of genic segments revealed potential causal genes for some agronomic traits including grain shattering, awnness, susceptibility to bacterial leaf bright, and salt tolerance. Analysis of unmapped sequences against the reference cultivar Nipponbare indicated existence of unique genes for pathogen and abiotic stress resistance in the NU varieties.

Conclusions: The results provide understanding of NU genomes for rice improvement for Africa reinforcing local capacity for food security and insights into molecular events in breeding of interspecific hybrid crops.

Keywords: Genome structure, *Oryza glaberrima*, Rice, Polymorphism, Upland NERICA, WAB56–104

* Correspondence: n.yamamoto@irri.org; a.kohli@irri.org
[1]International Rice Research Institute, Los Baños, Laguna, Philippines
Full list of author information is available at the end of the article

Background

Changing socio-economic conditions are driving the progressively increasing consumption of rice in the African countries. Hence, African countries have imported more than 10 million tons of rice annually in recent years (FAO 2013a). Besides, there is a sustained effort to supplement the imports with local rice cultivation in Africa. Namely, yield increase and harvested area expansion both were achieved to pull up the total rice production in Africa since 2000 (FAO 2013b). To further facilitate those, the improvement of the locally suited rice varieties is an important aspect in rice breeding.

Upland rice production is feasible with labor-saving management and can be a source of food and income in Africa. To integrate rice cultivation in poverty and hunger reduction, upland farmers should be able to grow rice. The development of the first generation upland NERICA (NU) varieties by the West Africa Rice Development Association (WARDA: at present the Africa Rice Center) in the 1990s was an important step in that direction. NU varieties are the products of surmounting above the reproductive barriers between Asian rice *Oryza sativa* and African rice *Oryza glaberrima* Steud. (Sano 1990). One of the first subsets of NU varieties (NU 1 to 11), derived from an upland tropical *japonica* cultivar WAB56–104, developed by WARDA as the recurrent parent, and the *O. glaberrima* accession CG14 as the donor parent, were released in 2000s (Sié et al. 2012). These NUs were adapted into the upland rice ecosystems in West African countries, exhibiting good yield potential (4–5 t/ha), short duration, no grain shattering and tolerance to lodging (Africa Rice Center 2008; Somado et al. 2008). Moreover, the NU varieties exhibit superior tolerance of biotic and abiotic stress factors such as rice yellow mottle virus (Attere and Fatokun 1983; Albar et al. 2003; Paul et al. 2003), rice blast (Somado et al. 2008), stem borders (Rodenburg et al., 2015), soil acidity (Somado et al. 2008), and soil salinity (Awala et al. 2010). Arguably, the NU rice grains exhibit high protein content (~ 10%) which makes for better nutritious value for the African consumers (Somado et al. 2008).

The NU varieties would be useful not only as maternal lines in rice breeding but also intermediary breeding materials that can transfer genes underlying useful agronomic traits from the parental lines into other varieties. The donor species *O. glaberrima* has attracted attention as a source of biotic and abiotic stress tolerance genes: i.e. genes for bacterial leaf blight, cyst nematode, iron toxicity, drought and weed competitiveness (Sano et al. 1984; Jones et al. 1997; Lorieux et al. 2003; Haefele et al. 2004; Singh et al. 2004; Sarla and Swamy 2005; Majerus et al. 2007; Vikal et al. 2007). In addition, the recurrent *O. sativa* parents seem to have useful genes for African rice breeding because those were selected from hundred lines under the local climate (Jones et al. 1997). Traore et al. (2011) reported that WAB56–104 showed long grain phenotype (~ 7.7 mm), low amylase content, and relatively short duration. Recently, a draft genome sequence of the donor parent of *O. glaberrima* accession CG14 was reported (Wang et al. 2014), however, there is a lack of genetic information on NU and the recurrent parental varieties.

Understanding on genome architecture of the NU varieties would promote their appropriate and informed utilization in rice breeding for African countries. However, it is rather vague if deciphered from the existing genomic profiles. Genotyping by the limited number of DNA markers provided evidence for macro-level similarity among NU genomes and also some non-parental alleles (Semagn et al. 2007; Fukuta et al. 2012; Pariasca-Tanaka et al. 2015). Recently, four NU varieties (1, 2, 8 and 9) were sequenced at 10 to 15X depth as a part of the rice 3000 genome project (Li et al. 2014; Alexandrov et al. 2015). An in-depth analysis of their genome sequences remains to be achieved.

In the present study, we conducted comparative resequencing analysis of the NU genomes using the Illumina high-throughput sequencing technique. We resequenced the genomes of four additional NU varieties (3, 4, 5, and 7) as well as of the two parental lines. Genome size and homozygosity were examined by k-mer analysis of the sequencing data. We then performed an integrated sequence analysis of all eight NU varieties for single nucleotide polymorphisms (SNPs), multi nucleotide polymorphisms (MNPs) and short insertion and deletions (InDels). Application of the whole genome polymorphisms allowed prediction and characterization of the NU genome structures at high resolution. Enrichment of Gene Ontology on common and variety/lineage-specific introgression segments from CG14 was tested for biased gene distributions. Analysis of chromosomal segment patterns identified gene candidates that are associated with phenotype. Gene information from WAB56–104 unmapped sequences, which were not mapped on the Nipponbare genome sequence properly, were analyzed and integrated with the four NU varieties sequenced in this study.

Methods

Whole genome sequencing

Rice seeds were distributed from Africa Rice Center (Côte d'Ivoire). Genomic DNA was prepared from etiolated seedlings cultured in the half strength of Murashige & Skoog liquid medium (Murashige and Skoog 1962). DNA extraction was performed using the method described by Doyle (1987) with a few modifications. Integrity of DNA was analyzed by electrophoresis in 2% agarose, and DNA samples were subjected to high-throughput sequencing by Illumina HiSeq 4000 in Macrogen Inc. (Seoul, Korea). In the sequencing analysis, DNA library was constructed from fragmented genomic DNA (ranging from approx. 0.2 to 10 kb) using TruSeq DNA PCR-FREE (350 bp) (Illumina), and 101 bp paired-end short reads were generated in fastq format. The obtained raw data were processed by the Trimmomatic software version 0.33 (Bolger et al. 2014) with the options of 'LEADING:15, TRAINLING:15, SLIDINGWINDOW:4:15, MINLEN:20' to trim adapter and low-quality sequences. The raw sequencing data were deposited to DDBJ Sequencing Read Archive (accession number: DRA006795).

k-mer analysis

The pre-processed high quality reads were applied to count k-mers by the software KMC 2 version 2.3.0 (Deorowicz et al. 2015). Genome size prediction was carried out using the software GCE version 1.0.0, which utilizes a Bayes model based method (Liu et al. 2013). Reference data were downloaded from NCBI Sequencing Read Archive for cv. Nipponbare (SRR545231, SRX179262, DRR028131, DRR028132, DRR083658, and SRR1043564), IR64 (SRR3098100), and Moroberekan (DRR003661).

Flow cytometry analysis

Genome DNA contents (pg/2C) were determined by a flow cytometry analyzer EC800 (Sony Biotechnology Inc., CA, USA). Nuclei of samples were prepared as follows: 1) fresh leaves were chopped into finely minced tissues using a razor blade and mixed with 600 μl of extraction buffer [50 mM Tris-HCl (pH 7.5), 0.5% polyvinypyrrolidone-K90, 0.01% Triton X-100, and 0.63% Na_2SO_4] at room temperature, 2) the supernatant was filtered via a single layer of nylon mesh with 25 μm pore size (NYTAL P-25, SEFAR AG, Ruschlikon, Switzaland), 3) the filtrate was subjected to RNase treatment and staining by propidium iodide (PI) with 3 μl of RNase A (10 mg/ml) and 6 μl of Propidium Iodide (PI) Solution (0.5 mg/ml, Sony Biotechnology) for 30 min at room temperature under dark. Signals of PI-stained nuclei were obtained by the flow cytometer with a band pass filter which transmits 570 ~ 620 nm excitation of PI at 40 μl/min of flow rate. At least 1000 nuclei were analyzed per measurement. Peak areas of given signals were used for calculation of DNA contents. Each measurement was with technical duplicates and four biological replicates. A linear standard curve was made based on *Arabidopsis thaliana* (Col-0), *Brachypodium distachyon* (L.) Beauv., *Solanum lycopersicum* (cultivar Micro-Tom), and *Zea mays* (cultivar Peter corn). DNA contents of these samples were defined as 0.32, 0.70, 2.0, 5.3 pg/2C, respectively, based on prior knowledge (Arumuganathan and Earle 1991; Bennett et al. 2003; Bennett and Leitch 2005). Determined DNA contents were converted into haploid

genome size using the conversion factor of 1 pg DNA = 978 million base pairs (Doležel et al. 2003). The rice reference cultivar Nipponbare was analyzed as a control for the African varieties.

Bioinformatics pipeline for polymorphism mining

In order to mine polymorphism candidates, we prepared a bioinformatics pipeline. Processed sequence reads were mapped against the rice genome (16 pseudomolecules of cultivar Nipponbare, chromosome (Chr) 1 to 12, ChrUn, ChrSy, plastid, and mitochondrial genome contigs of MSU 7.0: http://rice.plantbiology.msu.edu/) by Burrows-Wheeler Aligner MEM (Li and Durbin 2009) ver. 0.7.12 with option of '−O 4'. This mapping condition was determined after comparisons under conditions with different gap penalty in a small batch pre-analysis. Utilization of the full set of pseudomolecules made it possible to reduce mismapping between homologous loci between pseudomolecules. Generated mapping data file (SAM file) was processed for filtering out discordant paired reads using in-house Perl scripts. Filtered SAM file was converted into a BAM file and sorted by SAMtools ver. 0.1.19 (Li et al. 2009). Sorted BAM files of 10 genotypes were co-realigned by the Genome Analysis Tool Kit ver. 3.5 (McKenna et al. 2010). Resultant mapping data were converted into a pileup format file by SAMtools with the command 'mpileup' with '−Q 0' option. Sequence variants between the parental lines and NERICA to either parental line were called if the read depth was no less than 20 for WAB 56−104, CG 14, and NERICA 7, 25 for NERICA 5, 21 for NERICA 4, 6 for NERICA 9, 5 for NERICA 1, 18 for NERICA 3, 4 for NERICA 2 and NERICA 8 and coincidence within genotype was no less than 80% for homozygous polymorphism mining. SNPs and MNPs were called using an in-house Perl script 'SNiPer2'. Short Indel sites were called using the software VarScan2 version 2.3.9 with the command of somatic and the options of '−min-base-qual 0'.

Cleaved amplified polymorphic sequences (CAPS) assay

Genomic DNA surrounding SNP candidates were amplified by PCR using KAPA3G Plant PCR Kit (KapaBiosystems, Inc.). PCR products were digested by a restriction enzyme and electrophoresed in agarose gel with presence of FastStart SYBR Green Master (Roche Diagnostics GmbH) at 100 V for 60 min. Gels were imaged using a Molecular Imager Gel Dox XR System (Bio-Rad Laboratories, Inc.).

Survey for structural variations

A structural variant detection tool Manta version 1.1.1 (Chen et al. 2016) was employed for calling translocation breakends, inversions, tandem repeats, long insertions and long deletions with the default condition. Raw data files (VCF format) were converted into the BEDPE format using SVtools (https://github.com/ctsa/svtools) for analysis. Potential structural variants after QC filtering were selected. Copy number variations (CNVs) were searched using CNVnator version 0.3.3 (Abyzov et al. 2011).

Prediction of chromosomal segment type

All the SNPs, MNPs, and short InDel sites were aligned into a data matrix. Potential de novo mutation sites, which showed the same allele between the parental varieties but a different common allele in NU varieties, were excluded for reducing artifacts. Genetic imputation was carried out to infer chromosomal segments derived from WAB56–104, CG14, or non-parental varieties.

Gene set enrichment analysis

Gene Ontology enrichment was examined by the agriGO2 web tool (Tian et al. 2017). 'Singular Enrichment Analysis' was applied with the default parameters. Statistical analysis options of hypergeometric test with multi test adjustment method of Holm were selected.

De novo assembly

Full-length of unmapped paired end reads of WAB56–104 were collected and probable contaminant sequences were excluded based on results of BLASTN against human, *E. coli* and yeast genomic sequences (98% identity of 101 bp). Resultant paired-end reads were applied to de novo assembly using SOAPdenovo2 version 2.40 (Luo et al. 2012) with different k-mer from 21 to 81. The assembly result with the longest N50 length was obtained with k-mer of 67. Then, series of ambiguous sequence 'N' of the genomic scaffolds were filled using the software the GapCloser module version 1.12 (Luo et al. 2012).

Gene prediction and annotation

Repeat sequences on genome scaffolds were masked using RepeatMasker version open-4.0.6 with rmblastn version 2.6.0, RepBase Update 20,160,829 and RM database version 20,160,829. Then, structural gene annotations were given using Augustus web interface (http://bioinf.uni-greifswald.de/augustus/submission.php; Keller et al. 2011). Predicted protein sequences were annotated based on BLAST searches (Blast+ version 2.4.0) against the manually annotated protein database Swiss-Prot (Boeckmann et al. 2005), all proteins in rice (MSU7), in Arabidopsis (TAIR10, Lamesch et al., 2012), in maize (maizeGDB AGPv4, Andorf et al. 2015), in *Sorghum bicolor* (PlantGDB the number 79, Duvick et al. 2008), and in Brachypodium (PlantGDB the number 192). GO terms were assigned using BLAST2GO version 4.1.9 (Conesa and Götz 2008).

Mapping of unmapped short reads

Unmapped paired end reads were collected and mapped upon the genomic scaffolds of unmapped sequences from WAB56–104 using BWA mem with the default condition. After filtering out discordant and multi-mapped paired reads using in-house Perl scripts, averaged read depth on each gene transcribed region was calculated in each genotype.

Results and discussion

Whole genome re-sequencing

Re-sequencing data for NU3, 4, 5, and 7, and for WAB56–104 and CG14 were obtained at 32.5–53.8X coverage. In total 1.48 billion short reads with the length of 101 bp were generated (Additional file 1: Table S1). Analysis of the sequence reads revealed k-mer distributions suggesting homozygosity of the sequenced genomes (Additional file 2: Figure S1); a small peak (k-mer depth = 16) was observed in NERICA 5, and it could be contaminants in the sequencing steps. Genome sizes were predicted using the k-mer distributions; CG14 genome exhibited the smallest size of 399.8 Mbp, WAB56–104, NU3 and NU7 exhibited larger sizes ranging from 444.9 to 455.4 Mbp, and NU4 and NU5 genome sizes were intermediate between the parental lines (Table 1). To evaluate the reliability of those predictions, we analyzed publicly available sequencing data in cv. Nipponbare (temperate *japonica*), IR64 (indica), and an African variety Moroberekan (tropical *japonica*). The result of Nipponbare ranged from 371.1 to 421.3 Mbp, which was close to the actual genome size of 384.2 to 386.5 Mbp (Kawahara et al. 2013). The result of IR64 (355.6 Mbp) was smaller than that of Nipponbare, being consistent with the sizes of genome assemblies in both cultivars (Schatz et al. 2014). The result of Moroberekan was 446.8 Mbp, which was nearly identical to that of WAB56–104. These data supported the size diversity and the larger genome sizes of NU varieties and the parental variety compared to the Nipponbare genome.

It was notable that all the six varieties we sequenced exhibited larger genome sizes compared to Nipponbare. There are still controversial hypotheses on the origin of differences in genome size in different *O. sativa* genotypes. Burr (2002) reviewed the predicted size ranged from 403 to 430 Mbp, while Kawahara et al. (2013) mentioned the actual size of the sequenced Nipponbare genome was between 384.2 and 386.5 Mbp. Miyabayashi et al. (2007) observed that the DNA content of a wild rice *Oryza glumaepatula*, which has the AA genome, was 15.4% higher than that of the Nipponbare genome. To evaluate our k-mer-based estimation, nuclear DNA contents of the African varieties were determined by measuring PI-stained nuclei using flow cytometry. The estimated genome size by this way ranged from 392.

Table 1 Prediction of genome size and sequencing depth

Variety	k-mer analysis		Flow cytometry	
	Genome size (Mb)	Sequencing depth (X)	DNA content (pg/2C)	Genome size (Mb)
WAB 56–104	446.982	38.9	0.896 ± 0.016	438.27 ± 7.86
CG 14	399.771	43.7	0.802 ± 0.006	392.06 ± 2.83
NERICA 3	444.887	32.5	0.849 ± 0.016	415.16 ± 7.65
NERICA 4	425.708	45.5	0.878 ± 0.012	429.46 ± 5.91
NERICA 5	424.061	53.8	0.845 ± 0.019	412.96 ± 9.48
NERICA 7	455.363	37.8	0.914 ± 0.024	446.70 ± 11.87
Nipponbare	371.142–421.300	–	0.863 ± 0.014	421.76 ± 6.71
IR64	355.615	–	–	–
Moroberekan	446.806	–	–	–

k-mer was set as "17"

1 Mb of CG14 to 446.7 Mb of NU7 (Table 1). The genome size of Nipponbare (421.8 Mb) was larger than those of the latest genome assembly, but it was close to the range given from the *k*-mer analysis. The DNA content of Nipponbare was smaller than those of NU7 and WAB 56–104 with statistical significance (Student's *t*-test, at 10% level) and larger than that of CG14 with statistical significance (Student's *t*-test, at 1% level). A smaller genome size of *O. glaberrima* than those of *O. sativa* was previously reported by Martínez et al. (1994) and Uozu et al. (1997). Remaining three varieties did not show any clear difference although their values were numerically different from that of Nipponbare. Taken together, one common observation which was not inconsistent with previous studies was that the recurrent parent WAB 56–104 had a larger genome than Nipponbare, and it is likely to be inherited in NU7. If our relative estimation of WAB56–104 and Nipponbare genome size is trustable, the WAB56–104 genome could have unique genetic information that was not found on the Nipponbare genome assembly. We examined this possibility by de novo assembly of unmapped sequence reads later (section "Unmapped sequences from WAB56-104").

Genomic comparison in the parental varieties
Parental SNPs, MNPs, and short InDels were mined using a bioinformatics pipeline. To consider genome sequence diversity between *O. sativa* and *O. glaberrima* (Wang et al. 2014), we set the mapping parameters after optimization. Against the Nipponbare reference genome, 3,088,818 SNPs, 103,568 MNPs, and 404,070 potential InDels were found (Table 2). Approx. 87.8% and 11.9% of the parental SNPs represented common nucleotide alleles between Nipponbare and WAB56–104 and Nipponbare and CG14, respectively. The MNPs were comprised of four categories: di, tri, tetra, and penta-nucleotide polymorphisms, and di-MNPs were the most abundant (97.0%; Additional file 3: Table S2). Majority

(81.8%) of the MNPs represented the same nucleotide types as Nipponbare, but 6.2% were of CG14. For the short InDels, the length ranged from 1 to 45 bp, and 90% of these were 1–4 bp length (Additional file 4: Table S3). Most of the InDels (90.9%) were detected as changes in CG14 compared to the alleles of WAB56–104 and Nipponbare. These results are consistent with the knowledge that tropical *japonica* rice is genetically closer to temperate *japonica* than to *O. glaberrima*. More distant genetic relationship of WAB56–104 to *O. glaberrima* accessions was observed to Nipponbare (Semon et al. 2005). However, the actual similarity between Nipponbare and CG14 at polymorphic sites might be closer because a technical bias would exist when the sequence reads of the parental lines were aligned upon the Nipponbare genome sequence. Total of 42 polymorphic sites across all the chromosomes were tested by CAPS assays, and 40 of them were validated (Additional file 5: Table S4), suggesting more than 95% of mined polymorphisms in this study are genuine.

Hundred sites of structural variation between *O. sativa* and *O. glaberrima* were reported previously (Hurwitz et al. 2010). We searched potential structural variants, translocation breakends, inversions, tandem repeats and long InDels between WAB56–104 and CG14 *in silico*. By the reciprocal somatic calls, we predicted 8661 of translocation breakends, 329 of inversions, 320 of tandem duplications, and 4715 of long InDels (Table 3). The structrural variants were distributed across all the chromosomes (data not shown). The average length of structural variants was as follows: 1.5 Mbp for inversion, 59.4 kbp for tandem duplication, 101.3 kbp for long deletion, 1.7 kbp for long insertion. We also searched structural variations between Nipponbare and the two parental varieties separately. This computation revealed numerous structrural variants potentially existed between Nipponbare and the two parental varieties (Table 3). Number of structural variants in WAB56–104 ranged

Table 2 Polymorphic sites between WAB 56–104 and CG 14

Chr	Nucleotide polymorphism		Insertion		Deletion		Total
	SNP	MNP	WAB56–104	CG 14	WAB56–104	CG 14	
1	360,381	11,386	1972	20,132	2804	24,478	49,386
2	316,206	10,336	1564	17,498	2320	21,101	42,483
3	306,733	9299	1293	16,964	1879	20,338	40,474
4	270,022	9601	1370	14,362	2061	17,316	35,109
5	251,323	8021	1377	13,075	2055	15,560	32,067
6	271,081	9447	1413	14,354	1997	17,557	35,321
7	254,203	8756	899	14,537	1314	17,184	33,934
8	221,116	7317	1131	11,605	1632	14,365	28,733
9	190,775	6350	666	10,347	893	12,284	24,190
10	201,301	7146	1771	9125	2600	11,457	24,953
11	238,964	8611	1884	11,917	2360	14,295	30,456
12	203,812	7230	1527	10,421	2084	12,631	26,663
Un	538	9	3	20	5	32	60
Sy	2363	59	7	113	2	119	241
Total	3,088,818	103,568	16,877	164,470	24,006	198,717	404,070

from 32.9% (long insertion) to 48.7% (tandem duplication) of those in CG14. Average length of inversions and tandem duplications in CG14 (1.54 Mbp and 65.5 kbp, respectively) was close to those in WAB56–104 (1.35 Mbp and 55.2 kbp, respectively), while average length of long InDels in WAB56–104 (1.8 kbp and 7.0 kbp, respectively) was longer than those in CG14 (1.4 kbp and 5.9 kbp, respectively). This observation is due to much more InDels, which are of relatively shorter length, found in CG14. Experimental validation of those structural variations remains to be performed.

Polymorphisms in NU varieties
For polymorphisms among NU varieties, the new sequencing data of the four NU genomes 3, 4, 5 and 7 were used along with the publicly available sequences of the other four NU genomes 1, 2, 8, and 9 (Additional file 6: Table S5). Polymorphisms between each NERICA and both parents were called, and a total of 4,461,719 polymorphic sites were found for SNPs, MNPs and short InDels (Table 4). Population structure of the 10 varieties (8 NUs and 2 parents) was

analyzed using common polymorphic sites (1,026,367 sites). Principal component analysis (PCA) was used to dissect variance of allelic divergence of the 10 genomes (Fig. 1). PC1, which explained 70.9% variance, represented a large part of the difference between the two parental genomes. PC2 and PC3, which explained 14.7% and 8.22% variance, respectively, represented differences among the NU varieties. NU3 was plotted at the same location as NU4, and at a different site NU8 and NU9 were plotted at the same location. This could most likely be due to NU3 being a sister line of NU4 and NU8 being a sister line of NU9.

The number of candidate polymorphisms discovered revealed that the NU varieties contained the donor alleles in different proportions, which ranged from 1.32 to 9.03% (Table 4). For example, NU5 exhibited the highest proportion of the CG14-specific SNPs (9.81%). Comparatively lower proportions of the CG14-specific SNPs at 1.32 and 1.33% were noted in NU3 and NU4 respectively. These results reflected different amount of introgression segments from CG14. Frequency of the non-parental SNPs among the total SNPs ranged from

Table 3 Structural variants in the parental lines and Nipponbare

Comparison	Interchromosomal translocation breakend	Inversion[*1]	Tandem duplication[*2]	Long deletion[*3]	Long insertion[*3]
WAB 56–104 vs. CG14	6807	283	244	3741	584
CG14 vs. WAB56–104	1238	46	76	336	54
Nippobare vs. WAB 56–104	3346	165	154	2222	282
Nippobare vs. CG14	9176	381	316	6071	857

*1 Inversion between 100 bp to 10 Mbp
*2 Tandem duplication between 100 bp to 1 Mbp
*3 Insertion and deletion between 50 bp to 1 Mbp

Table 4 Number of polymorphisms in the upland NERICA

Genotype	Category	Allele type			CG 14 allele (%)		Non-parental allele (%)	
		WAB56–104	CG 14	Non-parental	Category	All	Category	All
NERICA 1	SNP	2464390	247641	101458	8.80	9.03	3.61	3.81
	MNP	79607	6336	8975	6.68		9.46	
	Short InDel	324348	43036	14852	11.26		3.89	
NERICA 2	SNP	1485308	148296	42301	8.85	8.68	2.52	2.65
	MNP	46795	3990	3566	7.34		6.56	
	Short InDel	171010	14367	5123	7.54		2.69	
NERICA 3	SNP	2989656	41101	39623	1.34	1.32	1.29	1.39
	MNP	98929	671	3991	0.65		3.85	
	Short InDel	333122	4601	5371	1.34		1.57	
NERICA 4	SNP	3008891	41574	40255	1.35	1.33	1.30	1.41
	MNP	99336	681	4059	0.65		3.90	
	Short InDel	322610	4520	5457	1.36		1.64	
NERICA 5	SNP	2637694	293878	79951	9.76	9.81	2.65	2.91
	MNP	85743	8884	7838	8.67		7.65	
	Short InDel	275398	34231	12349	10.63		3.84	
NERICA 7	SNP	2471034	117058	82482	4.38	4.32	3.09	3.36
	MNP	74904	1816	8946	2.12		10.44	
	Short InDel	272139	12842	11127	4.34		3.76	
NERICA 8	SNP	1476391	98444	96093	5.89	5.82	5.75	6.05
	MNP	46196	1787	8550	3.16		15.12	
	Short InDel	164605	11231	11142	6.01		5.96	
NERICA 9	SNP	1786210	120441	116935	5.95	5.88	5.78	6.10
	MNP	54940	2150	10599	3.18		15.66	
	Short InDel	203351	13928	14197	6.02		6.13	

1.39 to 6.10%. These numbers were compatible with the frequency of non-parental loci of 2.2% obtained by SSR maker-mediated genotyping (Semagn et al. 2007). Non-parental allele polymorphisms were the most frequent in NU9 and the least frequent in NU3 and NU4 (Table 4). Notably, the rates of CG14 allele and non-parental allele were at the same level, which implied that mutations, chromosomal rearrangements, or perhaps out-crossing contributed to the diversity of the NU genomes.

Chromosomal structure in NU varieties

The unique breeding processes of NU varieties could bring genomic alterations in various ways: de novo mutations, chromosomal rearrangement, translocation, chromosomal loss, etc., which might be required for stabilization of hybrid genomes and adaptive evolution (Baack and Rieseberg 2007; Rieseberg and Willis 2007; Morales and Dujon 2012). The chromosomal structure of the NU varieties was inferred using genetic imputation, by considering

all polymorphic sites to predict the recipient parent segments of WAB56–104, the introgression segments from CG14, and potential non-parental segments. For the proportion of CG14 segments among NU varieties a wide range of 1.40 in NU3 and 4 to 10.1% in NU2 was noted. Similarly, the potential non-parental segments ranged from 0.090 of NU3 and 4 to 2.74% of NU9. Graphical genotype revealed the chromosomal structures of the NU varieties (Fig. 2). Reiterating the result from PCA, NU2 and 5, NU3 and 4 and NU8 and 9 exhibited highly related genotypes, while NU2 and 5 revealed large differences in introgressions from CG14 on Chr 4. NU1 and 5 exhibited similarities at limited regions for example at the periphery of the centromeric region on the Chr 2, 3, 7 and 11, the upper edge of Chr 6, 7, and 9, and the bottom edge of Chr 4. These results suggested that the known NU varieties originated from a limited number of independent interspecific hybridization events but that sister lines were identified as independent events, adding to the number of apparently independent NU varieties. Frequent distribution of short

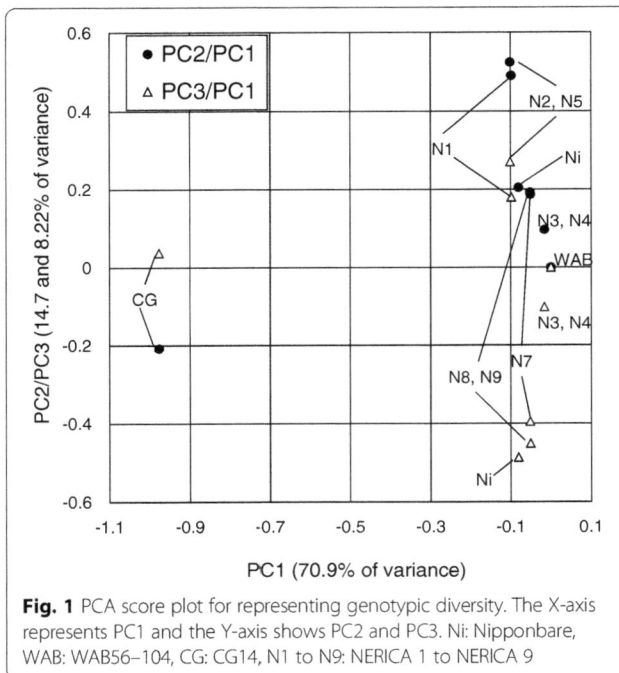

Fig. 1 PCA score plot for representing genotypic diversity. The X-axis represents PC1 and the Y-axis shows PC2 and PC3. Ni: Nipponbare, WAB: WAB56-104, CG: CG14, N1 to N9: NERICA 1 to NERICA 9

non-parental segments could be due to a combination of factors such as structural variants between Nipponbare and the parental lines, de novo mutations and genomic alterations at the chromosomal level. However, some large non-parental segments were observed. For example, in NU1 the middle part of Chr 8, which contained a fragrance gene *BADH2* from WAB638-1 (Asante et al. 2010). Chr 1 in NU5 and Chr 7 in NU7 also indicated non-parental segments. In the previous genotyping by microsatellite markers (Semagn et al. 2007), those non-parental segments were not detected.

Common introgression segments from CG14 across the eight NU varieties totaled 637,849 bp, composed from 3168 segments. These were frequent in Chr 6 (1608 segments) and 10 (1143 segments) (Additional file 7: Table S6). Also, the common introgression segments were closely distributed (within 1.4 to 1.8 Mbp) on each of these two chromosomes. For example on Chr 6, they occurred from 5,613,148 to 7,010,047 and on Chr 10 from 174,587 to 1,974,978. These fragmented segments might be due to structural variations between NERICA and Nipponbare/WAB56-104 or due to recombination between unlinked genes. The longest introgression segment was 24,125 bp on Chr 7 (3,885,715-3,909,839), which was followed by 23,541 bp on Chr 10 (1,501,250-1,524,790). To examine gene composition on the common introgression segments from CG14, enrichment of GO terms were tested. It appeared that 11 GO terms, including 'cell wall biogenesis' [GO:0042546] and 'transferase activity, transferring glycosyl groups' [GO:0016757], were

over-represented. That enrichment was due to five xyloglucan fucosyltransferase genes that were located at the same locus (Chr 6: 5,701,639-5,734,237) (Additional file 8: Table S7). Although this enrichment alone did not look biologically informative, this locus was co-localized with quantitative trait loci (QTLs) of typical agronomic traits and grain quality between *O. sativa* and *O. glaberrima* (Lorieux et al. 2000; Aluko et al. 2004). The facts might imply that the locus containing the xyloglucan fucosyltransferase genes was positively acted on during the selection process for developing the NU varieties. Since the region of Chr6: 5,613,148 to 7,010,047 contains 207 genes in the Nipponbare genome, we could not specify genes relevant to the QTLs in this study.

Xyloglucan fucosyltransferase catalyzes transfer of Fucose (Fuc) from GDP-Fuc to a galactosyl residue of xyloglucan in the process of fucogalactoxyloglucan biosynthesis. Xyloglucan is a key component in cell walls and can play a role in cell wall structure and function. Recently, fucogalactoxyloglucan, which is widely distributed across dicot plants, was surprisingly detected in root hairs and anthers in rice (Liu et al. 2015), however, the biological function of fucogalactoxyloglucan is still unclear. The five xyloglucan fucosyltransferase genes were categorized into three groups based on their protein sequence similarities (Additional file 9: Figure S2). *LOC_Os06g10910* is a candidate fucosyltransferase coexpressed with a xyloglucan glycosyltransferase *OsCSLC3* (Liu et al. 2015). *LOC_Os06g10920* was a probable ortholog of *AtFUT3*, of which overexpression altered cell wall composition and detrimental (Sarría et al. 2001). We observed several CG14-alleles that were introgressed into NU varieties caused amino acid substitutions in these genes (Additional file 10: Figure S3). Many CG14-alleles were also observed in the promoter regions. Cell wall in plants is the primary component that can determine the physical properties and protects them from against biotic and abiotic stress (Zabotina 2012). These genes from CG14 might affect on agronomic characteristics in the upland NU varieties.

Variety-specific/lineage-specific introgression segments were also predicted in all the NU varieties (Additional file 11: Table S8). The total nucleotide ranged from 81.9 kbp in NU4 to 6.85 Mbp in NU8 and 9 lineage. Number of genes with the segments varied from 218 in NU4 to 4906 in NU8 and 9 lineage. Various kinds of genes that would be from CG14 were found (Additional file 12: Table S9). Gene Ontology enrichment analysis (GOEA) indicated 109 over-represented GO terms, implying biased distribution of particular genes in NU1, 2, 5, 7 and the two lineage of NU2 and NU5, and NU8 and NU 9 (Additional file 13: Table S10). The listed GO categories might relate to their variety-specific characteristics, however, the biological meaning remains to be examined.

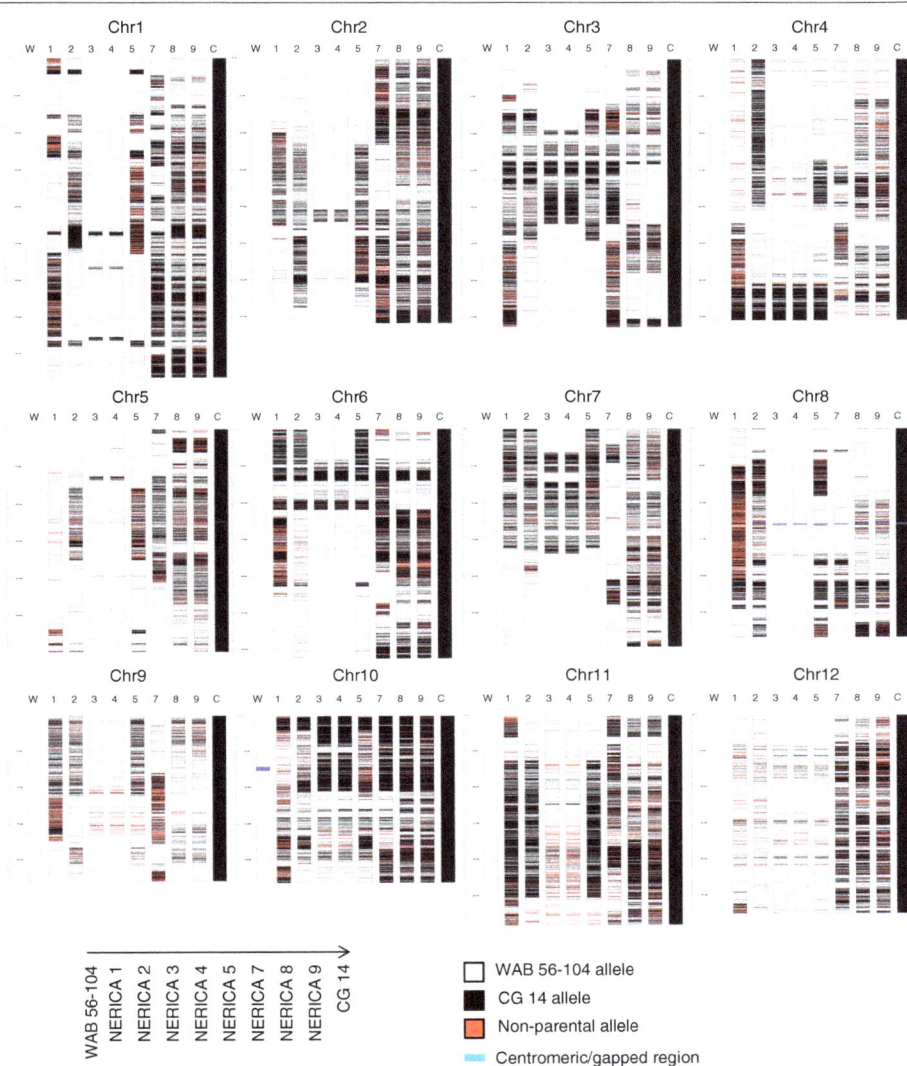

Fig. 2 Graphical genotype of the upland NERICA and the parental lines. W: WAB56–104, 1 to 9: NERICA 1 to NERICA 9 and, C: CG14. Introgression segments from CG14 are colored in black. Non-parental segments are represented in red. Centromeric and gap regions are shown in light blue. Scale bar is standardized as 100 bp

Gene composition relevant to traits

Grain shattering habit is the major constraint in *O. glaberrima* rice production (Africa Rice Center 2008). In agreement with disappearance of this trait in the NU varieties, key genes for grain shattering, *SHATTERING 3/SEED SHATTERING 4* (LOC_Os04g57530, Li et al. 2006), *qSH1* (LOC_Os01g62920, Konishi et al. 2006), and *SHAT1* (LOC_Os04g55560, Zhou et al. 2012), were on chromosomal segments of WAB56–104 (Fig. 3). In the case of *SHATTERING 3*, five of non-synonymous parental polymorphisms were on the coding region, and one of the polymorphisms was the identical allele associated with shattering habits (Asp for non-shattering, Lys for shattering) (Fix. 3a). The causal allele of *qSH1* was not polymorphic among the African varieties, however,

one non-synonymous SNP was found in CG14 (Fix. 3b). This SNP locates on the homeobox KN domain [PF05920] and substitutes Thr of WAB 56–104 into Pro of CG14. The structural and biochemical properties of the two amino acids might affect the function of the *qSH1* gene. *SHAT1* exhibited a mixed allelic pattern in the genic region while the probable promoter region of NU varieties consistently showed only of WAB 56–104 alleles (Fig. 3c). The corresponding CG14 segments, which are responsible for shattering, would be excluded in the processes of backcrossing followed by the artificial selection.

Discovery of trait-associated alleles helps in elucidating the genetic reasons underlying target trait variation. The high-resolution graphical genotype is useful to check

a

Chr.	Position	NI	WAB	NU1	NU2	NU3	NU4	NU5	NU7	NU8	NU9	CG14	Location	CDS posi.	A.A. posi.	Substitution type
Chr4	34231101	A	A	A	A		A	A	A	A		+AT	external	-	-	-
Chr4	34231570	G	G	G		G	G	G	G			C	intron	-	-	-
Chr4	34231586	A	A	A		A	A	A	A			C	intron	-	-	-
Chr4	34231673	A	A	A		A	A	A	A			T	intron	-	-	-
Chr4	34231674	C	C	C		C	C	C	C			T	intron	-	-	-
Chr4	34231690	A	A	A		A	A	A	A			G	intron	-	-	-
Chr4	34231812	T	T	T	T	T	T	T	T			C	intron	-	-	-
Chr4	34231847	C	C	C	C	C	C	C	C			G	intron	-	-	-
Chr4	34231896	T	T	T		T	T	T		T		+G	intron	-	-	-
Chr4	34231908	C	C	C		C	C	C	C	C		T	intron	-	-	-
Chr4	34231922	A	A	A		A	A	A	A	A		T	intron	-	-	-
Chr4	34231959	C	C	C		C	C	C	C	C		T	intron	-	-	-
Chr4	34231983	G	G	G		G	G	G	G			C	intron	-	-	-
Chr4	34232172	T	T	T	T	T	T	T	T		T	A	intron	-	-	-
Chr4	34232284	G	G	G			G	G	G	G		+C	intron	-	-	-
Chr4	34232292	A	A	A		A	A	A	A	A		G	intron	-	-	-
Chr4	34232350	T	T	T		T	T	T	T			C	intron	-	-	-
Chr4	34232493	C	C		C	C	C	C	C	C		G	coding	729	243	Glu to His
Chr4	34232542	G	G			G	G	G				-GCA	coding	680	226, 227	Leu-Pro to Pro
Chr4	34232931	G				G						A	coding	291	97	synonymous
Chr4	34232979	C	C			C	C	C				G	coding	243	81	synonymous
Chr4	34232985	A	A				A	A				C	coding	237	79	Asp to Lys
Chr4	34233097	C	C			C	C	C	C			G	coding	125	42	Gly to Ala
Chr4	34233166	A				A	A	A	A			G	coding	56	19	Val to Ala
Chr4	34233186	G	G			G	G	G	G			C	coding	36	12	synonymous
Chr4	34233251	T	T			T	T	T	T		T	C	5'UTR	-	-	-
Chr4	34233276	G	G		G	G	G	G	G		G	C	5'UTR	-	-	-
Chr4	34233315	G	G		G	G	G	G	G		G	T	5'UTR	-	-	-
Chr4	34233353	G	G	G	G	G	G	G	G	G	G	T	5'UTR	-	-	-
Chr4	34233366	T	T	T	T	T	T	T		T	T	+G	5'UTR	-	-	-
Chr4	34233383	A	A	A	A	A				A	A	+C	external	-	-	-

b

Chr.	Position	Ni	WAB	NU1	NU2	NU3	NU4	NU5	NU7	NU8	NU9	CG14	Location	CDS posi.	A.A. posi.	Substitution type
Chr1	36444961	C	C	C	C	C	C		C	C	C	+TACG	external	-	-	-
Chr1	36445166	G	G	G	G	G	G	G	G	G	G	A	3'UTR	-	-	-
Chr1	36445248	T	T	T	T	T	T	T	T	T	T	G	3'UTR	-	-	-
Chr1	36445340	G	G	G	G	G	G	G	G	G	G	A	3'UTR	-	-	-
Chr1	36445388	G	G	G	G	G	G	G	G	G	G	C	3'UTR	-	-	-
Chr1	36445991	A	A	A	A	A	A	A	A	A	A	G	coding	1458	486	synonymous
Chr1	36446177	T	T	T	T	T	T	T	T	T	T	C	coding	1272	424	synonymous
Chr1	36446195	C	C	C	C	C	C	C	C	C	C	T	coding	1254	418	synonymous
Chr1	36446291	G	G	G	G	G	G	G	G	G	G	T	intron	-	-	-
Chr1	36446575	A	A	A	A	A	A	A	A	A	A	G	intron	-	-	-
Chr1	36446592	C	C	C	C	C	C	C	C	C	C	G	intron	-	-	-
Chr1	36446835	C	C	C	C	C	C	C	C		C	T	intron	-	-	-
Chr1	36446947	T	T	T	T	T	T	T	T	T	T	C	intron	-	-	-
Chr1	36447147	A	A	A	A	A	A	A	A	A	A	G	intron	-	-	-
Chr1	36447151	A	A	A	A	A	A	A	A	A	A	C	intron	-	-	-
Chr1	36447471	T	T		T	T	T	T	T	T	T	-AAG	intron	-	-	-
Chr1	36447526	A	A	A	A	A	A	A	A	A	A	G	coding	1153	385	synonymous
Chr1	36448043	T	T	T	T	T	T	T	T	T	T	C	intron	-	-	-
Chr1	36448048	G	G	G	G	G	G	G	G	G	G	A	intron	-	-	-
Chr1	36448399	T	T	T	T	T	T	T	T	T	T	C	intron	-	-	-
Chr1	36448455	G	G	G	G	G	G	G	G	G	G	-AA	intron	-	-	-
Chr1	36449085	T	T		T	T	T	T	T	T	T	G	coding	418	140	Thr to Pro
Chr1	36449086	C	C		C	C	C	C	C	C	C	T	coding	417	139	synonymous
Chr1	36449508	G	G			G	G	G	G		G	A	5'UTR	-	-	-
Chr1	36449808	T	T	T		T	T	T	T	T	T	G	5'UTR	-	-	-
Chr1	36449883	G	G	G	G	G	G	G	G		G	A	5'UTR	-	-	-
Chr1	36450141	G	G	G	G	G	G	G	G	G	G	T	external	-	-	-
Chr1	36461451	C	C		C	C		C	C	C	C	+T	upstream	-	-	-
Chr1	36461792	T	G	G	G	G	G	G	G	G	G	G	qSH1	-	-	-
Chr1	36462072	A	A	A	A	A	A		A	A	A	-G	upstream	-	-	-

c

Chr.	Position	Ni	WAB	NU1	NU2	NU3	NU4	NU5	NU7	NU8	NU9	CG14	Location	CDS posi.	A.A. posi.	Substitution type
Chr4	33069949	T	T	T		T	T	T	T			+A	external	-	-	-
Chr4	33069997	A	+T	+T			+T		+T			A	upstream	-	-	-
Chr4	33070016	G	A	G		G	G	G	A			G	upstream	-	-	-
Chr4	33070086	G	G	G	G	G	G	G	G			A	upstream	-	-	-
Chr4	33070225	C	-GG	-GG		-GG	-GG	-GG	-GG		-GG	C	upstream	-	-	-
Chr4	33070763	T	T	T		T	T	T	T			A	upstream	-	-	-
Chr4	33070845	G	G	G	G	G	G	G	G		G	A	upstream	-	-	-
Chr4	33070872	C	C	C	C	C	C	C	C	C	C	T	upstream	-	-	-
Chr4	33070927	G	G	G	G	G	G	G	G	G		T	upstream	-	-	-
Chr4	33070938	C	C	C	C	C	C	C	C		C	T	upstream	-	-	-
Chr4	33070942	C	C	C	C	C	C	C	C		C	T	upstream	-	-	-
Chr4	33070945	T	C	C	C	C	C	C	C		C	C	upstream	-	-	-
Chr4	33070953	G	A	A	A	A	A	A	A		A	G	upstream	-	-	-
Chr4	33070972	G	G	G	G	G	G	G	G			A	upstream	-	-	-
Chr4	33071015	G	A	A	A	A	A	A	A	A		G	upstream	-	-	-
Chr4	33071043	C	T	T		T	T	T	T	T		C	upstream	-	-	-
Chr4	33071057	G	A	A	A	A	A	A	A	A		G	upstream	-	-	-
Chr4	33071069	C	C	C	C	C	C	C	C	C	C	+TAAG	upstream	-	-	-
Chr4	33071093	A	A	A	A	A	A	A	A	A	A	A	upstream	-	-	-
Chr4	33071314	C	A	A	A	A	A	A	A		A	C	upstream	-	-	-
Chr4	33071393	T	+C			+C	+C		+C	+C		A	upstream	-	-	-
Chr4	33071819	C	C	C	C	C		C			C	+CT	5'UTR	-	-	-
Chr4	33071867	C	C	C	C	C	C	C	C		C	T	5'UTR	-	-	-
Chr4	33072371	T	T	T		T	T	T	T	T	T	C	intron	-	-	-
Chr4	33072877	A	A	A	A	A	A	A	A	A	A	G	intron	-	-	-
Chr4	33073106	C	C	C	C	C	C	C	C	C	C	A	intron	-	-	-
Chr4	33073108	G	A	A	A	A	A	A	A	A	A	G	intron	-	-	-
Chr4	33073441	G	A	A	A	A	A	A	A	A	A	C	intron	-	-	-
Chr4	33073447	C	A	A	A	A	A	A	A	A	A	C	intron	-	-	-
Chr4	33073945	G	A	A	A	A	A	A	A	A	A	C	coding	940	314	Thr to Ala
Chr4	33074480	C	A			C	C	C	A	A	A	C	coding	1236	412	Gln to Pro
Chr4	33074493	C	C	C		C	C	C	C		C	A	coding	1278	426	synonymous
Chr4	33074673	G	G	G		G	G	G	G	G	G	+GA	3'UTR	-	-	-
Chr4	33074817	T	T	T	T	T	T	T	T	T	T	A	3'UTR	-	-	-
Chr4	33074818	G	G	G		G	G	G	G	G	G	T	3'UTR	-	-	-
Chr4	33074898	G	A	A	A	A	A	A	A	A	A	C	3'UTR	-	-	-
Chr4	33075124	T	C	C	C	C	C	C	C	C	C	T	3'UTR	-	-	-
Chr4	33075147	C	C	C	C	C	C	C	C	C	C	-TAT	3'UTR	-	-	-
Chr4	33075666	A	G	G	G	G		G	G	G	G	A	external	-	-	-

possibility of association with trait among the NU varieties and the parental varieties when genes, for which function was determined, showed allelic divergence. For example, Kishine et al. (2008) reported that the allele of *granule-bound starch synthase I* (*GBSSI*) from CG14 was associated with high grain amylose content in NU1, 2 and 5.Our graphical genotype indicated association of this gene with the corresponding CG14 introgression segment (Table 5, Additional file 14: Figure S4a). Also, a fragrant rice allele *BADH2*, which was found in NU1 (Asante et al. 2010), was located on the three non-parental segments of Chr8: 20379746–20,399,750 in NU1 genotype uniquely as expected (Table 5). Ikeda et al. (2009) documented the awnness in NU2 and 5, divergent apiculous color in NU1, 2 and 5, and distinct stigma color in NU1. It appeared that NU2 and 5 conserved CG14 alleles on *Regulator of Awn Elongation 1* (*RAE1*), which is one of genes involved in awn formation in CG14 (Furuta et al. 2015) (Table 5). We noted that *OsC1*, which is a determinant of anthocyanin accumulation in rice (Chin et al. 2016), was associated with apiculous color among the 8 NU varieties and the parental varieties (Table 5). Although these 'gene-trait relationships' need extensive analyses using other NU varieties or a rice diversity panel, the graphical genotype we have presented can be a starting point and also guides functional validation by molecular genetics.

Gene sequence information of the NU varieties is useful to explore divergent traits and their origin. Here we introduce one example each for biotic and *a*biotic stress tolerance (Table 5). *OsSWEET14* is a susceptibility gene to bacterial leaf blight in wild rice (Hutin et al. 2015). The corresponding locus in the graphical genotype showed that NU1, 2, and 5 had the CG14 alleles (Additional file 14: Figure S4b). Association analysis of those alleles and resistance to bacterial blight may lead to discover differences of NU varieties. Notably, Séré et al. (2005) reported the different susceptibility of NU1 (with

OsSWEET14 from CG14) and NU4 (without *OsSWEET14* from CG14).One of salt tolerance determinant gene *OsHKT1;5* (Platten et al. 2013), which encodes a Na$^+$ transporter, exhibited variation in the NU varieties. Namely, NU3 and NU4 were identical to WAB56–104, but other NU varieties contained non-parental segments in this gene region. This observation is consistent with the concept that salt tolerance in NU4 is derived from WAB56–104 as reported previously (Awala et al. 2010). Interestingly, Yamamoto et al. (2011) reported the similar physiological response of NU3 and NU4 to salt stress compared to other five NU varieties, while NU3 and NU4exhibited different profiles of total amino acids and polyamine. At the same time, Yamamoto et al. (2011) reported salt stress susceptibility of NU2 among the NU varieties. Those differences of NU varieties in salt stress response imply that multiple genetic components for salt stress adaptation are combined differently in their genomes. Platten et al. (2013) reported some *O. glaberrima* seem to have another salt exclusion mechanism that is independent on *OsHKT1;5*. Other seven Na$^+$ transporter genes (Platten et al. 2006) exhibited no difference in the chromosomal segments in NU3 and NU4 (Additional file 15: Table S11). Further investigation is required to specify the reason.

Unmapped sequences from WAB 56–104

More than 10% of the NU sequences were not mapped correctly on the rice reference genome of Nipponbare. To characterize those sequences, we performed de novo assembly of unmapped reads from WAB56–104. Total of 37,894 genomic scaffolds were constructed; those covered approx. 19 Mb of genomic scaffolds (Additional file 16: Table S12). The smaller value of total nucleotide of the genomics scaffolds than expected would be due to duplication of DNA segments. As evidence, a computational survey predicted 578 of chromosomal regions exhibiting higher copy number variation in WAB56–104 upon the

Table 5 Key genes segregating among the NU varieties

Gene abbreviation	Locus	Trait	Reference	Causal origin	Variety with the gene segment
GBSSI	LOC_Os06g04200	Grain amylose content	Kishine et al. (2008)	CG14	NU1, 2, and 5
BADH2	LOC_Os08g32870	Grain fragrance	Asante et al. (2010)	WAB638–1 (outcross)	NU1
RAE1	LOC_Os04g28280	Awnness	Furuta et al. (2015)	CG14	NU2 and 5
OsC1	LOC_Os06g10350	Purple pigment	Chin et al. (2016)	CG14	NU1, 2, and 5
OsSWEET14	LOC_Os11g31190	Bacterial leaf blight susceptibility	Hutin et al. (2015)	CG14	NU1, 2, and 5
OsHKT1;5	LOC_Os01g20160	Salt tolerance	Platten et al. (2013)	WAB56–104	NU3 and NU4

Table 6 Occurrence of WAB 56–104 specific genes in NU varieties and CG 14

Category	Gene identifier	Functional annotation		Gene occurrence					
		Gene of BLAST hit in Swiss prot	Species of BLAST hit in Swiss-prot	NU3	NU4	NU5	NU7	CG14	Remarks
Disease	g13.t1	Putative disease resistance protein RGA4	*Solanum bulbocastanum*	No	No	No	Yes	No	WAB 56–104 allele
	g51.t1	Putative disease resistance RPP13-like protein 3	*Arabidopsis thaliana*	Yes	Yes	Yes	No	No	WAB 56–104 allele
	g98.t1	Disease resistance protein RPP13	*Arabidopsis thaliana*	Yes	Yes	No	Yes	Yes	WAB 56–104 allele
	g108.t1	Putative disease resistance RPP13-like protein 1	*Arabidopsis thaliana*	No	No	No	No	Yes	Polymorphic in parents
	g342.t1	Chitin elicitor-binding protein	*Oryza sativa*	Yes	Yes	No	No	No	WAB 56–104 allele
	g632.t1	Putative disease resistance RPP13-like protein 2	*Arabidopsis thaliana*	Yes	Yes	No	Yes	No	WAB 56–104 allele
Receptor	g85.t1	Wall-associated receptor kinase-like 18	*Arabidopsis thaliana*	Yes	Yes	Yes	No	Yes	Both parental allele
	g115.t1	Wall-associated receptor kinase-like 18	*Arabidopsis thaliana*	Yes	Yes	Yes	No	Yes	Both parental allele
	g227.t1	Wall-associated receptor kinase 2	*Arabidopsis thaliana*	Yes	Yes	No	Yes	No	WAB 56–104 allele
	g162.t1	Receptor-like protein 12	*Arabidopsis thaliana*	Yes	Yes	Yes	No	Yes	WAB 56–104 allele in NU3 and NU4.CG 14 allele in NU5.
	g418.t1	G-type lectin S-receptor-like serine/threonine-protein kinase LECRK3	*Oryza sativa*	Yes	Yes	No	No	No	WAB 56–104 allele
Others	g173.t1	11-oxo-beta-amyrin 30-oxidase	*Glycyrrhiza uralensis*	No	No	Yes	No	Yes	Unspecified
	g268.t1	Uncharacterized protein	*Arabidopsis thaliana*	Yes	Yes	No	Yes	No	WAB 56–104 allele in NU3 and NU4. Unspecified in NU7.
	g338.t1	F-box/FBD/LRR-repeat protein	*Arabidopsis thaliana*	No	No	No	No	Yes	Polymorphic in parents
	g361.t1	Cytochrome P450 87A3	*Oryza sativa*	Yes	Yes	No	Yes	No	Both parental allele
	g427.t1	UPF0481 protein	*Arabidopsis thaliana*	Yes	Yes	No	No	No	WAB 56–104 allele
	g453.t1	Tetraspanin-8	*Arabidopsis thaliana*	Yes	Yes	No	No	No	WAB 56–104 allele
	g500.t1	Protein Brevis radix-like 2	*Arabidopsis thaliana*	Yes	Yes	Yes	No	Yes	WAB 56–104 allele
	g517.t1	UDP-glycosyltransferase 76E11	*Arabidopsis thaliana*	Yes	Yes	Yes	No	Yes	WAB 56–104 allele
	g597.t1	Basic blue protein	*Cucumis sativus*	Yes	Yes	Yes	No	Yes	WAB 56–104 allele
	g643.t1	Autonomous transposable element EN-1 mosaic protein	*Zea mays*	No	No	Yes	No	No	WAB 56–104 allele

Nipponbare genome (with E-value of 0.01), and notably, it was more than 460 sites in CG14. Hence, we assumed that some unmapped WAB56–104 sequence reads were mapped on highly related paralogous chromosomal regions in Nipponbare.

Structural annotation of long assembled genomic scaffolds (no less than 1 kb) by Augustus (Stanke and Morgenstern 2005) predicted 653 genes. A large part of the predicted gene proteins had homology with cereal proteins (Additional file 16: Table S12 and Additional file 17: Table S13). Notably, 95.6% (624) did not show strong sequence homology (less than 90% identity) with Nipponbare proteome sequences (Additional file 17: Table S13). Genes

with particular functional annotations showed frequent localization on the scaffolds with statistical significance (in Fisher's exact test at 5% level); 35 genes for disease resistance protein (p value = 0), 10 genes for wall-associated receptor kinase (WAK) (p value = 0), seven genes for leucine-rich repeat (LRR) receptor-like serine/threonine-protein kinase (p value = $5.6E^{-3}$), six genes for lectin receptor kinase (LRK) (p value = 0), and five genes for cytochrome P450 (p value = $4.4E^{-2}$). Potential *a*biotic stress-related genes (25) were obviously observed prominently. WAKs, LRR receptor-like serine/threonine-protein kinases, LRKs can be involved in pathogen resistance and *a*biotic stress response (Afzal et al. 2008; Kanneganti and Gupta 2008;

Vaid et al. 2013). A WAK was implicated as one of the NAM transcription factor regulated genes in drought tolerance (Dixit et al. 2015). Those maternal genes might be the basis for the characteristics in WAB56–104, perhaps for biotic and *a*biotic stress tolerance. Sakai et al. (2011) observed more frequent distribution of proteins with protein kinase domains, leucine-rich repeat domains, and disease resistance gene on unmapped sequences from *O. glaberrima*.

To analyze whether the maternal genes above exist in the NU genomes or not, we re-mapped unmapped sequence reads of NU3, 4, 5, and 7 upon the genomic scaffolds of unmapped WAB56–104 sequences. The result indicated that 56 genes exhibited segregation among the four NU varieties and CG14 (Additional file 17: Table S13). Fifty one of them were segregated among the four NU varieties. Total of 21 genes were functionally annotated by a BLAST search against the Swiss-Prot database (Table 6). Most of the genes exhibited WAB56–104 allele in NU varieties except two. It is unknown how those genes might contribute on divergent agronomic traits related to biotic and *a*biotic stress resistance in the NU and parental varieties. Association analysis using DNA markers of those genes may lead to discovery of useful genes in agronomy.

Conclusions

Interspecific hybrids NU varieties have a great potential for improvement of rice, especially for Africa. Due to the unique breeding processes, the NU genomes required in-depth analyses. The present study provided a draft whole picture of the NU genomes, including polymorphisms, introgression from the donor parent, some potential introgression segments by outcross and de novo mutations.

Currently, short length polymorphisms, especially SNPs have played a central role in rice breeding, genetics, and biology (Feltus et al. 2004). SNP is the ultimate DNA marker which can be detected by several methods such as SNP chip, next generation sequencing, restriction enzymes, and real-time PCR (Kim and Misra, 2007; Varshney et al. 2009). SNP genotyping data are useful for QTL mapping for agronomic traits, genome wide association study, and measurement of genetic distance (Yonemaru et al. 2010; Korte and Farlow 2013). Since SNPs can be relevant in terms of biochemical function of protein and gene regulation, nucleotide substitutions themselves could be causal of particular traits and thus be markers for breeding (Anderssen and Lübberstedt 2003; Gupta and Rustgi 2004). Approx. 81% of our polymorphic sites were on genic regions, and 43% were on promoter regions. In the NU varieties, divergence was observed in quantitative agronomic traits and apparent phenotypes (Ikeda et al. 2009; Sanni et al. 2009; Fukuta

et al. 2012; Saito et al. 2012; Rodenburg et al. 2015). Studies over the past years have also accumulated data that captures the physiological behavior of the NUs and their divergence in *a*biotic stress responses (Yamamoto et al. 2011; Atayese et al., 2016; Sikuku et al. 2012; Sakariyawo et al. 2015). In most cases the genetic reasons for divergence in traits are not known. The polymorphism panel resource we created is useful for identification of agronomically useful traits in rice.

In conclusion, our analyses revealed genome characteristics of the NU varieties and the parental varieties. The established whole genome polymorphic resource and knowledge are useful for addressing the genetic reasons of the prominent agronomic characteristics in the NERICA and promotes marker-assisted selection in the development of new rice cultivars. This is especially useful in the light of the critical information about the close relationship between pairs of some NERICA varieties. The genotyping matrix for the 8 NU and the parental lines is available upon request.

Additional files

Additional file 1: Table S1. Summary of whole genome sequencing data.

Additional file 2: Figure S1. *k*-mer analysis (*k*-merlength = 17). The X-axis shows *k*-mer depth, and the Y-axis shows proportion that represents the frequency at that *k*-mer depth divided by to total frequency of all *k*-mer depth. (a) WAB56–104, (b) CG14, (c) NERICA 3, (d) NERICA 4, (e) NERICA 5, (f) NERICA 7. Arrow indicates a peak of heterogeneous sequences.Peak between 0 to 8 of *k*-mer depth would be due to sequencing error or contaminants.

Additional file 3: Table S2. MNPs between WAB 56–104 and CG 14.

Additional file 4: Table S3. Distribution of InDel length.

Additional file 5: Table S4. Validation result of parental polymorphic sites.

Additional file 6: Table S5. Public whole genome sequence data in upland NERICA.

Additional file 7: Table S6. Common introgression segments from CG 14.

Additional file 8: Table S7. Location of xyloglucan fucosyltransferase genes on common long introgression segments from CG 14.

Additional file 9: Figure S2. Phylogeny of genes for fucogalactoxyloglucan biosynthesis. Predicted proteins sequences were aligned by 'CLUSTAL Ω' and a phylogenetic dendrogram was constructed by 'Simple Phylogeny' using the neighbor-joining method with the option of 'exclude gaps'. Twelve fucosyltransferase genes in Arabidopsis (AtFUT1 to AtFUT12), four candidate genes for galactosyltransferase in rice, four candidate genes for acetyltransferase genes in rice were analyzed together with the five fucosyltransferase genes with CG14-alleles. The five fucosyltransferase genes of interest were marked with a circle.

Additional file 10: Figure S3. Graphical representation of allelic patterns in the two fucosyltransferase. Substitution type indicates amino acids of WAB56–104-allele to CG14-allele. (a) LOC_Os06g10910 and (b) LOC_Os06g10920. Varietal names are abbreviated as follows: Nipponbare, Ni; WAB56–104, WAB; NERICA 1, NU1; NERICA 2, NU2; NERICA 3, NU3; NERICA 4, NU4; NERICA 5, NU5; NERICA 7, NU7; NERICA 8, NU8; NERICA 9, NU9. CDS posi. and A.A. posi. Represent positions in coding sequences and positions in deduced am no acid sequences, respectively. Alleles of CG14 are colored in gray.

Additional file 11: Table S8. Variety/lineage-specific introgression segments from CG 14.

Additional file 12: Table S9. Genes on variety/lineage-specific introgression segments from CG 14.

Additional file 13: Table S10. Enriched Gene Ontology on the introgression segments from CG 14.

Additional file 14: Figure S4. Graphical genotype of *GBSSI* and *OsSWEET14* at gene level. Polymorphic position is indicated by vertical line. WAB56–104 segment is presented by while rectangle, and CG14 allele is presented in gray rectangle. Arrow represent the position of gene. (a) *GBSSI*, (b) OsSWEET14. Representative polymorphic sites were used.

Additional file 15: Table S11. Allelic patterns of LOC_Os01g20160 (OsHKT1;5).

Additional file 16: Table S12. Summary of genome assembly and annotation of unmapped reads from WAB 56–104.

Additional file 17: Table S13. Genes on the genome assembly of unmapped sequences from WAB 56–104.

Abbreviations

CAPS: Cleaved amplified polymorphic sequences; Fuc: Fucose; GBSSI: Granule-bound starch synthase I; GO: Gene ontology; GOEA: Gene ontology enrichment analysis; InDel: Insertion and deletion; LRK: Lectin receptor kinase; LRR: Leucine-rich repeat; MNP: Multi nucleotide polymorphism; NERICA: New Rice for Africa; NU: Upland New Rice for Africa; PCA: Principal component analysis; PI: Propidium iodide; QTL: Quantitative trait locus; RAE1: Regulator of Awn Elongation 1; SNP: Single nucleotide polymorphism; SSR: Simple sequence repeat; WAK: Wall-associated receptor kinase; WARDA: The West Africa Rice Development Association

Acknowledgements

We thank Mr. Francisco V. Gulay and Mrs. Maria Elena Escosura for plant material preparation. We also thank Mrs. Blesilda Enriquez for preparation of genomic DNA. We also thank Dr. Kiyoshi Yamada in Faculty of Agriculture in Utsunomiya University for technical assistance of flow cytometry analysis.

Funding

This study was supported by the 'Rapid Mobilization of Alleles for Rice Cultivar Improvement in Sub-Saharan Africa' project of the Bill and Melinda Gates Foundation Grant Number OPP1080832 to AfricaRice.

Authors' contributions

NY created and analyzed the experimental and computational data. RG and CS assisted the experimental and computational procedures. TS arranged flow cytometry measurements and assisted data analysis. YT prepared standard plant materials for flow cytometry analysis and assisted data analysis. RV contributed to the design of this study and plant material preparation. AK conceived the idea of the study, organized inputs from all the authors and finalized the manuscript. All authors read and approved the final manuscript.

Authors' information

Strategic Innovation Platform, International Rice Research Institute, DAPO 7777, Metro Manila, 1226, Philippines.
Utsunomiya University, 350 Mine-machi, Utsunomiya, Tochigi, Japan.
Tokyo University of Technology, 1404–1 Katakura, Hachioji, Tokyo, Japan.
Africa Rice Center, 01 BP 4029, Abidjan 01, Côte d'Ivoire.

Competing interests

The authors declare that they have no competing interests.

Author details

[1]International Rice Research Institute, Los Baños, Laguna, Philippines. [2]Utsunomiya University, 350 Mine-machi, Utsunomiya, Tochigi, Japan. [3]Tokyo University of Technology, 1404-1 Katakura, Hachioji, Tokyo, Japan. [4]Africa Rice Center, 01 BP 4029, Abidjan 01, Côte d'Ivoire.

References

Abyzov A, Urban AE, Snyder M, Gerstein M (2011) CNVnator: an approach to discover, genotype, and characterize typical and atypical CNVs from family and population genome sequencing. Genome Res 21:974–984

Africa Rice Center (2008) Africa rice trends 2007. Africa Rice Center, Cotonou, Benin

Afzal AJ, Wood AJ, Lightfoot DA (2008) Plant receptor-like serine threonine kinases: roles in signaling and plant defense. MPMI 21:507–517

Albar L, Ndjiondjop MN, Esshak Z, Berger A, Pinel A, Jones M et al (2003) Fine genetic mapping of a gene required for Rice yellow mottle virus cell-to-cell movement. Theor Appl Genet 107:371–378

Alexandrov N, Tai S, Wang W, Mansueto L, Palis K, Fuentes RR, Ulat VJ, Chebotarov D, Zhang G, Li Z, Mauleon R, Hamilton RS, McNally KL (2015) SNP-seek database of SNPs derived from 3000 rice genomes. Nucleic Acids Res 43:D1023–D1027

Aluko G, Martinez C, Tohme J, Castano C, Bergman C, Oard JH (2004) QTL mapping of grain quality traits from the interspecific cross *Oryza sativa* x *O. glaberrima*. Theor Appl Genet 109:630–639

Andorf CM, Cannon EK, Portwood JLII, Gardiner JM, Harper LC, Schaeffer ML et al (2015) MaizeGDB update: new tools, data and interface for the maize model organism database. Nucleic Acids Res 44:D1195–D1201

Arumuganathan K, Earle ED (1991) Nuclear DNA content of some important plant species. Plant Mol Biol Rep 9:208–218

Asante MD, Kovach MJ, Huang L, Harrington S, Dartey PK, Akromah R et al (2010) The genetic origin of fragrance in NERICA1. Mol Breed 26:419–424

Atayese M, Olagunju S, Sakariyawo O, Oyekanmi A, Babalola O, Aderibigbe S et al (2016) Root response of some selected rice varieties to soil moisture stress at different phenological stages. Journal of agricultural science and. Environment 12:93–113

Attere AF, Fatokun CA (1983) Reaction of *Oryza glaberrima* accessions to rice yellow mottle virus. Plant Dis 67:420–421

Awala SK, Nanhapo I, Sakagami JI, Kanyomeka L, Iijima M (2010) Differential salinity tolerance among *Oryza glaberrima*, *Oryza sativa* and their interspecies including NERICA. Plant Prod Sci 13:3–10

Baack EJ, Rieseberg LH (2007) A genomic view of introgression and hybrid speciation. Curr Opin Genet Dev 17:513–518

Bennett MD, Leitch IJ (2005) Nuclear DNA amounts in angiosperms: progress, problems and prospects. Ann Bot 95:45–90

Bennett MD, Leitch IJ, Price HJ, Johnston JS (2003) Comparisons with *Caenorhabditis* (approximately 100 Mb) and drosophila (approximately 175 Mb) using flow cytometry show genome size in Arabidopsis to be approximately 157 Mb and thus approximately 25% larger than the Arabidopsis genome initiative estimate of approximately 125 Mb. Ann Bot 91:547–557

Boeckmann B, Blatter MC, Famiglietti L, Hinz U, Lane L, Roechert B, Bairoch A (2005) Protein variety and functional diversity: Swiss-Prot annotation in its biological context. C R Biol 328:882–999

Bolger AM, Lohse M, Usadel B (2014) Trimmomatic: a flexible trimmer for Illumina sequence data. Bioinformatics 30:2114–2120

Burr B (2002) Mapping and sequencing the rice genome. Plant Cell 14:521–523

Chen X, Schulz-Trieglaff O, Shaw R, Barnes B, Schlesinger F, Källberg M, Cox AJ, Kruglyak S, Saunders CT (2016) Manta: rapid detection of structural variants and indels for germline and cancer sequencing applications. Bioinformatics 32:1220–1222

Chin HS, Wu YP, HourAL HCY, Lin YR (2016) Genetic and evolutionary analysis of purple leaf sheath in Rice. Rice 9:8

Conesa A, Götz S (2008, 2008) Blast2GO: a comprehensive suite for functional analysis in plant genomics. Int J Plant Genomics:619832

Deorowicz S, Kokot M, Grabowski S, Debudaj-Grabysz A (2015) KMC 2: fast and resource-frugal k-mer counting. Bioinformatics 31:1569–1576

Dixit S, Biswal AK, Min A, Henry A, Oane RH, Raorane ML, Longkumer T, Pabuayon IM, Mutte SK, Vardarajan AR, Miro B, Govindan G, Albano-Enriquez B, Pueffeld M, Sreenivasulu N, Slamet-Loedin I, Sundarvelpandian K, Tsai Y-C, Raghuvanshi S, Hsing Y-IC, Kumar A, Kohli A (2015) Action of multiple intra-QTL genes concerted around a co-localized transcription factor underpins a large effect QTL. Scientific Reports 5(1)

Doležel J, Bartoš J, Voglmayr H, Greilhuber J (2003) Nuclear DNA content and genome size of trout and human. Cytometry A 51:127–128

Doyle JJ (1987) A rapid DNA isolation procedure for small quantities of fresh leaf tissue. Phytochem Bull 19:11–15

Duvick J, Fu A, Muppirala U, Sabharwal M, Wilkerson MD, Lawrence CJ et al (2008) PlantGDB: a resource for comparative plant genomics. Nucleic Acids Res 36:D959–D965

Feltus FA, Wan J, Schulze SR, Estill JC, Jiang N, Paterson AH (2004) An SNP resource for rice genetics and breeding based on subspecies indica and japonica genome alignments. Genome Res 14:1812–1819

Food and Agriculture Organization of the United Nations (FAO) (2013a) FAOSAT online statistical service. FAO, Rome

Food and Agriculture Organization of the United Nations (FAO) (2013b) Yearbook, FAO statistical. "World food and agriculture". FAO, Rome

Fukuta Y, Konisho K, Senoo-Namai S, Yanagihara S, Tsunematsu H, Fukuo A, Kumashiro T (2012) Genetic characterization of rainfed upland new Rice for Africa (NERICA) varieties. Breed Sci 62:27–37

Furuta T, Komeda N, Asano K, Uehara K, Gamuyao R, Angeles-Shim RB et al (2015) Convergent loss of awn in two cultivated rice species Oryza sativa and Oryza glaberrima is caused by mutations in different loci. G3 5:2267–2274

Gupta PK, Rustgi S (2004) Molecular markers from the transcribed/expressed region of the genome in higher plants. Funct Integr Genomic 4:139–162

Haefele SM, Johnson DE, M'Bodj D, Wopereis MCS, Miezan KM (2004) Field screening of diverse rice genotypes for weed competitiveness in irrigated lowland ecosystems. Field Crops Res 88:39–56

Hurwitz BL, Kudrna D, Yu Y, Sebastian A, Zuccolo A, Jackson SA et al (2010) Rice structural variation: a comparative analysis of structural variation between rice and three of its closest relatives in the genus Oryza. Plant J 63:990–1003

Hutin M, Sabot F, Ghesquière A, Koebnik R, Szurek B (2015) A knowledge-based molecular screen uncovers a broad-spectrum OsSWEET14 resistance allele to bacterial blight from wild rice. Plant J 84:694–703

Ikeda R, Sokei Y, Akintayo I (2009) Seed fertility of F1 hybrids between upland rice NERICA cultivars and Oryza sativa L. or O. glaberrima Steud. Breed Sci 59:27–35

Jones MP, Dingkuhn M, Aluko GK, Semon M (1997) Interspecific Oryza sativa L. x O. glaberrimaSteud. Progenies in upland rice improvement. Euphytica 94: 237–246

Kanneganti V, Gupta AK (2008) Wall associated kinases from plants—an overview. Physiol Mol Biol Plants 14:109–118

Kawahara Y, de la Bastide M, Hamilton JP, Kanamori H, McCombie WR, Ouyang S et al (2013) Improvement of the Oryza sativa Nipponbare reference genome using next generation sequence and optical map data. Rice 6:4

Keller O, Kollmar M, Stanke M, Waack S (2011) A novel hybrid gene prediction method employing protein multiple sequence alignments. Bioinformatics 27: 757–763

Kim S, Misra A (2007) SNP genotyping: technologies and biomedical applications. Annu Rev Biomed Eng 9:289–320

Kishine M, Suzuki K, Nakamura S, Ohtsubo KI (2008) Grain qualities and their genetic derivation of 7 new rice for Africa (NERICA) varieties. JAgric. Food Chem 56:4605–4610

Konishi S, Izawa T, Lin SY, Ebana K, Fukuta Y, Sasaki T, Yano M (2006) An SNP caused loss of seed shattering during rice domestication. Science 312:1392–1396

Korte A, Farlow A (2013) The advantages and limitations of trait analysis with GWAS: a review. Plant Methods 9:1

Lamesch P, Berardini TZ, Li D, Swarbreck D, Wilks C, Sasidharan R et al (2012) The Arabidopsis information resource (TAIR): improved gene annotation and new tools. Nucleic Acids Res 40:D1202–D1210

Li C, Zhou A, Sang T (2006) Rice domestication by reducing shattering. Science 311:1936–1939

Li H, Durbin R (2009) Fast and accurate short read alignment with burrows-wheeler transform. Bioinformatics 25:1754–1760

Li H, Handsaker B, Wysoker A, Fennell T, Ruan J, Homer N, Durbin R (2009) The sequence alignment/map format and SAMtools. Bioinformatics 25:2078–2079

Li JY, Wang J, Zeigler RS (2014) The 3,000 rice genomes project: new opportunities and challenges for future rice research. Gigascience 3:1

Liu B, Shi Y, Yuan J, Hu X, Zhang H, Li N, Li Z, Chen Y, Mu D, Fan W (2013) Estimation of genomic characteristics by analyzing k-mer frequency in de novo genome projects eprint arXiv:13082012

Liu L, Paulitz J, Pauly M (2015) The presence of fucogalactoxyloglucan and its synthesis in rice indicates conserved functional importance in plants. Plant Physiol 168:549–560

Lorieux M, Ndjiondjop MN, Ghesquière A (2000) A first interspecific Oryza sativa Oryza glaberrima microsatellite-based genetic linkage map. Theor Appl Genet 100:593 601

Lorieux M, Reversat G, Diaz SG, Denance C, Jouvenet N, Orieux Y et al (2003) Linkage mapping of Hsa-1Og, a resistance gene of African rice to the cyst nematode, Heteroderasacchari. Theor Appl Genet 107:691–696

Luo R, Liu B, Xie Y, Li Z, Huang W, Yuan J et al (2012) SOAPdenovo2: an empirically improved memory-efficient short-read de novo assembler. Gigascience 1:18

Majerus V, Bertin P, Lutts S (2007) Effects of iron toxicity on osmotic potential, osmolytes and polyamines concentrations in the African rice (Oryza glaberrima Steud.). Plant Sci 173:96–105

Martínez CP, Arumuganathan K, Kikuchi H, Earle ED (1994) Nuclear DNA content of ten rice species as determined by flow cytometry. Jpn J Genet 69:513–523

McKenna A, Hanna M, Banks E, Sivachenko A, Cibulskis K, Kernytsky A, DePristo MA (2010) The genome analysis toolkit: a MapReduce framework for analyzing next-generation DNA sequencing data. Genome Res 20:1297–1303

Miyabayashi T, Nonomura KI, Morishima H, Kurata N (2007) Genome size of twenty wild species of Oryza determined by flow cytometric and chromosome analyses. Breed Sci 57:73–78

Morales L andDujon B (2012) Evolutionary role of interspecies hybridization and genetic exchanges in yeasts. MMBR 76:721–739

Murashige T, Skoog F (1962) A revised medium for rapid growth and bio assays with tobacco tissue cultures. Physiol Plant 15:473–497

Pariasca-Tanaka J, Lorieux M, He C, McCouch S, Thomson MJ, Wissuwa M (2015) Development of a SNP genotyping panel for detecting polymorphisms in Oryza glaberrima/O. sativa interspecific crosses. Euphytica 201:67–78

Paul CP, Ng NQ, Ladeinde TA (2003) Mode of gene action of inheritance for resistance to rice yellow mottle virus. Afr Crop Sci J 11:143–150

Platten JD, Cotsaftis O, Berthomieu P, Bohnert H, Davenport RJ, Fairbairn DJ et al (2006) Nomenclature for HKT transporters, key determinants of plant salinity tolerance. Trends Plant Sci 11:372–374

Platten JD, Egdane JA, Ismail AM (2013) Salinity tolerance, Na^+ exclusion and allele mining of HKT1; 5 in Oryza sativa and O. glaberrima: many sources, many genes, one mechanism? BMC Plant Biol 13:32

Rieseberg LH, Willis JH (2007) Plant speciation. Science 317:910–914

Rodenburg J, Cissoko M, Kayeke J, Dieng I, Khan ZR, Midega CA et al (2015) Do NERICA rice cultivars express resistance to Striga hermonthica (Del.) Benth. andStriga asiatica (L.) Kuntze under field conditions? Field Crops Res 170:83–94

Saito K, Sokei Y, Wopereis MCS (2012) Enhancing rice productivity in West Africa through genetic improvement. Crop Sci 52:484–493

Sakai H, Ikawa H, Tanaka T, Numa H, Minami H, Fujisawa M, Shibata M, Kurita K, Kikuta A, Hamada M, Kanamori H, Namiki N, Wu J, Itoh T, Matsumoto T, Sasaki T (2011) Distinct evolutionary patterns of Oryza glaberrima deciphered by genome sequencing and comparative analysis. Plant J 66:796–805

Sakariyawo OS, Olagunju SO, Atayese MO, Okeleye KA, PAS S, Aderibigbe SG et al (2015) Physiological and yield response of some upland rice varieties to re-watering after imposed soil moisture stress. J Agric Sci & Env 15:93–111

Sanni KA, Ojo DK, Adebisi MA, Somado EA, Ariyo OJ, Sie M et al (2009) Ratooning potential of interspecific NERICA rice varieties (Oryza glaberrimax Oryza sativa). Int J Bot 5:112–115

Sano Y (1990) The genic nature of gamete eliminator in rice. Genetics 125:183–191

Sano Y, Sano R, Morishima H (1984) Neighbour effects between two co-occurring rice species, Oryza sativa and O. glaberrima. J Appl Ecol 21:245–254

Sarla N, andSwamy BM (2005) Oryza glaberrima: a source for the improvement of Oryza sativa. Curr Sci:955–963

Sarria R, Wagner TA, O'Neill MA, Faik A, Wilkerson CG, Keegstra K, Raikhel NV (2001) Characterization of a family of Arabidopsis genes related to xyloglucan fucosyltransferase1. Plant Physiol 127:1595–1606

Schatz MC, Maron LG, Stein JC, Wences AH, Gurtowski J, Biggers E et al (2014) Whole genome de novo assemblies of three divergent strains of rice, Oryza sativa, document novel gene space of aus and indica. Genome Biol 15:506

Semagn K, Ndjiondjop MN, Lorieux M, Cissoko M, Jones M, McCouch S (2007) Molecular profiling of an interspecific rice population derived from a cross between WAB56-104 (Oryza sativa) and CG14 (Oryzaglaberrima). Afr J Biotechnol 6:2014–2022

Semon M, Nielsen R, Jones MP, McCouch SR (2005) The population structure of African cultivated rice Oryza glaberrima (Steud.): evidence for elevated levels of linkage disequilibrium caused by admixture with O. sativa and ecological adaptation. Genetics 169:1639–1647

Séré Y, Onasanya A, Verdier V, Akator K, Ouedraogo LS, Segda Z, Coulibaly MM, Sido AY, Basso A (2005) Rice bacterial leaf blight in West Africa: preliminary

studies on disease in farmers' fields and screening released varieties for
resistance to the bacteria. Asian J Plant Sci 4:577–579

Sié M, Sanni K, Futakuchi K, Manneh B, Mandé S, Vodouhé R, Dogbe S, Dramé KN,
Ogunbayo A, Ndjiondjop MN, Traoré K (2012) Towards a rational use of African
rice (OryzaglaberrimaSteud.) for breeding in sub-Saharan Africa. G3 (6):1–7

Sikuku PA, Onyango JC, Netondo GW (2012) Physiological and biochemical
responses of five nerica rice varieties (*Oryza sativa* L.) to water deficit at
vegetative and reproductive stage. Agric Biol J North Am 3:93–104

Singh RK, Sharma RK, Singh AK, Singh VP, Singh NK, Tiwari SP, Mohapatra T
(2004) Suitability of mapped sequence tagged microsatellite site markers for
establishing distinctness, uniformity and stability in aromatic rice. Euphytica
135:135–143

Somado EA, Guei RG, Keya SO (2008) NERICA: the new rice for Africa–a
compendium: 2008 edition. Africa Rice Center (WARDA), Cononou, Benin.

Stanke M, Morgenstern B (2005) AUGUSTUS: a web server for gene
prediction in eukaryotes that allows user-defined constraints. Nucleic
Acids Res 33:W465–W467

Tian T, Liu Y, Yan H, You Q, Yi X, Du Z, Xu W, Su Z (2017) agriGO v2.0: a GO
analysis toolkit for the agricultural community, 2017 update. Nucleic Acids
Res 45(W1):W122–W129. https://doi.org/10.1093/nar/gkx382.

Traore K, McClung AM, Fjellstrom R, Futakuchi K (2011) Diversity in grain physico-
chemical characteristics of west African rice, including NERICA genotypes, as
compared to cultivars from the United States of America. IRJAS 1:435–448

Uozu S, Ikehashi H, Ohido N, Ohtsubo H, Ohtsubo E, Fukui K (1997) Repetitive
sequences: cause for variation in genome size and chromosome
morphology in the genus *Oryza*. Plant Mol Biol 35:791–799

Vaid N, Macovei A, Tuteja N (2013) Knights in action: lectin receptor-like kinases
in plant development and stress responses. Mol Plant 6:1405–1418

Varshney RK, Nayak SN, May GD, Jackson SA (2009) Next-generation sequencing
technologies and their implications for crop genetics and breeding. Trends
Biotechnol 27:522–530

Vikal Y, Das A, Patra B, Goel RK, Sidhu JS, Singh K (2007) Identification of new
sources of bacterial blight (*Xanthomonas oryzae* pv. Oryzae) resistance in wild
Oryza species and *O. glaberrima*. Plant Genet Resour 5:108–112

Wang M, Yu Y, Haberer G, Marri FR, Fan C, Goicoechea JL et al (2014) The
genome sequence of African rice (*Oryza glaberrima*) and evidence for
independent domestication. Nat Genet 46:982–988

Yamamoto A, Sawada H, Shim IS, Usui K, Fujihara S (2011) Effect of salt stress on
physiological response and leaf polyamine content in NERICA rice seedlings.
Plant Soil Environ 57:571–576

Yonemaru JI, Yamamoto T, Fukuoka S, Uga Y, Hori K, Yano M (2010) Q-TARO: QTL
annotation rice online database. Rice 3:194–203

Zabotina OA (2012) Xyloglucan and its biosynthesis. Front Plant Sci 3:134

Zhou Y, Lu D, Li C, Luo J, Zhu BF, Zhu J et al (2012) Genetic control of seed
SHATTERING in rice by the APETALA2 transcription factor SHATTERING
ABORTION1. Plant Cell 24:1034–1048

14

Genetic analysis for the grain number heterosis of a super-hybrid rice WFYT025 combination using RNA-Seq

Liang Chen[1,2†], Jianmin Bian[1,2,3*†] (iD), Shilai Shi[1,2†], Jianfeng Yu[1,2], Hira Khanzada[1,2], Ghulam Mustafa Wassan[1,2], Changlan Zhu[1,2,3], Xin Luo[1,2], Shan Tong[1,2], Xiaorong Yang[1,2], Xiaosong Peng[1,2,3], Shuang Yong[1], Qiuying Yu[1], Xiaopeng He[1,2], Junru Fu[1,2], Xiaorong Chen[1,2], Lifang Hu[1,2,3], Linjuan Ouyang[1,2] and Haohua He[1,2,3*]

Abstract

Background: Despite the great contributions of utilizing heterosis to crop productivity worldwide, the molecular mechanism of heterosis remains largely unexplored. Thus, the present research is focused on the grain number heterosis of a widely used late-cropping *indica* super hybrid rice combination in China using a high-throughput next-generation RNA-seq strategy.

Results: Here, we obtained 872 million clean reads, and at least one read could maps 27,917 transcripts out of 35,679 annotations. Transcript differential expression analysis revealed a total of 5910 differentially expressed genes (DG_{HP}) between super-hybrid rice Wufengyou T025 (WFYT025) and its parents were identified in the young panicles. Out of the 5910 DG_{HP}, 63.1% had a genetic action mode of over-dominance, 17.3% had a complete-dominance action, 15.6% had a partial-dominance action and 4.0% had an additive action. DG_{HP} were significantly enriched in carotenoid biosynthesis, diterpenoid biosynthesis and plant hormone signal transduction pathways, with the key genes involved in the three pathways being up-regulated in the hybrid. By comparing the DG_{HP} enriched in the KEGG pathway with QTLs associated with grain number, several DG_{HP} were located on the same chromosomal segment with some of these grain number QTLs.

Conclusion: Through young panicle development transcriptome analysis, we conclude that the over-dominant effect is probably the major contributor to the grain number heterosis of WFYT025. The DG_{HP} sharing the same location with grain number QTLs could be considered a candidate gene and provide valuable targets for the cloning and functional analysis of these grain number QTLs.

Keywords: Rice, Super-hybrid Rice, Heterosis, Grain number, RNA-seq

Background

Heterosis is a phenomenon in which hybrids exhibit superiority over their parental lines in economic traits, such as enhanced biomass production, development rate, stress tolerance and, most important, grain yield. Heterosis has been extensively used to increase crop productivity throughout the world. A major increase in rice yield was caused by the application of heterosis. Because of the key role of heterosis,

the molecular mechanisms should be elucidated. In the early twentieth century, dominance (Davenport 1908) and over-dominance (Shull 1908) were used to explain heterosis. However, with nothing about molecular concepts being covered, consequently, they cannot interpret the molecular genetic mechanisms of heterosis (Birchler et al. 2003). With the development of polymerase chain reaction (PCR), molecular markers have been widely used to identify the distance between the hybrid and its parents and to build the relationship between heterosis and genetic distance. However, marker PCR can only be used to classify heterotic groups and determine genetic diversity, but it cannot predict heterosis because the coefficient of the relationship between the genetic distance of SSR markers and yield

* Correspondence: jmbian81@126.com; hhhua64@163.com

†Liang Chen, Jianmin Bian and Shilai Shi contributed equally to this work.

[1]Key Laboratory of Crop Physiology, Ecology and Genetic Breeding, Ministry of Education, Jiangxi Agricultural University, Nanchang 330045, China

Full list of author information is available at the end of the article

heterosis is very small (Xu et al. 2009). Subsequently, molecular markers and hybrid genetic analysis have been used to locate QTLs for heterosis. A Pioneer study of the heterosis gene *qGY2–1* related to yield was reported in haplotype populations (He et al. 2006). To eliminate the epistasis effect among QTLs, Bian et al. (2011) used chromosome segment substitution lines (CSSLs) to study heterosis for yield traits in *indica* × *japonica* hybrid rice subspecies. With the advent of high-throughput sequencing technology, scientists conducted DNA sequencing of 1495 elite hybrid rice varieties and their inbred parental lines. Comprehensive analyses of heterozygous genotypes have revealed that heterosis mainly resulted from the accumulation of numerous superior alleles with positive dominant effects (Huang et al. 2015).

In addition, the association of heterosis with differentially expressed transcripts was also investigated at the RNA level. Wei et al. (2009) investigated differentially expressed transcripts from tissues at different growth development stages using super rice LYP9 and its parents and found that the differentially expressed transcripts were closely related to QTLs in response to heterosis. Huang et al. (2006) used 9198 unique sequence tags to study gene differential expression profiles of young panicles using the super rice SY63 combination and suggested that transcripts controlling DNA repair and replication were up-regulated and that the genes related to carbohydrate, energy and lipid metabolism, translation and protein degradation were down-regulated.

High-throughput RNA sequencing has been used to search for heterosis in rice to avoid defects of methods with low throughput, high cost, low sensitivity, clonal preference, and high background noise. RNA-seq was first used to compare the transcriptome profiles of reciprocal hybrids from Nipponbare and 93-11, along with their parents, at the seedling stage. In total, 2800 genes showed differential expression, and these transcripts were involved in energy metabolism, especially in the Calvin cycle, in which six key components were up-regulated (He et al. 2010). Later, Zhai et al. (2013) compared the transcriptome between super hybrid XY9308 and its parents through RNA-seq, which indicated that carbohydrate metabolism and plant hormone signal transduction were enriched in differentially expressed transcripts.

In this study, we focused on heterosis in the rice WFYT025, a widely used late-cropping *indica* super hybrid rice combination in China. The number of filled grains, one of the most important yield heteroses in yield contributing factors, showed great differences between WFYT025 and its female parent. Thus, we conducted transcriptome analysis using young panicles from the WFYT025 combination by high-throughput RNA-seq to detect the correlation of key transcripts with filled grain number heterosis. Some key transcripts were mapped in the QTL

interval related to grain number. Revealing the function of these transcripts may provide useful information for understanding the molecular mechanism underlying heterosis.

Results

Phenotype analysis for WFYT025 and its parents

In this study, we investigated the yield-related traits of WFYT025 and its parents. It was found that the panicles of WFYT025 and its male parent CHT025 were larger than those of the female parent WFB, and their grain number and primary branch number were also higher than those in WFB (Fig. 1a and b). However, no significant differences were observed between WFYT025 and parental line CHT025 for both grain number and primary branch number (Fig. 1b). Mid-parent heterosis (MPH) and higher parent heterosis (HPH) were estimated for the heterosis of panicles. The MPH for all of the traits except the seed setting ratio and tiller remained significant (Table 1). Traits such as primary branch number, secondary branch number, filled grain number, empty grain number and 1000-grain weight were significant for the MPH at $p < 0.05$, while traits such as spike length, total grain number and yield per plant were highly significant at $p < 0.01$. The MPH showed negative effects on the seed setting ratio. Apart from the seed setting ratio and empty grain number, the MPH values for all of the traits varied from 1.16 to 32.32%. In addition, HPH for yield per plant remained highly significant (22.99%) at the $p < 0.01$ level. Further analysis indicated that significant difference for yield per plant was mainly due to the large MPH range for filled grain numbers (20.01%) and 1000-grain weight (6.25%). This implied that compared to the 1000-grain weight, the yield heterosis was more likely to underlay the filled grain number between hybrid WFYT025 and maternal line WFB.

Identification of transcripts by sequencing

A total of 917 million raw reads were generated using the high-throughput Illumina HiSeq 2500 platform. The paired-end sequences with low-quality reads containing adapters were trimmed off. Finally, 87.2 million clean reads were obtained (Table 2). The correlation for the gene expression level from three biological replicates of each line was $0.97 < R^2 < 0.99$. (Additional file 1: Figure S1). We pooled the short reads and aligned them to the Nipponbare reference genome (IRGSP v1.0) to identify the transcripts. Out of 35,679 identified transcripts, 27,917 transcripts were mapped, covering 78.24% of the genome. In addition, the transcriptome profile of WFYT025 was similar to that of its female parent WFB (Fig. 2).

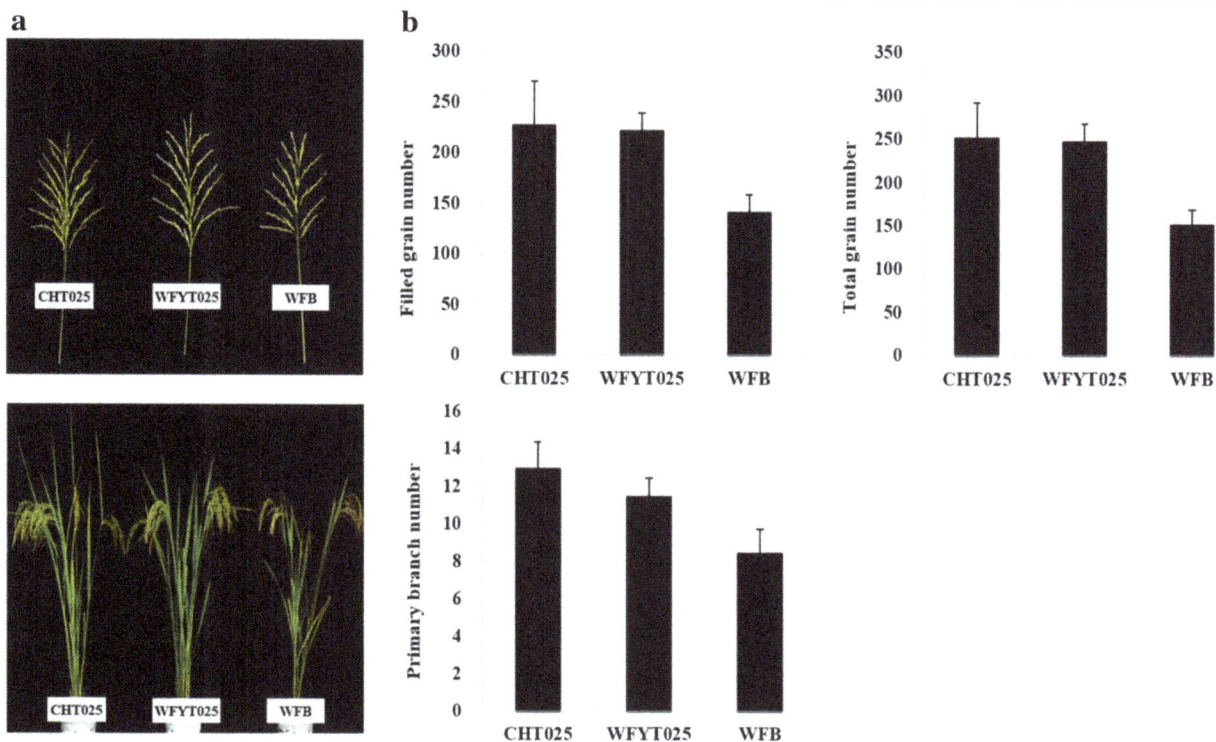

Fig. 1 Comparisons of super hybrid WFYT025 combination. **a** The upper panel illustrates the panicles from combination of super hybrid WFYT025. Left, CHT025; middle, WFYT025; right, WFB. The lower panel shows the combination of super hybrid WFYT025. Left, CHT025; middle, WFYT025; right, WFB. **b** Panicle traits of CHT025, WFYT025, and WFB

Validation of gene expression by quantitative real-time PCR (qRT-PCR)

To validate the results of mRNA sequencing data, the expression of a subset of 15 randomly selected DG_{HP} was determined by qRT-PCR. The list of primer sequences is presented in Additional file 2: Table S1. The results obtained from qRT-PCR and RNA-seq were compared, and expression trends were consistent for all transcripts in both analyses; the correlation coefficient (R^2) was 0.9339 (Fig. 3).

Analysis of differentially expressed genes (DEGs)

We adopted reads per kilobase million reads (FPKM) to measure gene expression levels. Two criteria were considered to identify putative DEGs: (1) the false discovery rate (FDR) should be ≤0.05 and (2) the fold change (FC)

Table 1 Phenotypic Analysis of Super Hybrid WFYT025 Combination

Traits	CHT025	WFYT025	WFB	MPH (%)	HPH (%)
Spike length(cm)	23.24 ± 1.75	24.34 ± 0.67	19.61 ± 1.71	13.64**	4.74
Primary branch number	12.89 ± 1.45	11.45 ± 1.00	8.46 ± 1.26	7.10*	− 11.21
Secondary branch number	49.26 ± 9.66	45.67 ± 5.44	26.51 ± 5.21	20.39*	− 7.28
Solid grain number	227.40 ± 42.72	221.43 ± 18.30	141.42 ± 17.60	20.01*	− 2.63
Total grain number	251.36 ± 40.39	247.08 ± 21.20	151.21 ± 17.89	22.64**	− 1.70
Empty grain number	23.95 ± 6.52	25.65 ± 10.68	9.78 ± 2.49	51.30*	7.08
Seed setting ratio (%)	89.65 ± 0.04	89.04 ± 4.29	93.05 ± 1.62	−2.54	−5.16
1000-grain weight (g)	18.64 ± 0.92	22.44 ± 0.63	23.60 ± 0.34	6.25*	− 5.15
Tiller	6.8 ± 0.92	8.7 ± 1.25	10.4 ± 2.50	1.16	−16.34
Yield per plant (g)	29.89 ± 7.02	42.8 ± 2.80	34.8 ± 6.89	32.32**	22.99**

**Significant difference with $p < 0.01$
*Significant difference with $p < 0.05$

Table 2 Number of Mapped Reads

Sample	Total Reads	Mapped Reads	Mapping Ratio (%)
CS	27,507,194	22,157,870	80.55
YS	21,214,278	17,312,734	81.61
BS	38,511,686	31,076,870	80.69
Total	87,233,158	70,547,474	80.95

CS, YS and BS stand for the samples from CHT025, WFYT025, WFB, respectively

should be ≥2. Following these criteria, 4160 DEGs have been identified between paternal line CHT025 and WFYT025. Of these, 2155 DEGs were up-regulated and 2005 were down-regulated. Additionally, 2809 DEGs were identified between maternal line WFB and WFYT025, of which 1463 DEGs were up-regulated and 1346 DEGs were down-regulated (Table 3). For a detailed comparison, the FPKM of all transcripts is presented in Additional file 3: Table S2. DEGs between parents are designated DG_{PP}, and DEGs among the hybrid and parents are designated DG_{HP}. DG_{HP} may be relevant to heterosis because differences in expression between hybrids and parents should underlie their phenotypic differences. While DG_{PP} only refers to the differences among the two parental lines (Song et al. 2010), there are still 3223 DG_{HP}s that overlapped with DG_{PP}, which indicates that these DG_{PP} are also associated with heterosis (Fig. 4). In addition, 1059 DG_{HP}s were shared between the hybrid and both of its parents.

The mode of inheritance for DG_{HP}

Using the method to evaluate the mode of inheritance, DG_{HP} were classified into four expression patterns: over-dominance (Hp ≤ − 1.2 or Hp > 1.2), dominance (− 1.2 < Hp ≤ − 0.8 or 0.8 < Hp ≤ 1.2), additive effect (− 0.2 < Hp ≤ 0.2), and partial dominance (− 0.8 < Hp ≤ − 0.2 or

0.2 < Hp ≤ 0.8) (Additional file 4: Table S3). As shown in Fig. 5, these data suggested that the over-dominant effect, dominant effect, partially dominant effect and additive effect accounted for 63.1%, 17.3%, 15.6% and 4.0%, respectively.

Functional classification of DG_{HP} by Gene Ontology (GO)

We applied Gene Ontology (GO) to classify the function of the mRNA. Using Web Gene Ontology Annotation Plot (WEGO) software (Ye et al. 2006), we distributed 5910 DG_{HP} into at least one term in the GO molecular function, cellular component, and biological process categories. Further analysis showed that 5910 DG_{HP} were present in 54 functional subcategories at a significance level of $p < 0.05$ (Fig. 6). In the cellular function category, cells and cell parts were mainly divided in the groups. For the molecular function category, DG_{HP} was enriched with binding and catalytic activity. With respect to biological processes, cellular and metabolic processes were highly enriched in DG_{HP}. We further analysed the GO terms of DG_{HP} enriched with the biological process subcategories. These GO terms, including response to stimulus, cell proliferation, carbohydrate metabolic process, organ formation, and gibberellin biosynthetic process, may underlie heterosis in the young panicle of WFYT025 (Tables 4 and 5).

DG_{HP} mapping Kyoto Encyclopedia of Genes and Genomes (KEGG) pathway

For the identification of metabolic pathways in which DG_{HP} were involved and enriched, the Kyoto Encyclopedia of Genes and Genomes pathway database was used. In total, 118 pathways were identified in 613 DG_{HP} (between paternal line CHT025 and hybrid line WFYT025). The top 20 most enriched pathways mainly

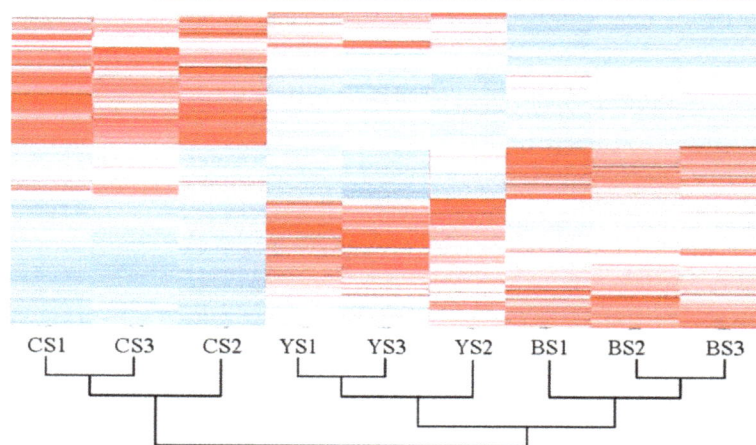

Fig. 2 Hierarchical clustering analysis of all gene models based on expression data. Each horizontal line refers to a gene. The color key represents RPKM normalized log2 transformed counts. With the color varied from blue to red, the expression of transcripts are from low to high. CS 1 to 3, YS 1 to 3 and BS 1 to 3 stand for the replicated samples from CHT025, WFYT025, WFB, respectively

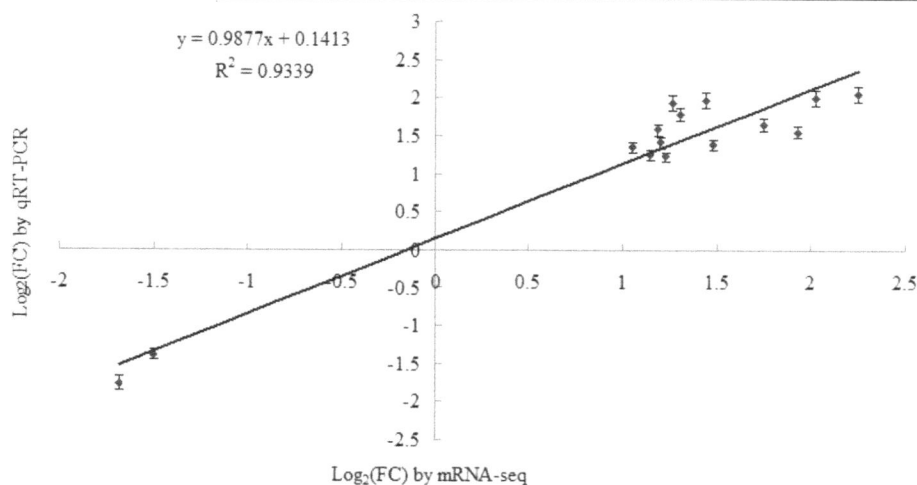

Fig. 3 Comparison of the log2 (FC) of 15 randomly selected transcripts using RNA-Seq and qRT-PCR

covered carbon fixation in photosynthetic organisms, DNA replication, fatty acid biosynthesis and metabolism, and phenylpropanoid biosynthesis (Fig. 7a). In contrast, 268 DG_{HP} between maternal line WFB and WFYT025 were classified into 107 pathways, and the top 20 most enriched pathways were mainly concentrated in plant hormone signal transduction, carotenoid biosynthesis, diterpenoid biosynthesis, zeatin biosynthesis, and cysteine and methionine metabolism with a significance level of $p < 0.05$ (Fig. 7b). This suggests that the considerable differences in young panicles between WFB and WFYT025 may be related to hormone regulation.

Comparison of DG_{HP} with grain yield-related genes (QTLs)

We were able to map the DG_{HP} that were significant in the KEGG analysis ($P < 0.05$) between WFYT025 and WFB for the QTLs associated with grain yield in the rice genome (http://www.gramene.org). As shown in Table 6, a total of 36 transcripts were mapped in the interval of 22 yield-related QTLs, including 15 grain number QTLs, 6 1000-grain weight QTLs and 1 yield per plant QTL. Most genes shared the same location with one yield-related QTL. However, Os03g0856700 corresponded to *qGP3–1* for grain number and *qSNP-3b* for spikelet number per panicle. Os04g0229100 was mapped to the same loci as *qGwt4a* for 1000-grain weight and *qSNP-4a for* spikelet number per panicle, while Os04g0578400 and Os04g0608300

shared the same chromosome segment with *qGPP-4* for grain number per panicle and *qSNP4–1* for spikelet number per panicle.

Discussion

Though heterosis has been extensively exploited in plant breeding and plays an important role in agriculture, the molecular and genetic mechanisms underlying this phenomenon remain poorly understood. Differential gene expression between a hybrid and its parents may be associated with heterosis (He et al. 2010; Kim et al. 2013; Zhang et al. 2008). Here, we investigated the relationship between transcriptional profiles and heterosis in super hybrid rice WFYT025 by RNA-Seq.

Comparative analysis of DG_{HP}

Using RNA-Seq analysis, 872 million high-quality paired-end reads of 150 bp were generated from the panicles of WFYT025 and its parental lines at the panicle differentiation stage, and 27,917 annotated transcripts were identified. Of these transcripts, 4160 DG_{HP} between hybrid WFYT025 and paternal line CHT025 and 2809 DG_{HP} between hybrid WFYT025 and maternal line WFB were identified.

The filled grain number heterosis exhibited significant differences between WFYT025 and WFB; however, there were no significant differences between WFYT025 and CHT025 (Fig. 1b, Table 1). Therefore, the results suggest that the expression of DG_{HP} between WFYT025 and WFB at the young panicle development stage may play an important role in grain number heterosis compared to that between WFYT025 and CHT025. Therefore, focusing on the expression of DG_{HP} between WFB and WFYT025 might find an association between DG_{HP} and heterosis for filled grain number.

Table 3 Number and Classification of DG_{HP}

Pattern	WFYT025 / CHT025	WFYT025 / WFB
Up	2155	1463
Down	2005	1346
Total	4160	2809

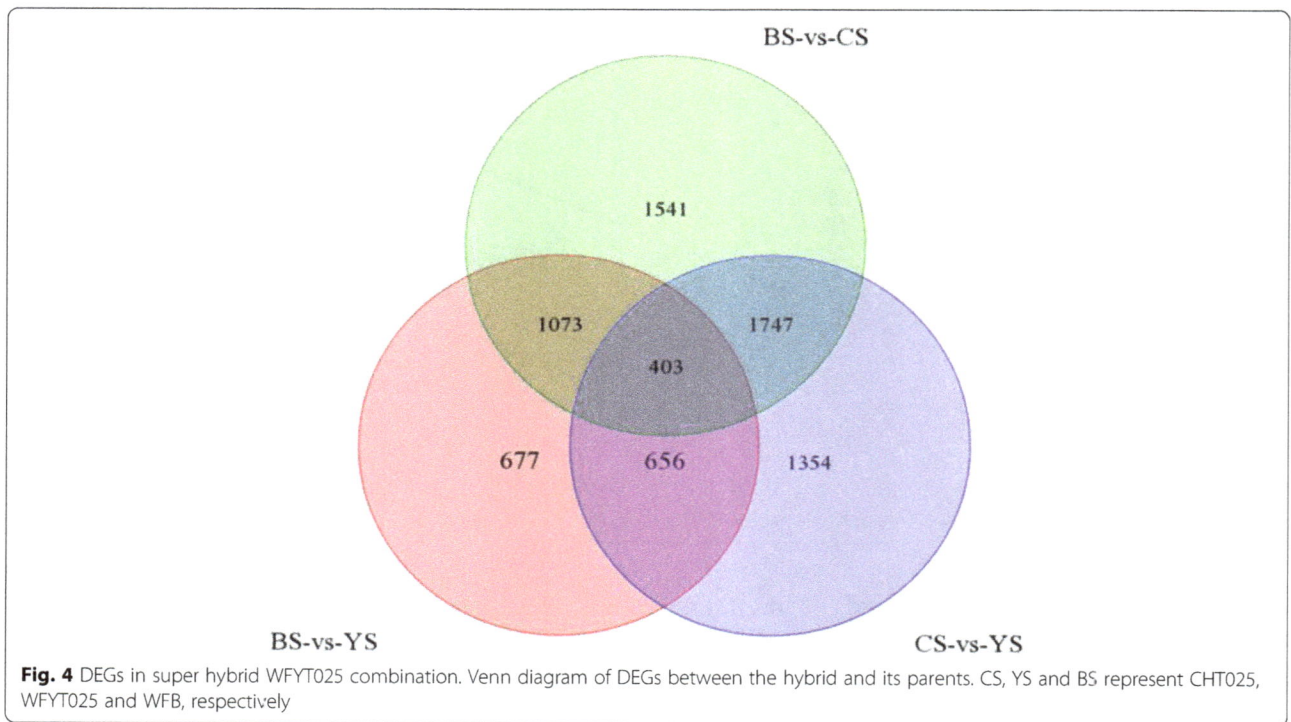

Fig. 4 DEGs in super hybrid WFYT025 combination. Venn diagram of DEGs between the hybrid and its parents. CS, YS and BS represent CHT025, WFYT025 and WFB, respectively

The genetic basis of heterosis

We have been able to identify a number of DG_{HP}s underlying grain number between hybrid WFYT025 and maternal line WFB, confirming the suggestion that heterosis is a polygenic phenomenon (Kusterer et al. 2007; Bian et al. 2011). Among the DG_{HP} 17.3% had a dominant effect, 15.6% had a partial dominant effect, 4% had an additive effect and the remaining 63.1% had an over-dominant effect. Thus, over-dominance was the major contributor to the heterosis of WFYT025.

Meanwhile, the expression differences of cloned yield trait genes have been investigated between the hybrid and its parents. Of the 143 genes related to grain yield traits,

11 genes, accounting for 7.7%, showed over-dominance; 12 genes, accounting for 8.3%, showed dominance; 71 genes, accounting for 49.6%, showed partial dominance; and 49 genes, accounting for 34.4%, showed partial dominance (Additional file 5: Table S4).

The role of hormone signal transduction in heterosis

It is well known that hormones act as signalling molecules in plants and can regulate physiological responses. Transcriptome analysis has uncovered many DG_{HP}s that are involved in the phytohormone response in young panicle tissue. For example, mRNA levels of Os12g0586100 encoding SNF1-related protein kinase2 (SnRK2), whose

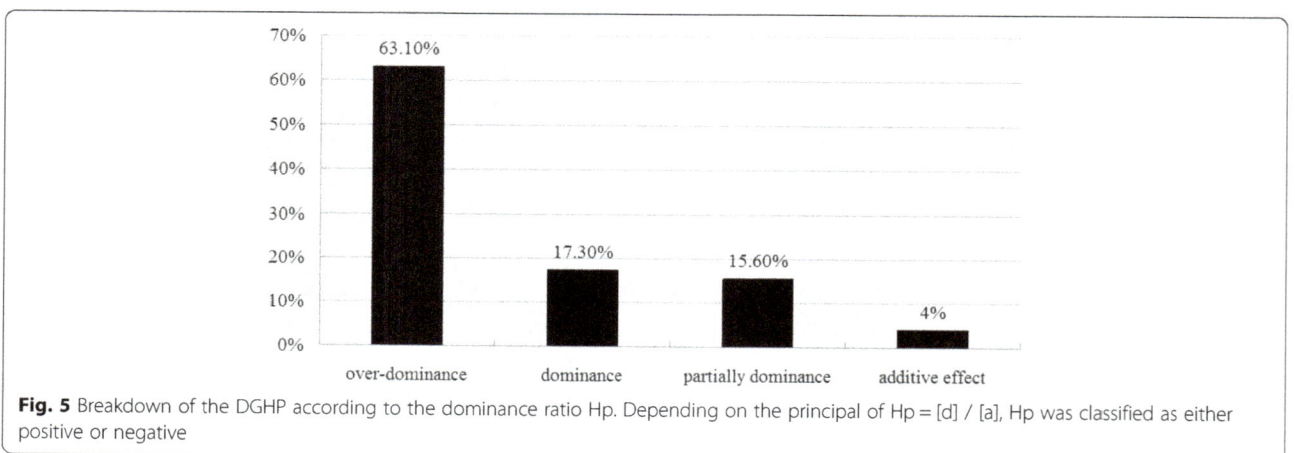

Fig. 5 Breakdown of the DGHP according to the dominance ratio Hp. Depending on the principal of Hp = [d] / [a], Hp was classified as either positive or negative

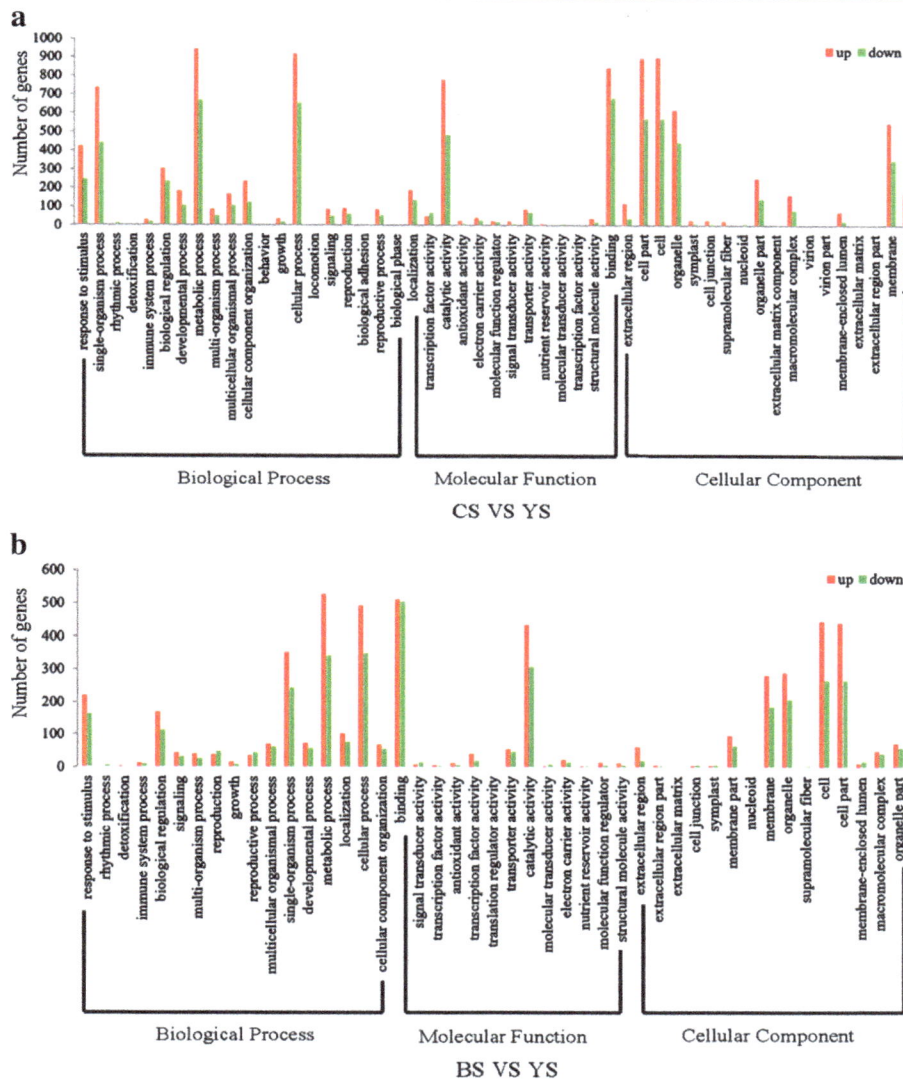

Fig. 6 Comparison of Gene Ontology (GO) classifications of DGHP. **a** CS and YS represent CHT025 and WFYT025 respectively. Red column and green column represent up-regulated and down-regulated transcripts respectively. **b** BS and YS represent WFB and WFYT025 respectively. Red column and green column represent up-regulated and down-regulated transcripts respectively

autophosphorylation is required for kinase activity towards downstream targets, were expressed poorly in WFYT025 compared to its parents. In addition, type-2C protein phosphatase (PP2C, a negative regulator) (Os01g0846300, Os05g0572700, Os01g0656200 and Os03g0268600) was up-regulated, and a similar observation was also reported by Merlot et al. (2001) and Zhai et al. (2013). These results are consistent with the negative-feedback regulatory mechanism in ABA signal transduction.

Moreover, transcripts involved in the gibberellin (GA) biosynthesis pathway were also differentially expressed between the hybrid and its two parents, in this study. GAs are a large family of diterpenoid compounds, some of which are bioactive growth regulators that control flower development

(Cowling et al. 1998). GAs are involved in the transformation of vegetative reproduction to reproductive growth (Poethig 1990; Evans and Poethig 1995). *OsGA20ox1* (Os03g0856700) encodes a GA20 oxidase, which is the key enzyme that catalyses the penultimate step reaction of gibberellin biosynthesis and enhances the grain number of rice by increasing the cytokinin activity in the rice panicle meristem (Wu et al. 2016). In this study, we observed that the expression level of *OsGA20ox1* in WFYT025 is up-regulated 2-fold higher than in WFB and showed over-dominance (Additional file 5: Table S4). This suggested that WFYT025 may possess strong potential for gibberellin biosynthesis compared to maternal line WFB, which promoted the amount of spikelet primordium in hybrid line WFYT025.

Table 4 Significant GO Terms of DG$_{HP}$ Between CS and YS in the Biological Process Category

GO ID	Description	p-value
GO:0042221	Response to chemical	0.000000
GO:0010035	Response to inorganic substance	0.000000
GO:0008283	Cell proliferation	0.000000
GO:0006260	DNA replication	0.000000
GO:0005975	Carbohydrate metabolic process	0.000004
GO:0006629	Lipid metabolic process	0.000088
GO:0009725	Response to hormone	0.000276
GO:0044550	Secondary metabolite biosynthetic process	0.000291
GO:0000281	Mitotic cytokinesis	0.000378
GO:0061640	Cytoskeleton-dependent cytokinesis	0.000378
GO:0051301	Cell division	0.001546
GO:0019344	Cysteine biosynthetic process	0.001784

The significant DG$_{HP}$ related to grain yield QTLs

We compared the significantly enriched DG$_{HP}$ to grain yield QTLs. As shown in Table 6, among the DG$_{HP}$-correlated QTLs, many QTLs were well characterized, including those for grain per panicle (e.g., q*GP-1a* (Yu et al. 1997), q*NG-1* (Lin et al. 1996), q*GP3–1* (Li et al. 2001), q*GPP-4* (Xiao et al. 1996), q*GP-6* (Hua et al. 2002), q*GP-7a* (Li et al. 2000)); number of spikelets on secondary branches per panicle (e.g., q*SSBP1–1* (Cui et al., 2002)); spikelet number per panicle (e.g., q*SNP-3b* (Xu et al. 2001), q*SNP-4a* (Mei et al. 2006), q*SNP4–1* (Takai et al. 2005), q*SP6–1* (Zhuang et al. 2001), q*SNP-6* (He et al. 2001), q*NFPB-11* (Yamagishi et al. 2004)); spikelet density (e.g., q*SD-15* (Li et al. 1998) and q*SSD-10* (Xiao et al. 1996)); 1000-grain weight (e.g., q*gw362* (Ishimaru 2003), q*GW3.1* (Thomson et al. 2003),

Table 5 Significant GO Terms of DG$_{HP}$ Between BS and YS in the Biological Process Category

GO ID	Description	p-value
GO:0006950	Response to stress	0.000012
GO:0050896	Response to stimulus	0.000047
GO:0048645	Organ formation	0.006863
GO:0071265	L-methionine biosynthetic process	0.008974
GO:0009686	Gibberellin biosynthetic process	0.009689
GO:0010160	Formation of organ boundary	0.011830
GO:0003156	Regulation of organ formation	0.016135
GO:0045596	Negative regulation of cell differentiation	0.016135
GO:0048497	Maintenance of floral organ identity	0.016135
GO:0010077	Maintenance of inflorescence meristem identity	0.017878
GO:2000027	Regulation of organ morphogenesis	0.030664
GO:0048586	Regulation of long-day photoperiodism, flowering	0.034206
GO:2000028	Regulation of photoperiodism, flowering	0.044460

q*Gwt4a* (Lin et al. 1995), q*Kw5* (Li et al. 1997), q*Gw-6* (Lu et al. 1996), and q*Gw7* (Li et al. 2000)); and yield per plant (e.g., *yd7a* (Li et al. 2000)).

The potential association between DG$_{HP}$ and QTLs was also suggested within many QTL regions, including putative protein phosphatase 2C (Os01g0846300) with q*SSBP1–1* for the number of spikelets on secondary branches per panicle and putative transketolase (Os05g0408900) with q*SD-15* for spikelet density. Interestingly, *OsGA20ox1* (Os03g0856700), which is related to gibberellin biosynthesis, is located in both q*GP3–1* for the number of grains per panicle and q*SNP-3b* for the spikelet number per panicle. Putative fatty acid hydroxylase (Os04g0578400), which is involved in carotenoid biosynthesis, and OsSAUR20-Auxin-responsive SAUR gene family member (Os04g0608300) was shared in both q*SNP4–1* for spikelet number per panicle and q*GPP-4* for number of grains per panicle. Except for a small number of cloned genes, such as Os01g0788400, Os02g0697400, Os02g0771600, *OsGA20ox1* (Os03g0856700), Os03g0760200, Os03g0645900, Os04g0474800, Os04g0522500, Os04g0556500, Os05g0380900, Os07g0154100, and Os07g0155600, the remaining genes(including Os01g0846300, Os05g0408900, Os04g0578400 and Os04g0608300), which have been located in grain yield QTLs (including grain number, 1000-grain weight, and yield), were not cloned. Studying the function of these candidate transcripts in these QTL regions may increase the knowledge of the molecular mechanisms underlying heterosis.

Transcription factors probably underlying heterosis

Since transcripts are always under different levels of regulation, such as transcription and splicing through genetic or epigenetic mechanisms, the detailed sequence comparisons and validations for different alleles of annotated DG$_{HP}$ are not suitable to display in this current report. Transcription factors (TFs) are certainly one of the causes of gene expression fluctuations. In this study, we indeed found that 51 TFs showed significant differential expression in the hybrid compared with the maternal line (Additional file 6: Table S5). It is a coincidence that a previous study also proposed that altered gene expression caused by interactions between transcription factor allelic promoter regions in hybrids was one reasonable mechanism underlying heterosis in rice (Zhang et al. 2008).

Furthermore, among the 51 TFs, we found that *LAX1*, which is the main regulator involved in the formation of axillary bud primordium in rice, is overrepresented in the hybrid (Komatsu et al. 2003). MADS-box 55 (*MADS50*) was up-regulated significantly, and MADS-box 56 (*MADS56*) was down-regulated in the hybrid compared to the maternal line (Additional file 6: Table S5). This is consistent with a previous study that suggests that *OsMADS50* and

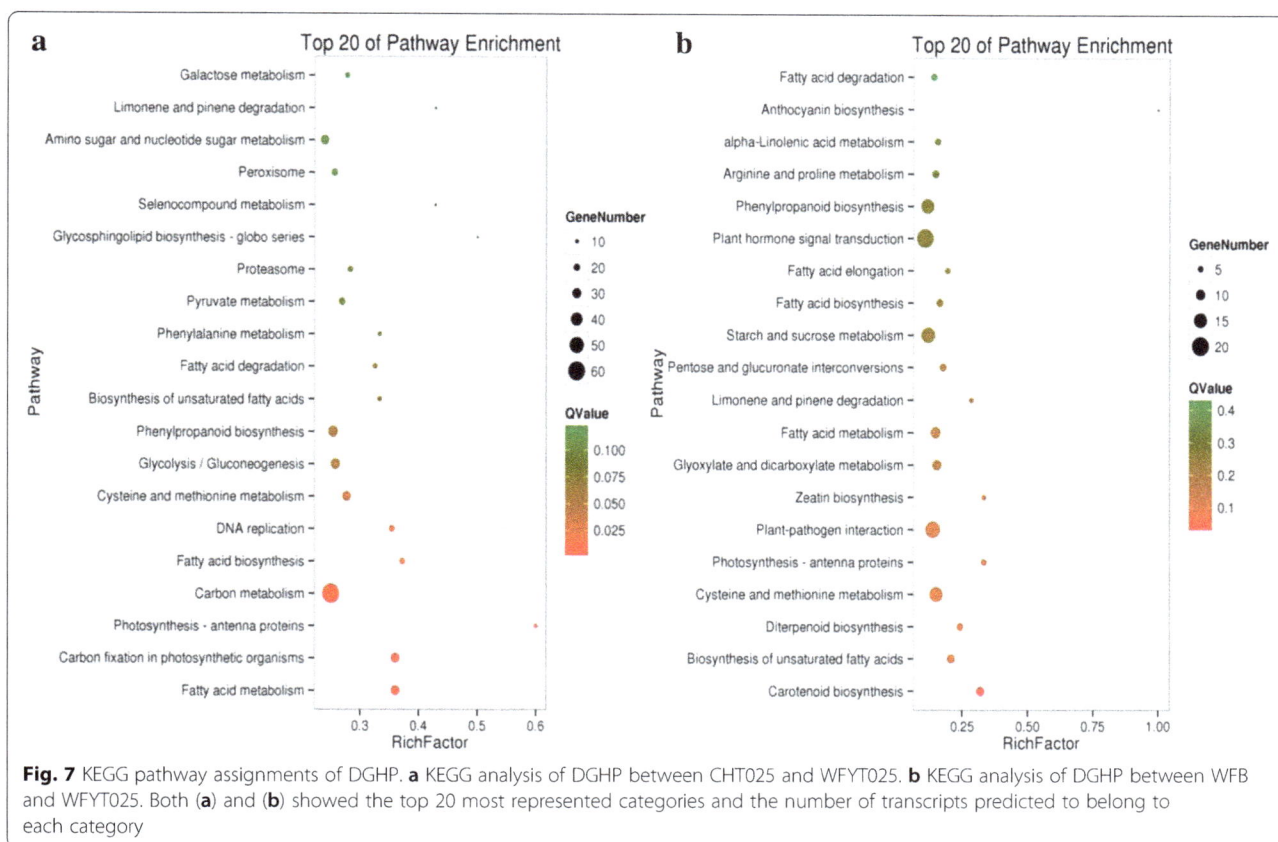

Fig. 7 KEGG pathway assignments of DGHP. **a** KEGG analysis of DGHP between CHT025 and WFYT025. **b** KEGG analysis of DGHP between WFB and WFYT025. Both (**a**) and (**b**) showed the top 20 most represented categories and the number of transcripts predicted to belong to each category

OsMADS56 function antagonistically in regulating LD-dependent flowering (Ryu et al. 2009). Certainly, except for 21 reported TFs, the remaining 30 novel TFs might play an important role in the young panicle and grain number heterosis.

Conclusions

In this study, we systematically investigated the transcriptome profiles from super-hybrid rice WFYT025 combinations for young panicles at the panicle differentiation stage by deep high-quality sequencing. We obtained a large amount of DG_{HP} and found that the over-dominance effect is the main mode of inheritance for DG_{HP}. Comparing the significantly enriched DG_{HP} ($P < 0.05$) between WFYT025 and WFB with QTLs in response to grain number, we found some candidate transcripts that may contribute to the increase in grain yield. Exploring these candidate transcripts will provide new opportunities for revealing the heterosis of grain yield.

Methods

Plant materials and growth conditions

The hybrid WFYT025 along with its parental lines Changhui T025 (CHT025) and Wufeng B (WFB) were planted in the experimental field of Jiangxi Agricultural University. WFYT025 is a super-hybrid rice combination derived from the cross between female parent WFB and male parent CHT025. WFYT025 and the two parents were sown at the experimental plot in Jiangxi Agricultural University in a completely randomized block design with three replications in autumn 2016. Each plot consisted of 50 rows, with each row consisting of 10 plants, each separated from its neighbour by 20 cm. Crop management followed normal procedures for rice. These three lines were selected in this study to measure phenotypic traits and conduct transcriptome analyses. At maturity time, panicles were selected with ten replicates for the estimation of heterosis. The young panicles at the differentiation stage were collected and stored at −80 °C for RNA-Seq analysis, and each sample had at least three biological replications to minimize systematic errors.

Panicle heterosis measurements

To determine 1000-grain weight, panicles were dried in an oven at 42 °C for 1 week. Panicle length, primary branch, secondary branch, number of filled grains and total grain number were measured manually. Mid-parent heterosis (MPH) and higher parent heterosis (HPH)

Table 6 Significant differentially Expressed Transcripts Mapped in each of the QTL Regions

Trait	QTL	Chr	Interval	DG$_{HP}$
GPP	qGP-1a	1	RM1-R753	Os01G0135700, Os01G0150800
NGP	qNG-1	1	RG374-RG394	Os01G0788400
NSP	qSSBP1-1	1	C86-C2340	Os01G0846300
GW	qgw362	2	C1445-C560	Os02G0697400, Os02G0771600
GPP	qGP3-1	3	G249-RG418	Os03G0760200, Os03G0762400, Os03G0797800, Os03G0856700
SNPP	qSNP-3b	3	RM227-RM85	Os03G0856700
GW	qGN3.1	3	RZ672-RZ474	Os03G0423300, Os03G0645900
GW	qGwt4a	4	RG788-RG190	Os04G0229100
SNPP	qSNP-4a	4	RM401-RM335	Os04G0229100, Os04G0474800, Os04G0486950,
GPP	qGPP-4	4	RZ569-RZ565	Os04G0492800, Os04G0498700, Os04G0518100, Os04G0522500 Os04G0535600, Os04G0556500, Os04G0565200, Os04G0578400, Os04G0608300, Os04G0611700, Os04G0611800, Os04G0618700
SNPP	qSNP4-1	4	RM303-RM255	Os04G0578400, Os04G0608300
GW	qKv/5	5	RG182-RG13	Os05G0374200, Os05G0380900
SD	qSD-15	5	RG13-RG346	Os05G0475400, Os05G0551700, Os05G0408900
GW	qGw-6	6	C235-G294	Os06G0347100, Os06G0486900
SP	qSP6-1	6	RG138-RZ398	Os06G0185100
GPP	qGP-6	6	RZ667-RG424	Os06G0347100
SSP	qSPN-6	6	C236-G294	Os06G0486900
GW	qGw7	7	R1440-RG128	Os07G0154100, Os07G0155600
YPP	yd7a	7	R1440-RG128	Os07G0154100, Os07G0155600
GPP	qGF-7a	7	R1440-RG128	Os07G0154100
SSD	qSSD-10	10	RG257-RZ583	Os10G0419400, Os10G0422200, Os10G0430200, Os10G0472900
NFPB	qNFPB-11	11	RM286-RM332	Os11G0141400, Os11G0152700

were calculated for these traits according to the following formulas: MPH = (F$_1$ – MP) / MP and HPH = (F$_1$ – BP) / BP, where F$_1$ is the performance of the hybrid, MP is the average performance of the two parents and BP is the performance of better parents. Hypothesis testing was performed using a t-test.

RNA extraction, cDNA library preparation and sequencing
Total RNA was extracted from rice panicles using Trizol reagent (Invitrogen, Carlsbad, CA, USA) and purified using an RNeasy Plant Mini Kit (Qiagen, Valencia, CA, USA) according to the manufacturer's instructions. The quality and integrity of RNA were tested using an Agilent Bioanalyzer 2100 system (Agilent, Santa Clara, CA, USA); RNA Integrity Number (RIN) values were greater than 8.5 for all samples. After total RNA extraction, eukaryotic mRNA was enriched by Oligo (dT) beads, while prokaryotic mRNA was enriched by removing rRNA using the Ribo-Zero TM Magnetic Kit (Epicentre). Then, the enriched mRNA was fragmented into 200-bp segments using fragmentation buffer and reverse transcribed into cDNA with random primers. Second-strand cDNA synthesis was subsequently performed using DNA polymerase I, RNase H, dNTP and buffer. Then, the cDNA fragments were purified with QIAquick PCR extraction kit, end repaired, poly (A) added, and ligated to Illumina sequencing adapters. The ligation product size was selected by agarose gel electrophoresis, PCR amplified, and sequenced with 100 cycles of paired-end sequencing (2 × 150 bp) using Illumina HiSeq TM 2500 by Gene Denovo Biotechnology Co. (Guangzhou, China). The processing of fluorescent images into sequences, base-calling and quality value calculations were performed using the Illumina data processing pipeline (version 1.8). The sequence reads were submitted to the NCBI Sequence.

Identification of differentially expressed mRNAs

Raw reads generated from high-throughput sequencing were treated as follows. First, to remove adapters that were added for reverse transcription and sequencing, sequences with too many unknown bases (>10%) and low-quality bases (>50% of the bases with a quality score ≤ 20) were removed. The reads mapped to the ribosome RNA (rRNA) database were removed with the read alignment tool Bowtie 2 (Langmead and Salzberg 2012). The remaining reads of each sample were then mapped to the Nipponbare reference genome (IRGSP build 1.0) by TopHat2 (version 2.0.3.12) (Kim et al. 2013). The parameters for alignment were set as follows: 1) the maximum read mismatch should be 2; 2) the distance between mate-pair reads should be 50 bp; 3) the error of distance between mate-pair reads should be ±80 bp. Differential expression was estimated and tested using the software package edgeR (R version: 2.14, edge R version: 2.3.52) (Robinson et al. 2010). We quantified gene expression levels in terms of fragments per kb for a million reads (FPKM) (Mortazavi et al. 2008), calculated the false discovery rate (FDR), and estimated the fold change (FC) and $\log_2$ values of FC. Transcripts that exhibited an FDR ≤ 0.05 and an estimated absolute $\log_2(\mathrm{FC}) \geq 1$ were considered to be significantly differentially expressed.

The mode of inheritance analysis

For statistical analysis, the analysis of variance (ANOVA) was usually by the model: $y = u + (GA) + (GD) + (SR) + e$, where y is the acquired gene expression, u is the overall mean, GA is the additive effect, GD is the dominant effect, SR is the replication effect, and e is the residual error (Lynch and Walsh 1998). $Hp = [d]/[a]$, referred to as the dominance ratio or potency (where $[a]$ and $[d]$ represent GA and GD, respectively), was also calculated to measure the non-additivity of the F_1 hybrid relative to its parents (Griffing 1990). Considering gene expression levels as quantitative traits, we adopted traditional quantitative genetic parameters, such as composite additive effect $[a]$ and composite dominance effect $[d]$, to estimate our expression profile. DG_{HP} were classified according to the dominance ratio Hp ($= [d]/[a]$), based on 99.8% confidence intervals constructed for $[d]$ - $[a]$ ($[d] > 0$) and $[d]$ + $[a]$ ($[d] < 0$). According to the value of Hp ($=[d]/[a]$), we considered that these genes belonged to partial dominance ($-0.8 < Hp \leq -0.2$ or $0.2 < Hp \leq 0.8$), over-dominance ($Hp \leq -1.2$ or $Hp > 1.2$), dominance ($-1.2 < Hp \leq -0.8$ or $0.8 < Hp \leq 1.2$) and additive effect ($-0.2 < Hp \leq 0.2$) (Stuber et al. 1987, Bian et al. 2011).

Cluster analysis

Cluster analysis of all annotated transcripts from the hybrid and its parents was performed. The FPKM-normalized expression counts for each transcript were clustered with the software Cluster 3.0, and the results were visualized using Treeview (Eisen et al. 1998).

Real-time quantitative PCR

The expression of genes with differential expression (DEGs) and results of RNA sequencing were validated by quantitative real-time PCR. Total RNA from nine samples (including three biological replicates) was extracted using the Prime Script™ RT reagent Kit with gDNA Eraser according to the manufacturer's instructions. SYBR-based qRT-PCR reactions (SYBR Green I, Osaka, Japan) were performed on an ABI VIIA@7 using the following thermal cycling conditions: 50 °C for 2 min; 95 °C for 5 min followed by 40 cycles at 95 °C for 15 s and 60 °C for 34 s. All qRT-PCR reactions were performed in triplicate samples, and the results were analysed with the system's relative quantification software (ver. 1.5) based on the ($\Delta\Delta CT$) method. The detection of the threshold cycle for each reaction was normalized against the expression level of the rice Actin1 gene with the primer sequences 5′-TGGCATCTCTCAGCACATT CC-3′ and 5′-TGCACAATGGATGGGTCAGA-3′.

Additional files

Additional file 1: Figure S1. Scatterplots comparing gene expression scores from biological replicates of WFYT025 and its parents. CS 1-3, YS 1-3, and BS 1-3 denote biological replicates from CHT025, WFYT025 and WFB, respectively.

Additional file 2: Table S1. Primer sequences for qRT-PCR expression analysis.

Additional file 3: Table S2. The FPKM of all transcripts.

Additional file 4: Table S3. Classification of DG_{HP} based on the dominance ratio H_P.

Additional file 5: Table S4. The mode of inheritance of cloned genes.

Additional file 6: Table S5. The DG_{HP} of all transcription factors between WFYT025 and WFB.

Abbreviations
ABA: Abscisic Acid; DEGs: The Genes with Different Expression; DG_{HP}: The Genes with Different Expression Between the Hybrid and Parents; DG_{PP}: The Genes with Different Expression Between Paternal Line and Maternal Line; FC: Fold Change; FDR: False Discovery Rate; FPKMs: Fragments Per kb for a Million Reads; GA: Gibberellins; GO: Gene Ontology; GPP: Grain per Panicle; GPP: Grains per Panicle; GW: Grain Weight; HPH: Higher Parent Heterosis; KEGG: Kyoto Encyclopedia of Genes and Genomes; MPH: Mid-Parent Heterosis; NFPB: Number of Florets per Branch; NGP: Number of Grains per Panicle; NSP: Number of Spikelets on Secondary Branches Per Panicle; PP2C: Type-2C Protein Phosphatase; qRT-PCR: Quantitative Real-Time Polymerase Chain Reaction; QTL: Quantitative Trait Locus; RNA-seq: RNA Sequencing Technology; SD: Spikelet Density; SNPP: Spikelet Number per Panicle; SnRK2: SNF1-Related Protein Kinase2; SP: Spikelet Number per Panicle; SSD: Spikelets Setting Density; SSP: Spikelet Number per Panicle; TFs: Transcription Factors; WEGO: Web Gene Ontology Annotation Plot Software; YPP: Yield per Plant

Acknowledgements
We thank the anonymous referees for their critical comments on this manuscript.

Funding

This research was supported by grant (2016YFD0101801) from The National Key Research and Development Program of China and Technology Department, grant (31560386) from National Nature Science Foundation of China and grant (201710410018) from National Undergraduate Training Program for Innovation and Entrepreneurship.

Authors' contributions

HH and JB conceived and designed the experiments. LC conceived and designed the experiments, and wrote the manuscript. SS and JY performed the experiments. CZ, LX, ST, XY, XP, QY, XH, JF, XC, LH, LO analyzed the data. HK and GMW revised the manuscript for the language. All authors read and approved the final manuscript.

Competing interests

The authors declare that they have no competing interests.

Author details

[1]Key Laboratory of Crop Physiology, Ecology and Genetic Breeding, Ministry of Education, Jiangxi Agricultural University, Nanchang 330045, China. [2]College of Agronomy, Jiangxi Agricultural University, Nanchang 330045, China. [3]Southern Regional Collaborative Innovation Center for Grain and Oil Crops in China, Changsha, China.

References

Bian JM, Jiang L, Liu LL, Xiao YH, Wang ZQ, Zhao ZG, Zhai HQ, Wan JM (2011) Identification of japonica chromosome segments associated with Heterosis for yield in Indica × japonica Rice hybrids. Crop Sci 50:2328–2337

Birchler JA, Auger DL, Riddle NC (2003) In search of the molecular basis of heterosis. Plant Cell 15:2236–2239

Cowling RJ, Kamiya Y, Seto H, Harberd NP (1998) Gibberellin dose-response regulation of GA4 gene transcript levels in Arabidopsis. Plant Physiol 117:1195

Cui KH, Peng SB, Xing YZ, Yu SB, Xu CG (2002) Genetic analysis of the panicle traits related to yield sink size of rice. Acta Genet Sin 29:144–152

Davenport CB (1908) Degeneration, albinism and inbreeding. Science 28:454

Eisen MB, Spellman PT, Brown PC, Botstein D (1998) Cluster analysis and display of genome-wide expression patterns. Proc Natl Acad Sci 95:14863–14868

Evans MM, Poethig RS (1995) Gibberellins promote vegetative phase change and reproductive maturity in maize. Plant Physiol 108:475

Griffing B (1990) Use of a controlled-nutrient experiment to test heterosis hypotheses. Genetics 126:753–767

He G, Luo XJ, Tian F, Li K, Zhu ZF, Su W, Qian XY, Fu YC, Wang XK, Sun CQ, Yang JS (2006) Haplotype variation in structure and expression of a gene cluster associated with a quantitative trait locus for improved yield in rice. Genome Res 16:618–626

He G, Zhu X, Elling AA, Chen L, Wang X, Guo L, Liang M, He H, Zhang H, Chen F, Qi Y, Chen R, Deng XW (2010) Global epigenetic and transcriptional trends among two rice subspecies and their reciprocal hybrids. Plant Cell 22:17–33

He P, Li JZ, Zheng XW, Shen LS, Lu CF, Chen Y, Zhu LH (2001) Comparison of molecular linkage maps and agronomic trait loci between DH and RIL populations derived from the same Rice cross. Crop Sci 41:1240–1246

Hua JP, Xing YZ, Xu CG, Sun XL, Yu SB, Zhang Q (2002) Genetic dissection of an elite rice hybrid revealed that heterozygotes are not always advantageous for performance. Genetics 162:1885

Huang X, Yang S, Gong J, Zhao Y, Feng Q, Gong H, Li W, Zhan Q, Cheng B, Xia J, Chen N, Hao Z, Liu K, Zhu C, Huang T, Zhao Q, Zhang L, Fan D, Zhou C, Lu Y, Weng Q, Wang ZX, Li J, Han B (2015) Genomic analysis of hybrid rice varieties reveals numerous superior alleles that contribute to heterosis. Nat Commun 6:6258

Huang Y, Zhang L, Zhang J, Yuan D, Xu C, Li X, Zhou D, Wang S, Zhang Q (2006) Heterosis and polymorphisms of gene expression in an elite rice hybrid as revealed by a microarray analysis of 9198 unique ESTs. Plant Mol Biol 62:579–591

Ishimaru K (2003) Identification of a locus increasing Rice yield and physiological analysis of its function. Plant Physiol 133:1083–1090

Kim D, Pertea G, Trapnell C, Pimentel H, Kelley R, Salzberg SL (2013) TopHat2: accurate alignment of transcriptomes in the presence of insertions, deletions and gene fusions. Genome Biol 14:R36

Komatsu K, Maekawa M, Ujiie S, Satake Y, Furutani I, Okamoto H, Shimamoto K, Kyozuka J (2003) LAX and SPA: major regulators of shoot branching in rice. Proc Natl Acad Sci 100:11765–11770

Kusterer BJ, Muminovic HF, Utz HP, Piepho S, Barth M, Heckenberger RC, Meyer T, Melchinger AE (2007) Analysis of a triple testcross design with recombinant inbred lines reveals a signifcant role of epistasis in heterosis for biomass-related traits in Arabidopsis. Genetics 175:2009–2017

Langmead B, Salzberg SL (2012) Fast gapped-read alignment with bowtie 2. Nat Methods 9:357–359

Li JX, Yu SB, Xu CG, Tan YF, Gao YJ, Li XH, Zhang QF (2000) Analyzing quantitative trait loci for yield using a vegetatively replicated F2 population from a cross between the parents of an elite rice hybrid. Theor Appl Genet 101:248–254

Li Z, Pinson SR, Park WD, Paterson AH, Stansel JW (1997) Epistasis for three grain yield components in rice (Oryza sativa L.). Genetics 145:453–465

Li Z, Pinson SRM, Stansel JW, Paterson AH (1998) Genetic dissection of the source-sink relationship affecting fecundity and yield in rice (Oryza sativa L.). Mol Breed 4:419–426

Li ZK, Luo LJ, Mei HW, Wang DL, Shu QY, Tabien R, Zhong DB, Ying CS, Stansel JW, Khush GS, Paterson AH (2001) Overdominant epistatic effect loci are the primary genetic basis of inbreeding depression and heterosis in rice. II. Grain yield components. Genetics 158:1737–1753

Lin HX, Qian HR, Zhuang JY, Lu J, Min SK, Xiong MZ, Huang N, Zheng KL (1995) Interval mapping of QTLs for yield and other related characters in rice. Rice Genet Newslett 12:251–253

Lin HX, Qian HR, Zhuang JY, Lu J, Min SK, Xiong ZM, Huang N, Zheng KL (1996) RFLP mapping of QTLs for yield and related characters in rice (Oryza sativa L.). Theor Appl Genetics 92:920–927

Lu C, Shen L, Tan Z, Xu Y, He P, Chen Y, Zhu L (1996) Comparative mapping of QTLs for agronomic traits of rice across environments using a doubled haploid population. TAG Theor Appl Genet 93:1211–1217

Lynch M, Walsh B (1998) Genetics Analysis of Quantitative Traits. Sinauer Associates Press, Sunderland

Mei HW, Xu JL, Li ZK, Yu XQ, Guo LB, Wang YP, Ying CS, Luo LJ (2006) QTLs influencing panicle size detected in two reciprocal introgressive line (IL) populations in rice (Oryza sativa L.). Theor Appl Genet 112:648–656

Merlot S, Gosti F, Guerrier D, Vavasseur A, Giraudat J (2001) The ABI1 and ABI2 protein phosphatases 2C act in a negative feedback regulatory loop of the abscisic acid signalling pathway. Plant J 25:295

Mortazavi A, Williams BA, McCue K, Schaeffer L, Wold B (2008) Mapping and quantifying mammalian transcriptomes by RNA-Seq. Nat Methods 5:621–628

Poethig RS (1990) Phase change and the regulation of shoot morphogenesis in plants. Science 250:923

Robinson MD, McCarthy DJ, Smyth GK (2010) edgeR: a Bioconductor package for differential expression analysis of digital gene expression data. Bioinformatics 26:139–140

Ryu CH, Lee S, Cho LH, Kim SL, Lee SY, Choi SC, Jeong HJ, Yi J, Park SJ, Han CD, An G (2009) OsMADS50 and OsMADS56 function antagonistically in regulating long day (LD)-dependent flowering in rice. Plant Cell Environ 32: 1412–1427

Shull GH (1908) The composition of a field of maize. Am Breed Assoc Rep 4:296–301

Song GS, Zhai HL, Peng YG, Zhang L, Wei G, Chen XY, Xiao YG, Wang L, Chen YJ, Wu B, Chen B, Zhang Y, Chen H, Feng XJ, Gong WK, Liu Y, Yin ZJ, Wang F, Liu GZ, Xu HL, Wei XL, Zhao XL, Ouwerkerk PB, Hankemeier T, Reijmers T, Heijden RVD, Lu C, Wang M, Greef JVD, Zhu Z (2010) Comparative transcriptional profiling and preliminary study on heterosis mechanism of super-hybrid rice. Mol Plant 3:1012–1025

Stuber C, Edwards M, Wendel J (1987) Molecular marker-facilitated investigations of quantitative trait loci in maize. II. Factors influencing yield and its component traits. Crop Sci 27(4):639–648

Takai T, Fukuta Y, Shiraiwa T, Horie T (2005) Time-related mapping of quantitative trait loci controlling grain-filling in rice (Oryza sativa L.). J Exp Bot 56:2107–2118

Thomson M, Tai T, McClung A, Lai X, Hinga M, Lobos K, Xu Y, Martinez C, McCouch S (2003) Mapping quantitative trait loci for yield, yield components and morphological traits in an advanced backcross population between Oryza rufipogon and the Oryza sativa cultivar Jefferson. Theor Appl Genet 107:479–493

Wei G, Tao Y, Liu G, Liu G, Chen C, Luo R, Xia H, Gan Q, Zeng H, Lu Z, Han Y, Li X, Song G, Zhai H, Peng Y, Li D, Xu H, Wei X, Cao M, Deng H, Xin Y, Fu X, Yuan L, Yu J, Zhu Z, Zhu L (2009) A transcriptomic analysis of superhybrid rice LYP9 and its parents. Proc Natl Acad Sci 106:7695–7701

Wu Y, Wang Y, Mi XF, Shan JX, Li XM, Xu JL, Lin HX (2016) The QTL GNP1 encodes GA20ox1, which increases grain number and yield by increasing Cytokinin activity in Rice panicle meristems. PLoS Genet 12:e1006386

Xiao J, Li J, Yuan L, Tanksley SD (1996) Identification of QTLs affecting traits of agronomic importance in a recombinant inbred population derived from a subspecific rice cross. Theo Appl Genet 92:230

Xu JL, Xue QZ, Luo LJ, Li ZK (2001) QTL Dissection of Panicle Number Per Plant and Spikelet Number Per Panicle in Rice (*Oryza sativa* L.). Acta Genet Sin 28:752–759

Xu ML, Jin ZX, Li XG, Zhang ZC, Liu HY, Zhang FZ, Zhao SY, Zhang HB (2009) Comparison of genetic distance among seven cultivars of japonica Rice based on SSR and SRAP and its relationship with Heterosis of yield traits. Molecular Plant Breeding: 2009-06

Yamagishi J, Miyamoto N, Hirotsu S, Laza RC, Nemoto K (2004) QTLs for branching, floret formation, and pre-flowering floret abortion of rice panicle in a temperate *japonica* × tropical *japonica* cross. Theor Appl Genet 109:1555

Ye J, Fang L, Zheng H, Zhang Y, Chen J, Zhang Z, Wang J, Li S, Li R, Bolund L, Wang J (2006) WEGO: a web tool for plotting GO annotations. Nucleic Acids Res 34:293–297

Yu SB, Li JX, Xu CG, Tan YF, Gao YJ, Li XH, Zhang QF, Maroot MAS (1997) Importance of epistasis as the genetic basis of heterosis in an elite rice hybrid. Proc Natl Acad Sci 94:9226–9231

Zhai R, Feng Y, Wang H, Zhan X, Shen X, Wu W, Zhang Y, Chen D, Dai G, Yang Z, Cao L, Cheng S (2013) Transcriptome analysis of rice root heterosis by RNA-Seq. BMC Genomics 14:19

Zhang HY, He H, Chen LB, Li L, Liang MZ, Wang XF, Liu XG, He GM, Chen RS, Ma LG, Deng XW (2008) A genome-wide transcription analysis reveals a close correlation of promoter INDEL polymorphism and heterotic gene expression in rice hybrids. Mol Plant 1:720–731

Zhuang JY, Fan YY, Wu JL, Xia YW, Zheng KL (2001) Comparison of the detection of QTL for yield traits in different generations of a rice cross using two mapping approaches. Acta Genet Sin 28:458

15

Dissection and fine-mapping of two QTL for grain size linked in a 460-kb region on chromosome 1 of rice

Qing Dong[1,2†], Zhen-Hua Zhang[1†], Lin-Lin Wang[1], Yu-Jun Zhu[1,2], Ye-Yang Fan[1], Tong-Min Mou[2], Liang-Yong Ma[1*] and Jie-Yun Zhuang[1*]

Abstract

Background: Grain size is a key determinant of grain weight and a trait having critical influence on grain quality in rice. While increasing evidences are shown for the importance of minor-effect QTL in controlling complex traits, the attention has not been given to grain size until recently. In previous studies, five QTL having small effects for grain size were resolved on the long arm of chromosome 1 using populations derived from *indica* rice cross Zhenshan 97///Zhenshan 97//Zhenshan 97/Milyang 46. One of them, *qTGW1.2c* that was located in a 2.1-Mb region, was targeted for fine-mapping in the present study.

Results: Firstly, the *qTGW1.2c* region was narrowed down into 1.1 Mb by determining genotypes of the cross-over regions using polymorphic markers newly developed. Then, one BC$_2$F$_9$ plant that was only heterozygous in the updated QTL region was identified. A total of 12 populations in generations from BC$_2$F$_{11:12}$ to BC$_2$F$_{15:16}$ were derived and used for QTL mapping. Two QTL linked in a 460-kb region were separated. The *qGS1-35.2* was delimited into a 57.7-kb region, containing six annotated genes of which five showed nucleotide polymorphisms between the two parental lines. Quantitative real-time PCR detected expression differences between near isogenic lines for *qGS1-35.2* at three of the six annotated genes. This QTL affected grain length and width with opposite allelic directions, exhibiting significant effect on ratio of grain length to width but showing little influence on yield traits. The other QTL, *qGW1-35.5*, was located within a 125.5-kb region and found to primarily control grain width and consequently affect grain weight.

Conclusions: Our work lays a foundation for cloning of two minor QTL for grain size that have potential application in rice breeding. The *qGS1-35.2* could be used to modify grain appearance quality without yield penalty because it affects grain shape but hardly influences grain yield, while *qGW1-35.5* offers a new gene recourse for enhancing grain yield since it contributes to grain size and grain weight simultaneously.

Keywords: Close linkage, Grain size, Minor effect, Quantitative trait locus, Rice

Background

Rice (*Oryza sativa* L.) is the staple food for more than half of the global population. Grain yield of rice depends on three components, i.e., number of panicles per plant, number of grains per panicle, and grain weight. Among them grain weight is mainly determined by grain size.

Grain size and shape is also an important quality trait that greatly influences the market value of grain products. In general, a short and bold rice grain is favored by consumers in Northern China, Japan and Korea, while a long and slender rice grain is preferred by consumers in the Africa, America and countries of Southeast Asia (Calingacion et al., 2014).

Grain size and shape are largely determined by grain length and width. All of them are complex traits controlled by a large number of quantitative trait loci (QTL). Up to date, a total of 14 QTL having large effect

* Correspondence: maliangyong@caas.cn; zhuangjieyun@caas.cn
†Qing Dong and Zhen-Hua Zhang contributed equally to this work.
[1]State Key Laboratory of Rice Biology and Chinese National Center for Rice Improvement, China National Rice Research Institute, Hangzhou 310006, China

for grain length and width in rice were cloned. One of them, *GL7/GW7*, has similar effects on grain length and width with opposite allelic directions, controlling grain shape but hardly influencing grain weight (Wang et al., 2015a; Wang et al., 2015b). The other 13 genes affect grain size and weight. Four of them mainly control grain width, including *GW2*, *GS5*, *qSW5/GW5*, and *GW8* (Li & Li, 2016). Eight others mainly control grain length, including *GS2/GL2*, *OsLG3*, *qLGY3/OsLG3b*, *GS3*, *GL3.1/qGL3*, *GL4*, *TGW6*, and *GLW7* (Li & Li, 2016; Wu et al., 2017; Yu et al., 2017; Liu et al., 2018; Yu et al., 2018). The remaining one, *GW6a*, has similar effects on grain length and width with the same allelic direction, and consequently exhibits a larger impact on grain weight (Song et al., 2015). It has been shown that these QTL regulate the proliferation and expansion of cells in spikelet hulls through diversified regulatory pathways. While most of them were involved in independent signaling pathways mediated by proteasomal degradation, plant hormones and G proteins, a number of genes were found to interact with each other (Yan et al., 2011; Wang et al., 2015a; Liu et al., 2018). These findings have greatly enriched our knowledge on the genetic control of grain size in rice, but much more efforts are needed to fill the gap in understanding the regulatory framework for this critical agronomical trait (Zuo & Li, 2014; Li & Li, 2016).

It has long been recognized that both major- and minor-effect QTL play important roles in the genetic control of complex traits (Mackay, Stone & Ayroles, 2009). QTL cloning in rice has been focused on those having large effects since the first success that was published in 2000 for heading date gene *Hd1* (Yano et al., 2000). Nevertheless, more and more attentions have been given to QTL with relative small effects in recent years. For heading date that has been taking the leading position in rice QTL studies, a number of minor-effect QTL were cloned or fine-mapped (Wu et al., 2013; Chen et al., 2014; Zhong et al., 2014; Chen et al., 2015; Shibaya et al., 2016). These QTL also showed important influences on the eco-geographical adaption and grain yield of rice, providing evidences for the importance of minor-effect QTL in controlling complex traits. For grain size and weight, QTL cloned are small in number and include no minor-effect QTL. While at least 546 QTL were detected in primary mapping and distributed over all regions of the 12 rice chromosomes (http://www.gramene.org), those that were cloned have very low genome coverage. None was located on chromosomes 1, 9, 10, 11 and 12, and on the long arm of chromosome 5 and short arms of chromosomes 2, 4, 6, 7 and 8. Isolation of QTL in these regions will be of great importance for establishing a gene network regulating grain size in rice.

In our previous studies, dissection of minor-effect QTL for grain weight and size was conducted using near isogenic lines (NILs) derived from a cross between *indica* rice cultivars Zhenshan 97 (ZS97) and Milyang 46 (MY46). Five QTL were resolved in an 8.2-Mb region on the long arm of chromosome 1 (Zhang et al., 2016). The present study aimed to fine-map one of the QTL, *qTGW1.2c* that was located in a 2.1-Mb interval (Wang et al., 2015c). Two linked QTL were separated in the target region, designated as *qGS1-35.2* and *qGW1-35.5*, respectively. The *qGS1-35.2* was delimited into a 57.7-kb region starting from the position of 35.2 Mb, affecting grain length and width with opposite allelic directions and showing little influence on grain weight. The *qGW1-35.5* was mapped in a 125.5-kb region starting from the position of 35.5 Mb, mainly controlling grain width and consequently affecting grain weight.

Methods

Plant materials

A total of 12 populations segregating in an isogenic background were used in this study. As described below and illustrated in Fig. 1, they were derived from a BC_2F_9 plant of the rice cross ZS97///ZS97//ZS97/MY46.

Firstly, new polymorphic markers were developed in the cross-over regions of *qTGW1.2c* (Wang et al., 2015c) and used to determine genotypes of a set of NILs that segregated this QTL. The *qTGW1.2c* region was narrowed down to be RM11807–RM11842 (details are presented in the first section of Results). Then, a BC_2F_9 plant that was only heterozygous in the RM11807–RM11842 interval was identified. The resultant BC_2F_{10} population consisting of 293 plants was genotyped using polymorphic markers in the target region. Two plants were selected, carrying heterozygous segments RM11807–RM11842 and RM265–RM11842, respectively. In the two resultant BC_2F_{11} populations consisting of 246 and 111 plants, respectively, homozygous non-recombinants (i.e., plants that were homozygous and showed no recombination in the corresponding segregating region) were identified and selfed. Two sets of NILs, namely L1 and L2, were developed and used for QTL analysis. The *qTGW1.2c* region was updated to be RM265–RM11842.

Another BC_2F_{10} plant carrying the RM265–RM11842 heterozygous segment was selected and selfed for two generations. A BC_2F_{12} population consisting of 259 individuals was genotyped. Five plants were selected, carrying sequential heterozygous segments extending from RM265 to RM11842. In the five resultant BC_2F_{13} populations consisting of 184, 188, 212, 190 and 206 plants, respectively, homozygous non-recombinants

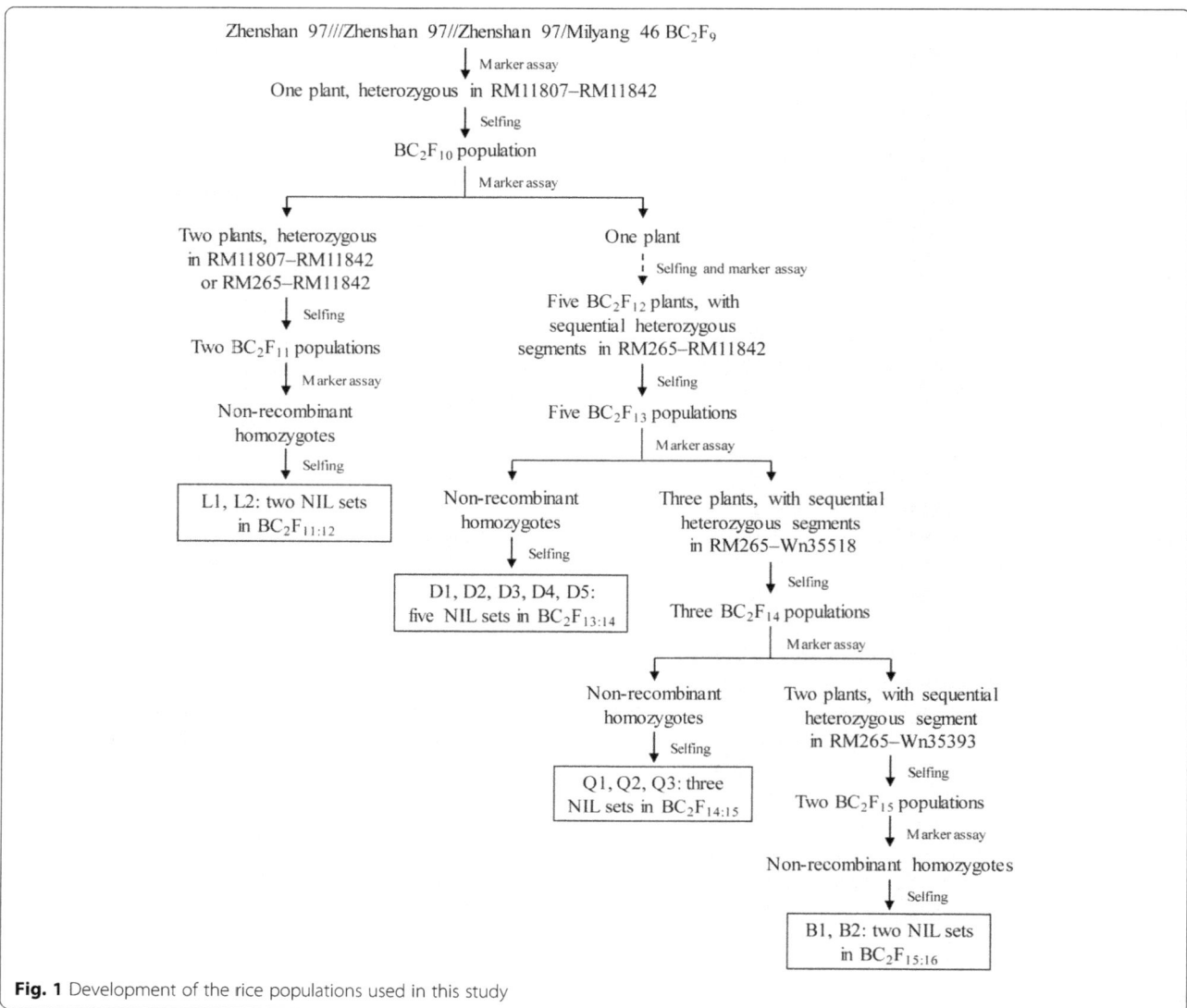

Fig. 1 Development of the rice populations used in this study

were identified and selfed. Five sets of NILs, namely D1, D2, D3, D4 and D5, were developed and used for QTL analysis. Two QTL were resolved, of which *qGS1-35.2* located in the upstream region was selected for further analysis.

Three plants were selected from the BC_2F_{13} populations, carrying sequential heterozygous segments covering *qGS1-35.2*. In the three resultant BC_2F_{14} populations consisting of 175, 166 and 187 plants, respectively, homozygous non-recombinants were identified and selfed. Three sets of NILs, namely Q1, Q2 and Q3, were developed and used for QTL analysis. The segregating region for *qGS1-35.2* was updated to be RM265–Wn35263.

Two other plants were selected from the BC_2F_{14} populations, carrying heterozygous segments RM265–Wn35263 and RM11824–Wn35393, respectively. In the

two resultant BC_2F_{15} populations consisting of 199 and 237 plants, respectively, homozygous non-recombinants were identified and selfed. Two sets of NILs, namely B1 and B2, were developed and used for fine-mapping of *qGS1-35.2*.

Field experiments and phenotyping

The rice populations were tested in the experimental stations of the China National Rice Research Institute located at either Hangzhou in Zhejiang Province or Lingshui in Hainan (Table 1). The experiments followed a randomized complete block design with two replications. In each replication, one line was grown in a single row of ten plants. Seedlings of about 25-day-old were transplanted with a planting density of 16.7 cm × 26.7 cm. Field management followed the normal agricultural practice. At maturity, five middle plants in each

Table 1 Rice populations and field experiments

Generation	Name	Segregating region		Number of lines[a]		Location and
		Marker	Physical position	NIL^{ZS97}	NIL^{MY46}	growing season[b]
BC$_2$F$_{11:12}$	L1	RM11807 – RM11842	34,722,600–35,694,183	41	37	HZ: May–Sep. 2014
	L2	RM265 – RM11842	35,197,724–35,694,183	21	22	HZ: May–Sep. 2014
BC$_2$F$_{13:14}$	D1	RM265 – Wn35518	35,197,724–35,518,354	25	27	HZ: May–Sep. 2016
	D2	RM265 – Wn35618	35,197,724–35,618,264	28	27	HZ: May–Sep. 2016
	D3	RM265 – RM11842	35,197,724–35,694,183	25	27	HZ: May–Sep. 2016
	D4	RM11828 – RM11842	35,315,714–35,694,183	23	25	HZ: May–Sep. 2016
	D5	Wn35518 – RM11842	35,518,508–35,694,183	25	28	HZ: May–Sep. 2016
BC$_2$F$_{14:15}$	Q1	RM265 – Wn35263	35,197,724–35,263,529	24	26	LS: Dec. 2016 – Apr. 2017
	Q2	RM265 – Wn35393	35,197,724–35,393,538	27	25	LS: Dec. 2016 – Apr. 2017
	Q3	Wn35263 – Wn35518	35,263,716–35,518,354	29	30	LS: Dec. 2016 – Apr. 2017
BC$_2$F$_{15:16}$	B1	RM265 – Wn35263	35,197,724–35,263,529	36	35	HZ: May–Sep. 2017
	B2	RM11824 – Wn35393	35,240,934–35,393,538	32	34	HZ: May–Sep. 2017

[a]NIL^{ZS97} and NIL^{MY46} are near isogenic lines with Zhenshan 97 and Milyang 46 homozygous genotypes in the segregating region, respectively
[b]HZ, Hangzhou, Zhejiang Province; LS, Lingshui, Hainan Province

row were harvested in bulk. Fully filled grain were selected and evaluated for grain weight and size following the procedure reported by Zhang et al. (2016). Four traits were measured for all populations, including 1000-grain weight (TGW, g), grain length (GL, mm), grain width (GW, mm) and ratio of grain length to width (RLW). One more trait for grain size, grain thickness (GT, mm), was measured for populations B1 and B2. The measurement was made over 20 fully filled grains using an electronic digital caliper with a precision of 0.001 mm (Shenzhen Star Instrument Co. Ltd., China). In addition, the two populations were evaluated for three other yield traits including number of panicles per plant (NP), number of grains per panicle (NGP) and grain yield per plant (GY, g).

Microscopy observation

For *qGS1-35.2* that was mapped within a 57.7-kb region in this study, NIL^{ZS97} and NIL^{MY46} were taken from population B1 and used for observation of outer glume epidermal cell. Young spikelet hulls were fixed with 2.5% glutaraldehyde for 24 h and then dehydrated by a graded series of ethanol. The dehydrated sample were coated with gold-palladium using ion sputter (Model E-1010, Hitachi, Japan) and observed using scanning electron microscope (Model TM-1000, Hitachi, Japan). Cell length and width of the outer glumes were measured, and cell number in the longitudinal direction was counted. For each NIL, 20 glumes from 20 plants were used.

DNA marker genotyping, sequence analysis and quantitative real-time PCR analysis

For population development and QTL mapping, total DNA was extracted using 2 cm-long leaf sample following

the method of Zheng et al. (1995). PCR amplification was performed according to Chen et al. (1997). The products were visualized on 6% non-denaturing polyacrylamide gels using silver staining. A total of 24 polymorphic DNA markers were used, including 11 simple sequence repeat (SSR) and 13 InDel markers (Additional file 1: Table S1). The SSR markers were selected from the Gramene database (http://www.gramene.org), and the InDel markers were designed according to the differences between ZS97 and MY46 detected by whole genome re-sequencing.

Sequence analysis was performed for six annotated genes located in the target QTL region. DNA was extracted using DNeasy Plant Mini Kit (QIAGEN, German) according to the manufacturer's instructions. The primers were designed according to the sequence of Nipponbare in RAP-DB (http://rapdb.lab.nig.ac.jp/:IRGSP-1.0) (Additional file 1: Table S2). Products amplified from the genomic DNA of ZS97 and MY46 were sequenced. Nucleotide sequence and the predicted amino acid sequence between ZS97 and MY46 were compared.

Panicles of 1 cm and 8 cm long were collected from NIL^{ZS97} and NIL^{MY46} in population B1. Total RNA was extracted using RNeasy Plus Mini Kit (QIAGEN, German). First-strand cDNA was synthesized using ReverTra AceR Kit (TOYOBO, Japan). Quantitative real-time PCR was performed on Applied Biosystems 7500 using SYBR qPCR Mix Kit (TOYOBO, Japan) according to the manufacturer's instructions. *Actin1* was used as the endogenous control. The data were analyzed according to the $2^{-\Delta\Delta Ct}$ method (Livak and Schmittgen, 2001). Three biological replicates and three technical replicates were used. The primers were listed in Additional file 1: Table S3.

Data analysis

Two-way analysis of variance (ANOVA) was performed to test the phenotypic differences between the two genotypic groups in each NIL set. The analysis was performed using the SAS procedure GLM (SAS Institute 1999) as described previously (Dai et al., 2008). Given the detection of a significant difference ($P < 0.05$), the same data were used to estimate the genetic effect of the QTL, including additive effect and the proportion of phenotypic variance explained (R^2). QTL were designated according to the rules recommended by McCouch and CGSNL (2008) with slight modification. Physical position of the first segregating marker in the QTL region was used as the unique identifier for the given QTL. For example, *qGW1-35.5* indicates that this QTL was associated with grain width (GW) and mapped in a region on chromosome 1 with the first segregating marker located at 35.5 Mb.

Cell length, width and number, as well as the expression level, were presented in mean ± s.e.m. Differences between NIL^{ZS97} and NIL^{MY46} were tested by student's *t*-test.

Results

Delimitation of *qTGW1.2c* from 2.1-Mb to 1.1-Mb by increasing marker density

In a previous study (Wang et al., 2015c), *qTGW1.2c* controlling grain weight in rice was located within a 2.1-Mb region between RM11800 and RM11885 on the long arm of chromosome 1. This interval included the segregating region RM11807–RM265 and two flanking cross-over regions, i.e., RM11800–RM11807 and RM265–RM11885. Based on sequence differences between ZS97 and MY46 detected by whole genome re-sequencing, seven polymorphic markers were developed. They were all located in one of the two cross-over regions RM265–RM11885. The original NIL population segregating *qTGW1.2c* was assayed using these markers. Four markers neighboring to RM11885 were homozygous, thus the downstream boundary of the QTL was moved from RM11885 to RM11844 (Fig. 2a). Therefore, *qTGW1.2c* was narrowed down to a 1.1-Mb region flanked by RM11800 and RM11844.

Dissection of *qTGW1.2c* into two QTL

Two NIL sets were developed following the updated location of *qTGW1.2c* (Fig. 2b). Highly significant genotypic effects ($P < 0.001$) were detected for TGW and GW in both populations, with the enhancing alleles all derived from MY46 (Table 2). In L1, the additive effects were 0.20 g for TGW and 0.027 mm for GW, explaining 17.6 and 49.3% of the phenotypic variance, respectively. In L2, the additive effects were 0.22 g for TGW and 0.025 mm for GW, contributing 21.6 and 44.7% to the

phenotypic variance, respectively. An opposite small effect was detected for GL, which was significant ($P = 0.0015$) in L1 only. The ZS97 allele increased GL by 0.017 mm, explaining 7.7% of the phenotypic variance. Obviously, *qTGW1.2c* affected grain weight mainly through grain width. For RLW, the ratio of GL to GW, highly significant effects ($P < 0.0001$) were detected in both populations (Table 2). The R^2 values were 57.2 and 58.6%, higher than the values estimated for GL and GW.

As described above, the effects detected in the two populations were similar, indicating that *qTGW1.2c* was located in the common segregating regions of L1 and L2. While the whole candidate region was segregated in L1, only a portion was segregated in L2. Thus, *qTGW1.2c* was located in the segregating region of L2, which was a 672.3-kb region flanked by Wn35060 and RM11844 (Fig. 2b). This result was used to develop five NIL sets with sequential segregating regions jointly covering the entire QTL region (Fig. 2c).

In all the five populations, significant effects were detected on GW with the enhancing alleles always derived from MY46 (Table 3). The additive effects were 0.009, 0.021, 0.023, 0.014 and 0.015 mm in D1, D2, D3, D4 and D5, respectively. Two alternative explanations could be given to the consistent allelic direction and varied magnitudes among the five populations. Firstly, there are two QTL for GW segregated in these populations. One was segregated in D1 but not in D4 and D5, and the other was segregated in D4 and D5 but not in D1. They were both segregated in D2 and D3, thus the additive effects were higher in the two populations than D1, D4 and D5. Secondly, one single QTL was segregated in these populations but the effect was not highly stable.

For GL, significant effects were detected in three populations (Table 3). The enhancing alleles were also derived from MY46 in D4 and D5, but from ZS97 in D1. These results indicate that two QTL for GL were located in the target region. One was segregated in D1 but not in D4 and D5, and the other was segregated in D4 and D5 but not in D1. The two QTL were both segregated in D2 and D3, thus the effect became nonsignificant due to opposite directions. Taking the results on GL and GW together, it could be concluded that two QTL simultaneously affecting the two traits were segregated in these populations. One was located in a region that was segregated in D1, D2 and D3 but homozygous in D4 and D5, with the allele from MY46 decreasing GL but increasing GW (Fig. 2c; Additional file 2: Figure S1). The other was located in a region that was segregated in D2, D3, D4 and D5 but homozygous in D1, with the allele from MY46 increasing GW and GL (Additional file 3: Figure S2).

The first QTL was located within the Wn35183–RM11828 interval that corresponds to a 132.4-kb region of the Nipponbare genome (Fig. 2c). It had little

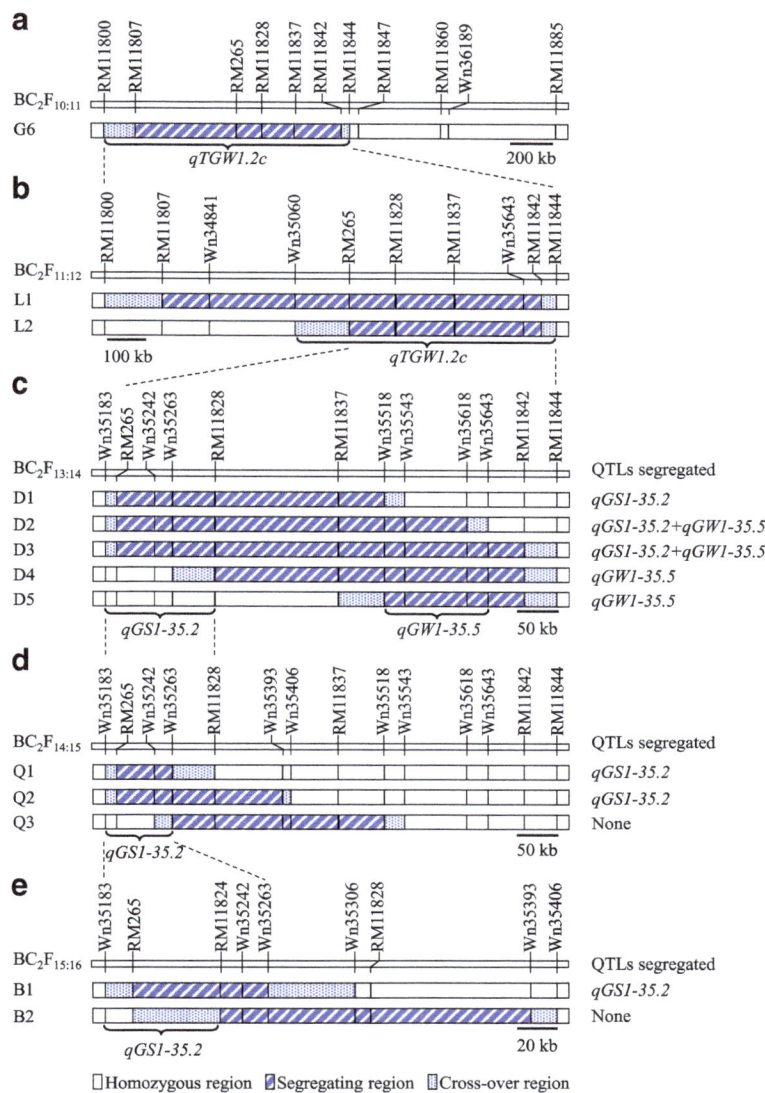

Fig. 2 Genotypic compositions of the near isogenic lines (NILs) in target regions. **a** Composition of NIL set G6 used by Wang et al. (2015c), updated in this study with more polymorphic markers. **b** Two sets of NILs in BC$_2$F$_{11:12}$. **c** Five sets of NILs in BC$_2$F$_{13:14}$. **d** Three sets of NILs in BC$_2$F$_{14:15}$. **e** Two sets of NILs in BC$_2$F$_{15:16}$

effect on TGW but significantly affected GL, GW and RLW. In the D1 population that segregated this QTL only, the MY46 allele decreased GL by 0.027 mm, increased GW by 0.009 mm, and decreased RLW by 0.017, having R^2 of 14.7, 11.6 and 39.0%, respectively (Table 3). Because this QTL primarily contributed to grain shape with the first segregating marker RM265 located at 35.2 Mb on chromosome 1, we designated it $qGS1$-35.2 (Fig. 2c).

The other QTL was located in a 125.5-kb region flanked by Wn35518 and Wn35643 (Fig. 2c). It affected GL and GW with the same allelic direction and exerted significant influence on TGW. In the D4 and D5 populations that segregated this QTL only, the MY46 allele

increased TGW by 0.20 and 0.14 g, GL by 0.018 and 0.014 mm, and GW by 0.014 and 0.015 mm, having R^2 of 14.3 and 7.2%, 8.3 and 5.5%, and 25.7 and 27.3%, respectively (Table 3). Because this QTL mainly contributed to grain width with the first segregating marker Wn35518 located at 35.5 Mb on chromosome 1, we designated it $qGW1$-35.5 (Fig. 2c).

Fine-mapping of $qGS1$-35.2

One QTL, $qGS1$-35.2, was selected for further analysis. Two more runs of NIL construction – QTL mapping were performed. The first run of QTL mapping was conducted using three NIL sets, Q1, Q2 and Q3. Significant effects for grain shape traits were detected in Q1 and Q2

Table 2 Validation of *qTGW1.2c* using two sets of near isogenic lines in BC$_2$F$_{11:12}$

Name	Trait[a]	Phenotype (mean ± sd)[b]		P	A[c]	R^2(%)[d]
		NIL^{ZS97}	NIL^{MY46}			
L1	TGW	28.67 ± 0.34	29.08 ± 0.32	< 0.0001	0.20	17.6
	GL	8.458 ± 0.040	8.423 ± 0.051	0.0015	−0.017	7.7
	GW	3.218 ± 0.021	3.272 ± 0.016	< 0.0001	0.027	49.3
	RLW	2.628 ± 0.017	2.575 ± 0.015	< 0.0001	−0.027	57.2
L2	TGW	28.86 ± 0.35	29.30 ± 0.35	0.0002	0.22	21.6
	GL	8.517 ± 0.044	8.502 ± 0.050	0.3107		
	GW	3.262 ± 0.017	3.311 ± 0.020	< 0.0001	0.025	44.7
	RLW	2.611 ± 0.012	2.568 ± 0.016	< 0.0001	−0.022	58.6

[a]*TGW* 1000-grain weight (g), *GL* grain length (mm), *GW* grain width (mm), *RLW* ratio of grain length to width
[b]NIL^{ZS97} and NIL^{MY46} are near isogenic lines with Zhenshan 97 and Milyang 46 homozygous genotypes in the segregating region, respectively
[c]Additive effect of replacing a Zhenshan 97 allele with a Milyang 46 allele
[d]Proprotion of phenotypic variance explained by the QTL effect

Table 3 Dissection of *qTGW1.2c* into two QTL using five sets of near isogenic lines in BC$_2$F$_{13:14}$

Name	Trait[a]	Phenotype (mean ± sd)[b]		P	A[c]	R^2(%)[d]
		NIL^{ZS97}	NIL^{MY46}			
D1	TGW	28.22 ± 0.50	28.24 ± 0.44	0.8500		
	GL	8.392 ± 0.042	8.338 ± 0.046	< 0.0001	−0.027	14.7
	GW	3.063 ± 0.022	3.080 ± 0.017	0.0023	0.009	11.6
	RLW	2.740 ± 0.018	2.707 ± 0.012	< 0.0001	−0.017	39.0
D2	TGW	27.10 ± 0.39	27.73 ± 0.37	< 0.0001	0.32	30.3
	GL	8.342 ± 0.039	8.332 ± 0.042	0.3473		
	GW	3.079 ± 0.023	3.121 ± 0.023	< 0.0001	0.021	33.1
	RLW	2.710 ± 0.020	2.670 ± 0.018	< 0.0001	−0.020	41.2
D3	TGW	27.22 ± 0.38	27.75 ± 0.49	< 0.0001	0.26	15.3
	GL	8.313 ± 0.051	8.298 ± 0.042	0.2498		
	GW	3.037 ± 0.020	3.084 ± 0.022	< 0.0001	0.023	39.8
	RLW	2.738 ± 0.016	2.691 ± 0.017	< 0.0001	−0.023	49.1
D4	TGW	27.71 ± 0.39	28.11 ± 0.40	0.0013	0.20	14.3
	GL	8.286 ± 0.050	8.321 ± 0.052	0.0205	0.018	8.3
	GW	3.087 ± 0.017	3.116 ± 0.021	< 0.0001	0.014	25.7
	RLW	2.684 ± 0.018	2.671 ± 0.017	0.0192	−0.007	8.9
D5	TGW	27.25 ± 0.36	27.53 ± 0.42	0.0126	0.14	7.2
	GL	8.339 ± 0.047	8.368 ± 0.049	0.0328	0.014	5.5
	GW	3.046 ± 0.015	3.075 ± 0.016	< 0.0001	0.015	27.3
	RLW	2.738 ± 0.019	2.721 ± 0.016	0.0015	−0.008	11.7

[a]*TGW* 1000-grain weight (g), *GL* grain length (mm), *GW* grain width (mm), *RLW* ratio of grain length to width
[b]NIL^{ZS97} and NIL^{MY46} are near isogenic lines with Zhenshan 97 and Milyang 46 homozygous genotypes in the segregating region, respectively
[c]Additive effect of replacing a Zhenshan 97 allele with a Milyang 46 allele
[d]Proprotion of phenotypic variance explained by the QTL effect

but not Q3. The QTL effects remained to be large on GL and RLW, and small or nonsignificant on TGW and GW (Table 4). The QTL location was delimited into an 80.4-kb region flanked by Wn35183 and Wn35263 (Fig. 2d). The second run was done using two NIL sets, B1 and B2. Significant effects for grain shape traits were detected in B1 but not B2. Finally, *qGS1-35.2* was mapped within a 57.7-kb region flanked by Wn35183 and RM11824 (Fig. 2e). The effect of this QTL was large on GL and RLW, small on GW, and nonsignificant on GT and yield traits including TGW, NP, NGP and GY (Table 4).

Length, width and number of the outer glume epidermal cells were compared between NIL^{ZS97} and NIL^{MY46} for *qGS1-35.2* (Additional file 4: Figure S3). Nonsignificant difference was detected on the cell length and width, but the cell number in the longitudinal direction was higher in NIL^{ZS97} than NIL^{MY46} ($P = 0.0027$). These results suggest that *qGS1-35.2* affects grain length by controlling cell division.

Candidate genes of *qGS1-35.2*

According to the Rice Annotation Project Database (http://rapdb.dna.affrc.go.jp/), there are six annotated genes in the 57.7-kb region for *qGS1-35.2*. Three of them encode proteins containing known functional domains. *Os01g823900* encodes the U-box E3 ubiquitin ligase OsPUB3 that regulates the response to abiotic stress (Byun et al., 2017). *Os01g0824600* produces two different transcripts, encoding a serine/threonine protein kinase domain containing protein or CBL-interacting protein kinase 11 that are involved in various biological processes (Sanyal et al., 2015). *Os01g0824700* encodes a member of the cyclin-like F-box domain containing proteins that are major components of E3 ubiquitin-protein ligase and participate in a large variety of biological processes including seed development (Somers & Fujiwara, 2009, Chen et al., 2013; Gupta, Garg & Bhatia, 2015). The remaining three annotated genes are *Os01g0823951*, *Os01g0824000* and *Os01g0824500* that encode hypothetical proteins.

Sequence comparisons of the six annotated genes were conducted between full-length genomic fragments of ZS97 and MY46 (Additional file 1: Table S4). Among the three genes encoding proteins of known functional domains, no difference was identified in *Os01g0824600* but single nucleotide polymorphisms (SNPs) were detected in the other two genes. For *Os01g0824700*, the G116A substitution resulted in a premature stop codon in MY46. For *Os01g823900*, four SNPs were found, of which two were synonymous and the other two were non-synonymous. SNPs resulting in non-synonymous mutation were also found for all the three genes encoding hypothetical proteins.

Table 4 Fine mapping of *qGS1-35.2* using five sets of near isogenic lines in $BC_2F_{14:15}$ and $BC_2F_{15:16}$

Generation	Name	Trait[a]	Phenotype (mean ± sd)[b]		P	A[c]	R²(%)[d]
			NIL^ZS97	NIL^MY46			
$BC_2F_{14:15}$	Q1	TGW	27.66 ± 0.20	27.61 ± 0.22	0.3588		
		GL	8.032 ± 0.033	7.989 ± 0.034	< 0.0001	−0.022	26.1
		GW	3.235 ± 0.015	3.242 ± 0.015	0.0917		
		RLW	2.483 ± 0.014	2.464 ± 0.010	< 0.0001	−0.010	18.2
	Q2	TGW	27.57 ± 0.26	27.60 ± 0.26	0.7044		
		GL	8.023 ± 0.034	7.964 ± 0.027	< 0.0001	−0.029	33.4
		GW	3.222 ± 0.014	3.235 ± 0.018	0.0084	0.006	7.5
		RLW	2.490 ± 0.011	2.462 ± 0.014	< 0.0001	−0.014	34.3
	Q3	TGW	28.00 ± 0.34	28.13 ± 0.28	0.0986		
		GL	8.015 ± 0.036	8.017 ± 0.026	0.8229		
		GW	3.246 ± 0.017	3.249 ± 0.015	0.4148		
		RLW	2.470 ± 0.014	2.468 ± 0.014	0.5968		
$BC_2F_{15:16}$	B1	TGW	28.80 ± 0.27	28.76 ± 0.29	0.6390		
		GL	8.386 ± 0.030	8.325 ± 0.036	< 0.0001	−0.030	29.0
		GW	3.128 ± 0.017	3.141 ± 0.021	0.0070	0.006	5.2
		GT	2.193 ± 0.014	2.197 ± 0.009	0.1490		
		RLW	2.681 ± 0.011	2.651 ± 0.018	< 0.0001	−0.015	36.5
		NP	16.54 ± 1.79	16.44 ± 2.19	0.8313		
		NGP	55.60 ± 5.84	57.12 ± 8.90	0.3960		
		GY	24.28 ± 1.28	24.67 ± 1.88	0.3065		
	B2	TGW	27.94 ± 0.44	28.04 ± 0.30	0.2894		
		GL	8.438 ± 0.047	8.443 ± 0.041	0.6192		
		GW	3.055 ± 0.027	3.057 ± 0.019	0.6095		
		GT	2.204 ± 0.011	2.209 ± 0.011	0.0843		
		RLW	2.762 ± 0.021	2.760 ± 0.014	0.7448		
		NP	15.16 ± 0.93	15.42 ± 1.13	0.3456		
		NGP	69.04 ± 7.68	67.82 ± 6.95	0.5098		
		GY	24.12 ± 3.38	24.11 ± 3.44	0.9599		

[a]*TGW* 1000-grain weight (g), *GL* grain length (mm), *GW* grain width (mm), *GT* grain thickness (mm), *RLW* ratio of grain length to width, *NP* number of panicles per plant, *NGP* number of grains per panicle, *GY* grain yield per plant (g)
[b]NIL^ZS97 and NIL^MY46 are near isogenic lines with Zhenshan 97 and Milyang 46 homozygous genotypes in the segregating region, respectively
[c]Additive effect of replacing a Zhenshan 97 allele with a Milyang 46 allele
[d]Proportion of phenotypic variance explained by the QTL effect

Transcript levels of the six annotated genes in panicle were compared between NIL^ZS97 and NIL^MY46 for *qGS1-35.2* (Additional file 5: Figure S4). For the two transcripts produced by *Os01g0824600*, significant difference was only detected on *Os01t0824600-2* encoding CBL-interacting protein kinase 11. In panicle of 1 cm and 8 cm long, the expression levels were 1.8 and 1.6 times higher in NIL^MY46 than NIL^ZS97, respectively. Two more genes were found to have significant expression differences between the two NILs. They were *Os01g0823951* and *Os01g082400* encoding hypothetical proteins. As compared with NIL^ZS97, the expression levels of NIL^MY46 in panicles of 1 cm and 8 cm were 6.2 and 7.6 times higher on *Os01g0823951*, and 1.6 and 0.8 times higher on *Os01g082400*, respectively. Nonsignificant difference was detected on other genes.

Discussion

In recent years, increasing attention has been paid to the cloning of minor QTL in rice, but none has been reported for traits determining grain size. In the present study, two minor QTL associated with grain size in rice were dissected and fine-mapped. They were located in the 460-kb interval Wn35183–Wn35643 on the long arm of chromosome 1. One of them, *qGS1-35.2*, was delimited into a 57.7-kb region flanked by Wn35183 and

RM11824, affecting grain length and width with opposite allelic directions and showing little influence on grain weight. The other one, *qGW1-35.5*, was mapped within a 125.5-kb region flanked by Wn35518 and Wn35643, primarily controlling grain width and consequently affecting grain weight. Our work lays a foundation for cloning the genes underlying these two minor QTL for grain size.

Clustering of genes for the same trait is frequently observed in plant genome. This has been evident for genes having large effects on grain size in rice. For example, a 3.2-Mb region on the short arm of chromosome 3 covers five genes, including *PGL1*, *BG1*, *OsLG3*, *OsLG3b/qLGY3*, and *TUD1* (Heang & Sassa, 2012a; Hu et al., 2013; Li & Li, 2016; Liu et al., 2018; Yu et al., 2018); a 4.3-Mb region on the short arm of chromosome 5 covers seven genes, including *APG*, *OsPPKL2*, *SRS3*, *GS5*, *GW5/qSW5*, *GSK2* and *OsCYP51G3* (Heang & Sassa, 2012b; Zhang et al., 2012; Huang et al., 2013; Xia et al., 2015; Li & Li, 2016). Clustering of QTL having small effect on grain size in rice has also been observed. In our previous studies, five minor QTL for grain size were dissected in a region on the long arm of chromosome 1 (Wang et al., 2015c; Zhang et al., 2016). One of them, *qTGW1.2c*, was separated into two QTL in the present study. Altogether, six minor QTL for grain size have been separated in a 7.1-Mb region using one *indica* rice cross, spanning from the upstream boundary marker Wn28447 for *qTGW1.1a* (Zhang et al., 2016) to the downstream boundary marker Wn35643 for *qGW1-35.5* reported here. These results suggest that grain size in rice is controlled by a large number of QTL, including a few loci with large effects and numerous loci with small effects, which is similar to the genetic architecture of heading date in rice (Hori et al., 2015).

Grain size and shape are both determined by grain length and width. While grain size is the major determinant of grain weight, grain shape is mainly related to consuming preference (Calingacion et al., 2014) and may not be associated with grain weight. For genes having similar effects on grain length and width with opposite directions, such as *GL7/GW7* (Song et al., 2015), the influence is usually exerted on grain shape rather than grain weight. One of the two QTL we identified, *qGS1-35.2*, affected grain shape without influencing grain weight and other yield traits. The MY46 allele decreased grain length but increased grain width, resulting in little effect on grain weight and enhanced effect on the ratio of grain length to width (Table 3). This type of QTL could be used to modify grain shape without yield penalty. For genes simultaneously controlling grain length and width with the same allelic direction, and those contributing to either grain length or width, such as most of the cloned genes conditioning these traits (Li and Li,

2016; Yu et al., 2017), the same direction of QTL effect is always simultaneously detected on grain size and grain weight. Another QTL we identified, *qGW1-35.5*, falls into this category (Table 3). The MY46 allele significantly increased both the grain length and width, and in the meantime enhanced grain weight. This type of QTL could be used for yield improvement. More and more choices for breeding utilization can be anticipated with the identification of new genes for grain size traits.

Three annotated genes encoding proteins with known functional domains were located in the *qGS1-35.2* region. They were involved in two important pathways that regulate grain size in plants (Zuo & Li, 2014; Li & Li, 2016). *Os01g0824700* and *Os01g0823900* encode two proteins that are important components of ubiquitin ligases (Xu et al., 2009; Byun et al., 2017), and *Os01g0824600* participates in plant hormone signaling pathway (Xiang, Huang & Xiong, 2007). For *Os01g0824700*, a premature stop mutation was found in MY46, which usually fully disrupt gene function. In addition, nonsignificant difference was detected between NIL[ZS97] and NIL[MY46] on the expression of *Os01g0824700*. It is unlikely that this is the gene for *qGS1-35.2*. For *Os01g823900*, two amino acid substitutions were identified, which may cause minor phenotypic change as suggested in previous studies (Matsubara et al., 2012; Wu et al., 2013; Shibaya et al., 2016). For *Os01g0824600*, no difference was identified in its coding region, but expression difference was found on one of the two transcripts of this gene (Additional file 5: Figure S4). Both *Os01g823900* and *Os01g0824600* are more likely to be the gene underlying *qGS1-35.2*.

Three other annotated genes, *Os01g0823951*, *Os01g0824000* and *Os01g0824500* encoding hypothetical proteins, were also located in the *qGS1-35.2* region. All of them showed amino acid substitutions. Expression differences were also detected on two of the genes, *Os01g0823951* and *Os01g0824000*. None of them could be ruled out from the candidate genes for *qGS1-35.2*. Therefore, more work is needed to clarify which gene is the one for QTL *qGS1-35.2*.

Conclusion

Two closely linked minor QTL for grain size in rice were separated on the long arm of chromosome 1. The *qGS1-35.2* was delimited into a 57.7-kb region in which six annotated genes were found. This QTL regulates grain length and width with opposite allelic directions, affecting grain shape but having little influence on grain weight and other yield traits, providing a potential gene resource for fine-tuning grain shape to modify grain appearance quality without yield penalty. The *qGW1-35.5* regulates grain width and length with the same allelic direction, simultaneously affecting grain shape, size and weight, offering a new gene resource for enhancing grain yield.

Additional files

Additional file 1: Table S1. Primers used for population development and QTL mapping. **Table S2.** Primers used for sequence analysis. **Table S3.** Primers used for quantitative real-time PCR. **Table S4.** Nucleotide and amino acid differences between Zhenshan 97 and Milyang 46.

Additional file 2: Figure S1. Grains of NIL^{ZS97} and NIL^{MY46} for qGS1-35.2. Scale bar, 20 mm.

Additional file 3: Figure S2. Grains of NIL^{ZS97} and NIL^{MY46} for qGW1-35.5. Scale bar, 20 mm.

Additional file 4: Figure S3. Characterization of the cells in outer glumes of NIL^{ZS97} and NIL^{MY46} for qGS1-35.2. **a** Scanning electron microscopic images of the cells. Scale bar, 200 μm. **b** Cell length, width and number. The cell numbers were measured in the longitudinal direction. Data are presented in mean ± s.e.m. ($n = 20$). A Student's t-test was used to generate the P values.

Additional file 5: Figure S4. Transcript levels of annotated genes in the qGS1-35.2 region. The experiment was performed using panicles of 1 cm (P1) and 8 cm (P8) collected from NIL^{ZS97} and NIL^{MY46} for qGS1-35.2. The expression levels were normalized to Actin1 and related to P1 of NIL^{ZS97}. Data are presented in mean ± s.e.m. ($n = 3$). A Student's t-test was used to generate the P values.

Abbreviations
ANOVA: Analysis of Variance; GL: Grain Length; GS: Grain Shape; GT: Grain Thickness; GW: Grain Width; GY: Grain Yield per Plant; MY46: Milyang 46; NGP: Number of Grains per Panicle; NIL: Near Isogenic Line; NP: Panicle per Plant; QTL: Quantitative Trait Locus; R^2: Proportion of Phenotypic Variance Explained; RAP-DB: Rice Annotation Project Database; RLW: Ratio of Grain Length to Width; SNP: Single Nucleotide Polymorphism; SSR: Simple Sequence Repeat; TGW: 1000-Grain Weight; ZS97: Zhenshan 97

Funding
This work was supported by the National Key Research and Development Program of China (2017YFD0100305), Zhejiang Provincial Natural Science Foundation of China (LQ18C130002), and a project of the China National Rice Research Institute (2017RG001–2).

Authors' contributions
JYZ, LYM and TMM conceived the study. JYZ and QD wrote the manuscript. QD, ZHZ and YYF performed the population development and data analysis. QD, LLW and ZHZ carried out the marker analysis, sequence analysis and quantitative real-time PCR analysis. QD, ZHZ and YJZ conducted the field experiments and microscopy observation. All authors read and approved the final manuscript.

Competing interests
The authors declare that they have no competing interests.

Author details
[1]State Key Laboratory of Rice Biology and Chinese National Center for Rice Improvement, China National Rice Research Institute, Hangzhou 310006, China. [2]State Key Laboratory of Crop Genetic Improvement and National Center of Plant Gene Research (Wuhan), Huazhong Agricultural University, Wuhan 430070, China.

References

Byun MY, Cui LH, Oh TK, Jung Y-J, Lee A, Park KY, Kang BG, Kim WT (2017) Homologous U-box E3 ubiquitin ligases OsPUB2 and OsPUB3 are involved in the positive regulation of low temperature stress response in rice (Oryza sativa L.). Front Plant Sci 8:16

Calingacion M, Laborte A, Nelson A, Resurreccion A, Concepcin JC, Daygon VD, Mumm R, Reinke R, Dipti S, Bassinello PZ, Manful J, Sophany S, Lara KC, Bao J, Xie L, Loaiza K, El-hissewy A, Gayin J, Sharma N, Rajeswari S, Manonmani S, Rani NS, Kota S, Indrasari SD, Habibi F, Hosseini M, Tavasoli F, Suzuki K, Umemoto T, Boualaphanh C, Lee HH, Hung YP, Ramli A, Aung PP, Ahmad R, Wattoo JI, Bandonill E, Romero M, Brites CM, Hafeel R, Lur H-S, Cheaupun K, Jongdee S, Blanco P, Bryant R, Lang NT, Hall RD, Fitzgerald M (2014) Diversity of global rice markets and the science required for consumer-targeted rice breeding. PLoS One 9:e85106

Chen J-Y, Guo L, Ma H, Chen Y-Y, Zhang H-W, Ying J-Z, Zhuang J-Y (2014) Fine mapping of qHd1, a minor heading date QTL with pleiotropism for yield traits in rice (Oryza sativa L.). Theor Appl Genet 127:2515–2524

Chen L, Zhong Z, Wu W, Liu L, Lu G, Jin M, Tan J, Sheng P, Wang D, Wang J, Cheng Z, Wang J, Zhang X, Guo X, Wu F, Lin Q, Zhu S, Jiang L, Zhai H, Wu C, Wan J (2015) Fine mapping of DTH3b, a minor heading date QTL potentially functioning upstream of Hd3a and RFT1 under long-day conditions in rice. Mol Breeding 35:206

Chen X, Temnykh S, Xu Y, Cho YG, McCouch SR (1997) Development of a microsatellite framework map providing genome-wide coverage in rice (Oryza sativa L.). Theor Appl Genet 95:553–567

Chen Y, Xu Y, Luo W, Li W, Chen N, Zhang D, Chong K (2013) The F-box protein OsFBK12 targets OsSAMS1 for degradation and affects pleiotropic phenotypes, including leaf senescence, in rice. Plant Physiol 163:1673–1685

Dai W-M, Zhang K-Q, Wu J-R, Wang L, Duan B-W, Zheng K-L, Cai R, Zhuang J-Y (2008) Validating a segment on the short arm of chromosome 6 responsible for genetic variation in the hull silicon content and yield traits of rice. Euphytica 160:317–324

Gupta S, Garg V, Bhatia S (2015) A new set of ESTs from chickpea (Cicer arietinum L.) embryo reveals two novel F-box genes, CarF-box_PP2 and CarF-box_LysM, with potential roles in seed development. PLoS One 10:e0121100

Heang D, Sassa H (2012a) An atypical bHLH protein encoded by POSITIVE REGULATOR OF GRAIN LENGTH 2 is involved in controlling grain length and weight of rice through interaction with a typical bHLH protein APG. Breeding Sci 62:133–141

Heang D, Sassa H (2012b) Antagonistic actions of HLH/bHLH proteins are involved in grain length and weight in rice. PLoS One 7:e31325

Hori K, Nonoue Y, Ono N, Shibaya T, Ebana K, Matsubara K, Ogiso-Tanaka E, Tanabata T, Sugimoto K, Taguchi-Shiobara F, Yonemaru J, Mizobuchi R, Uga Y, Fukuda A, Ueda T, Yamamoto S, Yamanouchi U, Takai T, Ikka T, Kondo K, Hoshino T, Yamamoto E, Adachi S, Nagasaki H, Shomura A, Shimizu T, Kono I, Ito S, Mizubayashi T, Kitazawa N, Nagata K, Ando T, Fukuoka S, Yamamoto T, Yano M (2015) Genetic architecture of variation in heading date among Asian rice accessions. BMC Plant Biol 15:115

Hu X, Qian Q, Xu T, Zhang Y, Dong G, Gao T, Xie Q, Xue Y (2013) The U-box E3 ubiquitin ligase TUD1 functions with a heterotrimeric G a subunit to regulate brassinosteroid-mediated growth in rice. PLoS Genet 9:e1003391

Huang R, Jiang L, Zheng J, Wang T, Wang H, Huang Y, Hong Z (2013) Genetic bases of rice grain shape: so many genes, so little known. Trends Plant Sci 18:218–226

Li N, Li Y (2016) Signaling pathways of seed size control in plants. Curr Opin Plant Biol 33:23–32

Liu Q, Han R, Wu K, Zhang J, Ye Y, Wang S, Chen J, Pan Y, Li Q, Xu X, Zhou J, Tao D, Wu Y, Fu X (2018) G-protein βγ subunits determine grain size through interaction with MADS-domain transcription factors in rice. Nat Commun 9:852

Livak KJ, Schmittgen TD (2001) Analysis of relative gene expression data using real-time quantitative PCR and the $2^{-\Delta\Delta Ct}$ method. Methods 25:402–408

Mackay TFC, Stone EA, Ayroles JF (2009) The genetics of quantitative traits: challenges and prospects. Nat Rev Genet 10:565–577

Matsubara K, Ogiso-Tanaka E, Hori K, Ebana K, Ando T, Yano M (2012) Natural variation in Hd17, a homolog of Arabidopsis ELF3 that is involved in rice photoperiodic flowering. Plant Cell Physiol 53:709–716

McCouch SR, CGSNL (Committee on Gene Symbolization, Nomenclature and Linkage, Rice Genetics Cooperative) (2008) Gene nomenclature system for rice. Rice 1:72⁻84

Sanyal S, Pandey A, Pandey GK (2015) The CBL–CIPK signaling module in plants: a mechanistic perspective. Physiol Plantarum 155:89–108

SAS Institute Inc (1999) SAS/STAT user's guide. Cary. SAS Institute

Shibaya T, Hori K, Ogiso-Tanaka E, Yamanouchi U, Shu K, Kitazawa N, Shomura A, Ando T, Ebana K, Wu J, Yamazaki T, Yano M (2016) Hd18, encoding histone acetylase related to Arabidopsis FLOWERING LOCUS D, is involved in the control of flowering time in rice. Plant Cell Physiol 57:1828–1838

Somers DE, Fujiwara S (2009) Thinking outside the F-box: novel ligands for novel receptors. Trends Plant Sci 14:206–213

Song XJ, Kuroha T, Ayano M, Furuta T, Nagai K, Komeda N, Segami S, Miura K, Ogawa D, Kamura T, Suzuki T, Higashiyama T, Yamasaki M, Mori H, Inukai Y, Wu J, Kitano H, Sakakibara H, Jacobsen SE, Ashikari M (2015) Rare allele of a previously unidentified histone H4 acetyltransferase enhances grain weight, yield, and plant biomass in rice. P Natl Acad Sci USA 112:76–81

Wang L-L, Chen Y-Y, Guo L, Zhang H-W, Fan Y-Y, Zhuang J-Y (2015c) Dissection of qTGW1.2 to three QTLs for grain weight and grain size in rice (Oryza sativa L.). Euphytica 202:119–127

Wang S, Li S, Liu Q, Wu K, Zhang J, Wang S, Wang Y, Chen X, Zhang Y, Gao C, Wang F, Huang H, Fu X (2015a) The OsSPL16-GW7 regulatory module determines grain shape and simultaneously improves rice yield and grain quality. Nat Genet 47:949–954

Wang Y, Xiong G, Hu J, Jiang L, Yu H, Xu J, Fang Y, Zeng L, Xu E, Xu J, Ye W, Meng X, Liu R, Chen H, Jing Y, Wang Y, Zhu X, Li J, Qian Q (2015b) Copy number variation at the GL7 locus contributes to grain size diversity in rice. Nat Genet 47:944–948

Wu W, Liu X, Wang M, Meyer RS, Luo X, Ndjiondjop M-N, Tan L, Zhang J, Wu J, Cai H, Sun C, Wang X, Wing RA, Zhu Z (2017) A single-nucleotide polymorphism causes smaller grain size and loss of seed shattering during African rice domestication. Nat Plants 3:17064

Wu W, Zheng X-M, Lu G, Zhong Z, Gao H, Chen L, Wu C, Wang H, Wang Q, Zhou K, Wang J-L, Wu F, Zhang X, Guo X, Cheng Z, Lei C, Lin Q, Jiang L, Wang H, Ge S, Wan J (2013) Association of functional nucleotide polymorphisms at DTH2 with the northward expansion of rice cultivation in Asia. P Natl Acad Sci USA 110:2775–2780

Xia K, Ou X, Tang H, Wang R, Wu P, Jia Y, Wei X, Xu X, Kang S-H, Kim S-K, Zhang M (2015) Rice microRNA Osa-miR1848 targets the obtusifoliol 14α-demethylase gene OsCYP51G3 and mediates the biosynthesis of phytosterols and brassinosteroids during development and in response to stress. New Phytol 208:790–802

Xiang Y, Huang Y, Xiong L (2007) Characterization of stress-responsive CIPK genes in rice for stress tolerance improvement. Plant Physiol 144:1416–1428

Xu G, Ma H, Nei M, Kong H (2009) Evolution of F-box genes in plants: different modes of sequence divergence and their relationships with functional diversification. P Natl Acad Sci USA 6:835–840

Yan S, Zou G, Li S, Wang H, Liu H, Zhai G, Guo P, Song H, Yan C, Tao Y (2011) Seed size is determined by the combinations of the genes controlling different seed characteristics in rice. Theor Appl Genet 123:1173–1181

Yano M, Katayose Y, Ashikari M, Yamanouchi U, Monna L, Fuse T, Baba T, Yamamoto K, Umehara Y, Nagamura Y, Sasaki T (2000) Hd1, a major photoperiod sensitivity quantitative trait locus in rice, is closely related to the Arabidopsis flowering time gene CONSTANS. Plant Cell 12:2473–2483

Yu J, Miao J, Zhang Z, Xiong H, Zhu X, Sun X, Pan Y, Liang Y, Zhang Q, Rashid MAR, Li J, Zhang H, Li Z (2018) Alternative splicing of OsLG3b controls grain length and yield in japonica rice. Plant Biotechnol J. https://doi.org/10.1111/pbi.12903

Yu J, Xiong H, Zhu X, Zhang H, Li H, Miao J, Wang W, Tang Z, Zhang Z, Yao G, Zhang Q, Pan Y, Wang X, Rashid MAR, Li J, Gao Y, Li Z, Yang W, Fu X, Li Z (2017) OsLG3 contributing to rice grain length and yield was mined by ho-LAMap. BMC Biol 15:28–45

Zhang H-W, Fan Y-Y, Zhu Y-J, Chen J-Y, Yu S-B, Zhuang J-Y (2016) Dissection of the qTGW1.1 region into two tightly-linked minor QTLs having stable effects for grain weight in rice. BMC Genet 17:98–107

Zhang X, Wang J, Huang J, Lan H, Wang C, Yin C, Wu Y, Tang H, Qian Q, Li J, Zhang H (2012) Rare allele of OsPPKL1 associated with grain length causes extra-large grain and a significant yield increase in rice. P Natl Acad Sci USA 109:21534–21539

Zheng KL, Huang N, Bennett J, Khush GS (1995) PCR-based marker-assisted selection in rice breeding: IRRI Discussion Paper Series No. 12. Los Banos: International Rice Research Institute

Zhong Z, Wu W, Wang H, Chen L, Liu L, Wang C, Zhao Z, Lu G, Gao H, Wei X, Yu C, Chen M, Shen Y, Zhang X, Cheng Z, Wang J, Jiang L, Wan J (2014) Fine mapping of a minor-effect QTL, DTH12, controlling heading date in rice by up-regulation of florigen genes under long-day conditions. Mol Breeding 34:311–322

Zuo J, Li J (2014) Molecular genetic dissection of quantitative trait loci regulating rice grain size. Annu Rev Genet 48:99–118

16

QTL analysis for chalkiness of rice and fine mapping of a candidate gene for *qACE9*

Yang Gao[1,2†], Chaolei Liu[1†], Yuanyuan Li[1], Anpeng Zhang[1], Guojun Dong[1], Lihong Xie[1], Bin Zhang[1], Banpu Ruan[1], Kai Hong[1], Dawei Xue[2], Dali Zeng[1], Longbiao Guo[1], Qian Qian[1] and Zhenyu Gao[1*] (iD)

Abstract

Background: An ideal appearance is of commercial value for rice varieties. Chalkiness is one of the most important appearance quality indicators. Therefore, clarification of the heredity of chalkiness and its molecular mechanisms will contribute to reduction of rice chalkiness. Although a number of QTLs related to chalkiness were mapped, few of them have been cloned so far.

Results: In this study, using recombinant inbred lines (RILs) of PA64s and 9311, we identified 19 QTLs associated with chalkiness on chromosomes 1, 4, 6, 7, 9 and 12, which accounted for 5.1 to 30.6 % of phenotypic variations. A novel major QTL *qACE9* for the area of chalky endosperm (ACE) was detected in Hainan and Hangzhou, both mapped in the overlapping region on chromosome 9. It was further fine mapped to an interval of 22 kb between two insertion-deletion (InDel) markers IND9-4 and IND9-5 using a BC₄F₂ population. Gene prediction analysis identified five putative genes, among which only one gene (*OsAPS1*), whose product involved in starch synthesis, was detected two nucleotide substitutions causing amino acid change between the parents. Significant difference was found in apparent amylose content (AAC) between NIL*qACE9* and 9311. And starch granules were round and loosely packed in NIL*qACE9* compared with 9311 by scanning electron microscopy (SEM) analysis.

Conclusions: *OsAPS1* was selected as a novel candidate gene for fine-mapped *qACE9*. The candidate gene not only plays a critical role during starch synthesis in endosperm, but also determines the area of chalky endosperm in rice. Further cloning of the QTL will facilitate the improvement of quality in hybrid rice.

Keywords: Chalkiness, Area of chalky endosperm, QTL analysis, Fine mapping, Rice

Background

Rice is one of the most important food crops, and fed more than half of the population in the world. In recent years, with the increase of living standard, more and more attention has been paid on rice quality, including the appearance quality, processing quality, nutritional quality, and cooking and eating quality, etc. Chalkiness is an important indicator of the appearance quality for rice. As the opaque part in endosperm, it is an optical character caused by the air-gap of loose arrangement between proteinoplast and amyloplast. Chalkiness can be evaluated by indexes, including area of chalky endosperm (ACE), degree of chalky endosperm (DCE) and percentage of grains with chalkiness (PGWC) when it is associated with high level of damage to the kernel during milling, and thus to a reduction in head rice recovery (Del Rosario et al. 1968). Furthermore, when chalky grain is steamed or boiled, cracks develop readily, reducing the palatability of the cooked product (Nagato and Ebata 1959; Cheng et al. 2005). Therefore, clarification of the heredity of chalkiness and its molecular mechanisms is of important significance to reduce the chalkiness, improve the appearance quality and the commercial value of rice (Tan et al. 2000).

In past decade, several rice mutants associated with chalkiness have been identified and a few genes have been cloned. The *OsPPDKB* gene, which control carbon flow into starch and lipid biosynthesis during grain filling and *starch synthase IIIa* (*SSIIIa*), whose product plays an

* Correspondence: zygao2000@hotmail.com

†Equal contributors

[1]State Key Laboratory of Rice Biology, China National Rice Research Institute, Hangzhou 310006, China

Full list of author information is available at the end of the article

important role in the elongation of amylopectin chains were cloned by Kang et al. (2005) and Fujita et al. (2007) subsequently. The abnormal growth and loose structure of starch grain in *gif1* mutant caused a significant rise in chalkiness, and the corresponding gene, *GIF1* was fine-mapped on chromosome 4, which encode a cell-wall invertase required for carbon partitioning during early grain filling (Wang et al. 2008). With map-based cloning strategy, She et al. (2010) identified *FLO2* gene on chromosome 4, whose product participated in production of storage starch and storage proteins in the endosperm.

Differences between cultivars in their responsiveness of *FLO2* expression during high-temperature stress indicated that *FLO2* may also be involved in heat tolerance during seed development. However, as a typical quantitative trait, chalkiness is vulnerable to environmental conditions, especially the temperature in the filling stage, when starch is accumulated in endosperm (Lanning et al. 2011; Siebenmorgen et al. 2013). To elucidate the effect of high temperature on grain-filling metabolism, Yamakawa et al. (2007) exposed caryopses of high temperature-tolerant and sensitive cultivars to high temperature (33 °C/28 °C) or control temperature (25 °C/20 °C) during the filling stage, and found that the starch synthesis-related genes, for example, *GBSSI*, were down-regulated at transcript level by high temperature, whereas those for starch-consuming α-amylases and heat shock proteins were up-regulated. In general, high temperature resulted in the occurrence of grains with various degrees of chalky appearance. Nevertheless, there were some varieties not influenced by the high temperature. Murata et al. (2014) developed an *Apq1*-NIL to evaluate the effect of temperature on various agronomic traits, and found that there is no significant difference in percentage of perfect grains (PPG) of the *Apq1*-NIL under high temperature and normal conditions, although PPG of the parent 'Koshihikari' is lower under high temperature.

The identification of quantitative trait loci (QTLs) for rice chalkiness and elucidation of the underlying genetic regulation mechanism are necessary for the development of markers for marker assisted selection (MAS) strategies in rice breeding (Yan and Bao 2014). Series of QTLs related to chalkiness have been mapped hitherto by different populations including DH, F_2 and RIL. Zeng et al. (2002) detected 9 QTLs for chalkiness on chromosomes 8, 11 and 12 respectively with 127 DH lines from transverse section, flank section and belly section. And the intervals and peaks of QTLs on 3 chromosomes were overlapped, some of them even uniform. A total of 22 QTLs for chalkiness were indentified with a population involving 66 chromosomal segment substitution lines (CSSLs) across eight environments, and 9 QTLs were consistently detected across 8 environments, which

indicated the 9 QTL alleles were more stable than other 13 QTL alleles (Wan et al. 2005). By using an F_8 recombinant inbred line (RIL) population consisting of 261 lines derived from a cross between Koshihikari and C602, Liu et al. (2012) detected three QTLs related to PGWC on chromosomes 5, 8 and 10. Zhao et al. (2015b) used the QK model to declare the usefulness of the targeted genes/QTLs. And *SSIIa* was the major gene for chalkiness and explained up to 17 and 21 % of variation of DEC and PGWC, respectively. In addition, the markers RMw513 and RM18068 were associated with DEC in 6 environments as well, and allelic combinations between *SSIIa*, RMw513 revealed more variations in DEC. Besides, many QTLs were fine mapped in limited region. Guo et al. (2011) narrowed down the *qPGWC-8* to a 142 kb region between two Indel markers 8G-7 and 8G-9. Recently, Sun et al. (2015) identified 10 common QTLs for the percentage of grain chalkiness and the degree of chalky endosperm using high-through-put single nucleotide polymorphism (SNP) genotyping of a CSSLs population, and validated the isoamylase gene (*ISA1*) residing on the *qPGC8-2* region, which preferentially expressed in the endosperm and revealed some nucleotide polymorphisms between the parents. Because chalkiness is controlled by multiple genes and its genetic mechanism is relatively complex, so far, only one major QTL for chalkiness, *Chalk5*, has been cloned in rice (Li et al. 2014).

In this study, the relationship between three indexes for rice chalkiness, ACE, DCE and PGWC were analyzed. Nineteen QTLs for chalkiness were identified using a RIL population derived from the cross PA64s × 9311 based on the high-density SNP based genetic map (Gao et al. 2013). A novel major QTL for chalkiness was fine mapped and one candidate gene was selected, which promote further cloning of the QTL and improvement of quality in hybrid rice.

Results

Phenotypic variation of the parents and RILs

The phenotypic differences between 9311 and PA64s are displayed and summarized in Fig. 1a and Table 1. The basic statistics for the RIL population are also shown in Table 1. Nearly normal distributions and bimodal distributions were observed in the RIL population for ACE, DEC and PGWC in Hangzhou and Hainan, respectively (Fig. 1b), indicating the three traits were controlled by multi-genes in Hangzhou and one or two major genes in Hainan.

Correlation analysis of three chalkiness related traits, heading date and grain shape

The correlations among the three chalkiness characteristics, ACE, DCE and PGWC, heading date (HD) and grain

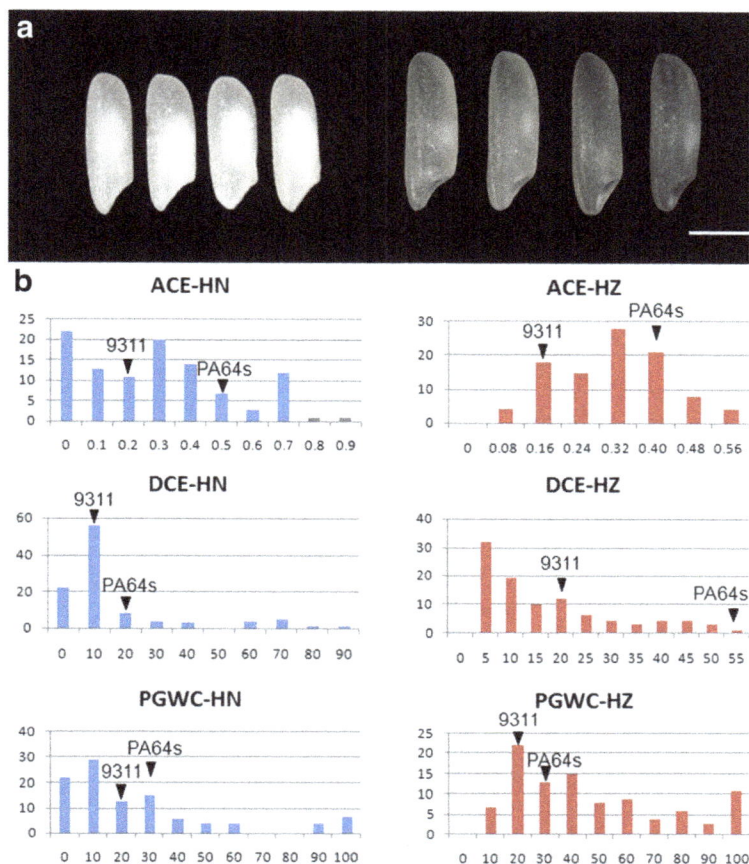

Fig. 1 Comparison of chalkiness between two parents and distribution of ACE, DCE and PGWC in the RIL population. **a** Seeds of PA64s (left) and 9311 (right). Bar =2.5 mm. **b** HN represents Hainan and HZ represents Hangzhou

shape traits, such as grain length (GL) and grain width (GW) are shown in Table 2. We identified significantly positive correlations in any couple of ACE, DCE and PGWC in both Hainan and Hangzhou. Meanwhile, the correlations between GW and PGWC, GW and DCE in Hainan and Hangzhou were positive at 5 % and 1 % significant level, respectively.

Table 1 Variations of phenotypes between parents and among RIL in Hainan and Hangzhou

Site	Variety/Population	ACE (cm²)	DCE (%)	PGWC (%)
Hainan	9311	0.18±0.03	2.8±0.6	16.0±2.8
	PA64s	0.44±0.04 [b]	11.7±0.8 [b]	27.0±4.2 [a]
	RIL	0.25±0.22	11.5±20.2	22.3±28.8
Hangzhou	9311	0.13±0.03	19.0±5.0	20.0±4.3
	PA64s	0.36±0.03 [b]	54.0±8.0 [b]	30.0±3.2 [a]
	RIL	0.27±0.11	14.4±13.7	42.6±28.1

Mean ± SD (n = 6 for parents and n = 104 for RIL)
[a] and [b] indicate the least significant difference at 0.05 and 0.01 probability level compared with 9311 in Hainan or Hangzhou, respectively

Detection of QTLs for ACE, DCE and PGWC

A total of 19 QTLs were detected for the traits of ACE, DCE and PGWC in both Hainan and Hangzhou, distributing on chromosomes 1, 4, 6, 7, 9 and 12 (Table 3; Fig. 2). Eight QTLs for ACE were identified, including 2 QTLs in Hainan and 6 QTLs in Hangzhou. Meanwhile, we detected qACE9 separately in Hainan and Hangzhou, which explained 12.6 and 13.6 % of the phenotypic variation and located within 7.59 ~ 23.65 cM on chromosome 9. Six QTLs for DCE were detected and each QTL explained 5.2 ~ 30.6 % of phenotypic variation. For the trait of PGWC, 5 QTLs were detected and each QTL explained 5.1 ~ 30.5 % of phenotypic variation. In Hangzhou, we identified qACE6-2, qDCE6 and qPGWC6, explained respectively up to 9.2, 30.6 and 30.5 % of the phenotypic variation and located within 10.92 ~ 22.12 cM on chromosome 6.

Among all the 19 QTLs detected with RILs, 6 QTLs were unreported at present, including qACE9. There were 16 QTLs distributed in the overlapping region on six chromosomes. A group of QTLs for all three traits

Table 2 Correlation coefficients between ACE, DCE, PGWC and GL, GW, HD

Traits in Hainan	ACE	DCE	PGWC
DCE	0.185 [a]		
PGWC	0.218 [b]	0.980 [b]	
GL	−0.101	0.055	0.071
GW	−0.155	0.224 [a]	0.204 [a]
HD	−0.222 [a]	−0.047	−0.085
Traits in Hangzhou	ACE	DCE	PGWC
DCE	0.428 [b]		
PGWC	0.460 [b]	0.984 [b]	
GL	−0.130	−0.002	−0.019
GW	0.188	0.298 [b]	0.301 [b]
HD	−0.035	−0.027	0.002

[a] and [b] indicate at 5 % and 1 % significant level, respectively

were detected in the overlapping region on chromosome 4, 6, 9 and 12. The QTLs for both DCE and PGWC were located in the overlapping interval on chromosomes 1 and 7.

Fine mapping of qACE9

For fine mapping of the novel major QTL *qACE9*, a line of RILs with PA64s genotype in the *qACE9* region was selected to backcross with recurrent parent 9311. Then phenotypic character was measured in F_2 population including 920 individuals derived from a BC_4F_1 line with 9311 genetic background exhibiting heterozygous across the entire *qACE9* region screened with markers SNP9-1 and SNP9-2. By comparing the sequences of the parents, four insertion-deletion (InDel) markers were developed. Combining the genotype and phenotype of individuals, the QTL was delimited between two InDel markers IND9-4 and IND9-5 in 22 kb interval (Fig. 3a). The target region contains 5 predicted genes (LOC_Os09g126620, LOC_Os09g126630, LOC_Os09g126640, LOC_Os09g126650 and LOC_Os09g12660) (Fig. 3b) based on Rice Genome Annotation Project Website (http://rice.plantbiology.msu.edu/). Among

Table 3 QTLs for ACE, DCE and PGWC detected in RIL population in Hainan and Hangzhou

Trait	QTL	Site	Chr.	LOD	P value	Genetic distance (cM)	PEV (%)	Subst. effect	Reported QTL
ACE	qACE1	Hainan	1	3.83	0.01	170.15–183.44	14.8	0.17	qPGWC-1a (Liu et al. 2011)
									qPGWC1d (Zhao et al. 2015a)
	qACE9	Hainan	9	3.36	0.01	7.59–13.04	12.2	0.16	
	qACE4	Hangzhou	4	2.01	0.03	158.65–164.02	8.5	0.05	
	qACE6-1	Hangzhou	6	4.01	0.01	0.00–10.92	16.2	−0.09	qPGWC-6 (Liu et al. 2011), qDEC6, qPGWC6 (Zhao et al. 2015a), qCA6-1 N-, qCA6-W+ (Peng et al. 2014)
	qACE6-2	Hangzhou	6	3.79	0.03	10.92–22.12	9.2	−0.09	qCR6-H+ (Peng et al. 2014)
	qACE7	Hangzhou	7	2.09	0.03	46.03–57.36	5.4	0.05	
	qACE9	Hangzhou	9	4.17	0.01	7.59–23.65	13.6	0.09	
	qACE12	Hangzhou	12	2.41	0.01	121.63–135.59	6.8	0.06	qWCR12-D- (Peng et al. 2014)
DCE	qDCE1	Hainan	1	3.43	0.01	208.31–225.92	11.2	−14.17	qPGWC.NH-1.2 (Bian et al. 2014)
	qDCE4	Hangzhou	4	5.97	0.01	154.53–165.67	12.9	8.61	
	qDCE6	Hangzhou	6	11.54	0.01	15.04–21.31	30.6	−13.27	qCR6-H+ (Peng et al. 2014)
	qDCE7	Hangzhou	7	2.72	0.03	70.43–79.87	5.6	−5.66	qPGWC-7 (Zhou et al. 2009), qWCA7-D+ (Peng et al. 2014)
	qDCE9	Hangzhou	9	2.62	0.01	0.36–12.63	5.2	5.46	qDEC9, qPGWC9a, qPGWC9b (Zhao et al. 2015a)
	qDCE12	Hangzhou	12	3.19	0.03	120.28–135.59	6.4	6.07	qWCR12-D- (Peng et al. 2014)
PGWC	qPGWC1	Hainan	1	9.14	0.03	208.31–225.48	11.5	−39.41	qPGWC.NH-1.2 (Bian et al. 2014)
	qPGWC4	Hangzhou	4	7.19	0.01	158.65–166.49	16.3	24.98	
	qPGWC6	Hangzhou	6	11.36	0.01	16.68–21.31	30.5	−34.21	qCR6-H+ (Peng et al. 2014)
	qPGWC7	Hangzhou	7	3.69	0.01	65.06–79.06	7.9	−17.38	qPGWC-7 (Zhou et al. 2009), qWCA7-D+ (Peng et al. 2014)
	qPGWC12	Hangzhou	12	2.51	0.01	121.63–135.59	5.1	13.92	qWCR12-D- (Peng et al. 2014)

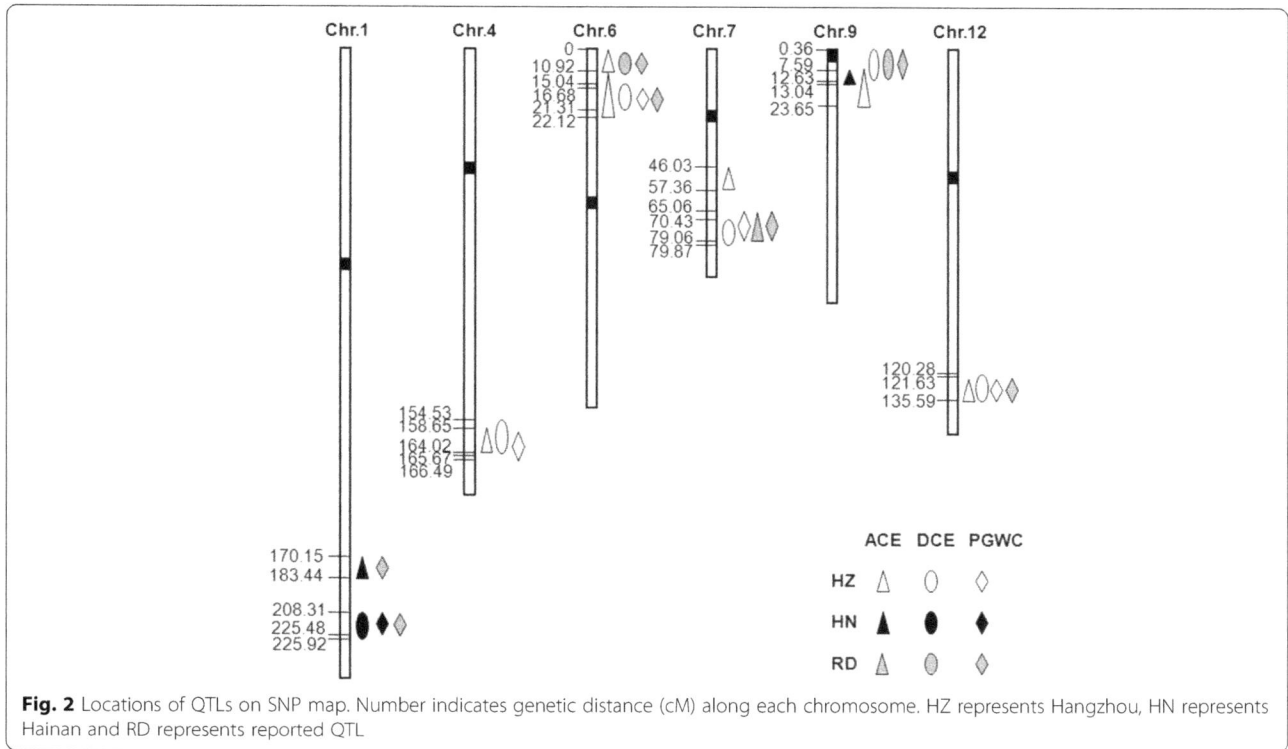

Fig. 2 Locations of QTLs on SNP map. Number indicates genetic distance (cM) along each chromosome. HZ represents Hangzhou, HN represents Hainan and RD represents reported QTL

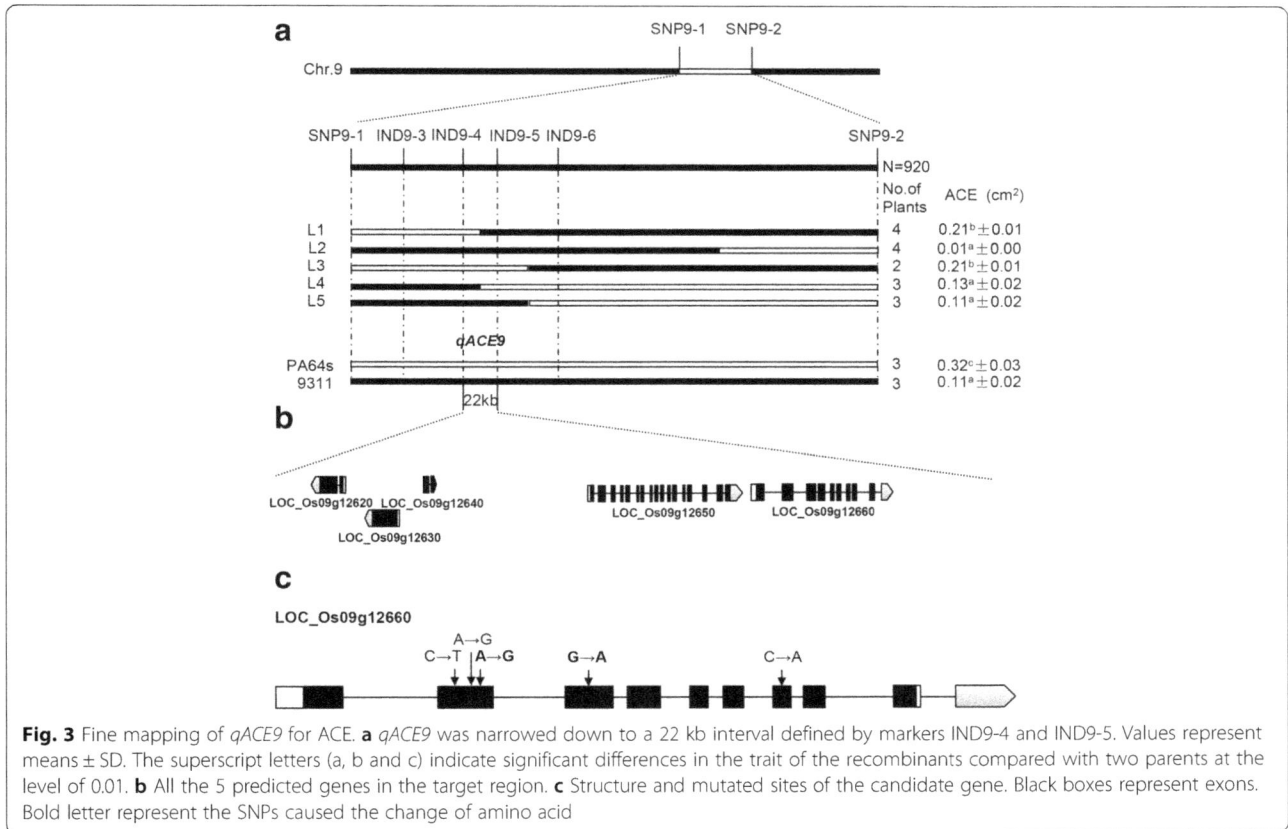

Fig. 3 Fine mapping of *qACE9* for ACE. **a** *qACE9* was narrowed down to a 22 kb interval defined by markers IND9-4 and IND9-5. Values represent means ± SD. The superscript letters (a, b and c) indicate significant differences in the trait of the recombinants compared with two parents at the level of 0.01. **b** All the 5 predicted genes in the target region. **c** Structure and mutated sites of the candidate gene. Black boxes represent exons. Bold letter represent the SNPs caused the change of amino acid

the 5 predicted genes, there were only 3 predicted genes (LOC_Os09g126620, LOC_Os09g126650 and LOC_Os09g12660) functional annotated. Sequence analysis of the 5 genes in 9311 and PA64s found two synonymous SNPs in LOC_Os09g126650 and 5 SNPs in LOC_Os09g12660 between two parents, among which two SNPs causing amino acid change (Fig. 3c). Therefore, LOC_Os09g12660 were finally selected as the candidate for qACE9. A real time PCR was performed for the five genes in the qACE9 locus (Fig. 4a, b). There were little

expression for LOC_Os09g126620, LOC_Os09g126630, LOC_Os09g126640 and LOC_Os09g126650 in two parents, and the expression level for LOC_Os09g12660 was higher in 9311 without significant difference with that in PA64s.

Characterization of the NILqACE9 with 9311 background

No significant differences were observed in grain size (GL, GW and length-width ratio (LWR)) between NILqACE9, a NIL carrying homozygous allele of PA64s between InDel markers IND9-3 and IND9-6 (approximately 321.8 kb) with 9311 background, and 9311. However, there was significant difference in apparent amylose content (AAC) between them (Fig. 4d), although no differences observed in amylopectin chain-length distributions (Fig. 4e). NILqACE9 had larger area of chalky endosperm, markedly different from that of 9311, though both had similar grain size (Fig. 5a–b). Scanning electron microscopy (SEM) analysis showed the NILqACE9 starch granules were round and loosely packed, and very different from polyhedral and densely packed starch granules of 9311 (Fig. 5c–f).

Discussion

Chalkiness is an important trait of rice appearance. High positive correlations were found between ACE, DCE and PGWC in both Hainan and Hangzhou demonstrated chalkiness related traits were significantly correlated. In addition, GW was found correlated positively with PGWC and DCE in both environments, which consistent with overlapping regions of qGW4, qPGWC4 and qDCE4 detected in Hangzhou (Gao et al. 2013). This conclusion was also proved by Bian et al. (2013), they use a segregating population derived from sgw (low GW) and cultivar 9311 (high GW) to detect qsgw7 associated with GW, and developed the NILqsgw7 which show lower grain width and chalkiness of brown rice than 9311.

Up to now, many rice mutants associated with chalkiness have been identified and a few genes have been cloned. Although only one QTL for chalkiness was cloned, many QTLs have been reported related to the trait. The qPGWC-6 and qPGWC-7 for PGWC located overlapped with our mapped region on chromosomes 6 and 7, and the latter was fine mapped to a 44 kb interval (Zhou et al. 2009). Using a set of CSSLs with 'Asominori' genetic background, Liu et al. (2011) detected 6 and 9 QTLs respectively for ACE and PGWC, and two QTLs, qPGWC-1a and qPGWC-6 were also in the same region mapped here. Peng et al. (2014) detected 79 QTLs associated with chalkiness traits using five populations across two environments, among which 5 QTLs in the overlapping region, qCA6-1 N-, qCA6-W+, qCR6-H+, qWCA7-D+ and qWCR12-D- also detected by us. With

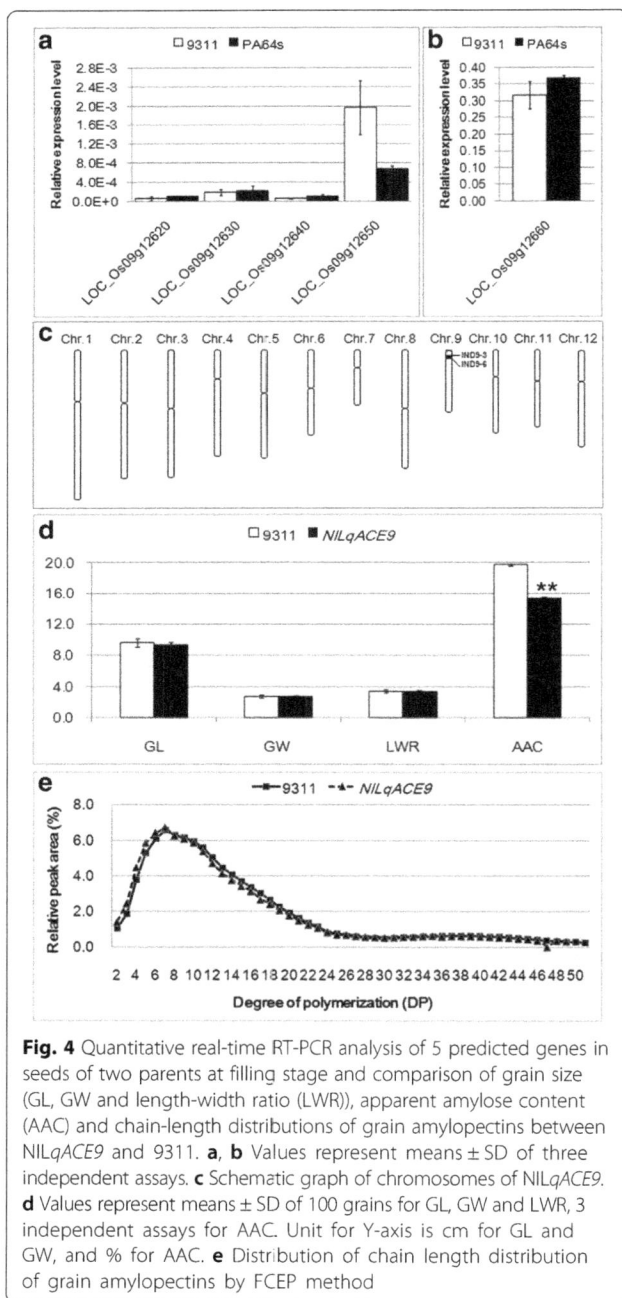

Fig. 4 Quantitative real-time RT-PCR analysis of 5 predicted genes in seeds of two parents at filling stage and comparison of grain size (GL, GW and length-width ratio (LWR)), apparent amylose content (AAC) and chain-length distributions of grain amylopectins between NILqACE9 and 9311. **a, b** Values represent means ± SD of three independent assays. **c** Schematic graph of chromosomes of NILqACE9. **d** Values represent means ± SD of 100 grains for GL, GW and LWR, 3 independent assays for AAC. Unit for Y-axis is cm for GL and GW, and % for AAC. **e** Distribution of chain length distribution of grain amylopectins by FCEP method

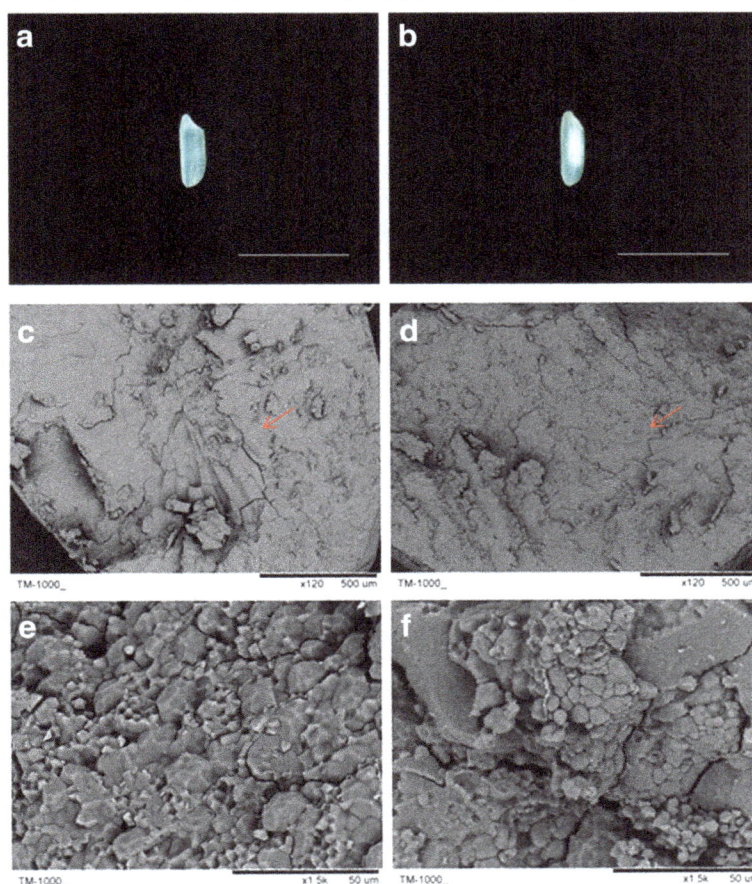

Fig. 5 Grain morphology and scanning electron microscopy (SEM) images of starch granule structure. **a**, **c** and **e** from 9311; **b**, **d** and **f** from NIL*qACE9*; The arrowhead in (**c**) and (**d**) represent the position of SEM images in (**e**) and (**f**); Bar represents 1 cm in (**a**) and (**b**)

a population composed of 37 introgression lines (ILs) of Habataki in the background of Sasanishiki, 54 QTLs were identified for grain quality across two different environments (Bian et al. 2014). Among them, the *qPGWC.NH-1.2* for percentage of grains with chalkiness was located in the same region indentified by us on chromosomes 1. Recently, Zhao et al. (2015a) used two sets of RILs derived from reciprocal crosses between Lemont and Teqing to study the genetic basis of chalkiness. A total of 53 and 68 QTLs were detected for DEC and PGWC respectively, among which *qDEC6, qDEC9, qPGWC6, qPGWC1d, qPGWC9a, qPGWC9b* were also identified here in the overlapping region. Because of no difference in CDS of *chalk5* between 9311 and PA64s by sequencing, there was no QTL detected in the region on chromosome 5 here.

In our study, a novel major QTL for ACE, *qACE9* was fine mapped to a 22 kb interval. Comparison of DNA sequence of 5 predicted genes between 9311 and PA64s found two SNPs in one candidate gene LOC_Os09g12660 (*OsAPS1*) caused amino acid changes, that is A_{427} to G_{427}

caused Serine to Aspartic, and G_{634} to A_{634} caused Glycine to Asparagine. They both located in the nucleotidyl transferase domain in small subunit of AGPase. There were two small subunits and four large subunits of AGPase, named OsAPS1, OsAPS2, OsAPL1, OsAPL2, OsAPL3 and OsAPL4, which compose one of four classes of enzymes for starch biosynthesis (Tian et al. 2009). Akihiro et al. (2005) cloned all six subunits in Nipponbare. Moreover, comparison of the deduced amino acid sequences of OsAPS1 and OsAPS2 showed high homology between them. Both of the small and large subunits were necessary to ensure the function of AGPase, which was essential for starch synthesis in the seed endosperm. To determine whether the *OsAGPS2* play a critical role during starch synthesis in developing rice endosperm, Lee et al. (2007) isolated mutant *osagps2* for OsAGPS2 by reverse genetic PCR screening of a rice T-DNA insertion library. The levels of AGPase activity and the starch content in *osagps2* were found to be remarkably reduced to 20 and 31 % of the wild type respectively in developing endosperms. Scanning electron microscopy showed that starch granules in the *osagps2* mutants are smaller in size and rounder in shape when compared to those from wild type endosperm. Because of high homology of *OsAPS1* and *OsAPS2* proteins, together with no difference in expression level for *OsAPS1* between the parents, therefore, the difference in chalkiness between 9311 and PA64s may be caused by *OsAPS1* at protein level rather than RNA level.

Conclusion

In the study, using high-density SNP linkage map, 19 QTLs for rice chalkiness were detected in Hangzhou and Hainan. With the BC$_4$F$_2$ population derived from a RIL and 9311, *qACE9*, a new major QTL for the area of chalky endosperm (ACE), was fine mapped within 22 kb physical interval on chromosome 9. One candidate gene, *OsAPS1*, whose product reported involved in starch synthesis, was finally selected based on difference in coding sequence causing amino acid change between the parents. There were significant differences in apparent amylose content (AAC) and starch granules structure in endosperm between NIL*qACE9* and 9311. It helps further cloning of the QTL and facilitates the improvement of rice quality.

Methods
Development of mapping population

A total of 104 RILs derived from the cross of the *indica* variety 9311 and the light-thermo-sensitive genic male sterile line PA64s were used in this study. The population was developed in the experimental fields at China National Rice Research Institute in Hangzhou, Zhejiang Province, and in Lingshui, Hainan Province, China. To develop a NIL containing the QTL for ACE, *qACE9* detected both in Hainan and Hangzhou on chromosome 9, a line of RILs with PA64s genotype in the *qACE9* region was selected to backcross with recurrent parent 9311. Two markers SNP9-1 and SNP9-2 (Table 4) were used for marker assisted selection (MAS) of each generation. As a result, a BC$_4$F$_1$ line, with 9311 genetic background exhibiting heterozygous across the entire *qACE9* region, was constructed. After self-crossing, a BC$_4$F$_2$ population was obtained for fine mapping of *qACE9*. A NIL carrying homozygous allele of PA64s in the target QTL region

between InDel markers IND9-3 and IND9-6 (Table 4), designated NIL*qACE9*, was also developed from one chromosome segment substitution line (CSSL) with 93–11 background (Fig. 4c).

Measurement of chalkiness related traits

The plot size was four rows of six plants with a 35 × 35 cm spacing. Mature seeds of each line were harvested 30 days after heading and dried in an electro-thermal incubator (ZXDP-A2160, Shanghai) at 30 °C for 72 h after harvest. The dried seeds (20 g) were dehulled and polished, and then intact seeds were selected for measurement of chalkiness related traits. The ACE, DEC and PGWC were evaluated according to He et al. (1999) and National Standard of People Republic of China (NSPRC 1999). To separate chalky grains from vitreous grains, 40 grains selected at random per entry were assessed on a chalkiness scanner. PGWC was calculated based on these photographs. ACE and the area of the whole endosperm for each grain were estimated visually by the software, and the values for both were averaged. DEC was the ratio of ACE to the area of the whole endosperm.

Scanning electron microscope (SEM) analysis of rice grains

For observation of starch granules, unbroken milled rice was cut transversely with a blade, and the pieces were stuck onto a 12-mm aluminum stub, and sputtered with gold on a polaron sputter coater. Samples were viewed with SEM and diameters of starch granules were estimated on the basis of the scale bar provided on the captured image.

Measurement of apparent amylose content (AAC) and chain-length distribution of starch

AAC was measured following the procedure of Perez and Juliano (1978) with some modifications. Absorbance of the starch solution was determined at 620 nm using the spectrophotometer. The method determining chain-length distribution of starch was essentially identical to the procedure described by Fujita et al. (2012) using the fluorescence capillary electrophoresis (FCEP) method of

Table 4 Primers for InDel markers and SNP markers developed

Primer	Forward (5'-3')	Reverse (5'-3')	Type
SNP9-1	AGCATAGTTGTAAAACATGCCAGAC	TGCCGGAAAATAAATTCACCC	SNP
SNP9-2	TTCGTATTTTATAGAACAGAGGG	TGTGTGCTAAGAACACAAAGG	SNP
IND9-3	CAGTATATGTGACGGAGCTATTTTC	ATTATCCTTGGTTATACACCG	InDel
IND9-4	CCAACCTCCAAGACTAGATGAAGTT	AACATTACTTGTGGGCTCTTG	InDel
IND9-5	TTTGATCGGACAATTTGTTT	AAAAACCGGAAAAAGAAAAG	InDel
IND9-6	TAGATGGGCCAGTTCAAATTG	ACCATATGTTTTTACATTTGATTGC	InDel

Table 5 Primers for real time PCR analysis

Primer	Forward (5'-3')	Reverse (5'-3')	Gene
RT-62	CTGCAGGCGAAGAAGGAT	GTGATCACCGTGTAGTTCGC	LOC_Os09g12620
RT-63	TACTACGCCTCGGTGGAGA	TCCGGGTAGACGTCGAAT	LOC_Os09g12630
RT-64	ACGTGGATTCAGCCAAATG	AATGGCAAGATCTCCGTAGG	LOC_Os09g12640
RT-65	ACATGCGCAAATATGGTTGT	CCAGAGAACACCACACCAAC	LOC_Os09g12650
RT-66	ATTCAGGCCCACAGAGAAAC	TGATCCTCCCTTCATCATCA	LOC_Os09g12660
Actin	CCATTGGTGCTGAGCGTTT	CGCAGCTTCCATTCCTATGAA	LOC_Os03g50885

O'Shea and Morell (1996) in a P/ACE MDQ Capillary Electrophoresis System (Beckman Coulters, CA, USA).

Statistical analyses and QTL analysis

All statistical analyses were completed using the SAS (Statistical Analysis System) v8.01. QTL analysis was performed with the MultiQTL package (www.multiqtl.com) using the maximum likelihood interval mapping approach for the RIL-selfing population. For major-effect QTLs, the LOD threshold was obtained based on a permutation test (1000 permutations, $P = 0.05$) for each dataset. QTLs were named according to Mccouch et al. (1997).

Design of markers for fine mapping

Primers were designed in *qACE9* region on the basis of insertions/deletions (InDels) and SNPs identified between 9311 and PA64s (Table 4) (Gao et al. 2013). Genotypes of SNP markers were screened by high-resolution dissociation curve analysis system (LightScanner 96, Idaho Technology Inc.).

Real time PCR analysis

Total RNA was isolated from panicles at filling stage with RNA extraction kit (Axygen). DNase treatment, cDNA synthesis, primer design and SYBR Green I real time PCR were carried out as described (Vandesompele et al. 2002) using a Rever Tra Ace® qPCR-RT kit (TOYOBA, Japan). Real time PCR amplification mixtures (10 μl) contained 50 ng template cDNA, 2 × SYBR Green PCR Master Mix (Applied Biosystems), and 200 nM forward and reverse primers. Reactions were conducted on an ABI PRISM_7900HT Sequence Detector (Applied Biosystems). The relative expression level of each transcript was obtained by comparing to the expression of the *Actin* gene. Primers for candidate genes and *Actin* are listed in Table 5.

Acknowledgments

This work was supported by Grants from the National Natural Science Foundation of China (Grant Nos. 31521064 and 31471167) and the National Key Research and Development Program of China (2016YFD0100902).

Authors' contributions

Z-YG and QQ conceived and designed the experiments. YG, C-LL, Y-YL, A-PZ, G-JD, L-HX, B-PR and KH performed the experiments. YG and C-LL analyzed the data. YG and Z-YG wrote the manuscript. D-WX, D-LZ, L-BG and QQ revised the manuscript. All authors read and approved the manuscript.

Authors' information

Dr. Gao Zhenyu, a plant geneticist, a member of Genetic Society of China. He specializes in plant genetics and molecular biology. He has engaged in the research on functional genomics and molecular breeding of rice quality and published series of related papers in eminent periodicals, such as PNAS, JIPB, TAG, Sci China et al.

Competing interests

The authors declare that they have no competing interests.

Author details

[1]State Key Laboratory of Rice Biology, China National Rice Research Institute, Hangzhou 310006, China. [2]College of Life and Environmental Sciences, Hangzhou Normal University, Hangzhou 310036, China.

References

Akihiro T, Mizuno K, Fujimura T (2005) Gene expression of ADP-glucose pyrophosphorylase and starch contents in rice cultured cells are cooperatively regulated by sucrose and ABA. Plant Cell Physiol 46(6):937–946

Bian JM, He HH, Li CJ, Shi H, Zhu CL, Peng X et al (2013) Identification and validation of a new grain weight QTL in rice. Genet Mol Res 12(4):5623–5633

Bian JM, Li CJ, He HH, Shi H, Yan S (2014) Identification and analysis of QTLs for grain quality traits in rice using an introgression lines population. Euphytica 195(1):83–93

Cheng FM, Zhong LJ, Wang F, Zhang GP (2005) Differences in cooking and eating properties between chalky and translucent parts in rice grains. Food Chem 90(1–2):39–46

Del Rosario AR, Briones VP, Vidal AJ, Juliano BO (1968) Composition and endosperm structure of developing and mature rice kernel. Cereal Chem 45(3):225–235

Fujita N, Yoshida M, Kondo T, Saito K, Utsumi Y, Tokunaga T, Nishi A, Satoh H, Park JH, Jane JL, Miyao A, Hirochika H, Nakamura Y (2007) Characterization of SSIIIa-deficient mutants of rice: the function of SSIIIa and pleiotropic effects by SSIIIa deficiency in the rice endosperm. Plant Physiol 144(4):2009–2023

Fujita N, Hanashiro I, Suzuki S, Higuchi T, Toyosawa Y, Utsumi Y, Itoh R, Aihara S, Nakamura Y (2012) Elongated phytoglycogen chain length in transgenic rice endosperm expressing active starch synthase IIa affects the altered solubility and crystallinity of the storage α-glucan. J Exp Bot 63(16):5859–5872

Gao ZY, Zhao SC, He WM, Guo LB, Peng YL, Wang JJ et al (2013) Dissecting yield-associated loci in super hybrid rice by resequencing recombinant inbred lines and improving parental genome sequences. Proc Natl Acad Sci U S A 110(35):14492–14497

Guo T, Liu XL, Wan XY, Weng XF, Liu SJ, Liu X et al (2011) Identification of a stable quantitative trait locus for percentage grains with white chalkiness in rice (*oryza sativa*). J Integr Plant Biol 53(8):598–607

He P, Li SG, Qian Q, Ma YQ, Li JZ, Wang WM, et al (1999) Genetic analysis of rice grain quality. Theor Appl Genet 98 (3-4):502–508

Kang HG, Park SH, Matsuoka M, An GH (2005) White-core endosperm *floury endosperm-4* in rice is generated by knockout mutations in the C4-type pyruvate orthophosphate dikinase gene (*OsPPDKB*). Plant J 42(6):901–11

Lanning SB, Siebenmorgen TJ, Counce PA, Ambardekar AA, Mauromoustakos A (2011) Extreme nighttime air temperatures in 2010 impact rice chalkiness and milling quality. Field Crops Res 124(1):104–136

Lee SK, Hwang SK, Han M, Eom JS, Kang HG, Han Y et al (2007) Identification of the ADP-glucose pyrophosphorylase isoforms essential for starch synthesis in the leaf and seed endosperm of rce (Oryza sativa L.). Plant Mol Biol 65(4):531–546

Li YB, Fan CC, Xing YZ, Yun P, Luo LJ, Yan B et al (2014) Chalk5 encodes a vacuolar H + −translocating pyrophosphatase influencing grain chalkiness in rice. Nat Genet 46(4):398–404

Liu XL, Wan XY, Ma XD, Wan JM (2011) Dissecting the genetic basis for the effect of rice chalkiness, amylase content, protein content, and rapid viscosity analyzer profile characteristics on the eating quality of cooked rice using the chromosome segment substitution line population across eight environments. Genome 54(1):64–80

Liu X, Wang Y, Wang SW (2012) QTL analysis of percentage of grains with chalkiness in Japonica rice (Oryza sativa). Genet Mol Res 11(1):717–724

Mccouch SR, Chen X, Panaud O, Temnykh S, Xu Y, Cho YG et al (1997) Microsatellite marker development, mapping and applications in rice genetics and breeding. Plant Mol Biol 35(1–2):89–99

Murata K, Iyama Y, Yamaguchi T, Ozaki H, Kidani Y, Ebitani T (2014) Identification of a novel gene (Apq1) from the indica rice cultivar 'Habataki' that improves the quality of grains produced under high temperature stress. Breeding Sci 64(4):273–281

Nagato K, Ebata M (1959) Studies on white-core rice kernel: II. on the physical properties of the kernel. Jpn J Crop Sci 28(1):46–50

NSPRC (National Standard of People Republic of China). High quality paddy, GB/T17891, Standards Press of China, Beijing; 1999. p. 1–3.

O'Shea MG, Morell MK (1996) High resolution slab gel electrophoresis of 8-amino-1, 3, 6-pyrenetrisulfonic acid (APTS) tagged oligosaccharides using a DNA sequencer. Electrophoresis 17(4):681–686

Peng B, Wang LQ, Fan CC, Jiang GH, Luo LJ, Li YB et al (2014) Comparative mapping of chalkiness components in rice using five populations across two environments. BMC Genet 15(1):49–63

Perez CM, Juliano BO (1978) Modification of the simplified amylase test for milled rice. Starch-Starke 30(12):424–426

She KC, Kusano H, Koizumi K, Yamakawa H, Hakata M, Imamura T et al (2010) A novel factor FLOURY ENDOSPERM2 is involved in regulation of rice grain size and starch quality. Plant Cell 22(10):3280–3294

Siebenmorgen TJ, Grigg BC, Lanning SB (2013) Impacts of preharvest factors during kernel development on rice quality and functionality. Annu Rev Food Sci Technol 4(4):101–116

Sun WQ, Zhou QL, Yao Y, Qiu XJ, Xie K, Yu SB (2015) Identification of genomic regions and the isoamylase gene for reduced grain chalkiness in rice. PLoS One. 10(3). doi:10.1371/journa.pone.0122013.

Tan YF, Xing YZ, Li JX, Yu SB, Xu CG, Zhang QF (2000) Genetic bases of appearance quality of rice grains in Shanyou 63, an elite rice hybrid. Theor Appl Genet 101(5):823–829

Tian ZX, Qian Q, Liu QQ, Yan MX, Liu XF, Yan CJ et al (2009) Allelic diversities in rice starch biosynthesis lead to a diverse array of rice eating and cooking qualities. Proc Natl Acad Sci U S A 106(51):21760–21765

Vandesompele J, De Paepe A, Speeman F (2002) Elimination of primer-dimer artifacts and genomic coamplification using a two-step SYBR green I real-time RT-PCR. Anal Biochem 303(1):95–98

Wan XY, Wan JM, Weng JF, Jiang L, Bi JC, Wang CM et al (2005) Stability of QTLs for rice grain dimension and endosperm chalkiness characteristics across eight environments. Theor Appl Genet 110(7):1334–1346

Wang E, Wang J, Zhu X, Hao W, Wang L, Li Q et al (2008) Control of rice grain-filling and yield by a gene with a potential signature of domestication. Nat Genet 40(11):1370–1374

Yamakawa H, Hirose T, Kuroda M, Yamaguchi T (2007) Comprehensive expression profiling of rice grain filling-related genes under high temperature using DNA microarray [OA]. Plant Physiol 144(1):258–277

Yan WG, Bao JS (2014) Rice-germplasm, genetics and improvement. InTech Publisher, p 239–78, ISBN 978-953-51-1240-2, doi:10.5772/51100.

Zeng DL, Qian Q, Ruan LQ, Tend S, Kunihiro Y, Fujimoto H et al (2002) QTL Analysis of chalkiness size in three dimensions. Chinese J Rice Sci 16(1):11–14

Zhao XQ, Daygon VD, McNally KL, Hamilton RS, Xie F, Reinke RF et al (2015a) Identification of stable QTLs causing chalk in rice grains in nine environments. Theor Appl Genet 129(1):141–153

Zhao XQ, Zhou LJ, Ponce K, Ye GY (2015b) The usefulness of known genes/qtls for grain quality traits in an incica population of diverse breeding lines tested using association analysis. Rice 8(1):1–13

Zhou LJ, Chen LM, Jiang L, Zhang WW, Liu LL, Liu X et al (2009) Fine mapping of the grain chalkiness QTL qPGWC-7 in rice (Oryza sativa L.). Theor Appl Genet 118(3):581–590

17

The Nipponbare genome and the next-generation of rice genomics research in Japan

Takashi Matsumoto[1,2]*, Jianzhong Wu[1,2], Takeshi Itoh[1,2], Hisataka Numa[1,2], Baltazar Antonio[1,2] and Takuji Sasaki[3]

Abstract

The map-based genome sequence of the *japonica* rice cultivar Nipponbare remains to date as the only monocot genome that has been sequenced to a high-quality level. It has become the reference sequence for understanding the diversity among thousands of rice cultivars and its wild relatives as well as the major cereal crops that comprised the food source for the entire human race. This review focuses on the accomplishments in rice genomics in Japan encompassing the last 10 years which have led into deeper understanding of the genome, characterization of many agronomic traits, comprehensive analysis of the transcriptome, and the map-based cloning of many genes associated with agronomic traits.

Keywords: Rice, *Oryza sativa*, Nipponbare, Genome, Annotation, Transcriptome, Agronomic traits

Introduction

The elucidation of the rice genome sequence is a major milestone in science as it paves the way for understanding the biology of a major cereal crop that feeds more than half of the world's population (International Rice Genome Sequencing Project 2005). Although roughly a hundred plant genome sequences have already been published to date, the map-based sequence of *Oryza sativa* ssp. *japonica* cv. Nipponbare remains as the only monocot genome that has been sequenced to a high-quality level. It has therefore become a reference for sequencing of other cereal crops with much larger genome sizes such as maize (Schnable et al. 2009), sorghum (Paterson et al. 2009), soybean (Schmutz et al. 2010), barley (International Barley Genome Sequencing Consortium 2012), and wheat (International Wheat Genome Sequencing Consortium 2014). More importantly, the rice genome sequence has become the most powerful tool in agriculture enhancing the ability of breeders to develop new cultivars with highly desirable traits such as high yield, resistance to biotic/abiotic stress, good eating quality, and cultivars that could adapt to an ever changing cultivation environment brought about by global warming. It is expected that subsequent sequencing of a wide array of rice germplasm throughout the world will be the platform for propelling the next green revolution to increase productivity under more sustainable conditions.

Although 90 % of rice is consumed mainly in Asia, it is also a major food source in many African and South American countries. Rice is a main staple in the Japanese diet with the current average per capita consumption of about 60 kg per year. It has been cultivated both as a staple and economic crop for more than 2000 years across the country and has been integrated in many aspects of the culture as well. Thousands of cultivars have been developed as a result of crossbreeding and selection conducted by farmers and breeders to suit the specific local conditions. Therefore, the complete rice genome sequence based on the cultivar Nipponbare led to the large-scale characterization of other *japonica* cultivars including the widely cultivated and elite cultivar Koshihikari (Yamamoto et al. 2010) known for good eating quality.

This review will focus on the accomplishments in rice genomics in Japan encompassing the last 10 years since the completion of the rice genome sequence. There is

* Correspondence: mat@affrc.go.jp
[1]National Institute of Agrobiological Sciences, 2-1-2 Kannondai, Tsukuba, Ibaraki 305-8602, Japan
[2]Present Address: National Agriculture and Food Research Organization, 2-1-2 Kannondai, Tsukuba, Ibaraki 305-8518, Japan
Full list of author information is available at the end of the article

no doubt however that a great deal of accomplishments has been achieved not only by the 10 participating countries in the international sequencing consortium but also by many rice researchers worldwide who have continuously engaged in understanding the rice biology based on the map-based Nipponbare genome sequence. In Japan, succeeding efforts in genome analysis from 2005 onwards have led to fine tuning of the genome assembly, deeper understanding of the structure of specific regions of the genome, characterization of many important traits across various cultivars, comprehensive profiling of the transcriptome, and the isolation and map-based cloning of many genes associated with agronomic traits.

Review

Enhancing the genome assembly and annotation

There have been continuous efforts to refine the genome assembly and enhance the annotation of the genes since the publication of the high-quality map-based sequence of the *japonica* cultivar Nipponbare. These efforts focused on gap-filling of the 12 chromosomes and characterization of the complex regions of the genome such as the centromeres, telomeres and nucleolar-organizing regions. Among the 12 chromosomes, the complex and highly repetitive centromere-specific DNA sequences were first reported in *Cen4* (Zhang et al. 2004), *Cen8* (Nagaki et al. 2004; Wu et al. 2004), and subsequently *Cen3* (Yan et al. 2006) which also complemented the previous extensive works on rice centromeres (Jiang et al. 1996; Cheng et al. 2002). We have continued to improve the quality of the Nipponbare genome pseudomolecules even after the completion of the IRGSP sequencing initiative. Using BAC sequence analysis, genome annotation, and FISH analysis, we characterized the nearly completed and high-quality genomic sequence of *Cen5* in chromosome 5 and revealed some striking differences among the centromeres in terms of the copy number and distribution pattern of the centromere-specific satellite repeat CentO as well as the distribution and expression of transcription units within the pericentromeric and centromeric regions (Mizuno et al. 2011). In the case of the telomeres, Fibre-FISH analysis revealed the presence of arrays of 730–1500 conserved copies of telomere-specific 5'-TTTAGGG-3' repeat sequence at the end regions of chromosomes 1S, 2S, 2L, 6L, 7S, 7L and 8S of Nipponbare (Mizuno et al. 2006). Gene annotation from the 500 kb subtelomere sequences clearly indicated that the rice chromosomal ends were gene-rich with high transcriptional expression. In addition, the subtelomere regions on these chromosome ends hardly contained TrsA, a subtelomeric repeat sequence of rice. On the other hand, clusters of TrsA have been observed in chromosomes 5L, 6S, 8L, 9L and 12L (Mizuno et al. 2008a). Sequence

comparison of these 14 telomere-ends and telomere-flanking regions also revealed the occurrence of deletions, insertions, or chromosome-specific substitutions of single nucleotides within the telomere specific repeats at the junction between the telomere and subtelomere, suggesting the telomeric variants in rice have arisen from the rapid expansion of a single mutation rather than from the gradual accumulation of random mutations (Mizuno et al. 2008b). More recently, the 14 telomere-ends from 12 chromosomes were successfully constructed from a fosmid library leading to the identification of telomere sequences and structure in rice (Mizuno et al. 2014). These additional sequenced regions of the genome have been incorporated into the genome assembly as we update the pseudomolecules on a regular basis. The most recent physical map of the genome covers almost 97 % of the entire genome with 62 remaining physical gaps (Fig. 1).

The latest genome assembly was constructed as a joint effort of the Rice Annotation Project Database (RAP-DB) of the National Institute of Agrobiological Sciences (NIAS) and the Michigan State University (MSU) Rice Genome Annotation Project to update and validate the reference IRGSP Nipponbare genome sequence and provide a unified set of pseudomolecules to the rice research community (Kawahara et al. 2013). The genome assembly was revised using the rice optical map (Zhou et al. 2007) to validate the minimal tiling path. The next-generation sequencing (NGS) data obtained by re-sequencing two individual Nipponbare plants using the Illumina Genome Analyzer II/IIx and Roche 454 GS FLX were used to check sequencing errors in the revised assembly. This resulted in the identification of 4886 sequencing errors and five insertions/deletions in the 321 Mb of the assembled genome corresponding to an error rate of 0.15 per 10,000 nucleotides in the original IRGSP assembly. The revised and unified genome assembly, Os-Nipponbare-Reference-IRGSP-1.0 (IRGSP-1.0), is now used to provide a common platform for genome annotation in the RAP-DB (http://rapdb.dna.affrc.go.jp, Rice Annotation Project 2008) and the MSU rice annotation database (http://rice.plantbiology.msu.edu/cgi-bin/gbrowse/rice/).

In line with the revision of the genome assembly, the RAP-DB has been enhanced further with the mapping of 154,579 transcript sequences from the genus *Oryza* and other monocot species (Sakai et al. 2013). In addition, literature-based manually curated data, transcriptome data, and NGS data of major rice cultivars were also incorporated into the database. The current release of RAP-DB consists of 37,869 loci including 1626 loci that correspond to literature-based manually curated annotation data, commonly used gene names, and gene symbols. Transcription data derived from Illumina RNA-seq

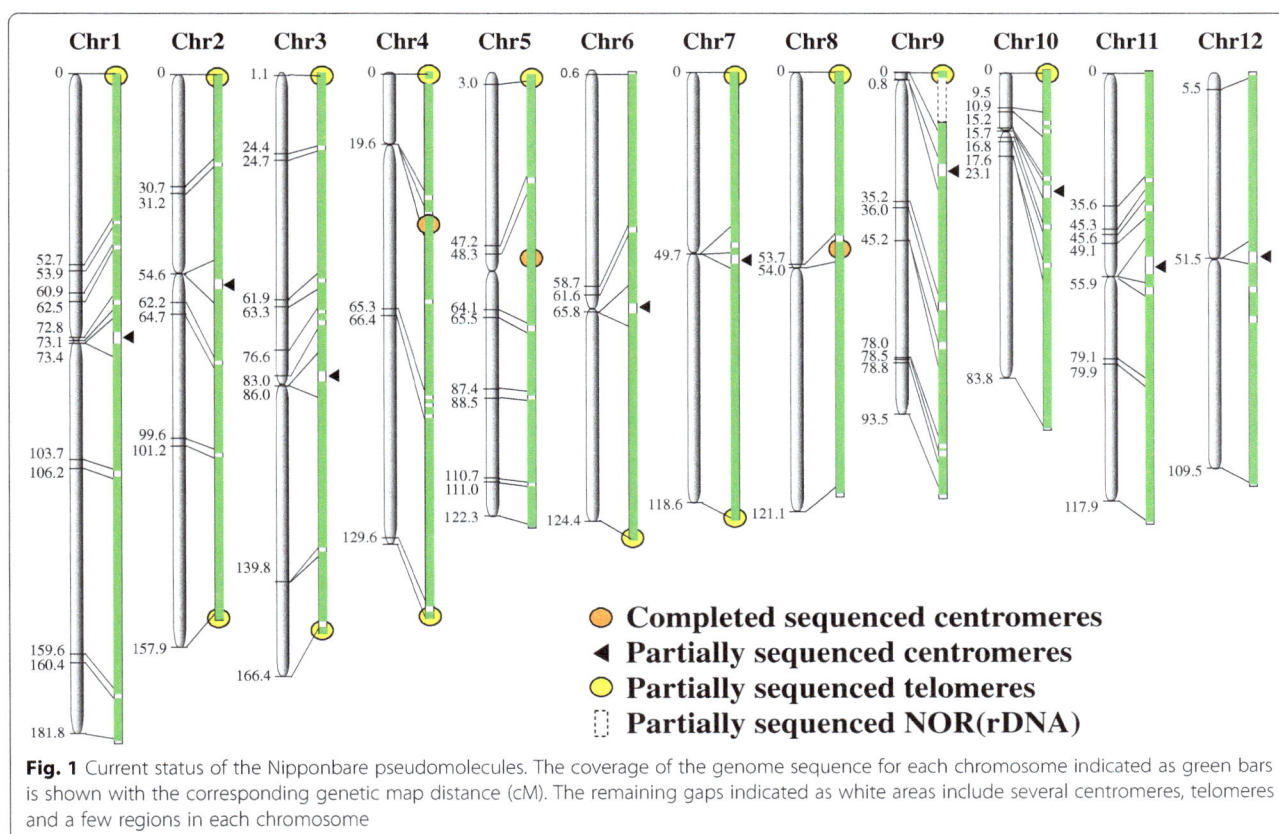

Fig. 1 Current status of the Nipponbare pseudomolecules. The coverage of the genome sequence for each chromosome indicated as green bars is shown with the corresponding genetic map distance (cM). The remaining gaps indicated as white areas include several centromeres, telomeres and a few regions in each chromosome

Legend within figure:
- ● Completed sequenced centromeres
- ◀ Partially sequenced centromeres
- ○ Partially sequenced telomeres
- ⬚ Partially sequenced NOR(rDNA)

analysis of various tissues under normal and stress conditions (Mizuno et al. 2010; Oono et al. 2011; Kawahara et al. 2012, 2016) have been added to enhance the utility of the database in understanding transcriptional regulatory networks. Links to gene families in rice, *Sorghum bicolor, Zea mays* and *Arabidopsis thaliana* are provided to facilitate analysis of how genes are conserved and evolved among plant species. An additional feature to RAP-DB is the Short-Read Assembly Browser (S-RAB) that provides a viewer for Illumina reads of the *japonica* cultivar Koshihikari and *indica* cultivar Guangluai-4 mapped to the Nipponbare genome, showing the alignments, single nucleotide polymorphisms (SNPs), and gene functional annotations. The RAP-DB is updated on a regular basis so that it can provide researchers with the latest information on characterization of rice genome structure and function. Recent advances in DNA sequencing technologies resulted in generation of massive genome sequencing data in a considerable number of rice cultivars and species. To facilitate efficient visualization of rapidly emerging large-scale sequencing data, a novel web-based browser, Tasuke (http://tasuke.dna.affrc.go.jp/), with various functions to show the variation and read depth of multiple genomes, as well as annotations and SNP data of hundreds of cultivars aligned to a reference genome at various scales, has been

developed for efficient utilization of emerging NGS data of rice cultivars (Kumagai et al. 2013).

Deeper perspectives on rice genetic resources

The reference Nipponbare genome facilitated the sequencing initiatives of rice germplasm aimed at understanding the genetic diversity which led to the domestication of rice as grown today. Foremost among these initiatives are the Oryza Map Alignment Project (OMAP) to clarify the diversity in the twelve wild rice genomes (Wing et al. 2007), and the international effort of resequencing a core collection of 3000 rice accessions from 89 countries to provide a foundation for large-scale discovery of novel alleles for important rice phenotypes (The 3,000 Rice Genomes Project 2014; Huang et al. 2012). To facilitate comprehensive understanding of the genome diversity in rice, we have also sequenced the African rice *O. glaberrima* known to be more resilient to water shortage as well as fungal or insect diseases than *O. sativa* (Sakai et al. 2011). The high-quality assembly and annotation of the *O. glaberrima* genome have also been reported, providing evidence for its independent domestication (Wang et al. 2014).

In contrast, successful efforts in Japan focus on understanding cultivars important to Japanese agriculture particularly those widely grown throughout Japan. To date, whole genome sequences from 16 varieties obtained by next-generation sequencers have been submitted in public databases by NIAS and other Japanese research organizations (Additional file 1: Table S2), and projects for sequencing other varieties and landraces are in progress.

The cultivar Koshihikari developed in 1953 is the most widely grown and favored cultivar in Japan occupying almost 80 % of total rice production including its relative cultivars. Many breeding efforts focus on further improvement of quality depending on the region where it is grown. The genome sequence of Koshihikari is therefore indispensable in breeding and designing rice to meet the demands of Japanese consumers. With the reference Nipponbare sequence, the next development was the sequencing of the Koshihikari genome with the Illumina sequencing technology (Yamamoto et al. 2010). A total of 67,051 SNPs between Koshihikari and Nipponbare, some of which derived from originating landraces and distributed through Koshihikari relatives, and 18 pedigree haplotype blocks which were artificially selected during breeding.

The Nipponbare pseudomolecule sequence was used as the template in the construction of a complete BAC-based physical map of the *O. sativa* ssp. *indica* cv. Kasalath (Kanamori et al. 2013). We also sequenced the centromere region of chromosome 8 in Kasalath (Wu et al. 2009). Comparative analysis with Nipponbare Cen8 revealed both collinearity and diversity in each ortholo-gous centromere. Subsequently, deep sequencing (>154-fold coverage) via the Roche GS-FLX Titanium or GS-FLX+ and Illumina GAIIx or HiSeq 2000 platforms and *de novo* assembly generated the 330.55 Mb Kasalath pseudomolecule sequence representing 91.1 % of the genome with 35,139 expressed loci annotated by RNA-Seq analysis (Sakai et al. 2014). Comparison of the Kasalath pseudomolecule with Nipponbare revealed 2,787,250 SNPs and 7393 large indel sites (>100 bp). On the other hand, comparison with the *indica* cultivar 93–11 showed 2,216,251 SNPs and 3780 large indels (Sakai et al. 2014). In particular, at least 14.78 Mb of indel sequences and 40.75 Mb of unmapped sequences were identified in the Kasalath genome in comparison with the Nipponbare genome suggesting that ~6.3 % of the total transcript loci in rice genome is presumably involved with gain or loss of genes.

Genotyping of the NIAS Genebank (https://www.gene.affrc.go.jp/index_en.php) rice accessions with 179 RFLP markers led to development of a rice diversity research set of germplasms (RDRS) in *indica*, *aus* and *japonica* accessions available for the detailed genetic studies and rice improvement (Kojima et al. 2005). Based on a result from screening 234 accessions of rice collected in Asia, the Americas, Africa, Europe and Oceania with 169 SSR (simple sequence repeats) markers,

moreover, current *O. sativa* cultivars and landraces can be classified in more detail into five genetically differentiated groups: *indica*, *aus*, *aromatic*, *temperate japonica*, and *tropical japonica* because of its deep genetic structure evolved during domestication and adaptation and its autogamous breeding system (Garris et al. 2005).

A series of comparative genomic studies among various species in the genus *Oryza* focused on a number of domestication or adaptation related genes such as the *sh4* gene region responsible for the reduction of grain shattering, the semi-dwarf1 (*sd-1*) gene (Wu et al. 2008; Asano et al. 2011), the major heading-date related genes such as *Hd1*, *Hd3a*, *Hd6*, *RFT1* and *Ghd7* (Fujino et al. 2010; Yamane et al. 2009; Ebana et al. 2011). Analysis of expression levels revealed clear association of the functional and nonfunctional alleles with early and late flowering, suggesting that *Hd1* is a major determinant of variation in flowering time of cultivated rice (Takahashi et al. 2009). Sequencing of BAC clones covering the chromosomal region of *Hd3a* and *RFT1* genes across the AA ~ GG genomes revealed that at least 89 % of the amino acid sequences encoded by *Hd3a* which promotes the transition to flowering under the short-day condition were conserved across the different *Oryza* species (Komiya et al. 2008). In comparison with *Hd1*, the *Hd3a* gene obviously showed much less genetic diversity with 95 ~ 100 % sequence identity among the accessions of *O. sativa* and *O. rufipogon* (Fig. 2a, b). Similarly, the *RFT1* gene which has been associated with late flowering of rice under long-day condition in a functional *Hd1* background (Ogiso-Tanaka et al. 2013) also showed low genetic diversity (Fig. 2c). Extremely high gene collinearity was also found in the surrounding region (~300kbp) of *Hd3a* and *RFT1* genes across the *Oryza* species despite the size differences caused mainly by transposable element insertions (Fig. 2d). Unlike *Hd3a*, the *RFT1* gene has only been found in the *Oryza* species that have the AA (including *O. sativa*) or BB genomes (*O. punctata*). These genomes diverged from a common ancestor only ~2 Mya. This result suggests that the *RFT1* gene may have originated from *Hd3a* by a recent duplication although a possible deletion of *RFT1* within the other species could not be ruled out. The Nipponbare reference sequence could contribute insights into the molecular mechanisms underlying genomic evolution and selection in rice which would benefit breeding programs to modify and control flowering time through efficient utilization of different genes or gene alleles.

Elucidating the molecular function of rice genes

The rice genome sequence of Nipponbare has been pivotal in the development of a system for discovering the biological functions of approximately 32,000 genes identified in rice. This has been addressed early on with the

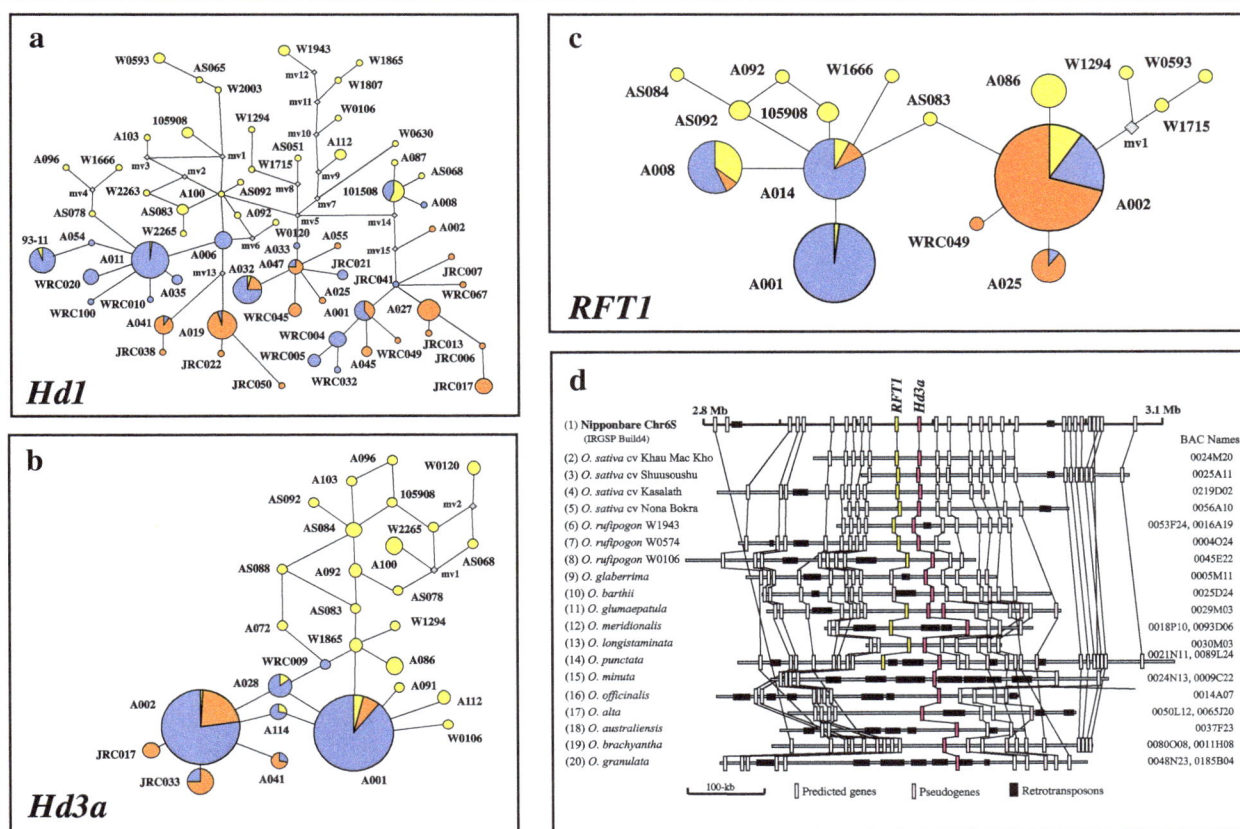

Fig. 2 Molecular and evolutionary analysis of *flowering* genes across rice accessions. The major flowering genes, namely, *Hd1*, *Hd3a* and *RFT1* genes were analysed using the exon sequences of 202 rice accessions. A-C: The resulting haplotype network of *Hd1* (**a**), *Hd3a* (**b**) and *RFT1* (**c**) constructed with the corresponding accession name and size proportional to the total number of samples from *O. rufipogon* (*yellow*), *O. sativa* ssp. *indica* (*blue*) and *O. sativa* ssp. *japonica* (*orange*). Lines between haplotypes represent the mutational steps between alleles. Hypothetical haplotypes (median vector) with discontinuous links are indicated as grey squares. **d**: Compositional and structural comparison of the *Hd3a* (*red box*), and *RFT1* region (*orange box*) across various Oryza species. The genes with loss-of-function (*white box*) and retrotransposons (*black box*) are also shown. Orthologous genes are linked by solid lines. The BAC sequences correspond to DDBJ accessions AP011450 ~ AP011476

development of resources for functional genomics such as the *Tos17* insertion mutant panel (Hirochika 2001) and the rice full-length cDNA collection (Kikuchi et al. 2003). The Tos17 mutant collection consisting of approximately 50,000 mutant lines characterized into flanking sequences have been used in full characterization of many genes characterized in Japan and nearly 50 *Tos17* mutant lines have been successfully used in tagging specific genes (See reference list at https://tos.nias.affrc.go.jp/doc/references.html).

Similarly, the information from approximately 28,000 fully-sequenced cDNA clones and the KOME Database (hhttps://dbarchive.biosciencedbc.jp/en/kome/desc.html) have been very useful for functional characterization of many genes. Both these resources are still widely used up to the present by researchers around the world particularly in systematically assigning functions to many predicted genes in the genome, in addition to other resources such as T-DNA insertion lines (Jung and An

2013), Ac/Ds tagging lines (Guiderdoni and Gantet 2012), nDART/aDart lines (Takagi et al. 2007). Analysis of the flanking sequences of these insertional mutants resulted in accumulation of nearly 448,000 gene-tagging sequence resources for characterization of gene functions (Wei et al. 2013).

The rice full-length cDNAs also serve as the main resources for large-scale gene expression profiling. Using these sequences as well as predicted gene information for plausible probes, we have designed 44 K Agilent oligonucleotide microarray to analyze the field transcriptome of field-grown rice (Fig. 3). A wide range of gene expression profiles based on organs and tissues at various developmental stages identified organ/tissue specific genes as well as growth stage-specific genes. Continuous transcriptome profiling of leaf from transplanting until harvesting stage uncovered two major drastic changes in the leaf transcriptional program (Sato et al. 2011). The rice transcriptome is well documented in two databases,

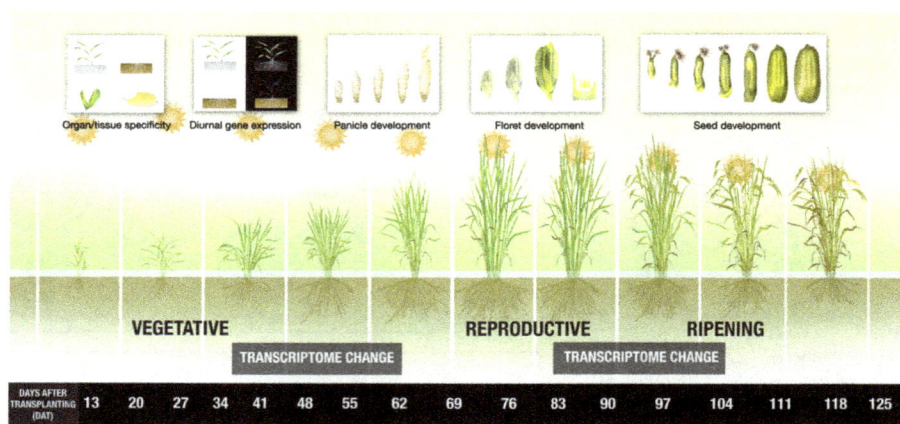

Fig. 3 Characterizing the field transcriptome of rice. Microarray analysis was used to characterize gene expression of rice cells and tissues at various stages of development from transplanting to harvesting in the field

namely, RiceXPro (http://ricexpro.dna.affrc.go.jp/) providing an overview of the transcriptional changes throughout the growth of the rice plant in the field, and RiceFREND (http://ricefrend.dna.affrc.go.jp/) for co-expression analysis of these genes. These resources are now widely used for deciphering gene functions and analysis of rice gene networks. Combining the massive field transcriptomic data and meteorological information with statistical model construction, we have also succeeded, to some extent, in predicting fluctuation of gene expression, or transcriptome dynamics (Nagano et al. 2012). These predictions may give insights for improving crop production, disease resistance and resilience to global stress. Information from transcription profiles can also predict the best cultivation conditions for a given variety in a given location.

With current advances in next-generation sequencing, we have embarked on analysis of gene expression using RNA-seq to measure the presence and quantity of RNA transcripts under specific growth conditions. As a result, unannotated salinity stress-inducible transcripts have been identified using the RNA-seq profile of seedlings treated with NaCl (Mizuno et al. 2010). Through RNA-seq analysis, previously unannotated salinity response genes, some of which might function in phosphate and cadmium stress tolerance were discovered. Simultaneous measurement of rice and rice blast fungus transcripts in infected plants using RNA-seq provided infection-responsive expression profiles for both rice and fungal transcripts. These profiles might indicate genes that may interact at different times and different tissues during infection (Kawahara et al. 2012). The same strategy has been used to characterize the transcriptome profile of *japonica* cultivar Nipponbare under phosphate starvation (Oono et al. 2011, 2013), as well as the diversity among the transcriptomes of rice cultivars including Nipponbare

with low tolerance to phosphate starvation stress, *japonica* cultivars IAC 25 and Vary Lava 701 with relatively higher tolerance, and *indica* cultivar Kasalath known to be highly tolerant to phosphate stress (Oono et al. 2013). Recently, we have also found that cadmium stress controls the expression of genes in drought stress signal pathways in rice based on genome wide transcriptome analysis (Oono et al. 2014).

Isolation and utilization of agronomically important genes

The genome sequence coupled with genomics tools and resources such as mutant lines, genetic populations and germplasm collection paved the way for marker-assisted selection, mapping of QTLs and map-based cloning of specific genes having agronomic properties. As a result, many QTLs have been detected as Mendelian factors in rice, including those responsible for increased yield, resistance to various insect pests and diseases, resistance to abiotic stress such as drought, salinity and submergence, good eating quality etc.

Heading date

Many QTLs involved in heading date, a key determinant of rice adaptation to different cultivation areas and cropping seasons, have been characterized. These include *Hd1*, *Hd2*, *Hd3a*, *Hd3b*, *Hd4*, *Hd5*, *Hd6*, *Hd8*, and *Hd9* (Yano et al. 2001). Subsequent studies focused on more detailed characterization of these QTLs based on the genome sequence. The *Hd1* contains a CCT domain with ~60 % identity to *Ghd7*, a long day dependent negative regulator of heading date. *Hd6* encodes casein kinase 2 alpha, and *Hd3a* is similar to an *Arabidopsis* FT-like protein (Takahashi et al. 2001; Kojima et al. 2002). Another major QTL, *Early heading date 1* (*Ehd1*), encodes a B-type response regulator that is suppressed under long-day conditions for which *Ghd7* is responsible (Doi et al. 2004). In more recent studies, it has been found that *Ehd3*, encoding a PHD finger-containing protein, is a critical promoter of rice flowering (Matsubara et al. 2011). *Hd17*, a homolog of Arabidopsis *EARLY FLOWERING 3* (*ELF3*), is involved in the photoperiodic flowering pathway by regulating the transcription level of a flowering repressor, *Grain number, plant height and heading date 7* (*Ghd7*) gene (Matsubara et al. 2012). The QTL *Hd16* is a gene for casein kinase I which is involved in the control of rice flowering time by modulating the day-length response (Hori et al. 2013).

Disease resistance

Developing cultivars with broad spectrum of resistance to diseases is a priority for rice breeding in Japan. Among the genes widely characterized for disease resistance, the rice *WRKY45* gene has been found to play a crucial role in resistance to bacterial and fungal blast induced by benzothiadiazole (BTH), a so-called plant activator that protects plants from diseases by activating plant innate immune system (Shimono et al. 2007). The *Pib* gene is the first cloned gene for resistance to blast induced by *Magnaporthe oryzae* (Wang et al. 1999) and it was reported much later that *Pi21* encodes a proline-rich protein with a heavy metal-binding domain and putative protein-protein interaction motifs (Fukuoka et al. 2009).

Domestication

The Nipponbare genome sequence has been instrumental in deciphering the evolutionary processes in the domestication of rice (Yang et al. 2012). Several genes that play key roles in selection and domestication have been analyzed. The seed shattering *qSH1* gene has been found to encode a BEL1-type homeobox gene and a SNP in the 5' regulatory region caused a loss of seed shattering owing to the absence of abscission layer formation (Konishi et al. 2006). More recently, *Kala4*, the gene responsible for the black color of rice grains (also referred to as purple rice) has been identified based on an extensive analysis of the genes associated with grain color in about 50 rice cultivars, tracing the origin to tropical *japonica* (Oikawa et al. 2015).

Abiotic tolerance

Map-based cloning of *qLTG3-1* which controls low-temperature germination in rice provides useful insights on cultivation in temperate as well as high altitude rice growing areas (Fujino et al. 2008). The molecular mechanism of deepwater response has been clarified through the identification of the genes *SNORKEL1* and *SNORKEL2*, which trigger deepwater response by encoding ethylene response factors involved in ethylene signaling (Hattori et al. 2009). With the molecular cloning of *Sdr4*, a seed dormancy QTL

in rice, the role of the gene as an intermediate regulator of dormancy in the seed maturation program has been clarified (Sugimoto et al. 2010). The *DEEPER ROOTING 1* (*Dro1*) has been recently discovered through screening and genetic analysis of the NIAS rice collection which have a great potential for improvement of rice yield under drought conditions by controlling the root system architecture in rice (Uga et al. 2013).

Yield

The major components that determine yield in rice have been widely characterized using the sequence information. The *qSW5* gene which corresponds to the QTL for seed width on chromosome 5 has been cloned and a deletion in the gene was found to be associated with larger grain size (Shomura et al. 2008). Furthermore, it has also been shown that this variant was selected during rice domestication for increased yields. Characterization of genes associated with productivity has also made significant progress. A loss-of-function mutation of rice *DENSE PANICLE 1* causes semi-dwarfness and slightly increased number of spikelets (Taguchi-Shiobara et al. 2011). A natural variant of *NARROW LEAF 1* (*NAL1*) gene selected in high-yield rice breeding programs increased the photosynthesis rate (Takai et al. 2013; Fujita et al. 2013). The *THOUSAND-GRAIN WEIGHT 6* (*TGW6*) gene limits endosperm cell number and grain length. Defective alleles lead to increase in grain size and yield (Ishimaru et al. 2013). Also researchers at Nagoya U. revealed the basic molecular strategy for construction good plant type for ideal yield performance, most of which related to metabolism or biosynthesis of plant hormones (Ueguchi-Tanaka et al. 2005; Ikeda et al. 2013)

A list of genes that have been characterized mainly or in collaboration with Japanese researchers in 2005–2014 is summarized in Additional file 2: Table S1. Significant contributions have been made in elucidating gene functions, identifying QTLs, characterizing molecular mechanisms, and establishing the DNA marker-assisted selection (MAS) as a precise and effective breeding strategy to produce novel varieties. There is no doubt that in the last 10 years since the completion of the rice genome, worldwide rice research has made significant output as evidenced in the number of rice related publications. An overview of the trend in rice research in the last 45 years is shown in Fig. 4 (data provided by Oryzabase, Kurata and Yamazaki 2006). Since 2005, the number of publications doubled in just a matter of 5 years (2010) and by the end of 2014, there are almost 2000 publications on rice alone (including 67 from Japan). In total, Japanese researcher contributed about 70–100 per year in the last 10 years since the completion of the rice genome sequence in 2004.

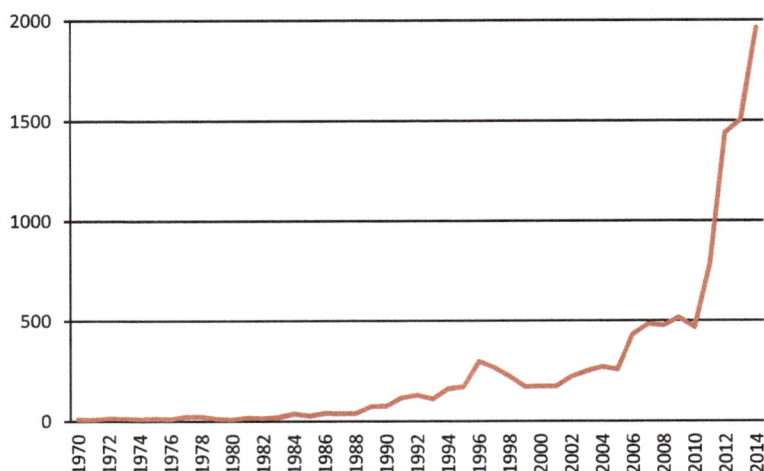

Fig. 4 The growth of rice publications before and after the completion of the rice genome sequence. The number of publications on rice research from 1970 to 2014 showed significant increase after the completion of the Nipponbare genome sequence in 2004

Conclusion

In Japan, rice genomics has been a part of major research programs of the Ministry of Agriculture, Forestry and Fisheries (MAFF) that address various issues in sustainable food production and agriculture. Foremost among these issues are the rapid aging of farm workers and depopulation of farming communities which would eventually affect agricultural production in the not so distant future. In breeding programs, various efforts are being initiated to integrate rice genomics technology to the development of novel varieties to reinvigorate Japanese agriculture. The new basic plan for food, agriculture and the rural areas which serve as the guideline for advancing the reform of measures and efforts by the entire nation so as to enable Japan's agriculture and rural areas to accurately respond to structural and other changes in the economy and society (http://www.maff.go.jp/e/basic_law/basiclaw_agri/basiclaw_agri.html). Rice being an integral part of Japanese agriculture is very much a part of these programs. Several ongoing MAFF-funded research projects focus on using the genome information of various crops for the development of various technologies to boost next-generation agriculture. As a part of that major project, the development of rice genomics resources and informatics tools are expected to contribute in such areas through attempts of various sectors to develop rice cultivars that address the specific needs of various rice producing regions in Japan considering the environmental changes in rice cultivation due to climate change.

As the rice research community embarks on various efforts in rice genomics, more rice genome sequence and transcriptome profiles will be generated in the very near future. Elucidating the molecular mechanisms controlling many biological processes will be supplemented by information to be obtained from other technologies such as proteomics, metabolomics (Okazaki and Saito 2016), epigenomics (Chen and Zhou 2013), and phenomics (Yang et al. 2013). Integration of all these data via advanced bioinformatics will elucidate the gene cascade or network in the whole rice plant that will serve as the platform on how to utilize and improve crop function. And Japanese researchers will continue to be a part of various initiatives for these advancements will most likely revolutionize rice breeding to circumvent future concerns of sustainable agriculture and food security.

Additional files

Additional file 1: Table S2. Rice varieties and landraces with whole-genome short-read sequences submitted by Japanese research organizations in the NCBI database.

Additional file 2: Table S1. Rice genes reported by Japanese researchers in various scientific journals in 2005–2014. Literatures were searched and obtained from PubMed with 'rice' and 'Oryza' as keywords in either the title or abstract, and further selected by natural language processing and manual curation. The data can be accessed from Oryzabase

Abbreviations
IRGSP, International Rice Genome Sequencing Project; NGS, next-generation sequencing technology; RAP-DB, Rice Genome Annotation Project Database

Acknowledgements
The authors wish to thank Dr. Yukiko Yamazaki (National Institute of Genetics) for providing the data on rice publications. We also thank Dr. Ben Burr for valuable comments and suggestions. The Ministry of Agriculture, Forestry and Fisheries of Japan provided the funding to help defray the costs of publication.

Authors' contributions
TM and TS developed the overall concept of this review. All authors contributed in organizing the content and writing the different sections of the manuscript. All authors read and approved the final manuscript.

Authors' information

TS was the director of the Japan Rice Genome Research Program (RGP) and chairman of the International Rice Genome Sequencing Project (IRGSP). TM, TI, JW and BA were members of RGP and are currently involved in various research on genomics of rice and other cereal crops.

Competing interests

The authors declare that they have no competing interests.

Author details

[1]National Institute of Agrobiological Sciences, 2-1-2 Kannondai, Tsukuba, Ibaraki 305-8602, Japan. [2]Present Address: National Agriculture and Food Research Organization, 2-1-2 Kannondai, Tsukuba, Ibaraki 305-8518, Japan. [3]Nodai Research Institute, Tokyo University of Agriculture, 1-1-1 Sakuragaoka, Setagaya, Tokyo 156-8502, Japan.

References

Asano K, Yamasaki M, Takuno S, Miura K, Katagiri S, Ito T, Doi K, Wu J, Ebana K, Matsumoto T, Innan H, Kitano H, Ashikari M, Matsuoka M (2011) Artificial selection for a green revolution gene during *japonica* rice domestication. Proc Natl Acad Sci U S A 108:11034–11039

Chen X, Zhou D-X (2013) Rice epigenomics and epigenetics: challenges and opportunities. Curr Opin Plant Biol 16:164–169

Cheng Z, Dong F, Langdon T, Ouyang S, Buell R, Gu M, Blattner F, Jiang J (2002) Functional rice centromeres are marked by a satellite repeat and a centromere-specific retrotransposon. Plant Cell 14:1691–1704

Doi K, Izawa T, Fuse T, Yamanouchi U et al (2004) *Ehd1*, a B-type response regulator in rice, confers short-day promotion of flowering and controls FT-like gene expression independently of *Hd1*. Genes Dev 18:926–936

Ebana K, Shibaya T, Wu J, Matsubara K, Kanamori H, Yamane H, Yamanouchi U, Mizubayashi T, Kono I, Shomura A, Ito S, Ando T, Hori K, Matsumoto T, Yano M (2011) Uncovering of major genetic factors generating naturally occurring variation in heading date among Asian rice cultivars. Theor Appl Genet 122:1199–1210

Fujino K, Sekiguchi H, Matsuda Y, Sugimoto K, Ono K, Yano M (2008) Molecular identification of a major quantitative trait locus, *qLTG3-1*, controlling low-temperature germinability in rice. Proc Natl Acad Sci U S A 105:12623–12628

Fujino K, Wu J, Sekiguchi H, Ito T, Izawa T, Matsumoto T (2010) Multiple introgression events surrounding the *Hd1* flowering-time gene in cultivated rice, *Oryza sativa* L. Mol Genet Genomics 284:137–146

Fujita D, Trijatmiko KR, Tagle AJ, Sapasap MV, Koide Y, Sasaki K, Tsakirpaloglou N, Gannaban RB, Nishimura T, Yanagihara S, Fukuta Y, Koshiba T, Slamet-Loedin IHS, Ishimaru T, Kobayashi N (2013) *NAL1* allele from a rice landrace greatly increase yield in modern *indica* cultivars. Proc Natl Acad Sci U S A 110: 20431–20436

Fukuoka S, Saka N, Koga H, Ono K, Shimizu T, Ebana K, Hayashi N, Takahashi A, Hirochika H, Okuno K, Yano M (2009) Loss of function of a proline-containing protein confers durable disease resistance in rice. Science 325:98–101

Garris AJ, Tai TH, Coburn J, Kresovich S, McCouch S (2005) Genetic structure and diversity in *Oryza sativa* L. Genetics 169:1631–1638

Guiderdoni E, Gantet P (2012) Ac-Ds solutions for rice insertion mutagenesis. Methods Mol Biol 859:177–187

Hattori Y, Nagai K, Furukawa S, Song X, Kawano R, Sakakibara H, Wu J, Matsumoto T, Yoshimura A, Kitano H, Matsuoka M, Mori H, Ashikari M (2009) The ethylene response factors *SNORKEL1* and *SNORKEL2* allow rice to adapt to deep water. Nature 460:1026–1030

Hirochika H (2001) Contribution of the Tos17 retrotransposon to rice functional genomics. Curr Opin Plant Biol 4:118–122

Hori K, Ogiso-Tanaka E, Matsubara K, Yamanouchi U, Ebana K, Yano M (2013) *Hd16*, a gene for casein kinase I, is involved in the control of rice flowering time by modulating the day-length response. Plant J 76:36–46

Huang X, Kurata N, Wei X, Wang ZX, Wang A, Zhao Q, Zhao Y, Liu K, Lu H, Li W, Guo Y, Lu Y, Zhou C, Fan D, Weng Q, Zhu C, Huang T, Zhang L, Wang Y, Feng L, Furuumi H, Kubo T, Miyabayashi T, Yuan X, Xu Q, Dong G, Zhan Q, Li C, Fujiyama A, Toyoda A, Lu T, Feng Q, Qian Q, Li J, Han B (2012) A map of rice genome variation reveals the origin of cultivated rice. Nature 490:497–501

Ikeda M, Miura K, Aya K, Kitano H, Matsuoka M (2013) Genes offering the potential for designing yield-related traits in rice. Curr Opin Plant Biol 16: 213–220

International Barley Genome Sequencing Consortium (2012) A physical, genetic and functional sequence assembly of the barley genome. Nature 491:711–716

International Rice Genome Sequencing Project (2005) The map-based sequence of the rice genome. Nature 436:793–800

International Wheat Genome Sequencing Consortium (2014) A chromosome-based draft sequence of the hexaploid bread wheat (*Triticum aestivum*) genome. Science 345:1251788

Ishimaru K, Hirotsu N, Madoka Y, Murakami N, Hara N, Onodera H, Kashiwagi T, Ujiie K, Shimizu B, Onishi A, Miyagawa H, Katoh E (2013) Loss of function of the IAA-glucose hydrolase gene *TGW6* enhances rice grain weight and increases yield. Nat Genet 45:707–711

Jiang J, Nasuda S, Dong F, Scherrer CW, Woo SS, Wing RA, Gill BS, Ward DC (1996) A conserved repetitive DNA element located in the centromeres of cereal chromosomes. Proc. Natl Acad Sci USA 93:14210–14213

Jung KH, An G (2013) Functional characterization of rice genes using a gene-indexed T-DNA insertional mutant population. Methods Mol Biol 956:57–67

Kanamori H, Fujisawa M, Katagiri S, Oono Y, Fujisawa H, Karasawa W, Kurita K, Sasaki H, Mori S, Hamada M, Mukai Y, Yazawa T, Mizuno H, Namiki N, Sasaki T, Katayose Y, Matsumoto T, Wu J (2013) A BAC physical map of aus rice cultivar 'Kasalath', and the map-based genomic sequence of 'Kasalath' chromosome 1. Plant J 76:699–708

Kawahara Y, Oono Y, Kanamori H, Matsumoto T, Itoh T, Minami E (2012) Simultaneous RNA-seq analysis of a mixed transcriptome of rice and blast fungus interaction. PLoS One 7–11, e49423

Kawahara Y, de la Bastide M, Hamilton J, Kanamori H, McCombie R, Ouyang S et al (2013) Improvement of the *Oryza sativa* Nipponbare reference genome using next generation sequence and optical map data. Rice 6:4

Kawahara Y, Oono Y, Wakimoto H, Ogata J, Kanamori H, Sasaki H, Mori S, Matsumoto T, Itoh T (2016) TENOR: Database for Comprehensive mRNA-Seq Experiments in Rice. Plant Cell Physiol 57(1), e7

Kikuchi S, Satoh K, Nagata T, Kawagashira N, Doi K, Kishimoto N et al (2003) Collection, mapping, and annotation of over 28,000 cDNA clones from *japonica* rice. Science 301:376–379

Kojima S, Takahashi Y, Kobayashi Y, Monna L, Sasaki T, Araki T, Yano M (2002) Hd3a, a rice ortholog of the Arabidopsis FT gene, promotes transition to flowering downstream of Hd1 under short-day conditions. Plant Cell Physiol 43:1096–1105

Kojima Y, Ebana K, Fukuoka S, Nagamine T, Kawase M (2005) Development of an RFLP-based rice diversity research set of germplasm. Breeding Sci 55:431–440

Komiya R, Ikegami A, Tamaki S, Yokoi S, Shimamoto K (2008) Hd3a and RFT1 are essential for flowering in rice. Development 135:767–774

Konishi S, Izawa T, Lin SY, Ebana K, Fukuta Y, Sasaki T, Yano M (2006) An SNP caused loss of seed shattering during rice domestication. Science 312:1392–1396

Kumagai M, Kim J, Itoh R, Itoh T (2013) Tasuke: a web-based visualization program for large-scale resequencing data. Bioinformatics 29:1806–1808

Kurata N, Yamazaki Y (2006) Oryzabase. An integrated biological and genome information database for rice. Plant Physiol 140:12–17

Matsubara K, Yamanouchi U, Nonoue Y, Sugimoto K, Wang ZX, Minobe Y, Yano M (2011) *Ehd3*, encoding a plant homeodomain finger-containing protein, is a critical promoter of rice flowering. Plant J 66:603–612

Matsubara K, Ogiso-Tanaka E, Hori K, Ebana K, Ando T, Yano M (2012) Natural variation in *Hd17*, a homolog of Arabidopsis *ELF3* that is involved in rice photoperiodic flowering. Plant Cell Physiol 53:709–716

Mizuno H, Wu J, Kanamori H, Fujisawa M, Namiki N, Saji S, Katagiri S, Katayose Y, Sasaki T, Matsumoto T (2006) Sequencing and characterization of telomere and subtelomere regions on rice chromosomes 1S, 2S, 2L, 6L, 7S, 7L and 8S. Plant J 46:206–217

Mizuno H, Wu J, Katayose Y, Kanamori H, Sasaki T, Matsumoto T (2008a) Characterization of chromosome ends on the basis of the structure of TrsA subtelomeric repeats in rice (*Oryza sativa* L.). Mol Genet Genomics 280:19–24

Mizuno H, Wu J, Katayose Y, Kanamori H, Sasaki T, Matsumoto T (2008b) Chromosome-specific distribution of nucleotide substitutions in telomeric repeats of rice (*Oryza sativa* L.). Mol Biol Evol 25:62–68

Mizuno H, Kawahara Y, Sakai H, Kanamori H, Wakimoto H, Yamagata H, Oono Y, Wu J, Ikawa H, Itoh T, Matsumoto T (2010) Massive parallel sequencing of mRNA in identification of unannotated salinity, stress-inducible transcripts in rice (*Oryza sativa* L.). BMC Genomics 11:683

Mizuno H, Kawahara Y, Wu J, Katayose Y, Kanamori H, Ikawa H, Itoh T, Sasaki T, Matsumoto T (2011) Asymmetric distribution of gene expression in the centromeric region of rice chromosome 5. Front Plant Sci 2:16

Mizuno H, Wu J, Matsumoto T (2014) Characterization of chromosomal ends on the basis of chromosome-specific telomere variants and subtelomeric repeats in rice (*Oryza sativa* L.). Subtelomeres 10:187–194

Nagaki K, Cheng Z, Ouyang S, Talbert PB, Kim M, Jones KM, Henikoff S, Buell CR, Jiang J (2004) Sequencing of a rice centromere uncovers active genes. Nat Genet 36:138–145

Nagano AJ, Sato Y, Mihara M, Antonio BA, Motoyama R, Itoh H, Nagamura Y, Izawa T (2012) Deciphering and prediction of transcriptome dynamics under fluctuating field conditions. Cell 151:1358–1369

Ogiso-Tanaka E, Matsubara K, Yamamoto S, Nonoue Y, Wu J, Fujisawa H, Ishikubo H, Tanaka T, Ando T, Matsumoto T, Yano M (2013) Natural variation of the *RICE FLOWERING LOCUS T 1* contributes to flowering time divergence in rice. PLoS One 8, e75959

Oikawa T, Maeda H, Oguchi T, Yamaguchi T, Tanabe N, Ebana K, Yano M, Ebitani T, Izawa T (2015) The birth of a black rice gene and its spread by introgression. Plant Cell 9:2401–2414

Okazaki Y, Saito K (2016) Integrated metabolomics and phytochemical genomics approaches for studies in rice. GigaScience 5:11

Oono Y, Kawahara Y, Kanamori H, Mizuno H, Yamagata H, Yamamoto M, Hosokawa S, Ikawa H, Akahane I, Zhu Z, Wu J, Itoh T, Matsumoto T (2011) mRNA-seq reveals a comprehensive transcriptome profile of rice under phosphate stress. Rice 4:50–65

Oono Y, Kawahara Y, Yazawa T, Kanamori H, Kuramata M, Yamagata H, Hosokawa S, Minami H, Ishikawa S, Wu J, Antonio B, Handa H, Itoh T, Matsumoto T (2013) Diversity in the complexity of phosphate starvation transcriptomes among rice cultivars based on RNA-Seq profiles. Plant Mol Biol 83:523–537

Oono Y, Yazawa T, Kawahara Y, Kanamori H, Kobayashi F, Sasaki H, Mori S, Wu J, Handa H, Itoh T, Matsumoto T (2014) Genome-wide transcriptome analysis reveals that cadmium stress signaling controls the expression of genes in drought stress signal pathways in rice. PLoS One 9, e96946

Paterson A, Bowers J, Bruggmann R, Dubchak I, Grimwood J, Gundlach H et al (2009) The *Sorghum bicolor* genome and the diversification of grasses. Nature 457:551–556

Rice Annotation Project (2008) The Rice Annotation Project Database (RAP-DB): 2008 update. Nucleic Acids Res 36:D1028–D1033

Sakai H, Ikawa H, Tanaka T, Numa H, Minami H, Fujisawa M, Shibata M, Kurita K, Kikuta A, Hamada M, Kanamori H, Namiki N, Wu J, Itoh T, Matsumoto T, Sasaki T (2011) Distinct evolutionary patterns of *Oryza glaberrima* deciphered by genome sequencing and comparative analysis. Plant J 66:796–805

Sakai H, Lee SS, Tanaka T, Numa H, Kim J, Kawahara Y, Wakimoto H, Yang CC, Iwamoto M, Abe T, Yamada Y, Muto A, Inokuchi H, Ikemura T, Matsumoto T, Sasaki T, Itoh T (2013) Rice Annotation Project Database (RAP-DB): an integrative and interactive database for rice genomics. Plant Cell Physiol 54(2), e6

Sakai H, Kanamori H, Arai-Kichise Y, Shibata-Hatta M, Ebana K, Oono Y, Kurita K, Fujisawa H, Katagiri S, Mukai Y, Hamada M, Itoh T, Matsumoto T, Katayose Y, Wakasa K, Yano M, Wu J (2014) Construction of pseudomolecule sequences of the aus rice cultivar Kasalath for comparative genomics of Asian cultivated rice. DNA Res 21:397–405

Sato Y, Antonio B, Namiki N, Motoyama R, Sugimoto K, Takehisa H, Minami H, Kamatsuki K, Kusaba M, Hirochika H, Nagamura Y (2011) Field transcriptome revealed critical developmental and physiological transitions involved in the expression of growth potential in *japonica* rice. BMC Plant Biol 11:10

Schmutz J, Cannon SB, Jessica Schlueter J, Ma J, Mitros T, Nelson W, Hyten DL et al (2010) Genome sequence of the palaeopolyploid soybean. Nature 463:178–183

Schnable P, Ware D, Fulton RS, Joshua C, Stein JC, Fusheng Wei F, Shiran Pasternak S (2009) The B73 maize genome: complexity, diversity, and dynamics. Science 326:1112–1115

Shimono M, Sugano S, Nakayama A, Jiang CJ, Ono K, Toki S, Takatsuji H (2007) Rice WRKY45 plays a crucial role in benzothiadiazole-inducible blast resistance. Plant Cell 19:2064–2076

Shomura A, Izawa T, Ebana K, Ebitani T, Kanegae H, Konishi S, Yano M (2008) Deletion in a gene associated with grain size increased yields during rice domestication. Nat Genet 40:1023–1028

Sugimoto K, Takeuchi Y, Ebana K, Miyao A, Hirochika H, Hara N, Ishiyama K, Kobayashi M, Ban Y, Hattori T, Yano M (2010) Molecular cloning of *Sdr4*, a regulator involved in seed dormancy and domestication of rice. Proc Natl Acad Sci U S A 107:5792–5797

Taguchi-Shiobara F, Kawagoe Y, Kato H, Onodera H, Tagiri A, Hara N, Miyao A, Hirochika H, Kitano H, Yano M, Toki S (2011) A loss-of-function mutation of rice *DENSE PANICLE 1* causes semi-dwarfness and slightly increased number of spikelets. Breeding Sci 61:17–25

Takagi K, Ishikawa N, Maekawa M, Tsugane K, Iida S (2007) Transposon display for active DNA transposons in rice. Genes Genet Syst 82:109–122

Takahashi Y, Shomura A, Sasaki T, Yano M (2001) Hd6, a rice quantitative trait locus involved in photoperiod sensitivity, encodes the alpha subunit of protein kinase CK2. Proc Natl Acad Sci U S A 98:7922–7927

Takahashi Y, Teshima KM, Yokoi S, Innan H, Shimamoto K (2009) Variations in Hd1 proteins, Hd3a promoters, and Ehd1 expression levels contribute to diversity of flowering time in cultivated rice. Proc Natl Acad Sci U S A 106:4555–4560

Takai T, Adachi S, Taguchi-Shiobara F, Sanoh-Arai Y, Iwasawa N, Yoshinaga S et al (2013) A natural variant of *NAL1*, selected in high-yield rice breeding programs, pleiotropically increases photosynthesis rate. Sci Rep 3:2149

The 3,000 Rice Genomes Project (2014) The 3,000 rice genomes project. GigaScience 3:7

Ueguchi-Tanaka M, Ashikari M, Nakajima M, Itoh H, Katoh E, Kobayashi M, Chow TY, Hsing YI, Kitano H, Yamaguchi I, Matsuoka M (2005) GIBBERELLIN INSENSITIVE DWARF1 encodes a soluble receptor for gibberellin. Nature 437:693–698

Uga Y, Sugimoto K, Ogawa S, Rane J, Ishitani M, Hara N, Kitomi Y, Inukai Y, Ono K, Kanno N, Inoue H, Takehisa H, Motoyama R, Nagamura Y, Wu J, Matsumoto T, Takai T, Okuno K, Yano M (2013) Control of root system architecture by *DEEPER ROOTING 1* increases rice yield under drought conditions. Nat Genet 45:1097–1102

Wang ZX, Yano M, Yamanouchi U, Iwamoto M, Monna L, Hayasaka H, Katayose Y, Sasaki T (1999) The Pib gene for rice blast resistance belongs to the nucleotide binding and leucine-rich repeat class of plant disease resistance genes. Plant J 19:55–64

Wang M, Yu Y, Haberer G, Marri PR, Fan C, Goicoechea JL, Zuccolo A, Song X, Kudrna D, Ammiraju JS, Cossu RM, Maldonado C, Chen J, Lee S, Sisneros N, Baynast A, Golser W, Wissotski M, Kim W, Sanchez P, Ndjiondjop MN, Sanni K, Long M, Carney J, Panaud O, Wicker T, Machado CA, Chen M, Mayer KF, Rounsley S, Wing RA (2014) The genome sequence of African rice (*Oryza glaberrima*) and evidence for independent domestication. Nat Genet 46:982–988

Wei FJ, Droc G, Guiderdoni E, Hsing YI (2013) International Consortium of Rice Mutagenesis: resources and beyond. Rice 6:39

Wing RA, Kim H, Goicoechea JL, Yu Y, Kudrna D, Zuccolo A, Ammiraju J, Luo M, Nelson W, Ma J, Sanmiguel P, Hurwitz B, Ware D, Brar D, Mackill D, Soderlund C, Stein L, Jackson S (2007) The oryza map alignment project (OMAP): A new resource for comparative genome studies within oryza. In: Rice Functional Genomics: Challenges, Progress and Prospects. Springer, New York, pp 395–409

Wu J, Yamagata H, Hayashi-Tsugane M, Hijishita S, Fujisawa M, Shibata M, Ito Y, Nakamura M, Sakaguchi M, Yoshihara R, Kobayashi H, Ito K, Karasawa W, Yamamoto M, Saji S, Katagiri S, Kanamori H, Namiki N, Katayose Y, Matsumoto T, Sasaki T (2004) Composition and structure of the centromeric region of rice chromosome 8. Plant Cell 16:967–976

Wu J, Mizuno H, Sasaki T, Matsumoto T (2008) Comparative analysis of rice genome sequence to understand the molecular basis of genome evolution. Rice 1:119–126

Wu J, Fujisawa M, Tian Z, Yamagata H, Kamiya K, Shibata M, Hosokawa S, Ito Y, Hamada M, Katagiri S, Kurita K, Yamamoto M, Kikuta A, Machita K, Karasawa W, Kanamori H, Namiki N, Mizuno H, Ma J, Sasaki T, Matsumoto T (2009) Comparative analysis of complete orthologous centromeres from two subspecies of rice reveals rapid variation of centromere organization and structure. Plant J 60:805–819

Yamamoto T, Nagasaki H, Yonemaru J, Ebana K, Nakajima M, Shibaya T, Yano M (2010) Fine definition of the pedigree haplotypes of closely related rice cultivars by means of genome-wide discovery of single-nucleotide polymorphisms. BMC Genomics 11:267

Yamane H, Ito T, Ishikubo H, Fujisawa M, Yamagata H, Kamiya K, Ito Y, Hamada M, Kanamori H, Ikawa H, Katayose Y, Wu J, Sasaki T, Matsumoto T (2009) Molecular and evolutionary analysis of the *Hd6* photoperiod sensitivity gene within genus Oryza. Rice 2:56–66

Yan H, Ito H, Nobuta K, Ouyang S, Jin W, Tian S, Lu C, Venu RC, Wang GL, Green PJ, Wing RA, Buell CR, Meyers BC, Jiang J (2006) Genomic and genetic characterization of rice Cen3 reveals extensive transcription and evolutionary implications of a complex centromere. Plant Cell 18:2123–2133

Yang CC, Kawahara Y, Mizuno H, Wu J, Matsumoto T, Itoh T (2012) Independent domestication of Asian rice followed by gene flow from *japonica* to *indica*. Mol Biol Evol 29:1471–1479

Yang W, Duan L, Chen G, Xiong L, Liu Q (2013) Plant phenomics and high-throughput phenotyping: accelerating rice functional genomics using multidisciplinary technologies. Curr Opin Plant Biol 16:180–187

Yano M, Kojima S, Takahashi Y, Lin H, Sasaki T (2001) Genetic control of flowering time in rice, a short-day plant. Plant Physiol 127:1425–1429

Zhang Y, Huang Y, Zhang L, Li Y, Lu T, Lu Y, Feng Q, Zhao Q, Cheng Z, Xue Y, Wing RA, Han B (2004) Structural features of the rice chromosome 4 centromere. Nucleic Acids Res 32:2023–2030

Zhou S, Bechner MC, Place M, Churas CP, Pape L, Leong SA, Runnheim R, Forrest DK, Goldstein S, Livny M, Schwartz DC (2007) Validation of rice genome sequence by optical mapping. BMC Genomics 8:278

Permissions

All chapters in this book were first published in RICE, by Springer; hereby published with permission under the Creative Commons Attribution License or equivalent. Every chapter published in this book has been scrutinized by our experts. Their significance has been extensively debated. The topics covered herein carry significant findings which will fuel the growth of the discipline. They may even be implemented as practical applications or may be referred to as a beginning point for another development.

The contributors of this book come from diverse backgrounds, making this book a truly international effort. This book will bring forth new frontiers with its revolutionizing research information and detailed analysis of the nascent developments around the world.

We would like to thank all the contributing authors for lending their expertise to make the book truly unique. They have played a crucial role in the development of this book. Without their invaluable contributions this book wouldn't have been possible. They have made vital efforts to compile up to date information on the varied aspects of this subject to make this book a valuable addition to the collection of many professionals and students.

This book was conceptualized with the vision of imparting up-to-date information and advanced data in this field. To ensure the same, a matchless editorial board was set up. Every individual on the board went through rigorous rounds of assessment to prove their worth. After which they invested a large part of their time researching and compiling the most relevant data for our readers.

The editorial board has been involved in producing this book since its inception. They have spent rigorous hours researching and exploring the diverse topics which have resulted in the successful publishing of this book. They have passed on their knowledge of decades through this book. To expedite this challenging task, the publisher supported the team at every step. A small team of assistant editors was also appointed to further simplify the editing procedure and attain best results for the readers.

Apart from the editorial board, the designing team has also invested a significant amount of their time in understanding the subject and creating the most relevant covers. They scrutinized every image to scout for the most suitable representation of the subject and create an appropriate cover for the book.

The publishing team has been an ardent support to the editorial, designing and production team. Their endless efforts to recruit the best for this project, has resulted in the accomplishment of this book. They are a veteran in the field of academics and their pool of knowledge is as vast as their experience in printing. Their expertise and guidance has proved useful at every step. Their uncompromising quality standards have made this book an exceptional effort. Their encouragement from time to time has been an inspiration for everyone.

The publisher and the editorial board hope that this book will prove to be a valuable piece of knowledge for researchers, students, practitioners and scholars across the globe.

List of Contributors

Pinky Agarwal and Swarup K. Parida
National Institute of Plant Genome Research (NIPGR), Aruna Asaf Ali Marg, New Delhi 110067, India

Akhilesh K. Tyagi
National Institute of Plant Genome Research (NIPGR), Aruna Asaf Ali Marg, New Delhi 110067, India
Interdisciplinary Centre for Plant Genomics and Department of Plant Molecular Biology, University of Delhi, South Campus, New Delhi 110021, India

Saurabh Raghuvanshi, Sanjay Kapoor, Paramjit Khurana and Jitendra P. Khurana
Interdisciplinary Centre for Plant Genomics and Department of Plant Molecular Biology, University of Delhi, South Campus, New Delhi 110021, India

Likai Chen, Weiwei Gao, Siping Chen, Liping Wang, Yongzhu Liu, Hui Wang, Zhiqiang Chen and Tao Guo
National Engineering Research Center of Plant Space Breeding, South China
Agricultural University, Guangzhou 510642, China

Jiyong Zou
National Engineering Research Center of Plant Space Breeding, South China Agricultural University, Guangzhou 510642, China
Guangdong Agricultural Technology Extension, Guangzhou 510520, China

Teresa B. De Leon and Prasanta K. Subudhi
School of Plant, Environmental, and Soil Sciences, Louisiana State University
Agricultural Center, Baton Rouge, LA, USA

Steven Linscombe
Rice Research Station, Louisiana State University Agricultural Center, Rayne, LA, USA

Wei Tong
Department of Plant Resources, College of Industrial Sciences, Kongju
National University, Yesan 32439, Republic of Korea

Tae-Sung Kim
Department of Plant Resources, College of Industrial Sciences, Kongju National University, Yesan 32439, Republic of Korea
Department of Agricultural Sciences, College of Natural Sciences, Korea National Open University, Seoul 03087, Republic of Korea

Yong-Jin Park
Department of Plant Resources, College of Industrial Sciences, Kongju National University, Yesan 32439, Republic of Korea
Center for Crop Genetic Resource and Breeding (CCGRB), Kongju National University, Cheonan 31080, Republic of Korea

Juan L. Reig-Valiente, Javier Terol, Manuel Talón and Concha Domingo
Centro de Genómica, Instituto Valenciano de Investigaciones Agrarias, Carretera CV 315 Km 10,7 (Carretera Moncada – Náquera Km 4.5), 46113 Moncada, Spain

Juan Viruel
Dpto. Biología Vegetal y Ecología, SGI Herbario – Universidad de Sevilla, Edif. Celestino Mutis, Av. Reina Mercedes s/n, 41012 Sevilla, Spain
Institut Méditerranéen de Biodiversité et d'Ecologie Marine et Continentale (IMBE), Aix Marseille Université, Chemin de la Batterie des Lions, 13007 Marseille, France

Ester Sales
Dpto. Ciencias Agrarias y del Medio Natural, Escuela Politécnica Superior, Universidad de Zaragoza, Ctra. Cuarte s/n, 22071 Huesca, Spain

Luis Marqués
Cooperativa de Productores de Semillas de Arroz, Avenida del Mar 1, 46410 Sueca, Spain

Marta Gut and Sophia Derdak
Centre Nacional d'Anàlisi Genòmica – Centre for Genomic Regulation (CNAG-CRG), Barcelona Institute of Science and Technology (BIST), Baldiri Reixac, 4, 08028 Barcelona, Spain
Universitat Pompeu Fabra (UPF), Barcelona, Spain

Honggen Zhang, Jianlan Che, Yongshen Ge, Yan Pei, Lijia Zhang, Qiaoquan Liu, Minghong Gu and Shuzhu Tang
Jiangsu Key Laboratory of Crop Genetics and Physiology/Co-Innovation Center for Modern Production Technology of Grain Crops, Key Laboratory of Plant Functional Genomics of the Ministry of Education, College of Agriculture, Yangzhou University, Yangzhou 225009, China

D. Shoba, M. Raveendran, S. Manonmani, S. Utharasu, D. Dhivyapriya, G. Subhasini, S. Ramchandar, R. Valarmathi and S. Robin
Tamil Nadu Agricultural University, Coimbatore 641003, India

Nitasha Grover, S. Gopala Krishnan and A. K. Singh
Division of Genetics, ICAR-Indian Agricultural Research Institute, New Delhi 110012, India

Pawan Jayaswal, Prashant Kale, M. K. Ramkumar, S. V. Amitha Mithra and N. K. Singh
ICAR-National Research Centre on Plant Biotechnology, Pusa, New Delhi 110012, India

T. Mohapatra
Indian Council of Agriculture Research, New Delhi 110 001, India

Kuldeep Singh
Punjab Agricultural University, Ludhiana 141004, India
ICAR-National Bureau of Plant Genetic Resources, Pusa, New Delhi 110012, India

N. Sarla
Indian Institute of Rice Research, Rajendranagar, Hyderabad 500030, India

M. S. Sheshshayee
University of Agricultural Sciences, Bengaluru 560065, India

M. K. Kar
National Rice Research Institute, Cuttack, Odisha 753006, India

R. P. Sharma
INSA Honorary Scientist, NRCPB, IARI, Pusa, New Delhi 110012, India

R. El-Namaky
Africa Rice Center (AfricaRice), P.B. 96, St. Louis, Senegal

Rice Research & Training Center, 33717, Sakha Kafr Sheikh, Egypt

P. A. J. van Oort
Africa Rice Center (AfricaRice), 01 B.P. 2551 Bouaké, Côte d'Ivoire Crop & Weed Ecology Group, Centre for Crop Systems Analysis, Wageningen University, AK, Wageningen, The Netherlands

Michael J. Thomson
Department of Soil and Crop Sciences, Texas A&M University, College Station, Houston, TX 77843, USA

Namrata Singh, Diane R. Wang, Francisco Agosto Perez and Susan R. McCouch
School of Integrative Plant Sciences, Plant Breeding and Genetics Section, Cornell University, Ithaca, New York 14853, USA

Mark H. Wright
School of Integrative Plant Sciences, Plant Breeding and Genetics Section, Cornell University, Ithaca, New York 14853, USA
Department of Genetics, Stanford School of Medicine, Stanford, California 94305, USA

Genevieve DeClerck
School of Integrative Plant Sciences, Plant Breeding and Genetics Section, Cornell University, Ithaca, New York 14853, USA
DeClerck Design, LLC, Freeville, NY, USA

Geraldine A. Malitic-Layaoen, Venice Margarette Juanillas, Ramil Mauleon and Tobias Kretzschmar
International Rice Research Institute, Los Baños, Philippines

Maria S. Dwiyanti
International Rice Research Institute, Los Baños, Philippines
Research Faculty of Agriculture, Hokkaido University, Sapporo, Hokkaido 060-8589, Japan

Joong Hyoun Chin
International Rice Research Institute, Los Baños, Philippines
Graduate School of Integrated Bioindustry, Sejong University, 209 Neungdong-ro, Gwangjin-gu, Seoul 05006, South Korea

Christine J. Dilla-Ermita
International Rice Research Institute, Los Baños, Philippines
Department of Plant Pathology, Washington State University, Pullman, Washington 99164, USA

Nagendra Kumar Singh
ICAR-National Research Centre on Plant Biotechnology, Pusa Campus, New Delhi 110012, India

Vijay Kumar Singh
ICAR-National Research Centre on Plant Biotechnology, Pusa Campus, New Delhi 110012, India
School of Biotechnology, Banaras Hindu University, Varanasi 221005, India

Ranjith Kumar Ellur and Ashok Kumar Singh
Division of Genetics, ICAR-Indian Agricultural Research Institute (ICAR-IARI), Pusa Campus, New Delhi 110012, India

M. Nagarajan
ICAR-IARI-Rice Breeding and Genetics Research Centre, Aduthurai, Tamil Nadu 612101, India

Brahma Deo Singh
School of Biotechnology, Banaras Hindu University, Varanasi 221005, India

Laiyuan Zhai, Shu Wang and Yun Wang
Rice Research Institute, Shenyang Agricultural University/Key Laboratory of Northern Japonica Rice Genetics and Breeding, Ministry of Education, Shenyang 110866, China

Tianqing Zheng and Yun Wang
Institute of Crop Sciences/National Key Facility for Crop Gene Resources and Genetic Improvement, Chinese Academy of Agricultural Sciences, 12# South Zhong-Guan-Cun Street, Haidain District, Beijing 100081, China

Xinyu Wang, Kai Chen and Jianlong Xu
Institute of Crop Sciences/National Key Facility for Crop Gene Resources and Genetic Improvement, Chinese Academy of Agricultural Sciences, 12# South Zhong-Guan-Cun Street, Haidain District, Beijing 100081, China
Agricultrual Genomics Institute at Shenzhen, Chinese Academy of Agricultural Sceinces, Shenzhen 518120, China

Zhikang Li
Institute of Crop Sciences/National Key Facility for Crop Gene Resources and Genetic Improvement, Chinese Academy of Agricultural Sciences, 12# South Zhong-Guan-Cun Street, Haidain District, Beijing 100081, China
Shenzhen Institute of Breeding and Innovation, Chinese Academy of Agricultural Sciences, Shenzhen 518120, China

Jie Wang, Kai Lu, Bowen Wu and Junjie Qian
Center of Applied Biotechnology, Wuhan Institute of Bioengineering, Wuhan 430415, China
National Key Laboratory of Crop Genetic Improvement, Huazhong Agricultural University, Wuhan 430070, China

Haipeng Nie, Qisen Zeng and Zhongming Fang
Center of Applied Biotechnology, Wuhan Institute of Bioengineering, Wuhan 430415, China
National Key Laboratory of Crop Genetic Improvement, Huazhong Agricultural University, Wuhan 430070, China

Naoki Yamamoto, Richard Garcia, Celymar Angela Solis and Ajay Kohli
International Rice Research Institute, Los Baños, Laguna, Philippines

Tomohiro Suzuki
Utsunomiya University, 350 Mine-machi, Utsunomiya, Tochigi, Japan

Yuichi Tada
Tokyo University of Technology, 1404-1 Katakura, Hachioji, Tokyo, Japan

Ramaiah Venuprasad
Africa Rice Center, 01 BP 4029, Abidjan 01, Côte d'Ivoire

Shuang Yong and Qiuying Yu
Key Laboratory of Crop Physiology, Ecology and Genetic Breeding, Ministry of Education, Jiangxi Agricultural University, Nanchang 330045, China

Liang Chen, Shilai Shi, Jianfeng Yu, Hira Khanzada, Ghulam Mustafa Wassan, Xin Luo, Shan Tong, Xiaorong Yang, Xiaopeng He, Junru Fu, Xiaorong Chen and Linjuan Ouyang
Key Laboratory of Crop Physiology, Ecology and Genetic Breeding, Ministry of Education, Jiangxi Agricultural University, Nanchang 330045, China
College of Agronomy, Jiangxi Agricultural University, Nanchang 330045, China

Jianmin Bian, Changlan Zhu, Xiaosong Peng, Lifang Hu and Haohua He
Southern Regional Collaborative Innovation Center for Grain and Oil Crops in China, Changsha, China

Zhen-Hua Zhang, Lin-Lin Wang, Ye-Yang Fan, Liang-Yong Ma and Jie-Yun Zhuang
State Key Laboratory of Rice Biology and Chinese National Center for Rice Improvement, China National Rice Research Institute, Hangzhou 310006, China

Qing Dong and Yu-Jun Zhu
State Key Laboratory of Rice Biology and Chinese National Center for Rice
Improvement, China National Rice Research Institute, Hangzhou 310006, China
State Key Laboratory of Crop Genetic Improvement and National Center of Plant Gene Research (Wuhan), Huazhong Agricultural University, Wuhan 430070, China

Tong-Min Mou
State Key Laboratory of Crop Genetic Improvement and National Center of Plant Gene Research (Wuhan), Huazhong Agricultural University, Wuhan 430070, China

Chaolei Liu, Yuanyuan Li, Anpeng Zhang, Guojun Dong, Lihong Xie, Bin Zhang, Banpu Ruan, Kai Hong, Dali Zeng, Longbiao Guo, Qian Qian and Zhenyu Gao
State Key Laboratory of Rice Biology, China National Rice Research Institute, Hangzhou 310006, China

Yang Gao
State Key Laboratory of Rice Biology, China National Rice Research Institute, Hangzhou 310006, China
College of Life and Environmental Sciences, Hangzhou Normal University, Hangzhou 310036, China

Dawei Xue
College of Life and Environmental Sciences, Hangzhou Normal University, Hangzhou 310036, China

Takashi Matsumoto, Jianzhong Wu, Takeshi Itoh, Hisataka Numa and Baltazar Antonio
National Institute of Agrobiological Sciences, 2-1-2 Kannondai, Tsukuba, Ibaraki 305-8602, Japan
National Agriculture and Food Research Organization, 2-1-2 Kannondai, Tsukuba, Ibaraki 305-8518, Japan

Takuji Sasaki
Nodai Research Institute, Tokyo University of Agriculture, 1-1-1 Sakuragaoka, Setagaya, Tokyo 156-8502, Japan

Index

www.ingramcontent.com/pod-product-compliance
Lightning Source LLC
Chambersburg PA
CBHW061258190326
41458CB00011B/3705